Agroforestry for sustainable agriculture

It is widely recognised that agriculture is a significant contributor to global warming and climate change. Agriculture needs to reduce its environmental impact and adapt to current climate change whilst still feeding a growing population, i.e. become more 'climate-smart'. Burleigh Dodds Science Publishing is playing its part in achieving this by bringing together key research on making the production of the world's most important crops and livestock products more sustainable.

Based on extensive research, our publications specifically target the challenge of climate-smart agriculture. In this way we are using 'smart publishing' to help achieve climate-smart agriculture.

Burleigh Dodds Science Publishing is an independent and innovative publisher delivering high quality customer-focused agricultural science content in both print and online formats for the academic and research communities. Our aim is to build a foundation of knowledge on which researchers can build to meet the challenge of climate-smart agriculture.

For more information about Burleigh Dodds Science Publishing simply call us on +44 (0) 1223 839365, email info@bdspublishing.com or alternatively please visit our website at www.bdspublishing.com.

Related titles:

Achieving sustainable cultivation of temperate zone tree fruits and berries
Volume 1: Physiology, genetics and cultivation
Print (ISBN 978-1-78676-208-5); Online (ISBN 978-1-78676-211-5, 978-1-78676-210-8)

Achieving sustainable cultivation of temperate zone tree fruits and berries
Volume 2: Case studies
Print (ISBN 978-1-78676-212-2); Online (ISBN 978-1-78676-215-3, 978-1-78676-214-6)

Achieving sustainable cultivation of tree nuts
Print (ISBN 978-1-78676-224-5); Online (ISBN 978-1-78676-227-6, 978-1-78676-226-9)

Achieving sustainable management of boreal and temperate forests
Print (ISBN 978-1-78676-292-4); Online (ISBN 978-1-78676-295-5, 978-1-78676-294-8)

Integrated management of diseases and insect pests of tree fruit
Print (ISBN 978-1-78676-256-6); Online (ISBN 978-1-78676-259-7, 978-1-78676-258-0)

Chapters are available individually from our online bookshop:
https://shop.bdspublishing.com

climate•SMART•publishing
IN AGRICULTURAL SCIENCE

BURLEIGH DODDS SERIES IN AGRICULTURAL SCIENCE
NUMBER 55

Agroforestry for sustainable agriculture

Edited by Professor María Rosa Mosquera-Losada, Universidade de Santiago de Compostela, Spain and Dr Ravi Prabhu, World Agroforestry Centre (ICRAF), Kenya

burleigh dodds
SCIENCE PUBLISHING

Published by Burleigh Dodds Science Publishing Limited
82 High Street, Sawston, Cambridge CB22 3HJ, UK
www.bdspublishing.com

Burleigh Dodds Science Publishing, 1518 Walnut Street, Suite 900, Philadelphia, PA 19102-3406, USA

First published 2019 by Burleigh Dodds Science Publishing Limited

Library of Congress Control Number: 2019941167

British Library Cataloguing in Publication Data
A catalogue record for this book is available from the British Library

ISBN 978-1-78676-220-7 (Print)
ISBN 978-1-78676-223-8 (PDF)
ISBN 978-1-78676-222-1 (ePub)
ISSN 2059-6936 (print)
ISSN 2059-6944 (online)

DOI 10.19103/AS.2018.0041

Typeset by Deanta Global Publishing Services, Chennai, India

Contents

Part 3 Agroforestry products

Series list

Title	Series number
Achieving sustainable cultivation of maize - Vol 1 From improved varieties to local applications Edited by: Dr Dave Watson, CGIAR Maize Research Program Manager, CIMMYT, Mexico	001
Achieving sustainable cultivation of maize - Vol 2 Cultivation techniques, pest and disease control Edited by: Dr Dave Watson, CGIAR Maize Research Program Manager, CIMMYT, Mexico	002
Achieving sustainable cultivation of rice - Vol 1 Breeding for higher yield and quality Edited by: Prof. Takuji Sasaki, Tokyo University of Agriculture, Japan	003
Achieving sustainable cultivation of rice - Vol 2 Cultivation, pest and disease management Edited by: Prof. Takuji Sasaki, Tokyo University of Agriculture, Japan	004
Achieving sustainable cultivation of wheat - Vol 1 Breeding, quality traits, pests and diseases Edited by: Prof. Peter Langridge, The University of Adelaide, Australia	005
Achieving sustainable cultivation of wheat - Vol 2 Cultivation techniques Edited by: Prof. Peter Langridge, The University of Adelaide, Australia	006
Achieving sustainable cultivation of tomatoes Edited by: Dr Autar Mattoo, USDA-ARS, USA & Prof. Avtar Handa, Purdue University, USA	007
Achieving sustainable production of milk - Vol 1 Milk composition, genetics and breeding Edited by: Dr Nico van Belzen, International Dairy Federation (IDF), Belgium	008
Achieving sustainable production of milk - Vol 2 Safety, quality and sustainability Edited by: Dr Nico van Belzen, International Dairy Federation (IDF), Belgium	009
Achieving sustainable production of milk - Vol 3 Dairy herd management and welfare Edited by: Prof. John Webster, University of Bristol, UK	010
Ensuring safety and quality in the production of beef - Vol 1 Safety Edited by: Prof. Gary Acuff, Texas A&M University, USA & Prof. James Dickson, Iowa State University, USA	011
Ensuring safety and quality in the production of beef - Vol 2 Quality Edited by: Prof. Michael Dikeman, Kansas State University, USA	012
Achieving sustainable production of poultry meat - Vol 1 Safety, quality and sustainability Edited by: Prof. Steven C. Ricke, University of Arkansas, USA	013
Achieving sustainable production of poultry meat - Vol 2 Breeding and nutrition Edited by: Prof. Todd Applegate, University of Georgia, USA	014

Acknowledgements

We wish to acknowledge the following for their help in reviewing particular chapters:

- Chapter 1: Dr James Chamberlain, US Forest Service, USA
- Chapter 3: Dr Federica Larcher, University of Turin, Italy; and Ms Sally Westaway, The Organic Research Centre, UK
- Chapter 5: Dr John Erik Hermansen, Aarhus University, Denmark; and Prof M. P. González Hernandez, Universidade de Santiago de Compostela, Spain
- Chapter 7: Dr Eike Luedeling, University of Bonn, Germany; and Dr Philip Smethurst, CSIRO, Australia
- Chapter 8: Prof Kevin O'Hara, University of California Berkely, USA
- Chapter 12: Dr Dan Dey, US Forest Service, USA
- Chapter 13: Dr Jens-Peter Barnekow Lillesø, University of Copenhagen, Denmark

We would also like to acknowledge Christian Dupraz of INRA for the cover image.

Introduction

The United States Department of Agriculture (USDA) defines agroforestry as: 'the intentional mixing of trees and shrubs into crop and animal production systems to create environmental, economic and social benefits'. Agroforestry seeks to balance the protection of forest resources, the exploitation of the ecosystem services that trees can contribute to agriculture and the role of agroforestry in diversifying the range of agricultural products and markets. The book provides a comprehensive review of ways of optimizing particular agroforestry practices, from riparian forest buffers and windbreaks to alley cropping, silvopasture and forest farming. In addition, the volume summarises current research on ecosystem services delivered by agroforestry in such areas as habitat conservation and soil health. The book also assesses research on optimising agroforestry products such as timber, nuts and fruit. The main focus of the book is on temperate agroforestry, but it also reviews particular issues facing agroforestry in the tropics.

Part 1 Agroforestry practices

The first part of the volume addresses agroforestry practices. The subject of Chapter 1 is riparian forest buffers and filter strips. Riparian forest buffers are planned combinations of trees, shrubs, grasses, forbs and bioengineered structures adjacent to or within a stream channel designed to mitigate the impact of land use on the stream or creek. At the landscape level, riparian forest buffers link the land and aquatic environment and perform vital ecological functions as part of the network of watersheds that connect forest, prairies, agricultural and urban lands. Following an introduction to the concept of riparian forest buffers, the chapter examines forest buffer design and function. The chapter discusses the distinct management zones in a riparian forest buffer, including undisturbed forest, managed forest and shrubs, and runoff control (grasses and forbs). The chapter concludes with sections on special design considerations and management, as well as ways of assessing buffer performance.

Chapter 2 moves on from riparian buffers to consider the role of windbreaks. Windbreaks, also known as vegetative environmental buffers (VEB) or shelterbelts, are a common feature of agricultural systems around the world. These strips of trees, shrubs and other perennial or annual vegetation perform a number of functions, including providing protection from the wind for homesteads, livestock and crops; reducing soil erosion; providing protection from drifting snow; providing wildlife habitat; and enhancing aesthetics in agricultural landscapes. The chapter offers an overview of the benefits of windbreaks and examines the process of tree and shrub selection and planting.

The chapter looks at designing windbreaks to reduce wind speed and snow drift, as well as describing the use of windbreaks for particulate capture, odor mitigation and wildlife conservation.

Chapter 3 shifts the focus to managing hedgerows in order to optimise ecosystem services in agroforestry systems. Hedgerows are composed of trees and/or shrubs and serve as natural fences, often between pastures and arable fields. The chapter discusses the importance of hedgerow management and maintenance to enhance multiple ecosystem services such as biomass production, wind and water protection, habitat provision and landscape aesthetics. It develops a categorisation of existing hedgerows, applicable to linear woody-features such as hedgerows, windbreaks and riparian buffer strips, and evaluates the condition of each category with respect to multiple ecosystem services. The chapter considers different measures for improving hedgerow conditions and provides guidelines for hedgerow restoration and management. The chapter includes a case study on hedgerow management, focusing on estimating the biomass potential of hedgerows and the implementation of a restoration and management strategy to selected hedgerows in two different scenarios.

The subject of Chapter 4 is temperate alley cropping systems. The implementation of modern agricultural practices has largely excluded trees from the rural landscape, causing negative environmental impacts. Alley cropping, an agroforestry practice where agricultural crops are grown simultaneously with long-term tree crops, helps mitigate negative environmental impacts and offers a promising land-use alternative to conventional farming for temperate regions. The chapter provides an overview of economic and ecological benefits, challenges, and major considerations of implementing these practices within North America. To illustrate the key issues, the chapter focuses on a system performance evaluation of a pecan-cotton system in the southern United States.

Chapter 5 moves on to consider silvopastoralism, a traditional agroforestry practice that is still used across the world for raising livestock, particularly in lands with a combination of grass understory and a sparse cover of trees and/or shrubs. The woody component plays multiple roles such as providing a forage resource, shade and shelter for livestock and delivering products such as timber, firewood or cork, for example. The chapter describes silvopastoral systems from around the world. The chapter also includes a section on the role of trees in promoting ecosystems services such as carbon sequestration, water quality and biodiversity conservation. The chapter concludes with a discussion of how to design and manage silvopastures.

Staying with the theme of diverse agroforestry, Chapter 6 examines forest farming. This is a relatively low-tech agroforestry practice for the cultivation of shade-tolerant non-timber forest products (NTFPs) such as medicinal plants, mushrooms, fruits, nuts, tree syrups and/or nursery stock. It is an ecologically sustainable way for forest owners to generate income, while maintaining forest health (also known as productive conservation). The chapter introduces the concept of forest farming with sections on both estimating yields and how to select sites for such practices. The chapter discusses the products with the greatest potential for income generation, such as shiitake mushrooms (*Lentinula edodes*), where logs begin yielding relatively soon after their inoculation, and American ginseng (*Panax quinquefolium*) which has greater income generating potential even though it requires many years before it begins to yield. Other NTFPs are also covered including medicinal plants, fruits, nuts and tree syrups. The chapter concludes with two case studies, one on the forest farming of tree nuts, and the other on the production of wild leeks (*Allium tricocchum*).

Moving from specific practices to an overview, Chapter 7 looks at the modelling of agroforestry systems, which can be highly complex in nature. The chapter examines the current state of agroforestry modelling before going on to describe two European agroforestry projects that involved modelling, the Silvoarable Agroforestry For Europe (SAFE) project and the AGroFORestry that Will Advance Rural Development (AGFORWARD) project. Looking towards the future, the chapter also considers current agroforestry modelling needs and potential trends.

The final chapter in the section, Chapter 8, deals with tree planting and management in agroforestry. The pattern of tree planting and tree management play an important part in the sustainability of an agroforestry system, as they determine the intensity of competition between trees and crop and the quality and quantity of wood production. The chapter focuses on the plantation and management of temperate agroforestry systems combining timber trees and herbaceous crops. The chapter covers choice of tree species, techniques of tree planting, plantation maintenance and approaches to tree pruning and thinning.

Part 2 Agroforestry ecosystem services

Opening the second part of the volume, which focuses on agroforestry ecosystem services, Chapter 9 turns to the relationship between trees, science and global society. Despite great advances in our understanding of the environmental, social and economic role of trees in farming systems, much work still remains to be done, especially regarding the wider adoption of agroforestry practices. The chapter offers an overview of tree agroecology, tree domestication, the commercialization of trees and the relevance of development studies to this field. The chapter assesses the potential for up-scaling the exploitation of trees as a sustainable resource and examines relevant issues connected with government policy and agribusiness.

Following this overview, Chapter 10 homes in on the role of agroforestry in habitat conservation and biodiversity. Agricultural biodiversity, commonly referred to as agrobiodiversity, focuses on agricultural habitats and food production landscapes. Adopting a multi-dimensional approach to their management is essential to ensure the sustainability of agricultural habitats. The chapter describes the application of the food system concept in providing a framework to enable such a multi-dimensional approach. The chapter outlines the multiple dimensions of food security and places food security in the context of global environment change. Finally, the chapter explains the concept of multifactor food security promotion.

The focus of Chapter 11 is on agroforestry as a system for improving soil health. Introducing agroforestry into agroecosystems can be an important method to help promote soil quality through its influence on soil physical, chemical and biological properties. The chapter highlights the benefits of agroforestry systems on soil properties important for soil quality. These include critical soil biological, physical and chemical properties important for maintaining and improving soil health. The chapter outlines the critical soil biological properties important for energy and nutrient transformations, as well as critical soil physical properties including soil density, porosity, water retention, pore-size distributions, hydraulic conductivity, infiltration, and thermal properties. The chapter shows that improving these parameters can reduce losses of sediment, nutrients and pesticides from land to water, and enhance soil water storage.

Part 3 Agroforestry products

The third part of the volume is dedicated to the diverse products of agroforestry. Chapter 12 examines the contribution of agroforestry to hardwood timber production. Managing hardwood trees for high-quality sawlogs within the agroforestry practices of alley-cropping, silvopasture, forested riparian buffers and upland (windbreaks) buffers means intensively managing relatively wide-spaced trees and a ground cover as a companion crop on the same unit of land. The chapter synthesizes available information on managing such trees for the production of veneer and high-quality sawlogs. The chapter includes sections on the impact of ground cover on tree growth and growing space requirements for hardwoods. The chapter also looks at pruning recommendations and practices, since pruning is essential for the production of high-quality logs of most species in any agroforestry practice. The chapter concludes with a section on log and wood quality and the general requirements of veneer logs.

The section then moves on to an assessment of the cultivation of nuts in Chapter 13. Row crop agriculture covers over 1.28 billion hectares of land globally. Though extremely productive in terms of yield, annual cropping systems rely on external inputs of energy, nutrients, and pesticides, leading to a suite of ecological consequences. The chapter focusses primarily on the opportunities and challenges associated with alley cropping practices involving overstory nut crops as one element of the solution to address global needs for food production which is both economically viable while enhancing ecosystem services. The chapter examines key challenges facing nut-based agroforestry systems, looking at the genetic improvement of nut trees as well as the challenge of managing temporal and spatial tree and crop interactions. The chapter addresses orchard design and management, pest management in nut tree alley cropping, and financial decision support tools.

The final chapter in Part 3 is Chapter 14 which reviews agroforestry for fruit trees. Although fruit trees are considered as high value for agroforestry and are the primary driver of agroforestry adoption worldwide, they are still underrepresented in agroforestry systems in temperate regions compared to the tropics. This chapter illustrates the large diversity of fruit tree-based agroforestry in Europe and in the Mediterranean North Africa. The chapter then describes the most common species-based (apple, olive) and place-based (e.g. oasis) agroforestry systems in these regions. Finally, the chapter discusses key biological and agronomical requirements of fruit trees that have to be considered when implementing successful fruit-tree based agroforestry systems. It reviews current trends in the design of agronomically and ecologically-sound fruit tree-based agroforestry systems.

Part 4 Tropical agroforestry

The volume's final section deals with agroforestry in tropical areas. Chapter 15 addresses the challenges involved in tropical agroforestry. From its early beginnings, agroforestry has moved from a 'technology in search of a problem' to a principal solution to critical global agendas, including climate smart agriculture, agroecological intensification, land rehabilitation, and provision of ecosystem services. The chapter addresses the challenges associated with agroforestry in agroecological intensification and sustainable landscapes. The chapter considers the challenge of developing policies in support of agroforestry, and the challenge of developing agroforestry at scale.

The subject of Chapter 16 is tropical tree domestication in agroforestry. The chapter examines the principles and techniques of tropical tree domestication, covering identification of species for domestication, selection of 'plus trees' and vegetative propagation methods. The chapter then provides examples of key tree species that have been targeted for domestication in the Amazon Basin (*Bactris gasipaes*, *Capirona*, *Guazuma crinita* and *Inga edulis*) and the Congo Basin (*Irvingia gabonensis*, *Irvingia wombolu*, safou, *Ricinodendron heudelotii*,

Cola acuminata, *Cola anomala*, *Cola nitida* and *Prunus africana*). The chapter concludes with a case study on participatory domestication of *Allanblackia floribunda*, a high value agroforestry tree species in Central Africa.

The volume's final chapter, Chapter 17, looks at trade-off analysis for better design strategies in tropical agroforestry and ecosystem services. A large body of research has documented a wide list of provisioning and regulating services that can be provided by tropical agroforestry systems (AFS). The chapter offers an overview of ecosystem services delivered by tropical AFS, and presents practical approaches for trade-off analysis between ecosystem services and plant biodiversity for better design (or re-design) and management of AFS. The chapter highlights the main provisioning and regulating services provided by tropical AFS (covering pest and disease regulation, nutrient cycling and soil quality, carbon sequestration and water regulation). The chapter gives an overview of practical approaches to assess trade-offs provides a case study of trade-off analysis in practice.

Part 1

Agroforestry practices

Chapter 1

Agroforestry practices: riparian forest buffers and filter strips

Richard Schultz, Thomas Isenhart, William Beck, Tyler Groh and Morgan Davis, Iowa State University, USA

1 Introduction

Intensive agriculture as practiced in much of the Temperate Zone around the world is not very friendly to the environment. Non-point source (NPS) pollution from this kind of agriculture has created major water quality issues for surface waters that originate or flow through these areas (Veum et al., 2009). In many landscapes in the Midwestern United States, more than 85% of the land is devoted to row crop agriculture or intensive grazing (Burkart et al., 1994). Small farms continue to be consolidated into larger farms in response to the need for economies of scale. In states east of the Rocky Mountains, vast areas are used to produce wheat, maize, soybeans and sorghum and to graze cattle (NRCS-USDA). Farm equipment that is operated by one person continues to become more sophisticated and able to cultivate and harvest larger and larger fields. The cost of the equipment and of labour further supports the continued expansion of large crop fields which are dependent on significant use of fertilizers and pesticides to optimize yields. In addition, to diversify income streams, farmers may fence off areas such as those along tightly meandering streams that are not suited to intensive crop production and graze livestock that usually have access to the streams within the fenced pastures. Livestock access to streams can do major damage to streambanks and stream water quality.

http://dx.doi.org/10.19103/AS.2018.0041.01

These trends of more intensive use of all available land in agricultural regions are likely to continue with the growing world population. Increased surface run-off laden with sediment and agrochemicals and streambank collapse continue to provide higher and more frequent peak stream flows. These are characterized by high sediment and agrochemical loads that result in more flooding, incision and widening of stream channels, reduction of base flows and reduction in water quality and the quality of aquatic ecosystems.

Despite our best efforts, it is unlikely that significant reduction in nutrient and sediment loading to surface waters will be achieved through voluntary, traditional in-field management alone (Dinnes et al., 2002). Increased use, by some farmers, of techniques such as cover crops, frequent side-dressing of small amounts of fertilizer, slow release fertilizer to promote soil and water quality, nutrient cycling efficiency and crop productivity have been studied as a way to reduce nutrient and sediment loading to surface waters and some farmers are using them (Snapp et al., 2005). However, they have some disadvantages, including increased farming costs, delay of spring soil warming and making it more difficult to predict nitrogen (N) mineralization, creating challenges to widespread adoption of such practices (Roesch-McNally et al., 2017).

The Natural Resources Conservation Service (NRCS) of the United States Department of Agriculture (USDA) National Water Quality Initiative is designed to provide both in-field and edge-of-field practices such as buffer and filter strips to promote soil health, reduce erosion and lessen nutrient run-off (https://www.nrcs.usda.gov/wps/portal/nrcs/detail/national/water/?cid=stelprdb1047761). One of the major NRCS Conservation Practice Standards is the Riparian Forest Buffer Standard (Practice 391). In those watersheds where the programme has been available, over 3600 farmers have taken advantage of the programme (Brewer, 2002). However, in the first 8 years of the programme, only 825 000 acres of farmland in the priority watersheds were enrolled with 11 impaired water bodies improved to the point of being removed from the US Environmental Protection Agency Impaired Waters List (USEPA 303(d) list). There are more than 390 million acres of cropland in the United States, many of them in need of conservation practices that protect surface waters (USDA 2012 Census of Agriculture).

2 Riparian forest buffers

Riparian forest buffers are an agroforestry practice that, when properly applied to the agricultural landscape, can enhance and diversify farm income opportunities, improve the environment and create wildlife habitat (Schultz et al., 2009). By developing an understanding of the interactions between the trees, shrubs and native prairie plants or introduced grasses, buffers can be designed to capture most surface run-off before it reaches the stream channel while also stabilizing the banks of the stream channel.

Riparian forest buffers are planned combinations of trees, shrubs, grasses, forbs and bioengineered structures adjacent to or within a stream channel designed to mitigate the impact of land use on the stream or creek. At the landscape level, riparian forest buffers link the land and aquatic environment and perform vital ecological functions as part of the network of watersheds that connect forests, prairies, agricultural and urban lands. By establishing and managing the trees, shrubs, grasses and forbs in the riparian zone, water quality and the aquatic ecosystems can be maintained or enhanced and the impact of floods can be mitigated. However, to be effective, riparian buffers must include plants that are adapted to the soils, topography and flood regime of the riparian zone and the stream as well as the long-term management by the landowner.

A well-established and maintained riparian forest buffer can:

- protect and improve water quality;
- stabilize eroding streambanks;
- help reduce flood impacts;
- recharge shallow groundwater;
- supply diverse food and cover for upland wildlife;
- enhance biodiversity of the landscape;
- improve carbon sequestration;
- improve aquatic habitats for fish and other organisms; and
- generate farm income from products harvested from the buffer.

An overview of the environmental benefits of riparian and other types of buffers has been provided by Lovell and Sullivan (2006) and Gundersen et al. (2010). The role of riparian and other agroforestry techniques in preventing nutrient run-off and NPS pollution is reviewed in Udawatta et al. (2002, 2006, 2011), Lee et al. (2003) and Simpkins et al. (2003). The impact of riparian buffers in enhancing water quality, preventing sediment trapping and streambank erosion is discussed in Zaimes et al. (2004), Liu et al. (2008), Klapproth and Johnson (2009a) and Udawatta et al. (2010). The broader potential contribution of riparian buffers and forests to carbon sequestration is reviewed in Dybala et al. (2019), while the biodiversity and broader social benefits they provide are discussed in Klapproth and Johnson (2009b,c).

Before designing and establishing a riparian buffer, it is critical to understand upland plant communities and their present management, the objectives of the landowner interested in installing a riparian forest buffer and their willingness and ability to manage that buffer.

Landowner concerns associated with establishment of buffers can include concerns such as:

- how much can buffers reduce sediment and nutrient movement into a stream;
- can buffers be used to heal gullies;
- can buffers reduce streambank erosion and slow stream meandering;
- what kind of buffer vegetation produces the best wildlife habitat and fishery;
- will trees in a riparian forest buffer fall into the stream and back up water into crop fields and field drainage tiles;
- are buffers a source of weed seeds;
- are cool-season grass filters as effective as riparian forest buffers;
- will forest buffers attract beavers that build dams that back up water;
- will deer become a problem for crops;
- how much maintenance is required to keep a buffer functioning properly;
- will a buffer be damaged by floods;
- is fencing needed to keep livestock out of a buffer;
- how much land will be taken out of crop production or pasture; and
- can specific products be harvested from the buffer to offset income losses from the land and similar other questions.

Riparian buffers can take different forms in response to landowner objectives and concerns as well as the regional location of the streams being buffered. Riparian forest buffers in agricultural landscapes in the Eastern United States, for example, may contain narrow corridors of remnant forests along streams with little else being needed to create an effective buffer of crop field run-off other than a grass filter lying between the crop field and the existing forest buffer. In the arid and semi-arid west, riparian forest buffers may consist of narrow strips of native flood plain species that often lie between grazed shrub and short grass communities and the stream. In the agricultural belt of the Midwestern United States, riparian forest buffers often need to be established from scratch.

Because riparian forest communities naturally evolved in the most fertile and moist position of the landscape, they are often easy to reestablish. However, in many agricultural landscapes, land uses have changed the hydrology so dramatically that the hydrology of these communities cannot be restored to their original condition. Stream channels have been incised and widened by higher discharge resulting from greater surface run-off from crop fields and heavily grazed pastures. Channelization of meandering streams, field tiling of some landscapes and urbanization also have contributed to higher stormflows and lower base flows. In many cases, water tables have been lowered to the point that the restored buffers require a plant community that did not naturally occur in that location. However, with proper planning and design the function of a healthy riparian forest community can be reestablished.

3 Riparian forest buffer design and function

Riparian forest buffers typically should be composed of three distinct management zones (Fig. 1):

- Zone 1: Undisturbed forest
- Zone 2: Managed forest and shrubs
- Zone 3: Run-off control (grasses and forbs)

These zones contain different kinds of plantings with different functions.

Zone 1 includes a zone of trees whose major function is to stabilize the streambank, provide a large long-term nutrient sink, help improve soil quality through annual leaf litterfall, provide vertical structure for wildlife habitat and potentially provide some shade to the stream channel to help stabilize daily stream water temperature especially if the desired fishery includes cold water demanding species such as trout (Table 1).

Trees should not be placed so close to the edge of the bank that they completely shade the stream throughout the whole day once they are mature.

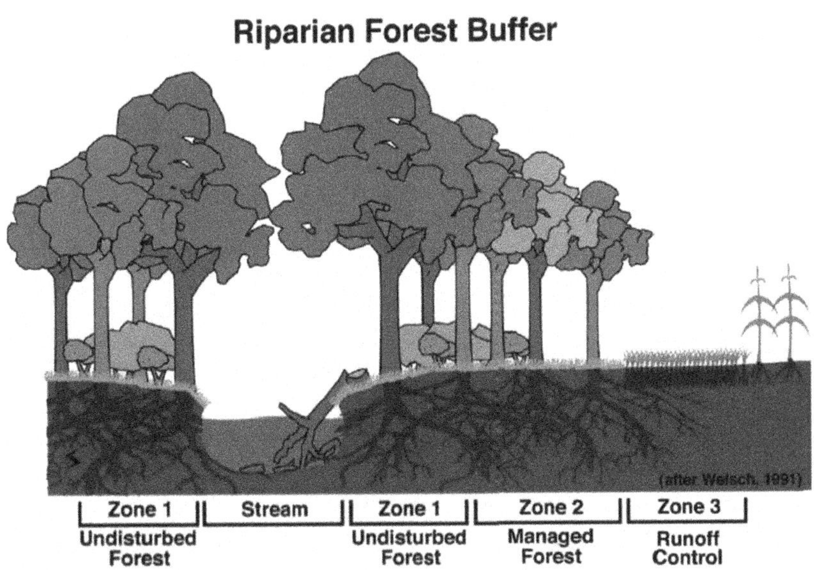

Figure 1 The traditional three-zone riparian buffer. Source: reprinted with permission from Schultz et al. (2004) by American Society of Agronomy.

Table 1 Functions of the grass, shrub and tree components of riparian buffers

Kind of plant	Functions
Prairie grasses/forbs	1. Slow water entering the buffer
	2. Trap sediment and associated chemicals
	3. Add organic carbon to a range of soil depth
	4. Added carbon improves soil structure
	5. Improve infiltration capacity of the surface soil
	6. Above-ground nutrient sink needs annual harvest
	7. Provide diverse wildlife habitat
	8. Do not significantly shade the stream channel
	9. Provide only fine organic matter input to stream
	10. Can provide forage and other products
Shrubs	1. Multiple stems act as a trap for flood debris
	2. Provide woody roots for bank stabilization
	3. Litter fall helps improve surface soil quality
	4. Above-ground nutrient sink needs occasional harvest
	5. Adds vertical structure for wildlife habitat
	6. Do not significantly shade the stream channel
	7. Provide only fine organic matter input to stream
	8. Can provide ornamental products and berries
Trees	1. Strong, deep woody roots stabilize banks
	2. Litter fall helps improve surface soil quality
	3. Long-lived, large nutrient sink needs infrequent harvest
	4. Adds vertical structure for wildlife habitat
	5. Vertical structure may inhibit buffer use by grassland birds
	6. Shade stream, lowering temperature and stabilizing dissolved oxygen
	7. Provide both fine organic matter and large woody debris to the channel
	8. Can provide a wide variety of fibre products

The reason is that if the streambank is completely shaded, grasses needed to stabilize the bank will be difficult to establish and grow and bank erosion will continue with the chance that trees could fall into the channel creating problems. If, however, streams are large enough and there is a desire to create natural in-stream habitat, trees could be placed closer to the bank so that some large mature trees could fall into the channel to provide large woody debris that helps create important pool and woody in-stream habitat for a myriad of organisms. For this to happen, trees in Zone 1 would not be harvested during their life time. If in-stream large woody debris is not desired because of concern for development of log dams and increased water levels harvesting with replacement would need to be done.

Tree spacing in Zone 1 should also be wide enough to allow grass or some other kind of cover crop to grow under the trees. If that does not happen, bare soil will exist under the trees after the litter from the previous fall leaf drop has either decomposed or been washed into the stream by surface run-off or by floodwaters from out-of-bank stream flow. This is an important consideration since the riparian buffer has been designed to slow surface run-off by providing a frictional surface of plants completely covering the soil. If there is no permanent ground cover under the trees, soil erosion and even gullies will develop carrying sediment into the stream and weakening streambanks.

One other consideration for row spacing is the direction that the buffers run. Rows should be wider for tree and shrub rows that run east and west to allow a longer time for the sun to actually shine directly down to the soil level. Rows running north and south can be narrower because the sun shines down between the rows unimpeded by the shade from trees.

Species of trees to be planted in Zone 1 would depend on channel incision. If the channel is a natural channel that is in contact with its flood plain, meaning that flood events typically would occur once every 2–3 years, true riparian tree species should be selected because the water table in such a situation would be relatively close to the surface. If on the other hand the channel is deeply incised as happens in watersheds where flow regimes have changed because of land-use changes or climate change, both true riparian tree species and upland tree species could be planted. Upland species are often slower growing and longer-lived trees. Trees in this zone might also include fruit trees or even shrubs if there was a desire to provide a short-term income generating plant community. If the buffer was established with assistance from a government conservation programme, it may not be possible to sell harvested fruits until the enrolled programme has lapsed. Zone 1 is usually the widest zone occupying up to two-thirds of the width of the buffer.

Zone 2 combines planting of trees and shrubs. It both helps to manage floodwater and allows run-off to infiltrate or percolate into the soil so that waterborne nutrients or pollutants are absorbed and cleansed by the soil and vegetation. Zone 2 would be used if Zone 1 consisted of trees. Zone 2 would have several rows of shrubs again spaced far enough apart so that sunlight can reach the soil at least during part of the day.

These rows of multiple woody stems provide an important barrier for slowing floodwater that is moving out into the agricultural field and trapping debris brought by the floodwater. The debris may consist of a wide variety of objects including large woody debris that, if not trapped by the shrubs in Zone 2, would end up in the adjacent crop fields, physically damaging the crops and requiring time and money for the farmer to remove. The shrub species in this zone can consist of edible berries and/or decorative woody florals such as red osier dogwood and curly willow. These are valuable components of the floral and decorating industries and can thus provide the farmer with income. The shrubs in Zone 2 also can provide a significant wildlife benefit to the buffer especially in attracting birds that may be important in helping to control pests in the adjacent crop fields. Bird species that are attracted to the shrub zone can be manipulated through the selection of shrub species that are planted.

Zone 3 is the zone adjacent to the crop field and the most important of the three zones. The zone is designed to provide high infiltration, sediment trapping and nutrient uptake ability while also dispersing any concentrated flow that runs into it. Native grasses and forbs provide the best buffering. They help to restore biological and physical soil quality to heavily used soils by adding large amounts of carbon to the profile from rapid turnover of roots that contain more than 70% of the total biomass of native prairie plant communities. This carbon plays a key role in redeveloping soil macro-aggregate structure that helps facilitate the high infiltration rates needed to get surface run-off into the soil profile. The carbon also serves as a substrate for increased soil microbial activity that is important both in building soil structure and processing some agricultural chemicals that move in the surface and groundwater (Dornbush et al., 2008).

Cool-season grasses are good at protecting the soil because, when water runs through the filter, they lie down, allowing the water to run over them and protecting the soil. However, they do not slow the flow of the water and are thus better adapted for use in grass waterways. Native grasses and forbs slow the water because their stiff stems seldom bend in response to surface run-off. This results in the majority of the sediment in surface run-off being dropped on the crop field at the edge of the buffer prior to the water moving through Zone 3. That sediment can be moved back upslope where it can be used by crops. The mix of native prairie grasses and forbs provides excellent habitat for prairie and forest edge wildlife such as pheasants and quail. In regions where hunting game is an activity, riparian buffers with a Zone 3 prairie strip and a Zone 2 shrub strip can provide excellent bird hunting opportunities that some landowners lease to hunters.

If a crop field has more than a 5–8% slope, a pure switch grass (*Panicum virgatum*) strip can be planted at the field edge of Zone 1 to slow the water. The deep rooting habit of the native prairie grasses and forbs creates a soil that has high infiltration rates. Even on soils that have been cultivated for many years and lost their surface soil structure, native prairie plants can recreate soil structure and porosity similar to that of the original undisturbed soil in 8–10 years (Marquez et al., 2004, 2019). Cool-season grasses such as fescue and brome take a significantly longer time to improve infiltration rates to the same depth as under a native plant buffer community.

Planning considerations during buffer design should include a strong focus on the landowners' desires and objectives while also retaining the buffer's ability to provide critical environmental benefits and services. Buffers designed for stabilizing collapsing streambanks in deeply incised channels should have the first rows of trees set back far enough from the edge of the bank to allow grass or other dense ground cover to grow both above and on the bank in full sunlight. In such cases, trees that have a propensity for producing large major roots with many smaller fibrous roots should be selected as these species' root systems can provide the reinforcing structures that hold the banks in place. Buffers designed to maximize capture and filtration of crop field surface run-off should consist of native prairie grasses and forbs that provide stiff stems to slow water at the field edge of the buffer, dropping much of the sediment and then providing high infiltration rates to significant depths that allow the potentially nutrient laden water to be filtered through an active plant community root system.

4 Special design considerations and management

Design guidelines and planning tools for riparian forest buffers are provided by Bentrup (2008) and MacFarland (2017). While it is relatively straightforward to design a three-zone riparian buffer based on the above standard design, it is critical to fit that buffer to the actual landscape which often requires additional conservation practices that must be integrated with the buffer to make the system function to its maximum potential. To accomplish this integration of various potential conservation practices, design planning should include an on-the-ground walk-through with the landowner as well as an aerial photograph of the site that shows other conservation practices such as grass waterways and other problem areas such as field drainage tiles, gullies and areas of severe bank erosion and collapse. Recent advances in remote sensing have the potential to help buffer zone planning and management significantly (Herring et al., 2006; Goetz, 2006). Techniques such as high-resolution imaging and laser-based techniques can provide detailed information on buffer zone properties such as topography, buffer length, width and vegetation structure as well as stream flow.

As mentioned earlier, where there is severe bank erosion, Zone 1 trees should be set back from the bank edge to allow enough sunlight to support the growth of dense grass or other cover on the bank. Shrubs could replace trees in the first row or two of Zone 1 to reduce the potential of shading the stream. This is an important consideration in some prairie landscapes where warm-water streams exist and lowering water temperature is not desired. Replacing the first row or two of trees with shrubs may also be appropriate where there is significant landowner concern about large woody debris falling into the stream which might raise water levels, thus backing up water into field drainage tiles.

If the region includes field drainage tiles such as those found on the Des Moines lobe and other recently glaciated regions in the Midwestern United States, a grass waterway of introduced cool-season grasses should be planted over the tile as it passes under the buffer unless the tile can be replaced by a section of solid tile that has no access holes or cracks that provide potential access to plant roots. The deep roots of the native plant community or of trees will access the field drainage tile and plug it to the point that it will no longer carry water from the wet areas of the upland crop field.

In areas where grass waterways from the upland intersect the riparian forest buffer, Zone 3 of the buffer should be expanded out into the crop field or at the expense of the other two buffer zones. Grass waterways are designed to carry surface run-off water safely downhill. When this fast-paced water approaches a buffer strip, it must be dramatically slowed to allow the water to infiltrate into the soil below the buffer. In such cases, Zone 3 should consist only of native grasses and forbs with a strip of native grasses without forbs right at the field edge of Zone 3. The grass waterway should be widened at the edge of the buffer, creating a pyramidal structure with the base against the buffer to allow water a place to slow and sit before it moves through Zone 3.

Buffer widths can vary depending on space available, soil and slope conditions and landowner objectives (Fig. 2). Dosskey et al. (2015) have developed AgBufferBuilder, a GIS tool to design buffer strips using digital elevation models and buffer area ratio relationships, to develop buffers that have a constant level of trapping efficiency along the extent of the buffer. Riparian buffers as narrow as 10–15 m can provide surface erosion control, but nitrogen (N) reduction in subsurface flow may require widths of 30–46 m depending on the soil type and the slope of the riparian zone. When working with the landowner, it is important to determine the number of up and down field passes a field operative can make with the equipment available. In a rectangular field, buffer widths should allow the farmer to end tillage and harvesting passes up and down field in a way that will bring them back to the end of the field that they use for access to the field. If the stream meanders along the edge of the field, the buffer will need to vary in width to create the rectangular shape of the field.

The length of a riparian buffer system should ideally include both sides of the channel beginning in the headwaters of the watershed and extending continuously as far downstream as possible. Natural buffers can be part of the total length of a buffer system as long as they fall within the required widths needed to capture surface run-off and reduce the nutrient content of the subsurface flow to the channel. Leaving significant lengths of the channel without riparian buffers or some other kind of perennial cover can actually

Figure 2 A riparian buffer applied to the landscape. Note the variation in width of Zone 3, the native grass and forb zone, to fit the landowners' objectives. Also note the gentle smooth edge of Zone 3 along the buffer to allow ease of cultivation with large equipment.

create more problems for channel stability. If bank collapse takes place along unbuffered reaches of the channel, channel widths increase in that zone which causes water depths in the channel to decrease until the water hits a narrow channel that is stable because it is buffered. The turbulence caused by forcing the water into a narrower channel can increase undercutting and scouring along the buffered bank, causing it to collapse especially if the riparian buffer is relatively young.

Long-term management of the buffer is required to maintain its design functionality. No grazing should be allowed in Zones 1 and 2. If properly managed, flash or rotational grazing in dry soil conditions can be undertaken in a Zone 3 sown with cool-season grass. Some harvesting could be done in Zone 1 if the species used are stump or root sprouting species. If they are root sprouting species, row definition will be lost and woody tree stems could sprout into the Zone 3 grasses and forbs because tree roots extend laterally away from the bole to an average distance of one tree height from the base. Berries and shrub stems, to be used for ornamental purposes, can be harvested from Zone 2. Zone 3 cool-season grass could be cut for hay once in the growing season – ideally done in the season with the least potential flooding. More importantly, if Zone 3 consists of native grasses and forbs, it must be burned every 3–5 years to maintain the biodiversity of the plant community. If that is not done, invasive weeds or grasses, such as reed canary grass (*Phalaris arundinacea*), will find their way into the zone over time.

Riparian forest buffers may need to be used in conjunction with other riparian management practices such as streambank bioengineering, in-stream boulder weirs or constructed wetlands. Both streambank bioengineering and in-stream boulder weirs (Fig. 3) are designed to stabilize the channel by creating steps in the channel evolution process. In-stream boulder weirs are designed to slow down-cutting in the channel by creating a series of weirs in the channel with 1:4 slopes upstream of a series of cross-channel crest stones, and 1:20 slopes of rock downstream of the crest stones. Weirs are placed so that the upstream pool behind the crest stones of one weir are backed-up to the gentle downstream side of the weir upstream of that weir during base-flow. The goal is to reduce down-cutting and pool development downstream of the weir crest stones.

Streambank bioengineering is designed to stabilize eroding streambanks usually associated with a channel that needs to adjust to changes in discharge.

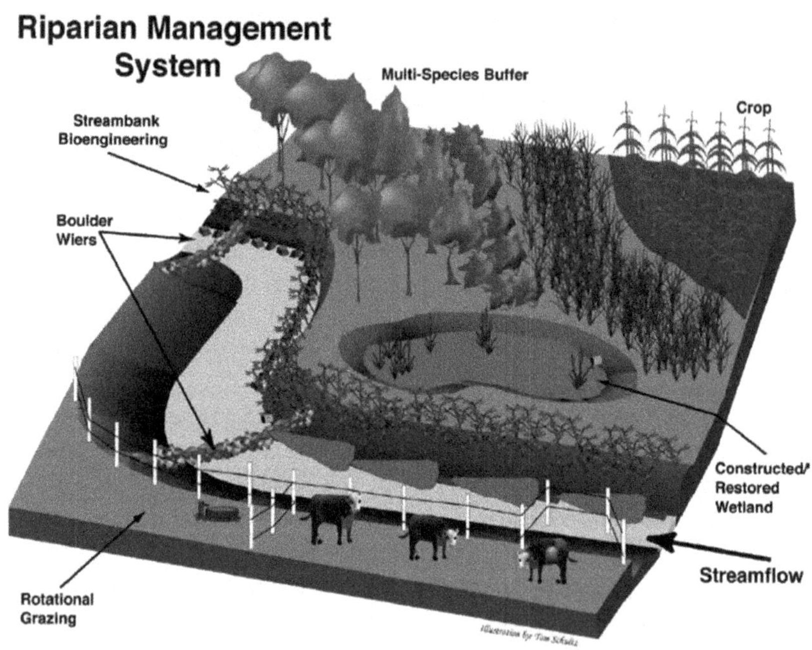

Figure 3 Riparian management system practices including from top right: streambank bioengineering, in-stream boulder weir structures, intensive rotational grazing, constructed wetlands and riparian forest buffer that could also be designed as a saturated buffer. Source: reprinted with permission from Schultz et al. (2004) by American Society of Agronomy.

Channels that are down-cutting, with the streambanks being taller than can be supported by the bank soil or parent material, can result in the banks collapsing, thus widening the channel. Bioengineering techniques include use of both hard engineering materials such as boulders for toe control and grasses and woody cuttings placed into the bank wall (Figs. 4 and 5).

Figure 4 Streambank bioengineering on streambanks that were severely eroded. Bear Creek in Central Iowa, USA - A USDA National Research and Demonstration Site - 1998.

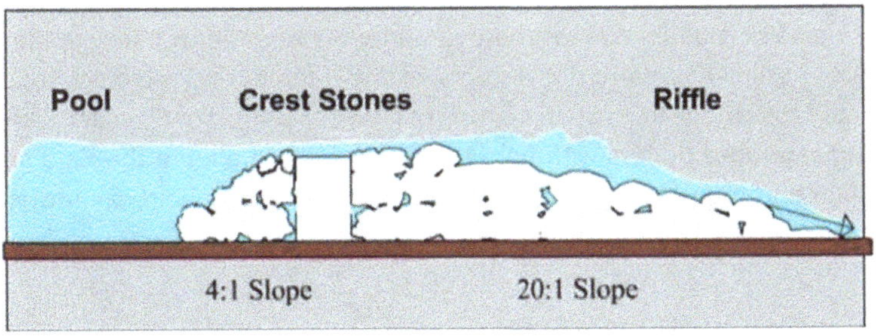

Figure 5 Boulder weir to control channel down-cutting by having a long gentle slope on the downstream side of the structure. Crest stones should be large stones that in the top layer allow fish to move over the crest. Boulder weirs are installed in a sequence with the upstream pool of the downstream weir backing water up to the long, gentle downstream rock sequence at base-flow conditions.

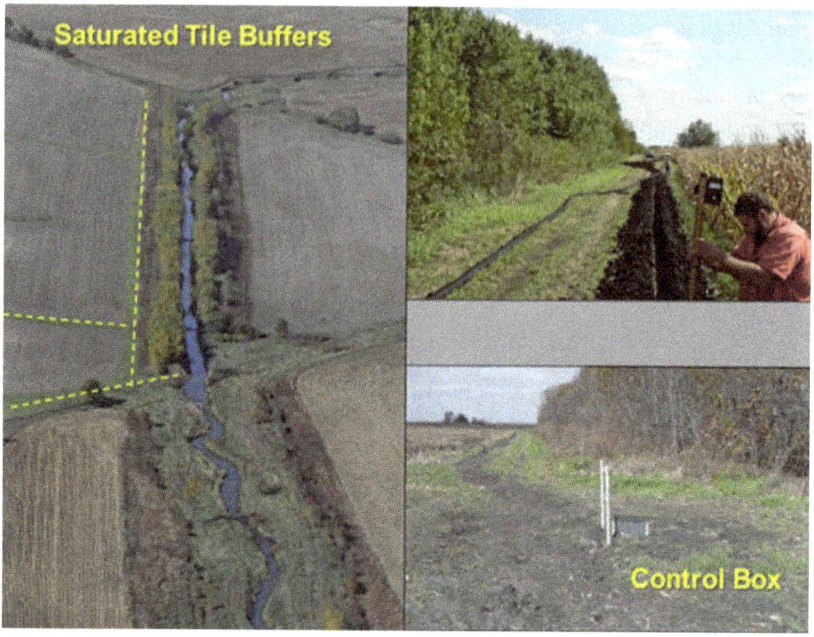

Figure 6 Saturated buffer. Field drain tile that intercepts tiles from upslope is laid parallel to the buffer. Water from the parallel tile now moves through the buffer instead of directly into the channel without any plant/soil treatment of the nutrient load.

Saturated riparian buffers are a relatively new addition to the buffer portfolio (Fig. 6). In this practice, a field tile that intercepts field tile draining upslope areas is laid parallel to the field edge of the buffer (Jaynes and Isenhart, 2014). Water flowing into this tile moves out of the tile and through the riparian buffer subjecting it to the treatment of the soil and plant community. Nitrate reduction is as high as 90% in the water flowing through the buffer before reaching the stream.

Streamside buffers cannot remove materials from field drainage tiles that exit directly into a stream. But an acre of tile-intercepting wetland has been calculated to remove from 20 to 40 tons of N over a period of 60 years. Likewise, creating saturated buffers with field tile that is laid parallel to the field side of the buffer and intercepts field tile that drains the adjacent field can reduce the nitrate content by 90% (Jaynes and Isenhart, 2014).

5 Assessing buffer performance

Tracer tests and isotope evidence shows that denitrification is the major groundwater nitrate removal mechanism in the buffer system (Schultz et al., 2004). Stratigraphy below buffers can determine the effectiveness of nutrient removal from shallow groundwater. With a shallow confining layer of till below a loamy root zone, buffers can remove up to 90% of the nitrate in groundwater. When the confining layer is found well below the rooting zone and porous sand and gravel are found between the till and the loam, residence time and contact with roots is dramatically reduced and buffers are unable to remove much nitrate from the groundwater. The difficulty in describing the stratigraphy below buffers makes it difficult to quantify the specific amount of remediation that a planned buffer might provide. To be able to measure in-stream water quality improvement, continuous buffers on both sides of the stream must extend at least 15 km.

Studies that have been conducted on the riparian forest buffers in the Bear Creek Watershed located on the Des Moines Lobe in Central Iowa in the United States have shown that a 7-m wide native grass filter strip on either side of the stream can reduce sediment loss to the stream by 95% and total nitrogen, phosphorous and nitrate and phosphate in the surface water by 60% (Schultz et al., 2004). This research suggests that adding a 9-m wide woody buffer to the grass filter results in removal of 97% of the sediment and 80% of the nutrients. There is also a 20% increase in removal of soluble nutrients with the added width (Simpkins et al., 2003; Lee et al., 2000, 2003). Riparian forest buffers can reach maximum efficiency for sediment removal in as little as 5 years and nutrient removal in as little as 10–15 years. Water can infiltrate in the soil up to five times faster in restored buffers in as little as 6 years after establishment compared to adjacent crop fields. Riparian buffer strips also have been shown to retain between 79% and 94% of the atrazine in run off from adjacent crop fields (Reungsang et al., 2005).

In terms of soil stability, it has been shown that buffered streambanks lose up to 80% less soil than row cropped or heavily grazed streambanks (Zaimes et al., 2004; Marquez et al., 2004). A study in the Central Claypan area of northeastern Missouri found that at the watershed scale, streambank erosion accounted for an average of 88% of the in-stream sediment and 23% of the N load on an annual basis suggesting the importance of using perennial vegetation to stabilize streambanks (Willett et al., 2012). Soils in riparian forest buffers contain up to 66% more total organic carbon in the top 50 cm than adjacent crop field soils (Tufekcioglu et al., 2003). *Populus* hybrids and switchgrass living and dead biomass sequester 3000 and 800 kg C ha^{-1} and immobilize 37 and 16 kg N ha^{-1}, respectively. Riparian forest buffers have more than eight times more below ground biomass than adjacent crop fields. Buffer soils show a 2.5-fold increase in soil microbial biomass and a fourfold increase in denitrification in the surface 50 cm of soil when compared with the adjacent crop field soil.

Bird species' use of buffers has shown that riparian forest buffers with a three-zone system of trees, shrubs and native grasses which provide a variety of habitat structure will support over 40 different bird species over the year in central Iowa compared to 8–10 species in non-buffer agricultural riparian zones with row crop culture to within 5 m of the channel (Berges et al., 2010). If properly designed, riparian buffers can protect a stream from chemical and sediment pollution while providing both terrestrial and aquatic wildlife habitat in agricultural landscapes that are dedicated to producing annual crops to feed humans and livestock or to create biofuels that replace fossil fuels.

6 References

Bentrup, G. 2008. Conservation buffers: design guidelines for buffers, corridors and greenways. USDA General Technical Report SRS-109. Available at: https://www.srs.fs.usda.gov/pubs/33522.

Berges, S. A., Schulte Moore, L. A., Isenhart, T. M. and Schultz, R. C. 2010. Bird species diversity in riparian buffers, row crop fields, and grazed pastures within agriculturally dominated watersheds. *Agroforest. Syst.* 79(1), 97–110. doi:10.1007/s10457-009-9270-6.

Brewer, M. 2002. Financial agents, water quality and riparian forest buffers. MS Thesis. Iowa State University, USA.

Burkart, M. R., Oberle, S. L., Hewitt, M. J. and Picklus, J. 1994. A framework for regional agroecosystems characterization using the national resources inventory. *J. Environ. Qual.* 23(5), 866-74. doi:10.2134/jeq1994.00472425002300050002x.

Dinnes, D., Karlen, D., Jaynes, D., Kaspar, T. C., Hatfield, J. L., Colvin, T. S. and Cambardella, C. A. 2002. Nitrogen management strategies to reduce leaching in tile-drained Midwestern soils. *Agron. J.* 94, 152–71.

Dornbush, M., Cambardella, C., Ingham, E. and Raich, J. 2008. A comparison of soil food webs beneath C3- and C4-dominated grasslands. *Biol. Fertil. Soils* 45(1), 73-81, doi:10.1007/s00374-008-0312-4.

Dosskey, M. G., Neelakantan, S., Mueller, T. G., Kellerman, T., Helmers, M. J. and Rienzi, E. 2015. AgBufferBuilder: A geographic information system (GIS) tool for precision design and performance assessment of filter strips. *J. Soil Water Conserv.* 70(4), 209-17.

Dybala, K. E., Matzek, V., Gardali, T. and Seavy, N. E. 2019. Carbon sequestration in riparian forests: a global synthesis and meta-analysis. *Glob. Change Biol.* 25(1), 57–67. doi:10.1111/gcb.14475.

Goetz, S. J. 2006. Remote sensing of riparian buffers: past progress and future prospects. *J. Amer. Water Res. Assoc.* 42(1), 133-43. doi:10.1111/j.1752-1688.2006.tb03829.x.

Gundersen, P., Lauren, A., Finer, L., Ring, E., Koivusalo, H., Saetersdal, M., Weslien, J. O., Sigurdsson, B. D., Högbom, L., Laine, J., et al. 2010. Environmental services provided from riparian forests in Nordic countries. *Ambio* 39(8), 555–66. doi:10.1007/s13280-010-0073-9.

Herring, J. P., Schultz, R. C. and Isenhart, T. M. 2006. Watershed-scale inventory of existing riparian buffers in northeast Missouri using GIS. *J. Amer. Water Res. Assoc.* 42(1), 145–55. doi:10.1111/j.1752-1688.2006.tb03830.x.

Jaynes, D. B. and Isenhart, T. M. 2014. Reconnecting tile drainage to riparian buffer hydrology for enhanced nitrate removal. *J. Environ. Qual.* 43(2), 631–8. doi:10.2134/jeq2013.08.0331.

Klapproth, J. and Johnson, J. 2009a. Understanding the science behind riparian forest buffers: effects on water quality. Virginia Cooperative Extension. Virginia Tech. Available at: https://pubs.ext.vt.edu/content/dam/pubs_ext_vt_edu/420/420-151/420-151_pdf.pdf.

Klapproth, J. and Johnson, J. 2009b. Understanding the science behind riparian forest buffers: effects on plant and animal communities. Virginia Cooperative Extension. Virginia Tech. Available at: http://pubs.ext.vt.edu/content/dam/pubs_ext_vt_edu/420/420-152/420-152_pdf.pdf.

Klapproth, J. and Johnson, J. 2009c. Understanding the science behind riparian forest buffers: benefits to communities and landowners. Virginia Cooperative Extension. Virginia Tech. Available at: https://pubs.ext.vt.edu/content/dam/pubs_ext_vt_edu/420/420-153/420-153_pdf.pdf.

Lee, K. H., Isenhart, T. M., Schultz, R. C. and Mickelson, S. K. 2000. Multispecies riparian buffers trap sediment and nutrients during rainfall simulations. *J. Environ. Qual.* 29(4), 1200–5. doi:10.2134/jeq2000.00472425002900040025x.

Lee, K. H., Isenhart, T. and Schultz, R. 2003. Sediment and nutrient removal in an established multi-species riparian buffer. *J. Soil Water Conserv.* 58, 1–8.

Liu, X., Zhang, X. and Zhang, M. 2008. Major factors affecting the efficacy of vegetated buffers on sediment trapping: a review and analysis. *J. Environ. Qual.* 37(5), 1667–74. doi:10.2134/jeq2007.0437.

Lovell, S. T. and Sullivan, W. C. 2006. Environmental benefits of conservation buffers in the United States: evidence, promise and open questions. *Agric. Ecosyst. Environ.* 112(4), 249–60. doi:10.1016/j.agee.2005.08.002.

MacFarland, K., Straight, R. and Dosskey, M. 2017. Riparian forest buffers: an agroforestry practice. Available at: https://www.fs.usda.gov/nac/documents/agroforestrynotes/an49rfb01.pdf.

Marquez, C. O., Garcia, V. J., Cambardella, C. A., Schultz, R. C. and Isenhart, T. M. 2004. Aggregate-size stability distribution and soil stability. *Soil Sci. Soc. Am. J.* 68(3), 725–35. doi:10.2136/sssaj2004.7250.

Marquez, C. O., Garcia, V. J., Schultz, R. C. and Isenhart, T. M. 2019. A conceptual framework to study soil aggregate dynamics. *Eur. J. Soil Sci.*, 1–14. doi:10.1111/ejss.12775.

NRCS Conservation Riparian Forest Buffer Practice Code 391. Available at: https://efotg.sc.egov.usda.gov/references/public/IA/Riparian_Forest_Buffer_391_STD_2014_05.pdf.

NRCS-USDA. The state of the land. Available at: https://www.nrcs.usda.gov/Internet/FSE_DOCUMENTS/nrcs143_012458.pdf.

Reungsang, A., Moorman, T. B. and Kanwar, R. S. 2005. Prediction of atrazine fate in riparian buffer strips soils using the root zone water quality model. *J. Water. Environ. Tech.* 3(2), 209–22. doi:10.2965/jwet.2005.209.

Roesch-McNally, G., Basche, A., Arbuckle, J., Tyndall, J. C., Miguez, F. E., Bowman, T. and Clay, R. 2017. The trouble with cover cropping: farmers' experiences with overcoming barriers to adoption. *Renew. Agric. Food Syst.* 33(4), 322–33.

Schultz, R. C., Isenhart, T. M., Simpkins, W. W. and Colletti, J. P. 2004. Riparian forest buffers in agroecosystems - lessons learned from the Bear Creek Watershed, Central Iowa, U.S.A. *Agrofor. Syst.* 63, 35–50.

Schultz, R. C., Isenhart, T. M., Colletti, J. P., Simpkins, W. W., Udawatta, R. P. and Schultz, P. L. 2009. Riparian and upland buffer practices. In: Garett, H. (Ed.), *North American*

Agroforestry: an Integrated Science and Practice. American Society of Agronomy, Madison, WI.

Simpkins, W., Wineland, T., Isenhart, T. and Schultz, R. 2003. Hydrologic setting control NO3-N removal in groundwater beneath multi-species riparian buffers. In: Kolpin, D. and Williams, J. (Eds), *Agricultural Hydrology and Water Quality: AWRA 2003 Spring Conference Proceedings*. American Water Resources Association, Middleburg, VA.

Snapp, S., Swinton, S., Labarta, R., Mutch, D., Black, J. R., Leep, R., Nyiraneza, J. and O'Neil, K. 2005. Evaluating cover crops for benefits, costs and performance within cropping system niches. *Agron. J.* 97(1), 322–32.

Tufekcioglu, A., Raich, J. W., Isenhart, T. M. and Schultz, R. C. 2003. Biomass carbon and nitrogen dynamics of multi-species riparian buffer zones within an agricultural watershed in Iowa, USA. *Agrofor. Syst.* 57(3), 187–98. doi:10.1023/A:1024898615284.

Udawatta, R. P., Krstansky, J. J., Henderson, G. S. and Garrett, H. E. 2002. Agroforestry practices, runoff and nutrient loss: a paired watershed comparison. *J. Environ. Qual.* 31(4), 1214–25.

Udawatta, R. P., Motavalli, P. P., Garrett, H. E. and Krstansky, J. J. 2006. Nitrogen losses in runoff from three adjacent agricultural watersheds with claypan soils. *Agric. Ecosyst. Environ.* 117(1), 39–48. doi:10.1016/j.agee.2006.03.002.

Udawatta, R. P., Garrett, H. E. and Kallenbach, R. L. 2010. Agroforestry and grass buffer effects on water quality in grazed pastures. *Agrofor. Syst.* 79(1), 81–7. doi:10.1007/s10457-010-9288-9.

Udawatta, R. P., Garrett, H. E. and Kallenbach, R. 2011. Agroforestry buffers for non-point source pollution reductions from agricultural watersheds. *J. Environ. Qual.* 40(3), 800–6. doi:10.2134/jeq2010.0168.

USEPA 303(d) List. https://www.epa.gov/tmdl/program-overview-303d-listing-impaired-waters.

Veum, K. S., Goyne, K. W., Motavalli, P. P. and Udawatta, R. P. 2009. Runoff and dissolved organic carbon loss from a paired watershed study of three adjacent agricultural watersheds. *Agric. Ecosyst. Environ.* 130(3–4), 115–22. doi:10.1016/j.agee.2008.12.006.

Willet, C. D., Lerch, R. N., Schultz, R. C., Berges, S. A., Peacher, R. D. and Isenhart, T. M. 2012. Streambank ersion in two watersheds of the Central Claypan Region of Missouri, United States. *J. Soil Water Conserv.* 67(4), 249–63. doi:10.2489/jswc.67.4.249.

Zaimes, G., Schultz, R. and Isenhart, T. 2004. Streambank erosion adjacent to riparian forest buffers, row-cropped fields and continuously-grazed pastures along Bear Creek in Central Iowa. *J. Soil Water Conserv.* 59, 19–27.

Chapter 2

Agroforestry practices: windbreaks

Gary Wyatt, Amanda Sames, and Diomy S. Zamora, University of Minnesota, USA

1 Introduction

Windbreaks, also known as vegetative environmental buffers (VEB) or shelterbelts, are a common feature of agricultural systems around the world. These strips of trees, shrubs, and other perennial or annual vegetation perform a number of functions including providing protection from the wind for homesteads, livestock, and crops; reducing soil erosion; providing protection from drifting snow; providing wildlife habitat; and enhancing aesthetics in agricultural landscapes (Brandle et al., 2004). More recently, windbreaks have been used to help mitigate dust and odor around swine and poultry facilities (Malone, 2004; Ullman et al., 2004; Lin et al., 2006; Colletti et al., 2006; Malone et al., 2006a; Tyndall and Colletti, 2007a; Adrizal et al., 2008; Tyndall, 2008; Burley et al., 2011).

Windbreaks are one of the five practices that underpin agroforestry. Agroforestry intentionally combines agriculture and forestry principles to create integrated and sustainable land-use systems. Agroforestry takes advantage of the synergies that can be derived from combining trees and/or shrubs with crops and/or livestock. Other forms of agroforestry practices are alley cropping, forest farming, riparian forest buffers, and silvopasture. A leading source of information on agroforestry in the United States is the US Department of Agriculture (USDA)–National Agroforestry Center in Lincoln, Nebraska (https://www.fs.usda.gov/nac/).

http://dx.doi.org/10.19103/AS.2018.0041.02

Several land grant universities have forestry and agroforestry classes and extension programs. The Center for Agroforestry at the University of Missouri (http://www.centerforagroforestry.org/) and the University of Minnesota offer resources for landowners (https://extension.umn.edu/forestry/agroforestry).

Windbreaks have a long history. An early national initiative dates back to the mid-1400s when the Scottish Parliament urged the planting of tree belts to protect agricultural production (Droze, 1977). From these beginnings, shelterbelts have been used extensively throughout the world to provide protection from the wind (Caborn, 1971; Brandle et al., 1988; Cleugh et al., 2002). In the United States, for example, as settlements moved west into the grasslands of the Great Plains, homesteaders planted trees to protect their homes, farms, and ranches.

A notable use of windbreaks in the United States came in the 1930s in response to the "Dust Bowl" conditions created by drought in the southern Plains, which led to severe soil erosion. The US Congress, then, authorized the Prairie States Forestry Project to plant windbreaks (Droze, 1977; Brandle et al., 2004) to address wind erosion and unemployment issues. In the span of about 7 years, foresters, the Civilian Conservation Corps and Works Progress Administration crews planted more than 200 million trees across six states. Of the more than 10000 original planting sites in Nebraska, more than 40% remain fully intact (Brocious, 2017).

2 Overview of windbreak benefits

Windbreaks have been shown to provide a range of benefits. They can improve crop yields, increase livestock health and comfort, reduce energy costs, and offer specialty products (nuts, fruits, or florals) or wood products (lumber, firewood). Windbreaks also increase soil health by preventing soil erosion (Brandle et al., 2004; Vacek et al., 2018), and can also prevent pesticide drift and pollution run-off from fields (Ucar and Hall, 2001). The microclimate created in the leeward side of windbreaks (in terms of air and soil temperature, radiation, humidity, and soil moisture) can increase crop yields in some circumstances (Cleugh and Hughes, 2002; Cleugh et al., 2002). In these spaces of decreased wind speed, temperatures tend to be slightly warmer and the air more humid than in the open field (Brandle et al., 2004). The exact effects of windbreaks on local microclimates, however, are highly contextual and may depend on such factors as time of day, orientation to the wind, windbreak internal and external structure, atmospheric conditions, and where measurements are taken (Brandle et al., 2004).

Research from around the world on crop yields around windbreaks have shown various results based on multiple variables including rainfall, temperature, and soil type (Kort, 1988). Documented crop yield increases include corn–12%, soybeans–13%, barley–25%, winter wheat–23%, spring wheat–8%, and hay–20% (Brandle et al., 1995, 2000). However, if the crop is planted close to the windbreak, there may be a crop yield decrease due possibly to shading and competition for belowground growth resources such as water and nutrients between the trees and the crops in that zone (Brandle et al., 2000). A grass and/or forbs buffer can be planted around the windbreak to improve sustainability and yields of the trees and crops. Windbreaks can also boost crop yields by playing a role in limiting the spread of airborne crop diseases (Gottwald and Timmer, 1995). Windbreaks are used, for example, in citrus orchards to prevent the spread of citrus canker bacteria caused when windblown rain splash transmits the bacteria into leaf and fruit tissue (Dewdney and Graham, 2014). As well as benefiting crops, properly placed windbreaks can also protect livestock in winter and summer weather conditions, either from the effects of wind, rain or snow in the winter, or from heat in the summer (Anderson and Bird, 1993).

Multi-row windbreaks can also reduce energy costs in rural homes from 10% to 25% (Ballasteros-Possu et al., 2017) due to the reduction of winter winds that can decrease the temperatures. Similarly, living snow fences (LSFs), which is a form of windbreaks, can reduce snow and drifting snow on farmland, residential homes, and on highways (Wight and Stuhr, 2002). LSFs are strategic plantings of trees and shrubs along highways for such purpose. Windbreaks are also used as visual screens to separate agricultural/industrial lands from residential areas, thereby increasing the aesthetic appeal of a landscape, and improving air quality by providing more oxygen from tree respiration. Windbreaks also filter dust particles thereby reduce odors from agricultural production (Raupach et al., 2001; Malone, 2004; Malone et al., 2006a,b; Colletti et al., 2006; Tyndall and Colletti, 2007a,b; Burley et al., 2011).

Windbreaks provide a habitat for wildlife and provide shelter, food, and travel corridors for insects, birds, and mammals, thus promoting pollination and the presence of beneficial species (e.g., those preying on crop pests) (Johnson and Beck, 1988; Hinsley and Bellamy, 2000). By planting nut and fruit bearing trees and shrubs, farmers can provide a habitat that can support many species of wildlife throughout the year (Wight and Stuhr, 2002). As with most tree plantings, windbreaks also store carbon in the biomass and in the soil, thus improving soil health (Ballesteros Possu, 2015).

3 Tree and shrub selection and planting

Tree and shrub selection is extremely important in windbreak design to ensure effective, long-lasting plantings (USDA, 2005). Plants need to be winter hardy and should have a successful history of planting in the selected area. Native plants should be the first choice. It is important to select multiple species of trees and shrubs to maximize the resilience of a windbreak. A mix of deciduous and coniferous plants is recommended. In selecting the plants, one must consider the potential of these plants to provide additional income to the farmer. Many plants offer potential sources of income such as fruits and nuts, timber, medicinal and specialty woods.

The number of rows and density of planting depends on the purpose of the windbreak. Windbreak density is the ratio of the solid portion of the windbreak to the total volume of the windbreak, while porosity is the ratio of the open portion of the windbreak to the total volume. The two terms are complementary in that their sum equals 1 or 100%. Wind flows through the open portions of a windbreak, thus as density increases, less wind passes through the barrier and wind speed reductions are greater (Brandle et al., 2009). Density depends on species used, planting spacing, and number of rows. Table 1 provides guidelines on planting densities for different kinds of windbreak function.

It is important to note that tree and shrub plant spacing affects active growth. Trees planted too close together may be stunted due to crowding, while planting too far apart from each other may cause gaps which reduce the effectiveness of the windbreak. It is important to note that conifers also need to be spaced apart to help prevent the spread of disease. Table 2 presents a general rule on establishing a windbreak based on species type.

Shrub willows are among other species suitable for windbreaks because willows can provide immediate wind protection and can provide multiple

Table 1 Guidelines on windbreak density by function

Density	25-50% density	50-65% density	65+% density
Protection	Crop, soil snow distribution	Air, snow accumulation	Farmstead, livestock, noise wildlife (ten rows)
Planting	• One row-deciduous shrub • Two row-deciduous tree and deciduous shrub	• Twin row-deciduous shrub • One row-evergreen tree • Two row-evergreen tree and deciduous tree • Three row-combination of deciduous trees/shrub	• Twin row-evergreen tree • Three or more row-combination of evergreen trees, deciduous trees/ shrubs

Table 2 General rule of windbreak establishment based on type of species

	Within the row (feet)	Between the row (feet)
Shrubs	3-6	6-10
Small trees (10-20 feet tall)	10-15	20
Deciduous trees	12-20	20-25
Conifer trees	20-25	20-25

Spruce trees may need to be spaced further apart if needle cast disease is a problem.

benefits (Volk et al., 2016). Willows can be cut and used as a biofuel or as decorative florals (e.g., pussy willow). Nurseries usually sell willows as bare root stock or sticks for larger plantings. In the United States, shrub willow stems can be cut from native willows in the local area or purchased as sticks or bare root plants from nurseries or Soil and Water Conservation District Offices. A recommended source of planting stock in the United States is Double A Willow (www.doubleawillow.com) who supplies cuttings of cultivars developed for rapid growth, plant form favorable to LSFs, and pest resistance. Willow sticks suitable for planting are 8-12 inches long.

Shrub willows planted for a LSF are typically planted in a single row one foot apart. An alternative approach is an offset double row with plants spaced 2 feet apart and 2.5 feet between rows. Bareroot plants and sticks can be planted with a pull behind tree planter or by hand. Soil preparation and weed control is critical to the survival and rapid growth of young willows. Perennial weeds/plants may be treated before and after tillage with systemic herbicides. Weeds may need to be controlled for the first 2 years. Biodegradable fabric weed mats can slow weed growth around the seedlings but need to be cut in an X pattern (around the willows/plants) so it does not girdle the willows/plants. Cutting or coppicing on a regular basis (every 2-5 years) can revitalize the plants but is not essential.

4 Windbreak design to reduce wind speed and snow drift

The primary purpose of windbreaks in agricultural settings is to reduce wind speed (Brandle et al., 2004; Tyndall and Colletti, 2007a). Both the internal and external structure of windbreaks affect the relationship to wind and wind speed. Internal structure is typically defined by porosity, which is the ratio of perforated area to total area, allowing air to move through windbreaks (Heisler and Dewalle, 1988; Tyndall and Colletti, 2007a). The greatest wind speed reduction over the greatest distance is achieved with a porosity of 40-60% (Tyndall and Colletti, 2007a). Less porous windbreaks can lead to greater reductions in wind speed in the immediate proximity of the windbreak, but the effect will last for a shorter horizontal distance away from the windbreak. Near-solid windbreaks can create a recirculation effect leading to turbulence within what should be the leeward side (Heisler and Dewalle, 1988).

The external structure of windbreaks includes such characteristics as the height and length of the windbreak, number of rows, species, windbreak shape, angle to the wind, placement on the landscape, and how the windbreak is managed. Height appears to have the most significant impact on wind speed. Wind speeds on the leeward side are reduced in proportion to height, typically for a horizontal distance between 10 and 30 times the height of the windbreak (Brandle et al., 2004; Ullman et al., 2004). For an average tree windbreak, the wind velocity reduction within a horizontal distance of ten times the height can be between 60% and 80%, and reduces to about a 20% reduction at a horizontal distance of twenty times the height (Fig. 1) (Nordstrom and Hotta, 2004). On the windward side, wind speed is also reduced beginning at a horizontal distance between two and five times the height ahead of the windbreak as air is forced upward and over the structure (Heisler and Dewalle, 1988).

Slope in multi-row windbreaks can also be an important factor. Windbreaks with steep vertical sides when viewed in profile tend to help reduce wind speeds at a greater amount over a greater distance than windbreaks with more streamlined, sloping sides (Heisler and Dewalle, 1988), while a wedge-shaped windbreak can help ramp air currents higher into the air, creating a larger zone of turbulence above the windbreak which may be beneficial for odor dilution (Tyndall and Colletti, 2007a,b).

As a plume of wind moves toward a windbreak in its path, some of the wind will pass through the vegetation and slow down, and some will be forced over the top or around the sides, creating a zone of turbulence above the windbreak (Brandle et al., 2004; Tyndall and Colletti, 2007a). To be most effective, windbreaks should be taller and wider than the approaching plume of wind (Heisler and Dewalle, 1988; Tyndall and Colletti, 2007a). Windbreak continuity is important as well. Any gaps in a windbreak can act like a funnel, creating higher velocity winds on the leeward side of the gap in the windbreak than on the windward side (Heisler and Dewalle, 1988; Brandle et al., 2004).

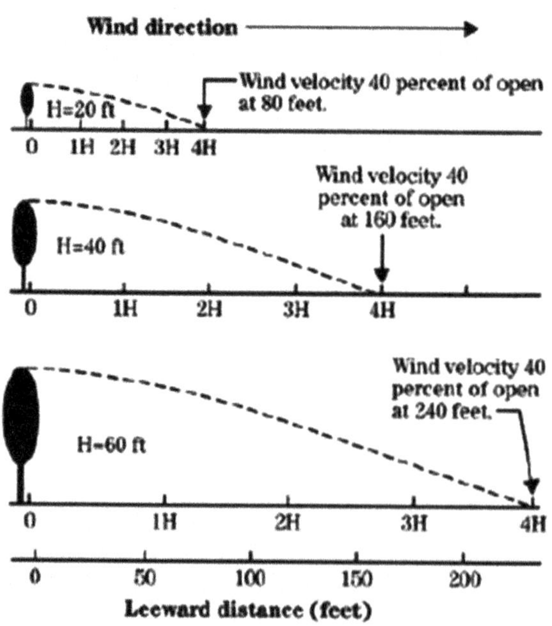

Figure 1 Leeward distance of wind protection is proportional to height of the barrier. Source: adapted from Brandle et al. (2004).

LSFs are shrubs, trees, or plants that form a wind barrier disrupting the wind, causing snow to be deposited on the upwind and downwind sides of the planting. LSFs have been shown to be economically viable solutions for mitigating blowing snow on roadways.

Typically snow drifts will collect ten times the height of the planting on the downwind side. Therefore a 10-foot willow windbreak could cause drifting downwind up to 100 feet (Fig. 2). It is important to design an LSF based on winter climate data. An example is the University of Minnesota's Climate Design Tool, which helps to determine width and number of rows to be planted

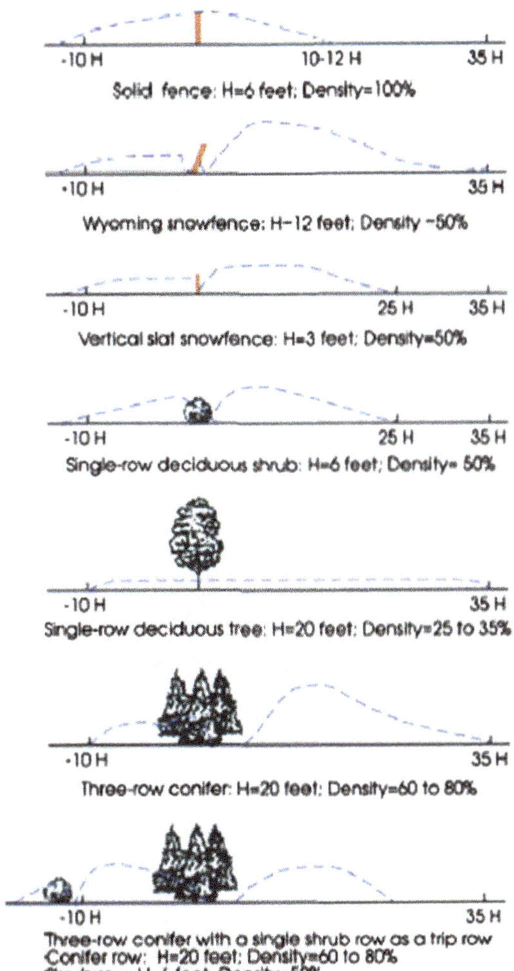

Figure 2 The height and density of a LSF will determine how much snow can be stored in the system. Source: adapted from Brandle et al. (2004).

(http://snowcontroltools.umn.edu). A setback range of 100–200 feet from a highway right of way is common. It may be necessary to extend the planting length of the LSF further out on one end (or both ends) to mitigate drifting problems around the end(s) (edge affect). This edge affect planting extension is determined by the Climate Design Tool. Downwind draft length is a function of the capacity/transport ratio. Capacity is primarily driven by height. Fast-growing willows can have massive snow catching capacity and reduced drift length shortly after planting.

5 Windbreaks and particulate capture

In addition to reducing wind speed, windbreaks can also affect anything the wind carries, including dust and particulate matter, snow, gasses, aerosols, microbes, and larger items such as feathers (Malone, 2004; Malone et al., 2006a; Tyndall and Colletti, 2007a; Tyndall, 2008). Typically, as wind penetrates through the windbreak, its speed decreases which allows gravity to act more effectively on solids and liquids carried along with the wind, causing them to settle out onto the ground or the vegetation (Tyndall and Colletti, 2007a). This effect has been observed in studies of propagule dispersion and pesticide drift (Burel, 1996; Tyndall and Colletti, 2007a; Ucar et al., 2003).

Windbreak vegetation and leaf surfaces can also directly intercept, trap, and even absorb or remove microbes, gasses, dust, and larger materials from the air (Raupach et al., 2001; Malone, 2004; Tyndall and Colletti, 2007a; Adrizal et al., 2008). Windbreaks have aerodynamically rough surfaces, which create small eddies in the air surrounding the vegetation. The resulting turbulence facilitates deposition of particles onto plant surfaces. However, if winds are strong enough, they can dislodge those impacted particles and cause them to become suspended in the air again (Tyndall and Colletti, 2007a).

Different species in windbreaks have different effects on wind-carried particles (Malone et al., 2006a; Tyndall and Colletti, 2007b). Conifers have a complex leaf shape, high circumference to area ratio, and hold their leaves all year round, making them the most effective species for particulate capture (Ucar et al., 2003; Tyndall and Colletti, 2007b) though hybrid willow has been shown to be a very effective trap species for particulate matter as well (Adrizal et al., 2008). Smaller leaves and complex leaf shapes tend to be more efficient traps, but stems, bark, and petioles also collect particulates, so windbreaks can and do still trap particulates even when no leaves are present (Tyndall and Colletti, 2007b; Adrizal et al., 2008). Deciduous trees, including hybrid poplar and hybrid willows, are less effective at trapping particulate matter, but appear to absorb more ammonia emissions than coniferous trees (Adrizal et al., 2008). Because different species appear to perform different tasks within a windbreak, planting multiple species and multiple rows and including multiple types of vegetation is recommended (Malone et al., 2006b; Tyindall and Colletti, 2007a,b; Adrizal et al., 2008).

The relationship between windbreak maturity and effectiveness at trapping dust and particulate matter is not quite as clear. Some studies have shown that for species that are already good at removing dust and particulate matter, younger trees may be even more effective because they tend to have higher foliage densities (Tyndall and Colletti, 2007a). But in a trial conducted by Burley et al. (2011), VEB did not show potential to reduce dust and virus transmission until the vegetation was older and thus taller and denser. It is likely the porosity of a windbreak, rather than age per say, that determines how effectively it can intercept or disrupt the flow of particulates (Raupach et al., 2001).

Species composition can also affect porosity at different heights within the windbreak. If the vegetation used in the windbreak leafs out several feet above the ground, for example, the windbreak may not intercept as much material at ground level as it does a few feet higher where leafy vegetation reduces porosity (Malone et al., 2006a; Tyndall and Colletti, 2007a). As a result, it is important to anticipate approximately where plumes will flow and intersect with a windbreak, and how the species composition and resulting height and porosity will change as the windbreak matures to ensure the windbreaks are effective over time. Yet, even with careful planning, ambient conditions such as temperature and humidity affect the process of both intercepting and trapping particulates, and many of these conditions cannot be controlled (Tyndall and Colletti, 2007a).

6 Windbreaks and odor mitigation

By trapping dust and particulate matter, windbreaks can also help to mitigate odor. Much, though not all, of the odor carried on a plume of wind is attached to dust and particulate matter (Tyndall and Colletti, 2007a). A number of studies have demonstrated this effect and the environmental and windbreak features that appear to affect the degree to which windbreaks mitigate odors emitted by poultry and swine barns (Lin et al., 2006, 2007) measured how much impact windbreaks had on in-field odor plumes from point odor sources. Using single row windbreaks with two deciduous species and two conifers, Lin et al. (2006, 2007) found that windbreaks do affect odor dispersion, and windbreaks with lower optical density, a proxy for porosity where higher optical density would indicate lower porosity, are more effective than those with higher density. However, single row deciduous poplar windbreak had the same level of optical porosity as the coniferous windbreaks but still was not as effective as the deciduous counterpart (Lin et al., 2006) theorize this is because coniferous species have stronger branches that are less flexible than poplar, and thus have a lower aerodynamic porosity. They also compared windbreak distance from the point source of odor by placing windbreaks either 15 or 60 m from point sources. Windbreaks located closer to the point source were more effective at dispersing odor plumes. Their trials also revealed that ambient air temperature can measurably influence odor dispersal, and the degree to which windbreaks affect the process. Warmer air temperatures resulted in better odor trapping. They suggest that this effect is a result of warmer temperatures creating more air turbulence and convective forces leading to greater movement of the plume while cold temperatures allow odor plumes to remain at ground level and are subject to fewer convective forces. Lin et al. (2007) also noted a significant reduction in the length of odor plumes when windbreaks were present regardless of windbreak porosity, though denser windbreaks produced slightly greater length reductions. Colletti et al. (2006) found windbreaks could also be effective in mitigating odor though they also noted the stress on trees from exposure to exhaust from poultry barn fans.

Others have found similar results. Adrizal et al. (2008) examined whether vegetative windbreaks could trap ammonia, dust, and other particulate matter which often harbor odor from commercial poultry barns. They found that conifers and hybrid willow hold the most dust and particulate matter mass while deciduous trees absorb the most ammonia. Their findings suggest that planting multiple rows of trees downwind of exhaust fans can effectively help trap and absorb airborne ammonia and dust. But other studies (Malone, 2004; Malone et al., 2006a) have shown only slight reductions in odor despite noticeable reductions in dust, particulate matter, and ammonia. Consequently, several researchers (Malone, 2004; Malone et al., 2006a; Tyndall and Colletti, 2007a; Lin et al., 2006, 2007; Adrizal et al., 2008) have suggested that the ability of windbreaks to capture or reduce odor is highly variable. Windbreaks do appear to become more effective in that role as they mature and when used in combination with other approaches (Tyndall and Colletti, 2007a), though this is variable depending on angle of the wind to the windbreak, wind speed and atmospheric conditions, including temperature (Malone, 2004; Lin et al., 2006, 2007). Damp conditions also appear to favor odor mitigation; when damp, the entire aerial surface of a plant is available for aerosol uptake to a far greater degree than when conditions are dry (Tyndall and Colletti, 2007a).

The ability of windbreaks to trap not only dust and particulates but odors as well has inspired questions about whether they can also prevent the spread of airborne diseases between commercial poultry and livestock operations. There is a wide range of research on the ways diseases like avian flu can be transmitted (Perkins and Swayne, 2003; Schofield et al., 2005; Lopez et al., 2006; Olsen et al., 2006; Sawabe et al., 2006, 2011; Tellier, 2006; Tiwari et al., 2006; Boon et al., 2007; Lowen et al., 2007; Barbazan et al., 2008; Weber and Stilianakis, 2008; Brown et al., 2009; Hanley and Borup, 2010; Wanaratna et al., 2011; Shriner et al., 2012, 2016; Zhong et al., 2014; APHIS-USDA, 2015; Flint et al.,

2015; Bui et al., 2016; Grear et al., 2017). A few studies have suggested ways vegetative buffers or windbreaks might reduce windborne virus transmission (Burley et al., 2011; Ssematimba et al., 2012; Jonges et al., 2015).

7 Windbreaks and wildlife conservation

Windbreaks can provide refuge, food, protection from wind and weather, breeding sites, and travel corridors for wildlife, primarily birds and mammals. They also shelter insects and serve as insect foraging sites for birds and small mammals that consume insects. In some cases, windbreaks may be the only site in the surrounding landscape where certain insects can be foraged. Windbreaks appear especially significant for off-ground nesting birds, as well as fox squirrels and eastern cotton tailed rabbits. However, many other species use windbreaks but would be present even if the shelterbelt was not there (Johnson and Beck, 1988).

Windbreaks can foster a range of wildlife species, with edge species particularly well suited to the linear nature of windbreaks (Brandle et al., 2004). Which species utilize a windbreak will depend on the windbreak's size, species composition, porosity, the spatial arrangement of the windbreaks on the landscape, time of year, and how they are managed (Yahner, 1983; Johnson and Beck, 1988; Hinsley and Bellamy, 2000).

As windbreaks increase in age and size, they tend to support greater species richness and abundance. There are many ways to measure size, with some studies indicating that width is the most important feature, and others suggesting height or volume matter more (Yahner, 1983; Johnson and Beck, 1988; Hinsley and Belamy, 2000). Shorter, blockier shelterbelts (12-15 rows) appear to be less supportive of wildlife than long, narrow ones (Johnson and Beck, 1988), and short, narrow windbreaks support the least species (Hinsley and Bellamy, 2000). Greater complexity in total vegetative species and types of species (grasses, forbs, trees, shrubs, etc.) that compose a windbreak can support a greater diversity of wildlife species. Complexity, like size, is often correlated with the age of the windbreak and many wild birds show a preference for windbreaks that are more than 5 years old (Johnson and Beck, 1988). Windbreaks are most likely to attract wildlife when there are limited natural or semi-natural habitats in the nearby surrounding landscape, as is common in many agricultural regions (Hinsley and Bellamy, 2000). Wild birds have been known to concentrate heavily in windbreaks when they are isolated on the landscape (Yahner, 1983; Burns, 2012).

Species using windbreaks may fluctuate depending on the time of year. In a study of avifauna in 3-9 row windbreaks in Rosemount Minnesota, USA, Yahner (1983) found evidence of this fluctuation. The greatest number of species was observed in the spring, followed by summer, then fall, and the least in winter. The most common species relative to population densities were common grackles (highest density in spring), house sparrows (lowest density in spring, but higher the rest of the year), European starlings (highest density summer and fall), American robins (highest density spring and summer), dark eyed juncos (highest density fall and winter), and mourning doves (highest density spring and summer). Overall, avifauna species richness and diversity in shelterbelts was higher in spring and summer than fall and winter.

8 Future trends and conclusion

In summary, windbreaks are an effective agroforestry practice that can be designed for multiple functions and benefits for a wide range of landscapes. Windbreaks offer many benefits, wind and snow drifting protection, soil erosion protection, carbon storage, crop protection, odor mitigation, energy savings, livestock protection, and wildlife habitat to mention a few. With our changing climate in many regions, it is important to be a visionary in selecting varieties of trees and shrubs that will grow healthy into the future. Native plants should be considered a priority; however, there are still selections of proven non-native plants that should be considered. Designing windbreaks with diversity of species in each planting while meeting the objectives of the planting is extremely important. Being aware of pests, invasive species, and diseases that are currently present or may be a future problem is also a valuable selection criteria. Windbreak design is very important to achieve the desired landscape protection. Field windbreaks can be extremely beneficial in protection crops and soil erosion. A grass or vegetative buffer is recommended between the trees/shrubs and the crop. The wildlife benefits of windbreaks are well documented and can be enhanced by diversified plantings and offering plants that offer edible nuts and fruits. Planting for wildlife habitat and shelter is a must for northern climates to help protect wildlife from snow and cold temperatures. Planting conifer varieties can deliver thermal protection for many wildlife species. Studies on windbreak plantings have already documented odor mitigation and visual screening. Future research is needed to see if dust and other particles that may be carrying diseases can be mitigated.

Windbreaks can be a very positive agroforestry planting in most landscapes if designed properly with the appropriate trees and/or shrubs that meet the objectives and purpose of the plantings.

9 Where to look further information

- *National Agroforestry Center*: https://www.fs.usda.gov/nac/
  The USDA National Agroforestry Center (NAC) accelerates the application of agroforestry through a national network of partners. NAC conducts

research, develops technologies and tools, coordinates demonstrations and training, and provides useful information to natural resource professionals on agroforestry.

- *The Center of for Agroforestry at the University of Missouri*: http://www.centerforagroforestry.org
  The Center for Agroforestry at the University of Missouri, established in 1998, is one the world's leading centers contributing to the science underlying agroforestry. The Center has a number of faculty from various disciplines doing research and outreach activities on agroforestry around the United States.
- *AGFORWARD*: http://www.agforward.eu
  FORWARD (AGroFORestry that Will Advance Rural Development) is an initiative in Europe dedicated to advancing agroforestry in the region,
- *Association for Temperate Agroforestry (AFTA)*: http://www.aftaweb.org/
  AFTA is a group of natural resource professionals, researchers, and farmers whose mission is to promote the wider adoption of agroforestry by landowners in temperate regions of North America. With members in the United States, Canada, and overseas, AFTA pursues its mission through activities such as networking, information exchange, public education, and policy development.

10 References

Adrizal, A., Patterson, P. H., Hulet, R. M., Bates, R. M., Myers, C. A. B., Martin, G. P., Shockey, R. L., van der Grinten, M., Anderson, D. A. and Thompson, J. R. 2008. Vegetative buffers for fan emissions from poultry farms: 2. Ammonia, dust and foliar nitrogen. *Journal of Environmental Science and Health. Part. B, Pesticides, Food Contaminants, and Agricultural Wastes* 43(1), 96–103. doi:10.1080/03601230701735078.

Anderson, V. L. and Bird, J. 1993. Effects of shelterbelt protection on performance of feedlot steers during a North Dakota winter. 1993 Beef Production Field Day, Carrington Res. Ext. Ctr., Livestock Unit. Vol 16: 19–21. North Dakota State University, Fargo, North Dakota.

APHIS USDA (Animal and Plant Health Inspection Service, United States of Department of Agriculture). 2015. Epidemiologic and other analyses of HPAI-affected poultry flocks. 9 September 2015 report.

Ballesteros Possu, W. 2015. Carbon storage potential of windbreaks on agricultural lands of the continental United States. Dissertations & Theses in Natural Resources, p. 117. Available at: http://digitalcommons.unl.edu/natresdiss/117.

Ballesteros Possu, W., Brandle, J. and Schoeneberger, M. 2017. Potential of windbreak trees to reduce carbon emissions by agricultural operations in the US. *Forests* 8(5), 138–45. doi:10.3390/f8050138.

Barbazan, P., Thitithanyanont, A., Misse, D., Dubot, A., Bosc, P., Luangsri, N. and Kittayapong, P. 2008. Detection of H5N1 avian influenza virus from mosquitoes collected in an infected poultry farm in Thailand. *Vector-borne and Zoonotic Diseases* 8(1), 105–10.

Boon, A. C., Sandbulte, M. R., Seiler, P., Webby, R. J., Songserm, T., Guan, Y. and Webster, R. G. 2007. Role of terrestrial wild birds in ecology of influenza A virus (H5N1). *Emerging Infectious Diseases* 13(11), 1720–4. doi:10.3201/eid1311.070114.

Brandle, J. R., Hintz, D. L. and Sturrock, J. W. 1988. *Windbreak Technology*. Elsevier Science Publishers, Amsterdam, 598p.

Brandle, J. R., Hodges, L. and Stuthman, J. 1995. Windbreaks and specialty crops for greater profits. In: Rietveld, W. (Ed.), *Agroforestry and Sustainable Systems: Symposium Proceedings*. USDA Forest Service Technical Report (RM-GTR-261).

Brandle, J. R., Hodges, L. and Wight, B. 2000. Windbreak practices. In: Garrett, E., Rietvald, W. and Fisher, R. (Eds), *North American Agroforestry: an Integrated Science and Practice*. American Society of Agronomy, Madison, WI.

Brandle, J. R., Hodges, L. and Zhou, X. H. 2004. Windbreaks in North American agricultural systems. *Agroforestry Systems* 61, 65–78.

Brandle, J. R., Hodges, L., Tyndall, J. and Sudmeyer, R. A. 2009. In: Garrett, H. E. (Ed), *North American Agroforestry: an Integrated Science and Practice*. Agronomy Society of America, Madison, WI, pp. 75–105.

Brocious, A. 2017. FDR's Great Wall of Trees: The Prairie States Forest Project. Nebraska PBS and NPR stations. Available at: http://netnebraska.org/article/news/1078024/fdrs-great-wall-trees-prairie-states-forestry-project.

Brown, J. D., Stallknecht, D. E., Berghaus, R. D. and Swayne, D. E. 2009. Infectious and lethal doses of H5N1 highly pathogenic avian influenza virus for house sparrows (*Passer domesticus*) and rock pigeons (*Columbia livia*). *Journal of Veterinary Diagnostic Investigation* 21(4), 437–45.

Bui, C. M., Gardner, L. and MacIntyre, C. R. 2016. Highly pathogenic avian influenza virus, Midwestern United States. *Emerging Infectious Diseases* 22(1), 138. doi:10.3201/eid2201.151053.

Burel, F. 1996. Hedgerows and their roles in agricultural landscapes. *Critical Reviews in Plant Sciences* 15(2), 169–90. doi:10.1080/07352689.1996.10393185.

Burley, H. K., Adrizal, A., Patterson, P. H., Hulet, R. M., Lu, H., Bates, R. M., Martin, G. P., Myers, C. A. B. and Atkins, H. M. 2011. The potential of vegetative buffers to reduce dust and respiratory virus transmission from commercial poultry farms. *The Journal of Applied Poultry Research* 20(2), 210–22. doi:10.3382/japr.2010-00298.

Burns, T. E., Ribble, C., Stephen, C., Kelton, D., Toews, L., Osterhold, J. and Wheeler, H. 2012.Use of observed wild bird activity on poultry farms and a literature review to target species as high priority for avian influenza testing in 2 regions of Canada. *The Canadian Veterinary Journal = la Revue Veterinaire Canadienne* 53(2), 158–66.

Caborn, J. M. 1971. The agronomic and biological significance of hedgerows. *Outlook on Agriculture* 6(6), 279–84. doi:10.1177/003072707100600609.

Cleugh, H. A. and Hughes, D. E. 2002. Impact of shelter on crop microclimates: a synthesis of results from wind tunnel and field experiments. *Australian Journal of Experimental Agriculture* 42(6), 679–701. doi:10.1071/EA02005.

Cleugh, H. A., Prinsley, R., Bird, P. R., Brooks, S. J., Carberry, P. S., Crawford, M. C., Jackson, T. T., Meinke, H., Mylius, S. J., Nuberg, I. K., et al. 2002. The Australian national windbreaks program: overview and summary of results. *Australian Journal of Experimental Agriculture* 42(6), 649–64. doi:10.1071/EA02003.

Colletti, J. P., Hoff, S., Thompson, J. and Tyndall, J. 2006. Vegetative environmental buffers to mitigate odor and aerosol pollutants emitted from poultry production sites. In:

Proceedings of the Workshop on Agricultural Air Quality: State of Science. Potomac Book Company, MD, pp. 5–8.

Dewdney, M. M. and Graham, J. H. 2014. Citrus canker. Florida Citrus Pest Management Guide, pp. 93–6.

Droze, W. H. 1977. *Trees, Prairies and People: A History of Tree Plantings in the Plains States*. Texas Woman's University Press, Denton, TX, p. 313.

Flint, P. L., Pearce, J. M., Franson, J. C. and Derksen, D. V. 2015. Wild bird surveillance for highly pathogenic avian influenza H5 in North America. *Virology Journal* 12, 151. doi:10.1186/s12985-015-0377-2.

Gottwald, T. R. and Timmer, L. W. 1995. The efficacy of windbreaks in reducing the spread of citrus canker caused by *Xanthomonas campestris* pv. citri. *Tropical Agriculture* 72, 194–201.

Grear, D. A., Dusek, R. J., Walsh, D. P. and Hall, J. S. 2017. No evidence of infection or exposure to highly pathogenic avian influenzas in peridomestic wildlife on an affected poultry facility. *Journal of Wildlife Diseases* 53(1), 37–45. doi:10.7589/2016-02-029.

Hanley, B. P. and Borup, B. 2010. Aerosol influenza transmission risk contours: a study of humid tropics versus winter temperate zone. *Virology Journal* 7, 98. doi: 10.1186/1743-422X-7-98.

Heisler, G. M. and Dewalle, D. R. 1988. Effects of windbreak structure on wind flow. *Agriculture, Ecosystems and Environment* 22–23, 41–69. doi:10.1016/0167-8809(88)90007-2.

Hinsley, S. A. and Bellamy, P. E. 2000. The influence of hedge structure, management and landscape context on the value of hedgerows to birds: a review. *Journal of Environmental Management* 60(1), 33–49. doi:10.1006/jema.2000.0360.

Johnson, R. J. and Beck, M. M. 1988. Influences of shelterbelts on wildlife management and biology. *Agriculture, Ecosystems and Environment* 22–23, 301–35. doi:10.1016/0167-8809(88)90028-X.

Jonges, M., Van Leuken, J., Wouters, I., Koch, G., Meijer, A. and Koopmans, M. 2015. Wind-mediated spread of low-pathogenic avian influenza virus into the environment during outbreaks at commercial poultry farms. *PLoS ONE* 10(5), e0125401. doi:10.1371/journal.pone.0125401.

Kort, J. 1988. Benefits of windbreaks to field and forage crops. *Agriculture, Ecosystems and Environment* 22–23, 165–90. doi:10.1016/0167-8809(88)90017-5.

Lin, X. J., Barrington, S., Nicell, J., Choiniere, D. and Vezina, A. 2006. Influence of windbreaks on livestock odour dispersion plume in the field. *Agriculture, Ecosystems and Environment* 116(3–4), 263–72. doi:10.1016/j.agee.2006.02.014.

Lin, X. J., Barrington, S., Nicell, J., Choiniere, D. and King, S. 2007. Livestock odour dispersion as affected by natural windbreaks. *Water, Air, and Soil Pollution* 182(1–4), 263–73. doi:10.1007/s11270-007-9337-9.

Lopez, J. C., McFarlane, R. and Scott, T. 2006. Environmental factors influence the prevalence of infectious bronchitis virus. In: *Proceedings of the 18th Australian Poultry Science Symposium*, pp. 127–30.

Lowen, A. C., Mubareka, S., Steel, J. and Palese, P. 2007. Influenza virus transmission is dependent on relative humidity and temperature. *PLoS Pathogens* 3(10), e151.

Malone, B. 2004. Using trees to reduce dust and odour emissions from poultry farms. In: *Proceedings of the 2004 Poultry Information Exchange*, pp. 33–8.

Malone, G., VanWicklen, G., Collier, S. and Hansen, D. 2006a. Efficacy of vegetative environmental buffers to capture emissions from tunnel ventilated poultry houses. In: *Proceedings of the Workshop on Agricultural Air Quality*, pp. 875–8.

Malone, G., Windsor, J., Abbott, D. and Collier, S. 2006b. Establishment of vegetative environmental buffers around poultry farms. In: *Proceedings of the Workshop on Agricultural Air Quality: State of the Science*, pp. 879–80.

Nordstrom, K. F. and Hotta, S. 2004. Wind erosion from cropland solutions in the USA: a review of problems, and prospects. *Geoderma* 121(3–4), 157–67. doi:10.1016/j.geoderma.2003.11.012.

Olsen, B., Munster, V. J., Wallensten, A., Waldenström, J., Osterhaus, A. D. and Fouchier, R. A. 2006. Global patterns of influenza A virus in wild birds. *Science* 312(5772), 384–8. doi:10.1126/science.1122438.

Perkins, L. E. L. and Swayne, D. E. 2003. Comparative susceptibility of selected avian and mammalian species to a Hong Kong-origin H5N1 high-pathogenicity avian influenza virus. *Avian Diseases* 47(3 Suppl.), 956–67. doi:10.1637/0005-2086-47.s3.956.

Raupach, M. R., Woods, N., Dorr, G., Leys, J. F. and Cleugh, H. A. 2001. The entrapment of particles by windbreaks. *Atmospheric Environment* 35(20), 3373–83. doi:10.1016/S1352-2310(01)00139-X.

Sawabe, K., Hoshino, K., Isawa, H., Sasaki, T., Hayashi, T., Tsuda, Y., Kurahashi, H., Tanabayashi, K., Hotta, A., Saito, T., et al. 2006. Detection and isolation of highly pathogenic H5N1 avian influenza A viruses from blow flies collected in the vicinity of an infected poultry farm in Kyoto, Japan 2004. *The American Journal of Tropical Medicine and Hygiene* 75(2), 327–32. doi:10.4269/ajtmh.2006.75.327.

Sawabe, K., Hoshino, K., Isawa, H., Sasaki, T., Kin, K. S., Hayashi, T., Tsuda, Y., Kurahashi, H. and Kobayashi, M. 2011. Blow flies were one of the possible candidates for transmission of highly pathogenic H5N1 avian influenza virus during the 2004 outbreaks in Japan. *Influenza Research and Treatment* 2011, 652652. doi:10.1155/2011/652652.

Schofield, L., Ho, J., Kournikakis, B. and Booth, T. 2005. Avian influenza aerosol sampling campaign in the British Columbia Fraser Valley, 9–19 April 2004. National Microbiology Laboratory Population and Public Health Branch, Health Canada. Available at: http://cradpdf.drdc-rddc.gc.ca/PDFS/unc92/p523877.pdf.

Shriner, S. A., VanDalen, K. K., Mooers, N. L., Ellis, J. W., Sullivan, H. J., Root, J. J., Pelzel, A. M. and Franklin, A. B. 2012. Low-pathogenic avian influenza viruses in wild house mice. *PLoS ONE* 7(6), e39206. doi: 10.1371/journal.pone.0039206.

Shriner, S. A., Root, J. J., Lutman, M. W., Kloft, J. M., VanDalen, K. K., Sullivan, H. J., White, T. S., Milleson, M. P., Hairston, J. L., Chandler, S. C., et al. 2016. Surveillance for highly pathogenic H5 avian influenza virus in synanthropic wildlife associated with poultry farms during an acute outbreak. *Scientific Reports* 6, 36237. doi:10.1038/srep36237.

Ssematimba, A., Hagenaars, T. J. and de Jong, M. C. M. 2012. Modelling the wind-borne spread of highly pathogenic avian influenza virus between farms. *PLoS ONE* 7(2), e31114. doi: 10.1371/journal.pone.0031114.

Tellier, R. 2006. Review of aerosol transmission of influenza A virus. *Emerging Infectious Diseases* 12(11), 1657–62. doi:10.3201/eid1211.060426.

Tiwari, A., Patnayak, D. P., Chander, Y., Parsad, M. and Goyal, S. M. 2006. Survival of two avian respiratory viruses on porous and nonporous surfaces. *Avian Diseases* 50(2), 284–7. doi:10.1637/7453-101205R.1.

Tyndall, J. 2008. The use of vegetative environmental buffers for livestock and poultry odor management. Natural Resource Ecology and Management Conference Papers, Posters and Presentations. Available at: http://lib.dr.iastate.edu/nrem_conf/20

Tyndall, J. and Colletti, J. P. 2007a. Air quality and shelterbelts: odor mitigation and livestock production a literature review. Forestry Publications Paper 1. Available at: http://lib.dr.iastate.edu/for_pubs/1.

Tyndall, J. and Colletti, J. 2007b. Mitigating swine odor with strategically designed shelterbelt systems: a review. *Agroforestry Systems* 69(1), 45–65. doi:10.1007/s10457-006-9017-6.

Ucar, T. and Hall, F. R. 2001. Windbreaks as a pesticide drift mitigation strategy: a review. *Pest Management Science* 57(8), 663–75. doi:10.1002/ps.341.

Ucar, T., Hall, F. R., Tew, J. E. and Hacker, J. K. 2003. Wind tunnel studies on spray deposition on leaves of tree species used for windbreaks and exposure of honey bees. *Pest Management Science* 59(3), 358–64. doi:10.1002/ps.650.

Ullman, J. L., Mukhtar, S., Lacey, R. E. and Carey, J. B. 2004. A review of literature concerning odors, ammonia, and dust from broiler production facilities: 4. Remedial management practices. *The Journal of Applied Poultry Research* 13(3), 521–31. doi:10.1093/japr/13.3.521.

USDA (United States Department of Agriculture). 2005. Technical Note 13: windbreak, shelterbelt and landscaping technology. United States Department of Agriculture - Natural Resources Conservation Service, Spokane, WA. Available at: https://www.nrcs.usda.gov/Internet/FSE_PLANTMATERIALS/publications/wapmctn6337.pdf.

Vacek, Z., Řeháček, D., Cukor, J., Vacek, S., Khel, T., Sharma, R. P., Kučera, J., Král, J. and Papaj, V. 2018. Windbreak efficiency in agricultural landscape of the Central Europe: multiple approaches to wind erosion control. *Environmental Management* 62(5), 942–54. doi:10.1007/s00267-018-1090-x.

Volk, T. A., Heavy, J. P. and Eisenbies, M. H. 2016. Advances in shrub-willow crops for bioenergy, renewable products and environmental benefits. *Food and Energy Security* 5(2), 97–106. doi:10.1002/fes3.82.

Wanaratana, S., Panyim, S. and Pakpinyo, S. 2011. The potential of house flies to act as a vector of avian influenza subtype H5N1 under experimental conditions. *Medical and Veterinary Entomology* 25(1), 58–63. doi:10.1111/j.1365-2915.2010.00928.x.

Weber, T. P. and Stilianakis, N. I. 2008. Inactivation of influenza A viruses in the environment and modes of transmission: a critical review. *Journal of Infection* 57(5), 361–73.

Wight, B. and Stuhr, K. 2002. *Windbreaks: an Agroforestry Practice, Agroforestry Notes, AF Note - 25*. National Agroforestry Center. United States Department of Agriculture - Forest Service.

Yahner, R. H. 1983. Seasonal dynamics, habitat relationships, and management of avifauna in farmstead shelterbelts. *The Journal of Wildlife Management* 47(1), 85–104. doi:10.2307/3808055.

Zhong, L., Wang, X., Li, Q., Liu, D., Chen, H., Zhao, M., Gu, X., He, L., Liu, X., Gu, M., et al. 2014. Molecular mechanism of the airborne transmissibility of H9N2 avian influenza A viruses in chickens. *Journal of Virology* 88(17), 9568–78. doi:10.1128/JVI.00943-14.

Chapter 3

Managing hedgerows to optimise ecosystem services in agroforestry systems

Penka Tsonkova and Christian Böhm, Brandenburg University of Technology Cottbus-Senftenberg, Germany; Rico Hübner, Technical University of Munich, Germany; and Julia Ehritt, Nature and Biodiversity Conservation Union Brandenburg, Germany

1 Introduction

Hedgerows composed of trees and/or shrubs are anthropogenically developed linear woody-features initially established as boundaries and managed regularly (Forman and Baudry, 1984; Baudry et al., 2000). They served as natural fences, often between pastures and arable fields, and also supplied timber and non-timber tree products such as fuel wood, small fruits, fodder and herbs for traditional medicine (Herzog, 2000). In temperate agricultural landscapes, woody perennials (trees or shrubs) are important for maintaining the equilibrium between a landscape's productive and ecosystem regulation functions (Herzog, 2000). Accordingly, hedgerows and associated elements (ditches and earth banks) were also developed to control water (for drainage or irrigation) or soil (to diminish wind and/or water erosion; Baudry

http://dx.doi.org/10.19103/AS.2018.0041.03

et al., 2000). The former function is provided by riparian buffer strips, while the latter is fulfilled by windbreaks. Hedgerows used along field borders are known in Germany as '*Wallhecken*' or '*Knicks*'. Hedgerows and windbreaks are considered the most important traditional agroforestry practices in the Atlantic region and Central Europe (Nerlich et al., 2013).

As a consequence of agricultural mechanisation, land consolidation and intensification, hedgerows have been increasingly removed from the landscape because they are perceived as obstacles to production objectives (Eichhorn et al., 2006; Nerlich et al., 2013). The agricultural landscape therefore came to be dominated by large fields which simplified and standardised landscapes and ecosystems. Consequently, many plant and animal species were eliminated and soil fertility declined (Nerlich et al., 2013). Due to reduced demand for firewood and the replacement of hedges with more efficient and economical post and wire fences, their economic value further diminished. The importance of historic functions such as boundaries, shelter and sources of products has shifted and hedgerows are primarily valued today for their ecological and cultural benefits (Baudry et al., 2000).

It is in this context that increased recognition of the ecological and amenity functions of hedgerows has been reflected in the Common Agricultural Policy (CAP) of the European Union since 1990. Currently, farmers can register hedges and wooded strips as landscape features which are recognised as part of cross-compliance and ecological focus areas (EFA) under the greening section of Pillar I of the CAP. In addition, several options in the Rural Development Regulation (Pillar II of the CAP) support the restoration and maintenance of existing hedgerows in the agricultural landscape (Mosquera-Losada et al., 2016; Santiago-Freijanes et al., 2018). However, within these options there is little emphasis on managing hedgerows for their productivity (Smith et al., 2012).

In contrast to the ecological and societal benefits, the direct economic benefit of hedgerows to the farmer has received less attention and their management is currently decoupled from agricultural land management. Existing woody-features, including those which are not subsidised by the direct payments of the CAP (Pillar I), are protected by local regulations. For example, the Hedgerows Regulations were introduced in England and Wales in 1997, to protect these characteristic elements of the countryside (HM Government, 1997). It is necessary for hedgerows to be preserved, managed and maintained continuously for adequate provision of ecosystem services (Schleyer and Plieninger, 2011).

Utilisation of the production function of hedgerows, as historically practised, could increase interest in hedgerow management and maintenance. In this context, a reliable estimate of the biomass potential of existing hedgerows will be necessary to determine how attractive their production function amongst other innovative uses linked to the bio-economy might be. Utilisation of the production function should be conducted in balance with the ecological and societal functions. The main objectives should be set according to the predominantly desired function in the landscape under consideration in order to reflect its heterogeneity (Baudry et al., 2000). For example, a management strategy aimed at utilising the production function of hedgerows should reflect local objectives while taking a landscape perspective into account.

The purpose of this chapter is to assess the condition of linear woody features in relation to relevant ecosystem services and discuss the opportunity for utilising existing hedgerows in the agricultural landscape. For this purpose, a new classification of existing linear woody-features applicable at the field level was developed. In this classification, features in the agricultural landscape consisting of both trees and shrubs, such as hedgerows, windbreaks and riparian buffer strips, are collectively referred to as hedgerows. Based on this classification, the condition of multiple ecosystem services provided by hedgerows is assessed and the effect of applying different measures of hedgerow management to these ecosystem services is discussed. This assessment is not limited to hedgerows having a width of up to 10 m, which are considered landscape features in the Regulation (EU) No 639/2014, but is also applicable to those features.

2 Methodological framework

The methodology essentially consists of three stages which are depicted by various shades of grey in Fig. 1. In the first stage, important ecosystem services provided by hedgerows are discussed. A classification of existing hedgerows, applicable at the field level, is developed. The link between hedgerow categories and ecosystem services is elaborated and main guidelines for hedgerow management in line with this assessment are provided. This first part is based on a review of relevant scientific literature and is applicable to hedgerows in general. In the following discussion of constraints to hedgerow management, the focus was narrowed to reflect those specific to Germany.

The lack of management of the majority of trees in German agricultural landscapes causes difficulties in applying existing comprehensive assessment methodologies, such as that provided by Larcher and Baudry (2013). In this study, the authors emphasise the importance of tree species and their management and propose a landscape grammar approach to deciphering the structural and functional aspects of hedgerows and suggesting rules for their design and management. Alternatively, the simple methodology proposed in this chapter may be used to assess the condition of hedgerows, independent of management type, when considering ecosystem services. When applied to a hedgerow network, the approach may be used to aid decision making for

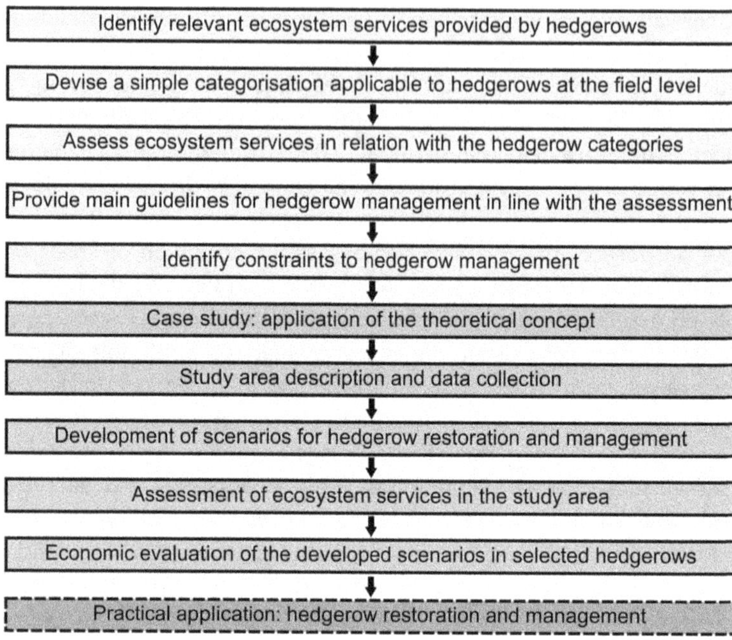

Figure 1 Methodological framework.

hedgerow management under consideration of multiple ecosystem services and possible trade-offs at the landscape level.

In the second stage, the application of the theoretical concept is demonstrated in a study area in the German federal state of Brandenburg. In this area, data is collected for hedgerow characterisation and assessment of ecosystem services. The assessment focussed on hedgerows for which farmers did not receive CAP subsidies, in order to exclude hedgerows which had been classed as landscape features. The biomass potential of existing woody features was determined in order to assess the attractiveness of the production function of hedgerows. General scenarios for restoration and management were developed and an assessment of ecosystem services provided by the hedgerows was conducted. This was followed by an economic evaluation of the developed scenarios in selected hedgerows.

The third stage is the practical application of the concept, which is beyond the scope of this study.

3 Ecosystem services associated with hedgerows

The most prominent ecosystem functions and related ecosystem services associated with hedgerows can be grouped as follows: production (biomass), regulation (wind and water protection), ecological enhancement (habitat provision) and amenity (landscape aesthetics).

3.1 Biomass production

The production function of hedgerows may be utilised in terms of cultivating fast-growing trees, grown in short rotation periods (for biomass feedstock) or by growing slow-maturing trees in longer rotation periods (for timber) (Eichhorn et al., 2006; Christen and Dalgaard, 2013). Recent studies have shown that existing woody-features can be used as a source of biomass for energy production (Wiegmann et al., 2007; Conrady et al., 2014; Peters et al., 2014). The average technical biomass potential of existing woody-features in the German federal state of Brandenburg was estimated at 3 t dry matter ha^{-1} $year^{-1}$ (Peters et al., 2014). The evaluation of technical biomass potential considers restrictions, such as the technical limits on biomass collection or legal regulations which could restrict biomass use (Brosowski et al., 2016). Existing nature conservation regulations also impose restrictions.

In this context, scenarios modelling different intensities of biomass extraction from hedgerows were simulated in a landscape in the federal state Hesse in Germany (Sauerbrei et al., 2017). The results suggested that increasing biomass extraction rates may reduce the suitability of hedgerows as a habitat for birds. For this reason, management strategies should be carefully designed. Practical trials in southern England suggest that introducing coppice management of hedges for wood-fuel may be beneficial for the wildlife of individual hedges and for biodiversity at the landscape scale (Westaway et al., 2016). In this study, a biodiversity protocol was developed to assess the likely impacts.

Similarly, riparian buffer strips with slow-maturing or fast-growing trees could be used to provide biomass along with the traditionally important functions of surface water protection and landscape aesthetics (Borin et al., 2010; Christen and Dalgaard, 2013). Assuming that buffer strips are established on fertile soils and depending on the combination of species and the growing conditions, an annual yield of 5–8 t dry matter ha^{-1} $year^{-1}$ for short rotation forestry and up to 16 t dry matter ha^{-1} $year^{-1}$ for short rotation coppices with willow (*Salix* spp.) and poplar (*Populus* spp.) was estimated, based on literature data by Christen and Dalgaard (2013). The average annual yield or estimations of theoretical biomass potential reflect the total available biomass in the area considered.

The biomass volume of trees can be determined by using allometric equations (Zianis et al., 2005; Böhm et al., 2011) or by conducting tree volume measurements (Kramer and Akça, 2008). Marketable tree volume is calculated from the tree diameter at breast height (DBH) and the tree height, using a form factor which converts the total volume to the marketable quantity (Kramer and

Akça, 2008). While the shrub layer biomass in hedgerows is of comparatively low economic value, additional benefits can be gained from hedgerow management. For example, in the United Kingdom, actively managed hawthorn (*Crataegus monogyna*) and blackthorn (*Prunus spinosa*) hedgerows exhibited increased carbon storage in above- and below-ground biomass (Axe et al., 2017). The shrub layer can also be used for the production of alternative hedgerow products such as berries. Due to the large variety of such products, they were not considered in this study.

3.2 Wind protection

The major benefits of windbreaks in North and Eastern Europe are protection and associated modification of the microclimate (Cleugh, 1998; Nuberg, 1998; Brandle et al., 2004; Torita and Satou, 2007). The shelter effect depends on characteristics such as height, permeability, orientation and location in the landscape (Nuberg, 1998). The sheltered area on the leeward side of a windbreak extends to around 25 times its height (Deumlich et al., 2006). The vertical distribution of density within a windbreak also influences wind flow and therefore its effectiveness (Brandle et al., 2009). Windbreaks having greater porosity in their lower levels will funnel wind through these areas, increasing the wind speed in the lee of the windbreak and thus decreasing the level of leeward protection (Brandle et al., 2009). Hence, the presence of a shrub layer below the tree layer is beneficial in improving wind protection at ground level close to the hedgerow.

For optimum efficiency, hedgerows should be aligned perpendicular to the prevailing wind direction. Accordingly, hedgerows planted for wind protection in Germany were predominantly established in a north-south direction (Reif and Achtziger, 2000). The preferred species in East Germany are fast-growing trees such as poplar and black alder (*Alnus glutinosa*). Their fast growth ensures rapid protection and allows for regular harvesting. Böhm et al. (2014) suggested that short rotation coppice strips of poplar and black locust (*Robinia pseudoacacia*), planted in a north-south direction in an alley-cropping system in southern Brandenburg, are suitable for reducing wind speed from a height of 2 m. Thus for effective wind protection over the area of 25 times the hedgerow height, the width between such low hedgerows should not exceed 50 m.

Typically, the minimum alley width allowing for agricultural machinery use in the temperate region is about 25 m. However, alleys which are too narrow may reduce the overall crop yield in an agroforestry system, due to the increased proportion of the competition zone (usually 1–2 times the hedgerow height). The effectiveness of hedgerows for wind protection at the landscape scale can be evaluated by applying a GIS-approach to estimate the density of agricultural landscape structures.

3.3 Water protection

Riparian buffer strips can improve both surface- and groundwater quality (Pavlidis and Tsihrintzis, 2018). They function as barriers and filter nonpoint source pollutants from agricultural areas, especially those carried by surface run-off (Bach, 2000; Lee et al., 2003; Schultz et al., 2004; Ryszkowski and Kedziora, 2007; Borin et al., 2010; Christen and Dalgaard, 2013; Nerlich et al., 2013; Wolton et al., 2014). The deeper root system of trees may provide a safety net for retrieving nutrients which have leached below the rooting zone of crops (Jose et al., 2004). According to a recent review by Pavlidis and Tsihrintzis (2018), tree roots in agroforestry systems are able to reduce nitrogen and phosphorus residues in soils from 20% up to 100% and have the potential to reduce pesticide leaching and run-off considerably (up to 90% in the case of run-off). The efficacy of these strips depends on characteristics such as width and composition. According to Lee et al. (2003), combinations of dense, stiff, native warm-season grass and woody vegetation improve the removal of nonpoint source pollutants from agricultural areas.

Tree shading further counteracts temperature-induced eutrophication in neighbouring water bodies by reducing solar radiation and water temperature, thus mitigating oxygen deficiency in summer (Bach, 2000; Herdt et al., 2007). On the landscape scale, hedgerows change the local hydrology (Ryszkowski and Kedziora, 2007; Christen and Dalgaard, 2013). They control evapotranspiration rates, surface run-off and percolation of water across the soil profile (Ryszkowski and Kedziora, 2007). When planted along the contours of a terrain, hedgerows can effectively reduce erosion by water and consequent related soil loss and surface run-off (Nerlich et al., 2013; Wolton et al., 2014). Due to their capacity for improving water infiltration rates and reducing surface run-off, hedgerows can be effective in mitigating flooding events (Carroll et al., 2004).

3.4 Habitat provision

Hedgerows benefit biodiversity by providing habitats, refuges, corridors and barriers (Baudry et al., 2000). Intersecting hedgerows in a landscape form a network providing pathways for the movement of species, and also subdivide the landscape into fields of varying size (Forman and Baudry, 1984). Hedgerows provide a habitat corridor for forest plants (McCollin et al., 2000; Wehling and Diekmann, 2009) and for carabids (Burel, 1992), birds and pollinators. Their flowers are also a source of nectar for pollinating insects (Jacobs et al., 2009).

In addition to early flowering plants such as hazel (*Corylus avellana* L.), willow (*Salix* spp.), blackthorn (*Prunus spinosa*) and wild cherry (*Prunus avium*), indispensable for the spring development of honeybee and bumblebee populations, late flowering plants such as wild roses (*Rosa* spp.), raspberries

(*Rubus idaeus*), blackberries (*Rubus fruticosus*) and ivy (*Hedera helix*) are of great value to insect biodiversity and ensure the late summer and winter development of bee colonies (Jacobs et al., 2009). In apiculture, components of an extract from walnut leaves (*Juglans regia* L.) can be useful against *Ascosphaera apis*, the fungal cause of the honey bee larval disease chalkbrood (Garbaczewska et al., 2014). Hedgerows provide habitat and food resources for overwintering invertebrates, mammals and birds (Best et al., 1995; Hinsley and Bellamy, 2000; LfL, 2005; DVL, 2006; Staley et al., 2013; Dondina et al., 2016). Riparian buffer strips and associated ditches provide habitat for species-rich aquatic, amphibian and terrestrial biocenosis (Handke, 1999; Bach, 2000).

Due to the structural diversity created by their vertical stratification, hedgerows are inhabited by a variety of species (Forman and Baudry, 1984; Knauer, 1993; Weber, 2003, LfL, 2005, DVL, 2006). The vegetation community within a hedgerow will differ widely according to its orientation (north-south, east-west) and the light availability and microhabitat heterogeneity (e.g. warm and dry conditions on the southern side of the hedgerow and cool and wet conditions on its northern side) (Forman and Baudry, 1984; Knauer, 1993; Weber, 2003). However, a resurveying of hedgerows in northern Germany after 48 years by Litza and Diekmann (2017) found a decrease in overall plant species from 182 to 164, which was also negatively influenced by the height of the shrub layer and an increase in nutrient-demanding species. Neophyte species increased from two to five.

On the landscape scale, the distribution of hedgerows and their degree of connection in forming a biotope network are essential for habitat provision. Preserving patches of open habitats in the agricultural landscape is also important to provide habitat for breeding birds, such as the skylark (*Alauda arvensis*; Lutze et al., 2010).

3.5 Landscape aesthetic

The structuring of the countryside by trees, bushes, orchards, hedges, copses and forest edges increases the recreational value of a landscape. Landscapes with many individual features contain more information than those which have been cleared and are thus more interesting for recreation and leisure (Schulze, 1982). Evaluation of the aesthetic perception of hedgerows is beyond the scope of this chapter: *'The visual characteristics of a landscape are some of the most widely experienced but also most difficult, and controversial, [. . .] to define'* (Sang et al., 2008). The number of aesthetically relevant aspects is high as are the corresponding possibilities and categories of evaluation.

For example, according to Augenstein (2002), the aesthetic quality of a landscape should be considered, while according to the German Environmental Protection Act, its important aspects are beauty, peculiarity and diversity. Other important aspects include the consideration of visual relationships; assessment of the level of intersection, urban sprawl (anthropogenic structures, roads and highways), the degree of naturalness, closeness to nature, originality and wildness. Contrasting elements such as noise versus silence, 'soundscapes' and light versus shadow may also be used. Hübner (2016) proposed the assessment of openness/enclosure in a landscape to determine the effect on landscape aesthetics of planting additional woody-features.

The references to evaluation criteria for landscape aesthetics made in this chapter, selectively draw on our observations, verified in the Brandenburg study area in East Germany, but also apply (with restrictions) to other areas in the temperate zone. Landscape aesthetics are especially enhanced by the presence of old trees, known as Methuselah or biotope trees. A greater tree and shrub height also affects a wider area aesthetically and imparts a more effective structure to the landscape. The density of woody structures within a distinct area is important for landscape aesthetics. However, a saturation effect can occur and adding further woody structures to a landscape may counteract the expectation that more trees/bushes will improve its aesthetics.

Hübner et al. (2016) used landscape metrics analyses in the study region followed by a landscape aesthetics judgement made by experts. According to this analysis, improving landscapes through planting additional woody structures was generally possible for a value below 0.4 (58% of the area), whereas above this value, aesthetic considerations had to be taken into account (41% of the area). Above a threshold of 0.7, any measure for planting additional woody structures is counterproductive. However, this was only observed in 1% of the area. In terms of design, the (re-)planting of hedgerows may enhance impaired landscapes and disguise structures such as highways. A negative perception may occur when important visual relationships, for example along valley-bottoms or towards monuments in a landscape, are interrupted.

4 Categorisation of existing hedgerows

Whatever their function, hedgerows can be grouped according to their common characteristics in order to assess their condition. In addition to width, height and associated elements such as ditches, important features recorded in field surveys and mapping of hedgerows reflect their vegetation (tree and shrub layer coverage, density and tree and shrub species present; Roßkamp, 1999; DEFRA, 2007; Müller, 2013). They may therefore be described by the following characteristics (1) hedgerow type (proportion of trees and shrubs), (2) hedgerow structure (continuity) and (3) degree of naturalness (the proportion of woody species native to the area).

4.1 Hedgerow type

As with forests, hedgerows are composed of different vertical layers, that is tree layer (>5 m in Germany), shrub layer (0.5–5 m) and herbaceous layer (<0.5 m) (LfL, 2005). In this chapter, hedgerow type refers to the proportion of woody components (trees and shrubs). In line with Wolton et al. (2013), the tree layer is defined as the woody part of the hedge which is typically more than 5 m high, but may be also represented by substantial shorter pollards and coppice stools. The shrub layer is the woody part of the hedge consisting of shrubs and bushes, which are usually multi-stemmed, branched and less than 5 m in height. Nine combinations can be distinguished in the tree and shrub layers and may be combined into five main hedgerow types: tree hedgerow (TH), predominantly tree hedgerow (PTH), equal proportion of trees and shrubs (EPTSH), predominantly shrub hedgerow (PSH) and shrub hedgerow (SH). The combinations are shown in percentage in Table 1.

The categorisation of hedgerows depends on the predominant vegetation and reflects disturbances related to management activities within the hedgerow, as well as habitat quality. It is useful in accounting for the dynamics and diversity of available habitats and for different levels of soil protection (Schmucki et al., 2002). For example, hedgerows were systematically established in northern Germany between the seventeenth and nineteenth centuries, and during this period, most hedgerows were cut every 8–12 years to obtain fuel wood, thus helping to maintain a dense structure (Huwer and Wittig, 2012). Such hedgerows in a good condition consist of a dense shrub layer with a maximum of 25% tree coverage equally distributed along the hedgerow (Roßkamp, 1999). Due to reduced utilisation of wood and fodder resources and the lack of regular management, tree layer development predominated and the shrub layer diminished, due to increased shading (Roßkamp, 1999; Kurz et al., 2001; Huwer and Wittig, 2012). It can be seen therefore, that hedgerow management is required to enhance the structural complexity which results in improved hedgerow condition.

The variety of vegetation layers found in a hedgerow has been related to both floral and faunal diversity (Forman and Baudry, 1984). Hedgerow size (height/width/volume) and the presence/abundance of trees, as well as the provision of cover and the botanical and structural complexity of the vegetation, were the most important factors positively associated with species richness and abundance of breeding birds in hedgerows (Hinsley and Bellamy, 2000). Similarly, diversely structured riparian strips consisting of native species planted in multiple rows provide habitat for flora and fauna (Bach, 2000). The different layers of grass, shrub and tree in riparian buffers are also important for various functions. For example, while both tree and shrub layers provide vertical structure for wildlife habitat, the tree layer provides better shading,

Table 1 Hedgerow type according to woody vegetation cover

Hedgerow type	Shrub layer (%)	Tree layer (%)	Illustrative example
Tree hedgerow (TH)	0-33	66-100	
Predominantly tree hedgerow (PTH)	0-33	33-66	
	33-66	66-100	
Equal proportion of trees and shrubs (EPTSH)	0-33	0-33	
	33-66	33-66	
	66-100	66-100	
Predominantly shrub hedgerow (PSH)	33-66	0-33	
	66-100	33-66	
Shrub hedgerow (SH)	66-100	0-33	

while the shrub layer has to be more frequently harvested (Schultz et al., 2004). A dense shrub layer is important to prevent wind from funnelling under the taller tree layer and to improve protection of nearby crops.

A high proportion of shrubs within hedgerows receives positive responses in landscape aesthetic evaluations. This also applies to an increase in structure by accompanying trees, especially from a long-distance perspective, provided that the stand does not become monotonous in appearance (Hübner et al., 2018). A pronounced tree layer (characterised by densely growing individual trees, in a very compact formation and a height of more than 10 m) may stand out negatively in an otherwise flat landscape, especially where no transition to the surrounding vegetation is maintained by a softened edge. It is important to integrate these concepts into the landscape when designing a new hedgerow, for example by planting a more pronounced shrub layer or an herbaceous fringe which merges into the surrounding vegetation and crops.

4.2 Hedgerow structure

Hedgerow structure refers to the continuity of the canopy along a hedgerow, whether consisting of tree or shrub. Gaps in hedgerows may occur for various reasons, including fallen trees or mismanagement of the hedgerow over time (Gelling et al., 2007), as hedgerows require periodic management and replanting with trees and/or shrubs to maintain their dense woody structure (Croxton et al., 2004; Staley et al., 2015). The hedgerow structure can be classified as closed (C), with small gaps (SG), and with big gaps (BG; Table 2).

Disruption of habitat continuity usually results in an increase in the length of the boundary between fragments and their surrounding habitats (Collinge, 1996). Newly created edges experience shifts in microclimatic characteristics, which may significantly modify plant and animal communities (Saunders et al., 1991; Collinge, 1996). In this context, gaps in continuity alter the habitat conditions and microclimate, causing temperature fluctuations and increased light availability. The latter may cause changes in the composition of the hedgerow as well-developed hedgerows receive sufficient light only in their outer parts and not in the inner closed areas (Weber, 2003; Litza and Diekmann, 2017). Gaps along hedgerows which disrupt continuity are also likely to

Table 2 Hedgerow structure

Hedgerow structure	Description	Illustrative example
Closed (C)	Dense canopy which fully covers the hedgerow along the entire hedgerow length	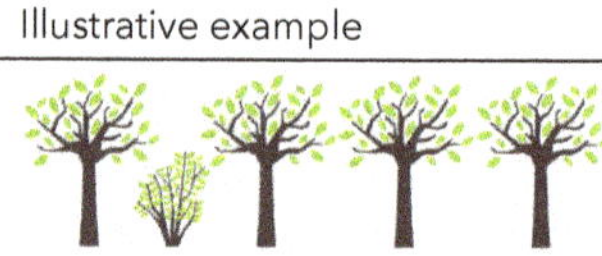
Small gaps (SG)	Relatively dense canopy which almost fully covers the hedgerow along the entire hedgerow length	
Big gaps (BG)	Gappy canopy, there are big gaps along the hedgerow (at least 33% from the hedgerow length is vegetation free)	

constrain movement along habitat corridors (Bright, 1998; Dondina et al., 2016). The continuity of a hedgerow canopy is especially important regarding this corridor function which ensures the movement of plants and animals which thrive in shady conditions.

Furthermore, effective protection against the drift of pesticides and fertilisers can only be expected from riparian strips which have more dense and closed vegetation than hedgerows which are further away from bodies of water (Bach, 2000). A mixture of grass, shrubs and trees is beneficial, as grass strips can intercept pollutants carried by surface run-off, although trees are more effective due to their deeper root system. The continuity of a windbreak also influences its efficiency because wind flow will be concentrated through an opening, creating a zone leeward of the gap in which wind speeds exceed those in the open field (Brandle et al., 2004). However, a mixture of hedgerows having both closed structures and gaps is considered more natural and increases the attractiveness of the landscape (Hübner et al., 2018). In order to provide continuity and a diverse structure, greater width is required, and space may be scarce. Structure appears less important from a distance, making similar or repetitive structures seem harmonious.

4.3 Degree of naturalness

The degree of naturalness, which is often assessed by the similarity of the studied biocenosis to the presumed natural state prior to human intervention, is an important criterion used to assess forests in nature conservation (Reif and Walentowski, 2008). The degree of naturalness of hedgerows in this study refers to the percentage of native species planted in them and can be grouped as near natural (NN), mixed (M) and exotic (E; Table 3).

This criterion is crucial for the ecological and amenity functions of hedgerows. In Germany between 1950 and 1980, priority was given to the establishing of hedgerows with a large number of species, irrespective of their origin. Since the 1980s, the focus has changed to planting hedgerows with fewer, but native, species (Reif and Achtziger, 2000). The main reason for the

Table 3 Degree of naturalness

Degree of naturalness	Description
Near natural (NN)	Hedgerows consist predominantly of tree and shrub species which are native to the area. More than 75% of the species are native.
Mixed (M)	Hedgerows consist of a mixture of native and non-native species. The amount of native species is between 25% and 75%.
Non-native, exotic (E)	Hedgerows consist predominantly of tree and shrub species which are non-native to the area. Less than 25% of the species are native.

earlier practice was the pursuit of a higher conservation value through planting a high number of species. However, this overlooked the importance of other nature conservation criteria, such as the unique character of a locality and the possibility of endangering the local genotypes (Reif and Achtziger, 2000).

Native species in a riparian buffer strip are recommended over non-native species and hybrids as they are usually better adapted to site conditions over the long term (Schultz et al., 2004). Planting native tree and shrub species also ensures a near natural development of the watercourse (Herdt et al., 2007). Hedgerow management which allows the growth of native species with a complex physical structure, such as common alders, can improve the effectiveness of hedgerows as corridors for mammals (Dondina et al., 2016). Wild bees, and in some cases managed bees, preferred to forage on native plants in hedgerows over co-occurring weedy, exotic plants (Morandin and Kremen, 2013).

With regard to landscape aesthetics, diversity in tree and shrub species is attractive, especially from a close-up perspective, where flowering and fruiting aspects or pleasant odours may be experienced (Hübner, 2016). A richness of species with different flowering and fruit maturation times guarantees year-round attractiveness and provides high ecological value and pollination ecosystem services. The most important seasonal aspect is colourful autumn foliage. However, the degree of naturalness measured by the proportion of neophytes may be inversely related to habitat provision. From a purely aesthetic point of view, many deciduous neophytes present positive aspects. Coniferous trees are often viewed negatively and a degree of naturalness is generally associated with a non-monotonous appearance.

Combining the classes of five hedgerow types (TH, PTH, EPTSH, PSH and SH), three hedgerow structures (C, SG and BG) and three degrees of naturalness (NN, M and E) resulted in 45 categories which describe existing hedgerows in the agricultural landscape (Fig. 2).

5 Condition of ecosystem services according to hedgerow category

5.1 Classification of conditions

Coupling the 45 categories obtained by combining the classes of hedgerow types, structure and degree of naturalness described above with the ecosystem functions and related services associated with hedgerows, that is production (biomass), regulation (wind and water protection), ecological enhancement (habitat provision) and amenity (landscape aesthetics), described in Section 3, offers insights into the current condition of a hedgerow. For this purpose, the hedgerow condition of each category and ecosystem service was qualitatively

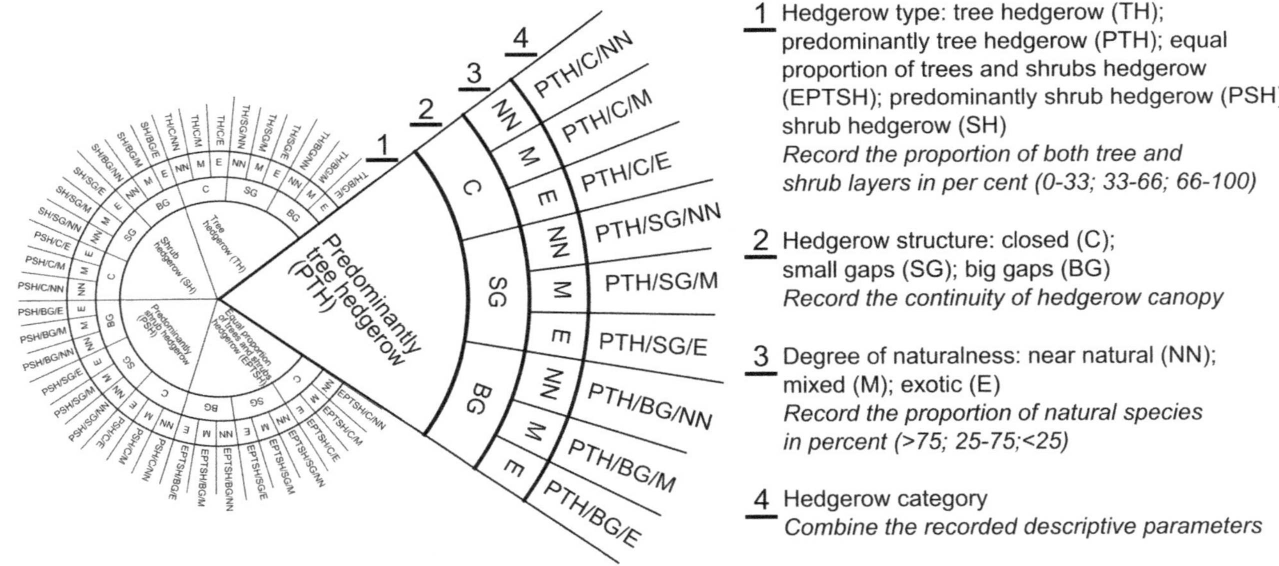

Figure 2 Systematic derivation of 45 hedgerow categories from five hedgerow types, three hedgerow structures and three degrees of naturalness.

assessed by the authors, based on the reviewed literature, and expressed in three classes: 1: 'poor' - the hedgerow category is not suitable for provision of the ecosystem service; 2: 'fair' - the hedgerow category is moderately suitable for the provision of the ecosystem service and 3: 'good' - the hedgerow category is suitable for the provision of the ecosystem service.

5.2 Multiple services

Figure 3 shows the results of a polychoric-based principal component analysis, describing the condition of multiple ecosystem services, along the 45 hedgerow categories. The largest variation of the categories along the PC1 axis was determined by the regulation (water and wind protection) and to some extent production (biomass) functions. The PC2 axis was the second most important direction determined by the ecosystem services, landscape aesthetics and habitat provision.

While it is not possible for all services to be simultaneously in a good condition, multiple services may be enhanced in several categories (shown in green in Fig. 3). In this context, TH, PTH, EPTSH and PSH with closed structure and M or NN vegetation could be associated with a good condition of ecosystem services with the regulation, and to some extent, production functions, predominating. On the other hand, PTH, EPTSH and PSH with SG and NN vegetation were associated with a good condition for both ecological and amenity functions. A poor condition of multiple ecosystem services was suggested for hedgerows of all types with BG and E vegetation (shown in red in Fig. 3). In these hedgerows, at least three ecosystem services were simultaneously found to be in a poor condition.

5.3 Hedgerow management and ecosystem services

The qualitative assessment was subsequently used to emphasise the link between hedgerow management and ecosystem services and to demonstrate how the application of different measures affects the condition of these services. The assessment of the initial state of each ecosystem service according to hedgerow category and impact of the measures (1) filling gaps and (2) increasing the proportion of native species, is visible in Fig. 4. For example, to improve water quality in all hedgerow types, both measures are shown as beneficial. Thus filling gaps with native species would have the highest impact on improving water quality, and would also lead to an improved habitat and to some extent, to improved landscape aesthetics. On the other hand, improving the production function by filling gaps in a TH would be beneficial, while increasing the proportion of native species would

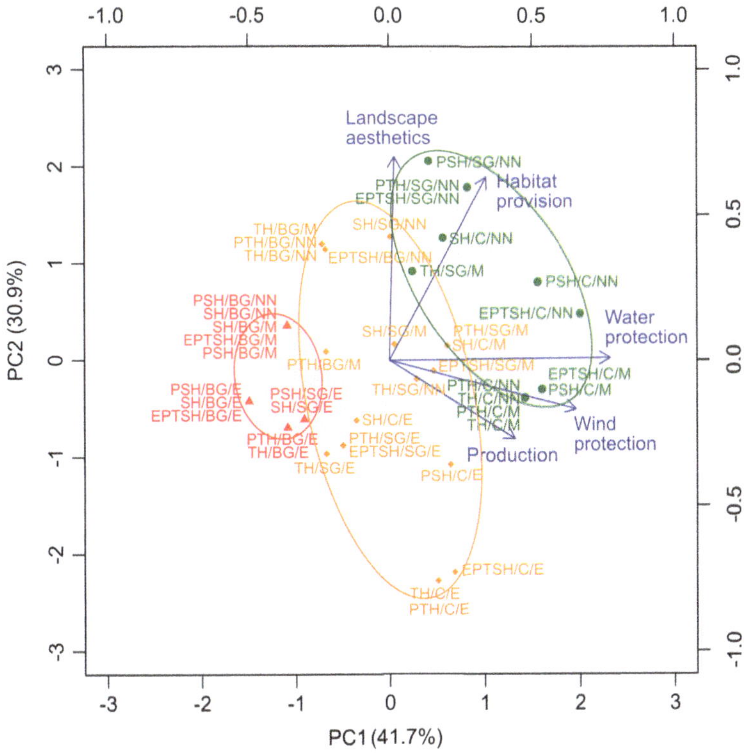

Figure 3 Result of a polychoric-based principal component analysis depicting the condition of multiple *ecosystem services* (biomass production, wind protection, water protection, habitat provision and landscape aesthetics) along the *45 hedgerow categories* obtained by combining the characteristics *hedgerow type*: tree hedgerow (TH), predominantly tree hedgerow (PTH), equal proportion of trees and shrubs (EPTSH), predominantly shrub hedgerow (PSH) and shrub hedgerow (SH)/*hedgerow structure*: closed (C), small gaps (SG) and big gaps (BG)/*degree of naturalness*: near natural (NN), mixed (M) and exotic (E). The assessment of landscape aesthetics captured only the close-up effect. The protection effect of each category was assessed by 1 - poor, 2 - fair and 3 - good condition for the ecosystem services according to the literature. Aligned categories have a similar contribution. *Coloured groups* reflect simultaneously poor (red), fair (yellow) or good (green) condition for multiple services. PC1 is determined by the ecosystem services water protection, wind protection and production, PC2 by landscape aesthetics and habitat provision.

not affect productivity. The biomass production function is not affected by filling the gaps in a SH.

The effect on ecosystem services of altering specific hedgerow characteristics during restoration is further depicted in Table 4. For example, to enhance the production function of hedgerows, increasing tree proportion,

Hedgerow type	Degree of naturalness	Measure A: Filling gaps — From 'Big gaps' to 'Small gaps': Production	Wind protection	Water protection	Habitat provision	Landscape aesthetics	Measure A: Filling gaps — From 'Small gaps' to 'Closed': Production	Wind protection	Water protection	Habitat provision	Landscape aesthetics
TH	NN	↑	→	↑	→	↓	→	↑	↑	→	→
	M	↑	→	↑	→	→	→	↑	↑	→	↓
	E	↑	→	→	→	→	→	↑	↑	→	↓
PTH	NN	→	↑	↑	↑	→	↑	→	↑	↓	↓
	M	→	↑	↑	→	→	↑	→	↑	→	→
	E	→	↑	→	→	→	↑	→	↑	→	↓
EPTSH	NN	↑	↑	↑	→	↑	→	↑	↑	→	↓
	M	↑	↑	↑	→	→	→	↑	↑	→	→
	E	↑	↑	→	→	→	→	↑	↑	→	↓
PSH	NN	→	↑	↑	↑	↑	↑	↑	↑	↓	→
	M	→	↑	↑	→	→	↑	↑	↑	→	→
	E	→	↑	→	→	→	↑	↑	↑	→	→
SH	NN	→	↑	↑	→	↑	→	→	↑	→	→
	M	→	↑	↑	→	→	→	→	↑	→	→
	E	→	↑	→	→	→	→	→	↑	→	→

Hedgerow type	Hedgerow structure	Measure B: Increase proportion of natural species — From 'Exotic' to 'Mixed': Production	Wind protection	Water protection	Habitat provision	Landscape aesthetics	Measure B: Increase proportion of natural species — From 'Mixed' to 'Near natural': Production	Wind protection	Water protection	Habitat provision	Landscape aesthetics
TH	C	→	→	↑	↑	↑	→	→	→	→	→
	SG	→	→	↑	↑	↑	→	→	→	→	↓
	BG	→	→	→	↑	↑	→	→	→	→	→
PTH	C	→	→	↑	↑	↑	→	→	→	→	→
	SG	→	→	↑	↑	→	→	→	→	↑	↑
	BG	→	→	→	↑	→	→	→	→	→	↑
EPTSH	C	→	→	↑	↑	↑	→	→	→	↑	→
	SG	→	→	↑	↑	→	→	→	→	↑	↑
	BG	→	→	→	↑	→	→	→	→	↑	→
PSH	C	→	→	↑	↑	→	→	→	→	→	↑
	SG	→	→	↑	↑	→	→	→	→	↑	↑
	BG	→	→	→	↑	→	→	→	→	→	→
SH	C	→	→	↑	↑	→	→	→	→	→	↑
	SG	→	→	↑	↑	→	→	→	→	→	↑
	BG	→	→	→	↑	→	→	→	→	→	→

Figure 4 Effect of filling gaps (Measure A) and increasing the proportion of natural species (Measure B) on the condition of ecosystem services, assessed according to the scientific literature, reviewed by the authors. The background colour depicts the initial state (red: poor, yellow: fair, green: good). Arrows indicate the change in condition by one class with respect to the ecosystem service after applying the specific measure (↑ improvement, → no change, ↓ decrease). *Hedgerow type*: tree hedgerow (TH), predominantly tree hedgerow (PTH), equal proportion of trees and shrubs (EPTSH), predominantly shrub hedgerow (PSH) and (SH) shrub hedgerow. *Hedgerow structure*: closed (C), small gaps (SG) and big gaps (BG). *Degree of naturalness*: near natural (NN), mixed (M) and exotic (E).

Table 4 Important characteristics describing the provision of ecosystem services by hedgerows

	Ecosystem service				
Hedgerow characteristic	Production	Wind protection	Water protection	Habitat provision	Landscape aesthetics
Tree proportion	↑	↑	↑	↑↓	↑↓l
Shrub proportion	↓	↑	↑	↑↓	↑l
Hedgerow continuity	↑	↑	↑	↑	↑↓c, l
Proportion of native species			↑	↑	↑c, l
Number of tree species				↑	↑c
Tree height	↑	↑	↑	↑↓	↑l
Tree DBH (-age)	↑			↑	↑c, l
Number of stems per m^2	↑↓	↑	↑		↓c
Shrub height		↑			↑c
Number of shrub species				↑	↑c
Density of woody-features (m ha^{-1}; m km^{-2})	↑	↑	↑	↑↓	↑↓c, l

↑ - An increase in the hedgerow characteristic improves the ecosystem service.
↓ - An increase in the hedgerow characteristic diminishes the ecosystem service.
↑↓ - An increase in the hedgerow characteristic may improve or diminish the ecosystem service.
l = long-distance effect; c = close-up effect.

hedgerow continuity, tree height and DBH are beneficial. Increasing stem density is not desired when the objective is growing wood for timber. On the other hand, a high number of stems can be advantageous in the case of producing biomass feedstock. However, multiple thin stems which are considered 'impenetrable' reduce hedgerow attractiveness (see also Reppin and Augenstein, 2018). A trade-off was suggested with regard to trees with larger DBH which may be attractive for their production function, while retaining old trees (Methuselah or biotope trees) is important to improve the aesthetic value of the landscape and habitat provision.

Reducing shrub proportion is beneficial in increasing the production of biomass, although it reduces the level of wind protection. A highly diverse landscape with a variety of hedgerow types and structures is beneficial for habitat provision as species differ in their environmental requirements. Overall, habitat provision and landscape aesthetics are associated with a high degree of controversy. The perception of hedgerow characteristics related to landscape aesthetics also changes depending on the viewing distance.

The density of woody-features measured by the length of hedgerows in an area (m ha^{-1} or m km^{-2}) reflects the link between hedgerow effectiveness regarding ecosystem services and their distribution in the landscape. In regions with low forest cover, the risk for wind erosion increases when the minimal density of woody-features is below 5 km km^{-2} (BMVEL, 2001). A dense network of linear woody structure is also required for a functioning biotope network. For example, a hedgerow density of 80 m ha^{-1} was encouraged in the federal state of Schleswig-Holstein (Knickel et al., 2001). While a high structural density is beneficial for many species, it should also be kept in mind that retaining open habitats in the agricultural landscape are important for breeding birds, such as the skylark (Lutze et al., 2010).

5.4 Prioritising services

In summary, Fig. 4 and Table 4 can be used as guidelines for hedgerow restoration and management with respect to multiple ecosystem services while considering potential conflicts between services. The criteria employed in decision making should be structured according to their importance in the local conditions. For example, in an agricultural landscape prone to wind erosion, priority should be given to wind protection, while in a landscape of high tourist value, aesthetics may be given priority. An objective view requires prioritising ecosystem services according to their condition and relevance in the area. For example, in an intensively managed agricultural area intersected by multiple ditches, water protection should be a priority, as opposed to an extensively managed landscape, where water quality is already in a good condition.

6 Key constraints to contemporary hedgerow management

6.1 Legal

The management of hedgerows in the German agricultural landscape is difficult because hedges, tree rows or copses are not legally considered as forests. They can, within size limits, either be part of the agricultural area of a farm and thus entitle the farmer for direct payments or they are located on field margins and do not belong to the eligible agricultural area and are therefore not legally considered as part of a field. Irrespective of their status, hedgerows are regarded as protected landscape components which cannot be removed permanently and cannot be used without permission from the responsible authorities. Their exact legal status is further complicated as they are subject to different regulations. As so-called landscape features within the meaning of the German agricultural regulation 'Agrarzahlungen-VerpflichtungenVerordnung' (AgrarZahlVerpflV), which is based on EU Regulations linked to conditionality and greening in Pillar I, and to rural development in Pillar II, the permanent removal of woody plants which are recognised as administrative landscape features is prohibited. Woody plants are also cross-compliance relevant, that is payments under the EU agricultural subsidy are tied to the preservation of these hedgerows. Landscape features are formally part of agricultural land and are mostly located directly on or adjacent to fields.

Hedgerows which do not have the legal status of a landscape feature do not belong to the eligible agricultural area and are therefore not legally considered as part of a field. Such hedges are subject to the tree preservation regulations created by districts and independent cities or municipalities. The background of the ruling is the request formulated in the Federal Nature Conservation Act for preserving trees outside forests and hedgerows. The regulations for the protection of hedgerows may vary depending upon the regional requirements. Most of the regulations prohibit the removal or damage of woody plants, especially thicker trees and shrub hedges. As a result, the use of hedgerows is prevented or severely limited. Before using the wood of these hedgerows, it is necessary or at least advisable to obtain permits from the competent authorities in order to avoid penalties.

It is necessary to clarify whether the use of trees in the course of hedgerow management, as described in an exemplary manner in Section 7 of this chapter, is considered as damaging interference. If this is the case, the illustrated approach to their management would at best be possible only to a very limited extent. According to the Tree Protection Regulation of the Elbe Elster district (Landkreis Elbe-Elster, 2013), which is binding for the presented case study in Section 7, no trees with a diameter larger than 32 cm in 1 m of height (for oak, beech and rowan only 16 cm) should be removed. Hedges with a minimum height of 1.50 m and 200 m^2 base area are protected, as are trees of smaller diameter planted as replacements in accordance with the tree protection ordinance.

6.2 Profitability

The historic production function and thus the related profitability of hedgerows are currently of less importance. In both Pillars I and II of the CAP, measures related to supporting isolated trees or hedgerows in agricultural landscapes are considered to be non-productive investments (Santiago-Freijanes et al., 2018). The application of such non-productive EFA elements in Pillar I by farmers in Germany was much lower than the use of measures related to productivity such as nitrogen-fixing crops and catch crops (Pe'er et al., 2017).

The maintenance of linear woody structures is labour intensive as they occupy a small area and are dispersed in the landscape. In Germany, landscape maintenance is usually carried out by Landcare Associations (LCA) *'Landschaftspflegeverbände'* which describe the measures to be taken, calculate the costs, develop project proposals and apply for subsidies from the regional government through various agri-environment schemes (Bluemlein, 2009). However, existing schemes supporting farmland trees in the German federal state of Saxony were often characterised by a narrow range of measures, the reluctant participation of farmers and other land users and by limited performance (Schleyer and Plieninger, 2011). The premiums offered in the funding schemes co-financed by the EU have been generally perceived as insufficient to compensate for the costs incurred (Schleyer and Plieninger, 2011).

Neglect of the productivity function of hedgerows, associated with the integrated management of hedgerows and agricultural crops which is typical for agroforestry, hinders the provision of potential benefits to farmers (Tsonkova et al., 2018a). The condition of hedgerows is also deteriorating, aggravating conflict between farmers and conservationists. For example, in East Germany, windbreaks which were established during the socialist period and consist predominantly of non-native and fast-growing trees, particularly hybrid poplar, are widely spread.

These farmed woody-features have been largely neglected, both in terms of management and of awareness among land users and society. The main conservation strategy is to prohibit their destruction, while little attention has been given to actively managing them for conservation (Plieninger and Schaich, 2014). Further problems arise for farmers due to breakage of branches and stumps lying on the ground (DVL, 2006). Increased shading of neighbouring agricultural land and an increase in the width of hedgerows are further perceived as reducing profitability. The sustainable utilisation of hedgerows is necessary to improve their condition and the provision of ecosystem services.

7 Case study: opportunities for utilisation and further development of existing hedgerows to enhance multiple ecosystem services

7.1 Data collection in the study area

The study area is located within the administrative district of Kleine Elster and the municipalities of Sonnewalde and Finsterwalde in the southern part of the eastern German federal state of Brandenburg (Fig. 5). The average temperature and precipitation in this area during the period 1981–2010 was 9.3°C and 600 mm year^{-1}, respectively. The predominant soil textures are sand and loamy sand.

In 2015, the hedgerows adjacent to agricultural land in the study area, which were neither classified as forest and nor registered as landscape features, were mapped based on the crown visible in digital orthophotos (40 cm - resolution) obtained from the Brandenburg Surveying and Geoinformation Office (LGB). Woody-features having a length of at least 20 m were digitised, while no minimal or maximal rule for width was set. Subsequently, a representative study area amounting to 4 km^2 (~1% of the total area), and comparatively rich in woody-features, was selected for a detailed analysis. In 2016, the hedgerows were classified on site according the characteristic type and structure described in Section 4. Up to three hedgerows representing each combination were randomly selected out of these classes, giving a total of 37 hedgerows (Fig. 5).

In these 37 hedgerows, measurements of tree height, diameter of >7 cm at breast height, 1.3 m above the ground (DBH), together with the tree and shrub species were recorded. Measurements of smaller trees with DBH <7 cm were not conducted and were excluded from further assessment. The shrub species were recorded according to their relative distribution in the hedgerow at two heights: higher than 1.5 m (shrub layer 1) and below 1.5 m (shrub layer 2). Hedgerows longer than 100 m were measured in five plots of 20 m in length, equally distributed throughout the total hedgerow length. Where the length was less than 100 m, the entire hedgerow was recorded.

The measurements of tree height and DBH were used to determine the theoretical maximum biomass potential of these hedgerows. The marketable tree volume (V) in m^3 is the product of tree basal area (g [m^2]), tree height (h [m]) and a form factor (f [-]) which converts the total tree volume to marketable volume (Kramer and Akça, 2008). The form factor was extracted from yield tables for the main tree species in the area (Schober, 1995).

$$V = g \times h \times f \tag{1}$$

$$g = \frac{\pi}{4} \times DBH^2 \tag{2}$$

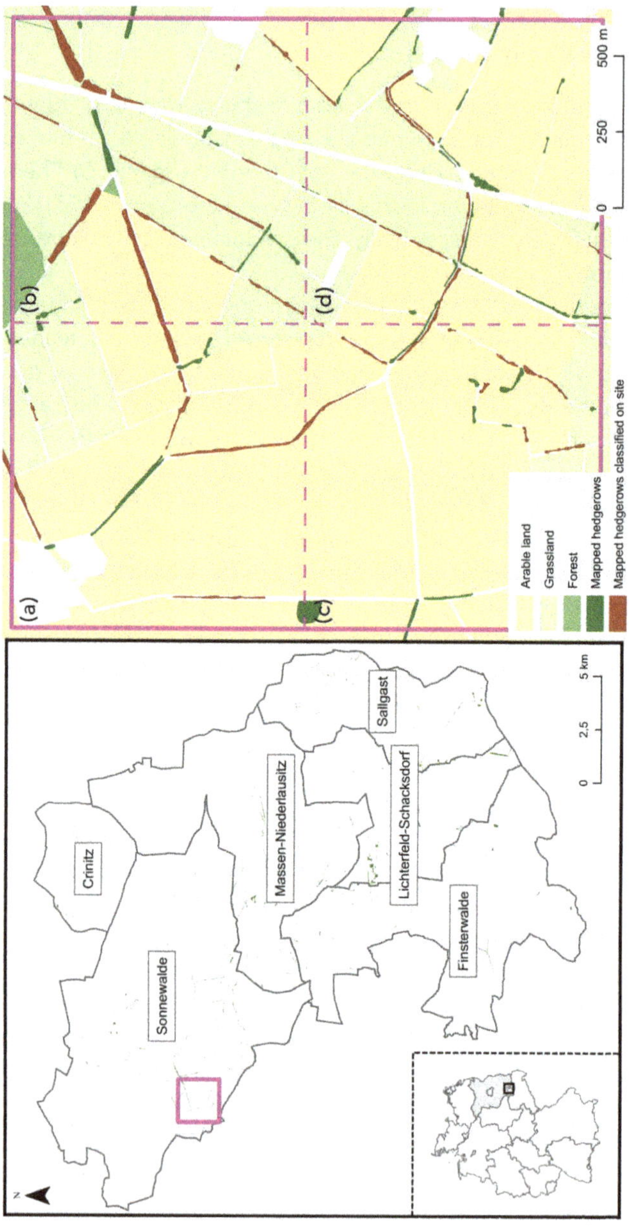

Figure 5 The study area in the German federal state of Brandenburg, main land use and the mapped hedgerows (n = 83), highlighting those randomly selected and investigated in detail on site (n = 37).

The ecosystem services biomass production, wind and water protection, habitat provision and landscape aesthetics were assessed, based on the hedgerow categories defined in Section 4 and coupled with ecosystem services in Section 5, according to their distribution in the landscape. Results on the landscape scale were presented for an area of 1 km^2 marked by A, B, C, D in Fig. 5.

7.2 Development of example scenarios for hedgerow restoration and management in the study area

Under the assumption that the legal status of hedgerows does not obstruct their management, an economic assessment was conducted for two example scenarios. Scenario I focussed on cultivating fast-growing trees (managed as short rotation coppices) for the production of biomass feedstock, and slow-maturing trees for the production of trunk wood. Scenario II focussed solely on growing slow-maturing trees for the production of trunk wood. The developed scenarios are examples for utilising hedgerow productivity while maintaining hedgerows in a good condition for the balanced provision of ecosystem services. The decision-making process for each hedgerow should be conducted individually, taking into account the management objectives and ecosystem services of interest.

The management strategy aims to restore the entire hedgerow over a period of 20 years with management interventions every 5 years, thus 20% of the hedgerow is renewed in one intervention. In order to reduce disturbance in hedgerow continuity, the management should take place in 20 m sections distributed over the entire hedgerow length. The planting design and management of a 20 m section, according to hedgerow types for both scenarios, are shown in Fig. 6. According to guidelines for hedgerow establishment, tree height should be between 1.0 m and 1.5 m and shrub height between 50 cm and 70 cm at the time of planting (DVL, 2000; Bioland et al., 2011). The selected species should be traditional for the area in accordance with species found in the hedgerows. In each 20 m section, one tree of high ecological value, for example *Quercus robur*, should be planted and retained. Shrubs are planted in groups with one or two predominating species in each hedgerow (DVL, 2006).

The interventions for hedgerow restoration and management considered in this chapter include harvesting mature trees, removing their roots so new plantings can be made, clearing shrubs, preparing the soil for replanting, planting and fencing trees and shrubs, mulching trees and coppicing trees and shrubs. The investment costs for tree and shrub plants were made in accordance with the recent price list of F.-O. Lürssen Baumschulen GmbH & Co. KG (2017). The labour time required for hedgerow management described in Table 5, including the fixed and variable costs of the machinery used, were calculated based on standardised costs (Ackermann et al., 2005; Joswig et al.,

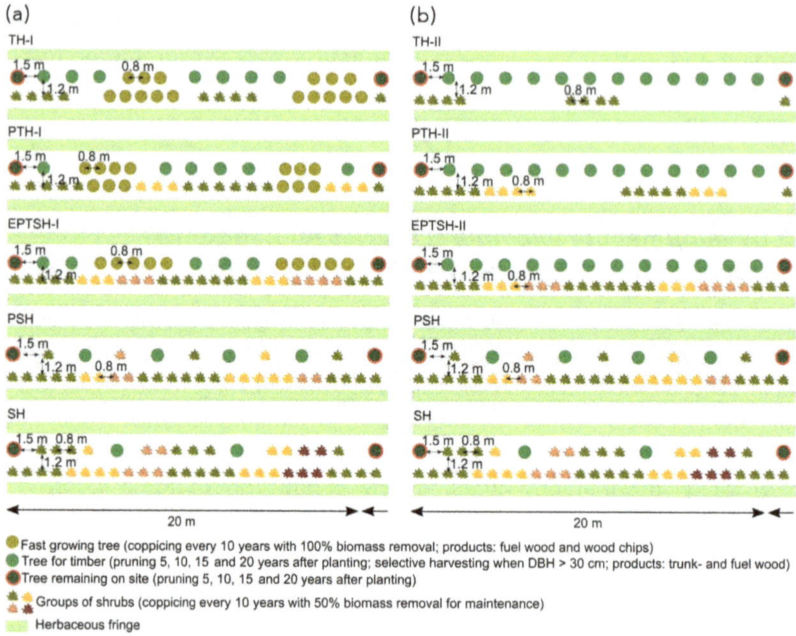

Figure 6 Planting design and management of a 20 m long section for (a) Scenario I. Production of biomass feedstock and trunk wood and (b) Scenario II. Production of trunk wood, according to *hedgerow type*: tree hedgerow (TH), predominantly tree hedgerow (PTH), equal proportion of trees and shrubs (EPTSH), predominantly shrub hedgerow (PSH) and shrub hedgerow (SH).

2012). The hourly labour costs were taken from the latest tariff register of Berlin and Brandenburg (SIAS, 2017).

The maintenance of fast-growing trees in Scenario I requires coppicing every 10 years. The entire biomass can be removed and used for the production of wood chips and fuel wood. Timber trees require maintenance in the first 20 years of growth when they are regularly pruned. When their diameter reaches 30 cm, they can be harvested and their biomass sold as trunk- and fuel wood. Revenues are obtained by selling trunk-, fuel wood and wood chips at the current prices of 80 € per solid cubic metre, 40 € per stacked cubic metre and 22 € per loose cubic metre, respectively, assuming that the products are purchased by small consumers for their own use and who would therefore pay higher prices.

The uniform series of equal investments (A; hereafter called annuity of hedgerow management) to be made under consideration of a defined time period was calculated by Eq. (3; Schmidt, 2011). Net present value is the difference between the present value of cash inflows and outflows (R_t) over

Table 5 Assumed average costs of investment materials and labour efforts including labour time and machinery used

Action			Costs	Unit
	Investment			
Planting	Tree		2.50-6.00	€ $tree^{-1}$
	Shrub		1.50-2.50	€ $shrub^{-1}$
	Mesh fence		2.50	€ $plant^{-1}$
	Labour effort			
Planting	Soil preparation by using a rotary cultivator		0.10	€ m^{-2}
	Establishment according to hedgerow type	Shrub hedgerow	0.95	€ $plant^{-1}$
		Predominantly shrub hedgerow, equal proportion of trees and shrubs, predominantly tree hedgerow	1.00	€ $plant^{-1}$
		Tree hedgerow	1.10	€ $plant^{-1}$
	Mulching of trees with solid manure according to hedgerow type	Shrub hedgerow	2.40	€ $tree^{-1}$
		Predominantly shrub hedgerow	1.70	€ $tree^{-1}$
		Equal proportion of trees and shrubs, predominantly tree hedgerow, tree hedgerow	1.40	€ $tree^{-1}$
	Installing mesh fencing		0.73	€ $plant^{-1}$
Removing trees	Manually harvesting trees by using a chainsaw and cutting the branches into 1 m stems	Tree diameter between 30 and 50 cm	56.00	€ $tree^{-1}$
		Tree diameter between 50 and 70 cm	78.00	€ $tree^{-1}$
		Tree diameter >70 cm	120.40	€ $tree^{-1}$
	Removal of tree roots by using a stump grinder	Diameter of tree base <55 cm	41.70	€ $tree^{-1}$
		Diameter of tree base <70 cm	100.10	€ $tree^{-1}$
		Diameter of tree base <85 cm	116.70	€ $tree^{-1}$
	Shrub clearing by using a brush cutter	Shrub coverage 30%	4.40	€ m^{-2}
		Shrub coverage 50%	6.40	€ m^{-2}
		Shrub coverage 70%	8.50	€ m^{-2}
Coppicing	Coppicing fast-growing trees by a chainsaw and chopping the branches by wood chopper	100% biomass removal	49.10	€ m^{-1}
		50% biomass removal	26.10	€ m^{-1}
		25% biomass removal	13.20	€ m^{-1}
Maintaining trees and shrubs	Coppicing shrubs by a chainsaw	50% biomass removal	1.35	€ m^{-1}
	Maintenance of trees for stem wood by pruning shears		4.90	€ $tree^{-1}$

Own calculations based on Ackermann et al. (2005), F.-O. Lürssen Baumschulen GmbH & Co. KG (2017), Joswig et al. (2012), SIAS (2017).

a period of time (t), at a specified interest rate (i). The capital recovery factor converts a present value into a stream of equal annual payments over a specified time period (t), at a specified interest rate (i). In this study, annuity for the time period (t) of 60 years was calculated at an assumed interest rate (i) of 3%.

$$A = NPV \times CRF \quad (3)$$

$$NPV = \sum \frac{R_t}{(1+i)^t} \quad (4)$$

$$CRF = \frac{(1+i)^t \times i}{(1+i)^t - 1} \quad (5)$$

7.3 Categorisation of hedgerows and assessing the condition of their ecosystem services in the study area

Within the study area, agricultural land use was practised in more than 89% of each km^2 with a prevalence of arable land, except in area B, where grassland use predominated (Table 6). The overall hedgerow length in areas A–D (density of hedgerows) was below 5 km km^{-2}, and was therefore mostly insufficient for wind protection. The proportion of hedgerows from the agricultural land varied between 1.82% and 4.74% in areas C and B, respectively. This was higher than the average proportion of hedgerows in the entire study area which was 1.74%.

Table 6 Main land use, proportion of hedgerow and hedgerow density in the study area

Area	Land use	Proportion agricultural land per km^2 (%)	Proportion of hedgerows from agricultural land (%)	Density of hedgerows per hectare agricultural land (m ha^{-1})	Density of hedgerows (m km^{-2})
A	Arable land	73	2.45	27.75	2497
	Grassland	17			
B	Arable land	17	4.74	40.93	3643
	Grassland	72			
C	Arable land	80	1.82	19.24	1866
	Grassland	17			
D	Arable land	58	2.67	41.32	3719
	Grassland	32			
Study area average[a]	Arable land	33	1.74	16.78	780
	Grassland	12			

Areas A, B, C and D according to Fig. 5.
[a] For areas with at least 5% agricultural land per km^2 (n = 313).

The proportion of hedgerows therefore complied with the recommendation for good agricultural practices in which the minimum area of woody-features in the landscape should amount to 1-2% of the agricultural area (Knickel et al., 2001). The average proportion of registered landscape features in 2015 amounted to an additional area of 0.7%. The lower proportion of landscape features agrees with the overall trend of lower registration for these features. For example, in 2015, only between 15% and 41% of the existing woody-features and buffer strips were registered as EFA in Germany (Pe'er et al., 2017).

7.3.1 Distribution and structure

The total length of woody-features in the study area was 11 725 m in 4 km^2. The predominant hedgerow type was EPTSH with 34% of the total length, followed by PTH and TH which together accounted for more than half of the hedgerows (Table 7). Assuming an equal distribution in the landscape for the five hedgerow types, approximately 2300 m should have been allocated to each type. On this basis, with a total length of 1400 m, both PSH and SH were under-represented. Regarding structure, the highest proportion of hedgerows were those with SG

Table 7 Relative distribution of the hedgerow length in % in the four 1 km^2 areas (A, B, C, D) according to hedgerow category

	Area				
	A	B	C	D	Total
Characteristics	Hedgerow distribution (%)				
Hedgerow type					
Tree hedgerow (TH)	3	10	3	8	23
Predominantly tree hedgerow (PTH)	4	10	6	12	31
Equal proportion of trees and shrubs (EPTSH)	14	6	6	8	34
Predominantly shrub hedgerow (PSH)	0	1	0	2	3
Shrub hedgerow (SH)	1	4	1	2	9
Hedgerow structure					
Closed (C)	1	5	1	5	13
Small gaps (SG)	13	18	7	14	52
Big gaps (BG)	7	8	8	12	36
Degree of naturalness[a]					
Near natural (NN)	6	19	8	17	50
Mixed (M)	8	10	5	8	30
Exotic (E)	7	3	3	7	20

[a] The tree and shrub species were determined on site in 37 hedgerows with a total length of 7079 m and the resulting distribution was used to upscale the degree of naturalness for the remaining hedgerows in the study area (n = 83).

at 52%, followed by those having BG and closed structures with 36% and 13%, respectively. Assuming that BG occupy at least one third of the hedgerow length, about 1400 m of the existing hedgerows are currently vegetation free and need replanting. Regarding the degree of naturalness, 50% of the hedgerows were considered NN, followed by M and E with 30% and 20%, respectively.

7.3.2 Tree and shrub species

The main tree species in the area were black alder (*Alnus glutinosa* (L.) Gaertn.), followed by hybrid poplar (in particular *Populus × canadensis*) and common oak (*Quercus robur* L.), composing 52%, 17% and 13%, respectively, of the species recorded (Fig. 7). The majority of young trees were black alder, 46% of which had a DBH <30 cm. Black alder is typically grown along ditches and the highest proportion of mapped woody-features were riparian buffer strips, followed by hedgerows and windbreaks. The majority of older trees were hybrid poplars which are non-native to the area. The non-native species *Quercus rubra* accounted for less than 1% of the tree species.

In the shrub layer, those categorised as layer 1 (higher than 1.5 m) clearly predominated, more than 75% of the shrub coverage being recorded in this layer (Fig. 8). The predominant shrub species in this layer were *Sambucus nigra* and *Salix* spp. Of the shrubs found, *Symphoricarpos albus* was not native to the area; however, it was found only in one hedgerow. The non-native trees (*Populus × canadensis*, *Quercus rubra*) and shrub species (*Symphoricarpos albus*) should be foremost in consideration for removal.

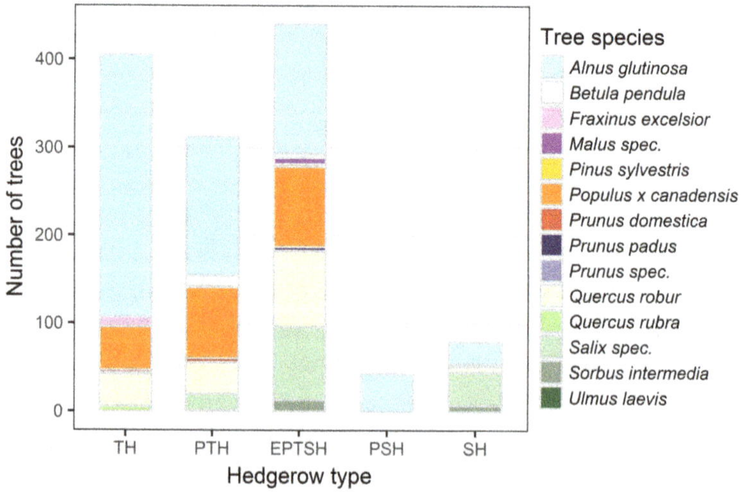

Figure 7 Distribution of tree species recorded according to hedgerow type (n = 1277).

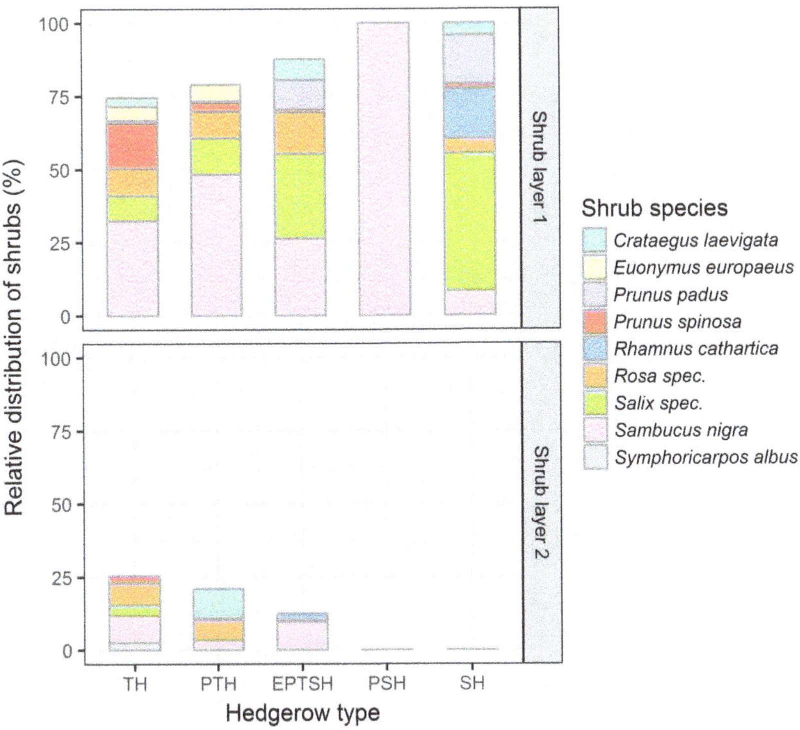

Figure 8 Relative distribution of shrub species according to hedgerow type. The sum of shrub layer 1 (shrubs higher than 1.5 m) and shrub layer 2 (shrubs lower than 1.5 m) in each hedgerow type amounts to 100% of the shrub layer recorded.

7.3.3 Production function

Of the 45 categories identified in Section 3, 27 were found in the study area (Fig. 9). Overall, the longest category with 1200 m was TH with SG and mixed vegetation. The predominant TH, PTH and EPTSH hedgerow types suggest high biomass availability in the study area, since these categories were associated with the highest estimated theoretical biomass potential (Table 8). The marketable tree volume ranged between 240 and 690 m^3 ha^{-1} according to hedgerow type. The highest potential estimated for TH gradually reduced with tree proportion and reached its minimum for SH. The potential of EPTSH increased with tree proportion (Tsonkova et al., 2018b).

The calculated potential for TH and PTH in the study area was higher than the average stocks of biomass in German forests of 330 m^3 ha^{-1} (considering all age groups and all tree species), reported by the Third National Forest Inventory (Thünen Institut, 2012). This is due to trees receiving more light than is the case under forest conditions and their crowns consequently being

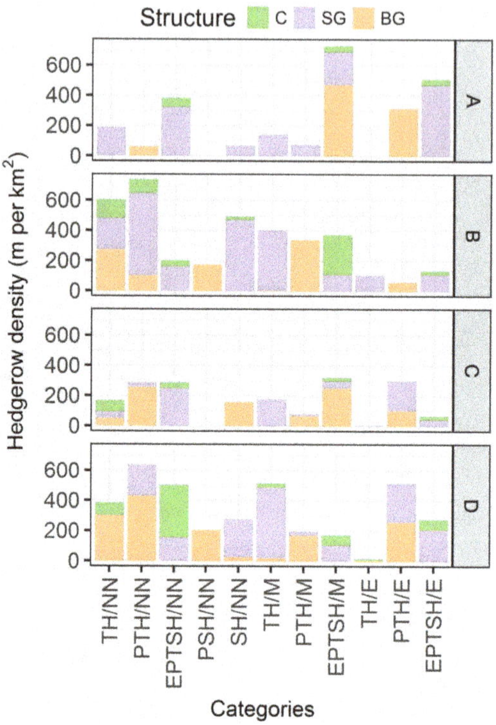

Figure 9 Distribution of hedgerows (m km^{-2}) according to the characteristics *hedgerow type*: tree hedgerow (TH), predominantly tree hedgerow (PTH), equal proportion of trees and shrubs (EPTSH), predominantly shrub hedgerow (PSH) and shrub hedgerow (SH)/*hedgerow structure*: closed (C), with small gaps (SG) and with big gaps (BG)/*degree of naturalness*: near natural (NN), mixed (M) and exotic (E) (n = 83).

better developed. They are also planted in higher densities. For example, the hedgerow density was up to 1500 stems ha^{-1}, compared to forests, where the highest density of ~900 stems ha^{-1} was found in Brandenburg.

Table 8 Estimated theoretical biomass potential (mean ± SE) according to hedgerow type (n = 37)

Hedgerow type	n	Merchantable tree volume (m^3 ha^{-1}) mean (±SE)
Tree hedgerow (TH)	7	690 (±190)
Predominantly tree hedgerow (PTH)	10	540 (±110)
Equal proportion of trees and shrubs (EPTSH)	13	330 (±38)
Predominantly shrub hedgerow (PSH)	1	345 (NA)
Shrub hedgerow (SH)	6	240 (±110)

7.3.4 Condition of multiple services

According to the assessment of hedgerow condition based on the categories and their link to ecosystem services described in Section 5, the production function was in the range of fair to good, with 54% and 29% of the hedgerows, respectively (Fig. 10). 35% of the hedgerows were considered good in regard to landscape aesthetics and 18% regarding habitat provision. 64% of the hedgerows were considered fair in terms of landscape aesthetics and 62% in habitat provision. The lowest overall condition was found for the services of wind and water protection with 54% and 47% of the total length, respectively, being in a poor condition. The lowest condition for wind and water protection, quantified by the weighted average calculated over the total length for each area, was found in areas A and C (Table 9).

Areas A and C were characterised by predominately arable land use and the lowest proportion of hedgerows with closed structure (Tables 6 and 7). Coupling the predominant arable land use with the mainly sandy soil texture and the poor condition of wind protection, suggests there would be a high risk of wind erosion in this landscape. Improved water protection is required as most of the hedgerows were classified as riparian buffer strips, planted along ditches. Improvement in the condition of these hedgerows can be achieved by

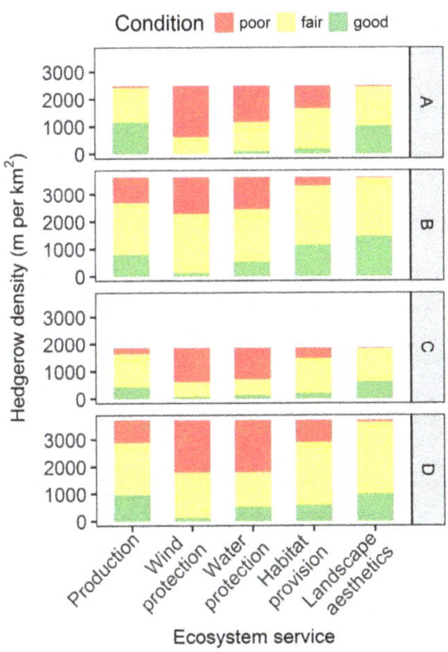

Figure 10 Distribution of hedgerows (m km^{-2}) according to the condition of ecosystem services (production, wind- and water protection, habitat provision and landscape aesthetics; n = 83).

Table 9 Weighted average of the hedgerow condition according to the length allocated to each category

Area	Production	Wind protection	Water protection	Habitat provision	Landscape aesthetics
A	1.436	0.250	0.509	0.743	1.391
B	0.960	0.664	0.824	1.232	1.393
C	1.114	0.377	0.461	0.906	1.323
D	1.035	0.517	0.628	0.941	1.247

increasing the proportion of shrub species, closing the BG and replacing the E species (Section 5).

7.4 Hedgerow utilisation and further development

To demonstrate opportunities for hedgerow utilisation and further development, several hedgerows covering multiple categories presented in Fig. 11 were selected for a detailed analysis. A general description of each hedgerow is provided in Table 10.

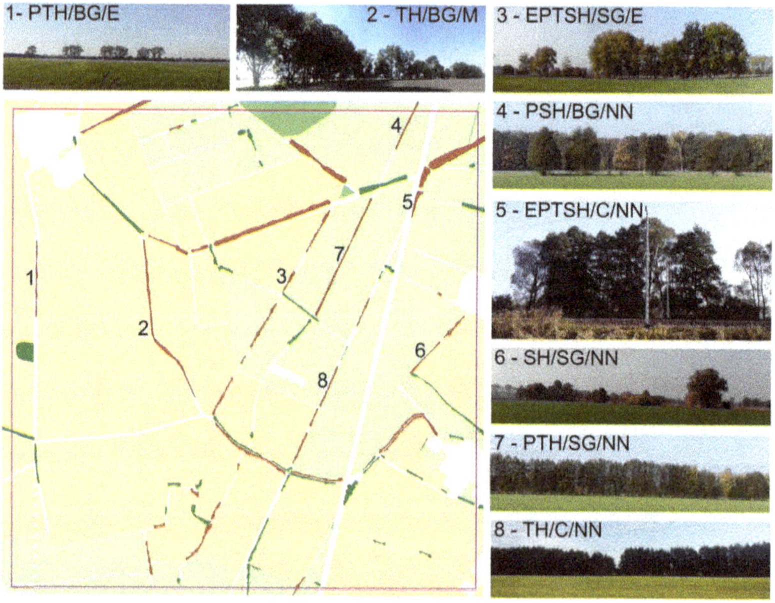

Figure 11 Overview of selected hedgerows for the case study according to the characteristics *hedgerow type*: tree hedgerow (TH), predominantly tree hedgerow (PTH), equal proportion of trees and shrubs (EPTSH), predominantly shrub hedgerow (PSH) and shrub hedgerow (SH)/*hedgerow structure*: closed (C), small gaps (SG) and big gaps (BG)/*degree of naturalness*: near natural (NN), mixed (M) and exotic (E).

Table 10 Main characteristics of selected hedgerows

Hedgerow	1	2	3	4	5	6	7	8
Main function	Windbreak	Windbreak	Hedgerow	Riparian strip	Hedgerow[a]	Hedgerow	Riparian strip	Riparian strip
Category	PTH/BG/E	TH/BG/M	EPTSH/SG/E	PSH/BG/NN	EPTSH/C/NN	SH/SG/NN	PTH/SG/NN	TH/C/NN
Tree layer (%)	33–66	66–100	33–66	0–33	33–66	0–33	66–100	66–100
Shrub layer (%)	0–33	0–33	33–66	33–66	33–66	66–100	33–66	0–33
Continuity of hedge-row canopy	Gappy	Gappy	Relatively dense	Gappy	Dense	Relatively dense	Relatively dense	Dense
Proportion of native species (%)	<25	25–75	<25	>75	>75	>75	>75	>75
Length (m)	305	748	112	280	110	175	566	181
Width (m)	5	6	4	3	8	3	2	3
Sampled area (%)	33	13	89	36	91	57	18	55
Length of gaps (m)	91	224	11	84	0	17	57	0
Tree average height (m)	23	14	11	13	12	9	14	10
Tree average BHD (cm)	103	33	21	23	20	28	20	12
Total merchantable tree volume[b] (m^3)	109	154	9	20	20	11	82	19

[a] Near a railway line.
[b] Theoretical biomass potential.

7.4.1 Management

It was assumed that hedgerow management is conducted every 5 years according to the scenarios in Fig. 6 and the interventions described in Table 11. During the first year, gaps should be filled to create hedgerows which are closed or have SG. The main tree species remains black alder and one common oak tree is planted and retained in each 20 m section. In both scenarios, trees are harvested when they reach maturity (in this study DBH >30 cm). Trees with initial DBH <30 cm are harvested in year 40 when it was assumed that they had reached maturity. Trees in hedgerow 8, with DBH <30 cm, are first harvested in year 40. It was assumed that trees planted in year 1 will reach maturity in year 60, when they will be harvested, their aboveground biomass sold and their roots removed to make room for new planting. In Scenario I, fast-growing trees planted in year 1 are coppiced every 10 years with 100% biomass removal. The biomass of fast-growing trees was derived from average growth rates of trees with a short life cycle, such as black alder in Germany (Thünen Institut, 2012). In Scenarios I and II, shrubs are coppiced every 10 years with 50% biomass removal for maintenance purposes.

7.4.2 Annuities

The annuity of hedgerow management calculated by Eqs. (3)-(5) and the average costs in Table 5 are presented for filling gaps and for hedgerow restoration and management conducted in 20 m long sections for each hedgerow type in Scenarios I and II. The results are presented in intervals of 10 years. Accordingly, tree pruning in years 5, 15, 45 and 55 is accounted for in years 10, 20, 50 and 60, respectively.

The annuity of hedgerow management for filling gaps varied between −8 € m^{-1} $year^{-1}$ for a tree hedgerow in Scenario II (TH-II) and −15 € m^{-1} $year^{-1}$ for a hedgerow consisting of equal proportion of trees and shrubs in Scenario I (EPTSH-I) in year 1 (Fig. 12), which suggests that the initial costs increase with hedgerow complexity. Over the period of 60 years, EPTSH-I had the lowest annuity value of −2.4 € m^{-1} $year^{-1}$, while TH-II along with PSH and SH had the highest value of −1.0 € m^{-1} $year^{-1}$, indicating that overall, the management of complex hedgerows such as EPTSH is more cost-intensive than that of hedgerows for the production of timber or for the maintenance of SH.

In the selected hedgerows, the annuity of hedgerow management in year 1 varied between −1.8 € m^{-1} $year^{-1}$ for hedgerow 1 and −18.5 € m^{-1} $year^{-1}$ for hedgerow 6 in Scenario I, and between +1 and −18.5 € m^{-1} $year^{-1}$ for the same hedgerows in Scenario II (Table 12). Hedgerow 1 was estimated to have the highest economic potential, due to having the highest average DBH. In this hedgerow, harvesting tree biomass and selling it as trunk- and fuel wood was

Table 11 Overview of hedgerow management conducted in each hedgerow in 5-year intervals in 20 m long sections

	Hedgerows							
Year	1-PTH/BG/E	2-TH/BG/M	3-EPTSH/SG/E	4-PSH/BG/NN	5-EPTSH/C/NN	6-SH/SG/NN	7-PTH/SG/NN	8-TH/C/NN
1	FG[a]	FG[a]	FG[a]	FG[a]		FG[a]	FG[a]	
								CF50[a]
	R	R	R	R	R	R	R	
5	R	R	R	R	R	R	R	
	P	P	P	P	P	P	P	
10,	R	R	R	R	R	R	R	
20	CF100	CF100	CF100		CF100		CF100	CF25[a]
	CS50[a]	CS50[a]	CS50[a]	CS50[a]	CS50[a]	CS50[a]	CS50[a]	
	P	P	P	P	P	P	P	
15	R	R	R	R	R	R	R	
	CF100	CF100	CF100		CF100		CF100	
	CS50[a]	CS50[a]	CS50[a]	CS50[a]	CS50[a]	CS50[a]	CS50[a]	
	P	P	P	P	P	P	P	
25,	CF100	CF100	CF100		CF100		CF100	
35	CS50[a]	CS50[a]	CS50[a]	CS50[a]	CS50[a]	CS50[a]	CS50[a]	
30	CF100	CF100	CF100		CF100		CF100	CF25[a]
	CS50[a]	CS50[a]	CS50[a]	CS50[a]	CS50[a]	CS50[a]	CS50[a]	
40	CF100	CF100	CF100		CF100		CF100	
	CS50[a]	CS50[a]	CS50[a]	CS50[a]	CS50[a]	CS50[a]	CS50[a]	
	H	H	H	H	H	H	H	R

(Continued)

Table 11 *(Continued)*

	Hedgerows							
Year	1-PTH/BG/E	2-TH/BG/M	3-EPTSH/SG/E	4-PSH/BG/NN	5-EPTSH/C/NN	6-SH/SG/NN	7-PTH/SG/NN	8-TH/C/NN
45	CF100	CF100	CF100		CF100		CF100	
	CS50[a]	CS50[a]	CS50[a]	CS50[a]	CS50[a]	CS50[a]	CS50[a]	
	P	P	P	P	P	P	P	P
50, 55	CF100	CF100	CF100	CS50	CF100		CF100	CF100
	CS50[a]	CS50[a]	CS50[a]	CS50[a]	CS50[a]	CS50[a]	CS50[a]	CS50[a]
	P	P	P	P	P	P	P	P
60	CF100	CF100	CF100		CF100		CF100	CF100
	CS50[a]	CS50[a]	CS50[a]	CS50[a]	CS50[a]	CS50[a]	CS50[a]	CS50[a]
	HR	HR	HR	HR	HR	HR	HR	
	P	P	P	P	P	P	P	P

[a] Over the entire hedgerow.

CS50 = Coppicing shrubs with 50% biomass extraction; CF100 = Coppicing fast-growing trees with 100% biomass extraction; FG = Filling gaps; H = Harvesting and replacing trees with DBH <30 cm in year 1; HR = Harvesting all trees in the section planted in year 1 and removing their roots; P = Pruning; R = Restoration management in 20 m long sections (Includes harvesting trees with DBH >30, removing tree roots, clearing shrubs, preparing soil, planting and fencing trees and shrubs according to the panting scheme for each scenario, mulching trees).

In Scenario II, no fast-growing trees are planted, hence no coppicing of fast-growing trees takes place.

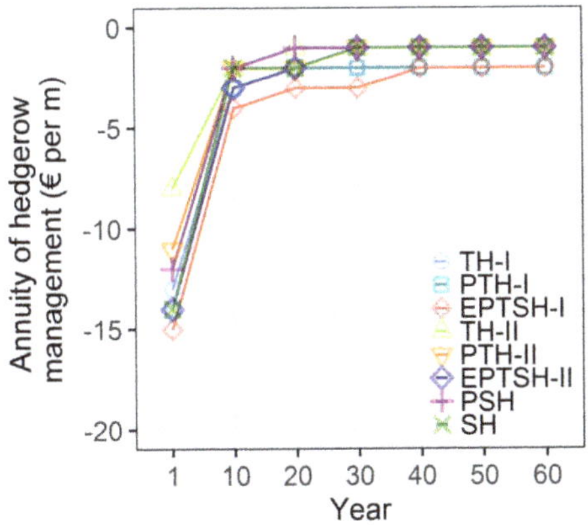

Figure 12 Annuity of hedgerow management over 60 years for *hedgerow types*: tree hedgerow (TH), predominantly tree hedgerow (PTH), equal proportion of trees and shrubs (EPTSH), predominantly shrub hedgerow (PSH) and shrub hedgerow (SH) according to Scenario I and II.

sufficient to cover the initial planting costs in Scenario II. However, filling the BG of this hedgerow which occupy about 91 m (Table 10) requires an additional 8 € m^{-1} during the first year, according to Scenario II (Fig. 13). The planting costs for filling gaps varied between 8 and 14 € m^{-1} in Scenario II and between 13 and 15 € m^{-1} in Scenario I (Fig. 13). The investment costs for hedgerow restoration were in general around 50% lower than that of new planting to fill gaps. Over a period of 60 years, the annuity of hedgerow management in Scenario I varied

Table 12 Annuity of hedgerow management of selected hedgerows (€ m^{-1})

	Scenario I		Scenario II	
	Time period			
Hedgerow	t = 1 (€ m^{-1})	t = 60 (€ m^{-1})	t = 1 (€ m^{-1})	t = 60 (€ m^{-1})
1 - PTH/BG/E	−1.85	−1.36	1.04	−0.68
2 - TH/BG/M	−6.69	−1.64	−1.78	−0.59
3 - EPTSH/SG/E	−11.14	−1.27	−9.90	−1.07
4 - PSH/BG/NN	−12.63	−1.25	−12.63	−1.25
5 - EPTSH/C/NN	−17.50	−1.51	−16.26	−1.31
6 - SH/SG/NN	−18.54	−1.23	−18.54	−1.23
7 - PTH/SG/NN	−6.59	−1.28	−3.55	−0.92
8 - TH/C/NN	−17.71	−1.63	−17.71	−1.40

Hedgerow type: tree hedgerow (TH), predominantly tree hedgerow (PTH), equal proportion of trees and shrubs (EPTSH), predominantly shrub hedgerow (PSH), and shrub hedgerow (SH)/Hedgerow structure: closed (C), small gaps (SG), and big gaps (BG)/Degree of naturalness: near natural (NN), mixed (M), and exotic (E).

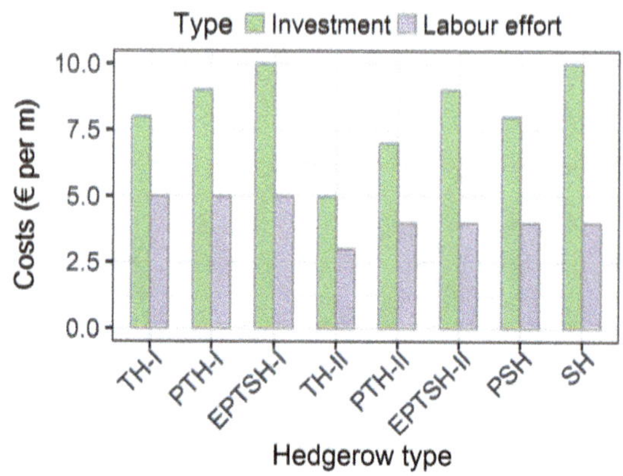

Figure 13 Costs of planting hedgerows including investment (plants and fence) and labour effort (preparing soil, planting, fencing and mulching trees) for *hedgerow types*: tree hedgerow (TH), predominantly tree hedgerow (PTH), equal proportion of trees and shrubs (EPTSH), predominantly shrub hedgerow (PSH) and shrub hedgerow (SH) according to Scenario I and II.

between −1.2 and −1.6 € m^{-1} $year^{-1}$ and in Scenario II between −0.6 and −1.4 € m^{-1} $year^{-1}$ (Table 12), indicating Scenario II to be the better alternative.

7.4.3 Restructuring

Hedgerow types can be restructured during restoration activities in order to better reflect the landscape objectives and improve the provision of ecosystem services. In the study area, TH predominated and very few SH were found. In this context, a comparison between further management of hedgerows 1 and 7 as PTH under Scenario I and II and restructuring them to PSH and SH was conducted.

According to the assessment in Fig. 4, the ecosystem services production and landscape aesthetics of hedgerow 1 in its current category (PTH/BG/E) are in a fair condition, while wind and water protection as well as habitat provision are in a poor condition. Over a period of 60 years, continued management of hedgerow 1 as PTH under Scenario II results in annuity of −0.7 € m^{-1} $year^{-1}$, while converting it to PSH results in annuity of −0.6 € m^{-1} $year^{-1}$ (Fig. 14). This hedgerow is located in an agricultural landscape and converting it to PSH would improve the provision of ecosystem services at a lower cost in the long term. Restructuring this hedgerow to PSH with a closed structure and NN vegetation (PSH/C/NN) would raise the provision of wind and water protection, as well as landscape aesthetics to meet the requirements of a good condition (Fig. 4). The habitat provision would be improved to a fair condition. Hence, the further management of this hedgerow as PSH is recommended.

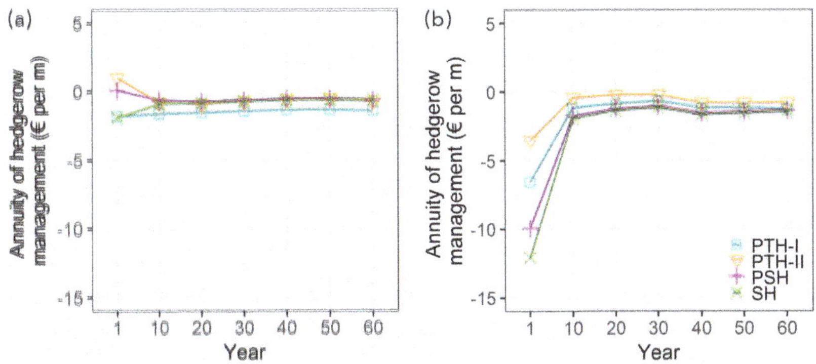

Figure 14 Annuity of hedgerow management over 60 years for the hedgerows (a) 1 - PTH/BG/E and (b) 7 - PTH/SG/NN with continued management as predominantly tree hedgerow (PTH) under Scenario I and II or conversion to predominantly shrub hedgerow (PSH) and shrub hedgerow (SH).

According to the assessment in Fig. 14, hedgerow 7 is in an overall fair condition. The ecosystem services, wind and water protection and production are in a fair condition in the current hedgerow category (PTH/SG/NN). The services of habitat provision and landscape aesthetics are also in a good condition. For this hedgerow, the best economic option is therefore to continue managing the hedgerow as PTH under Scenario II with annuity of −0.9 € m^{-1} $year^{-1}$ over 60 years (Fig. 14). The hedgerow is located in an area relatively rich in woody-features which is predominantly used as grassland. Restoration and management according to Scenario II can therefore be used for sustainable biomass production without adversely affecting other ecosystem services. Retaining the SG in this hedgerow should be considered, as closing them to the category (PTH/C/NN) would improve the ecosystem services of production and water protection to a good condition, but would also reduce the ecosystem services of habitat provision and landscape aesthetics to the category of a fair condition.

7.4.4 Biomass utilisation and profitability

The overall results of these example scenarios for hedgerow management suggest a financial loss. The calculated values are specific for the selected hedgerows because together with the hedgerow type, they are determined by the biomass availability, the type of trees present and their age as well as by management assumptions. Managing trees for the production of trunk wood in Scenario II demonstrated the lowest financial loss over the entire period, which suggests that growing wood for timber under the considered conditions was a more viable alternative. However, labour-intensive coppicing of fast-growing trees, in combination with the small number of trees and the low prices of biomass feedstock, resulted in a lack of profitability for this scenario.

Because the management of these woody-features is labour intensive and on a small scale, developing regional concepts of biomass utilisation is crucial in increasing profitability. In this context, trees can be grown for longer periods and the wood sold as veneer wood at higher prices, resulting in positive annuity values as demonstrated by Romer et al. (2016) in two example hedgerows in southern Brandenburg. Taking into account a tripling of wood prices the annuity of management in hedgerow types TH and PTH over 60 years in both scenarios is a positive value, reaching +0.7 € m^{-1} $year^{-1}$ for hedgerow 1 and +1.3 € m^{-1} $year^{-1}$ for hedgerow 2 under Scenario II. To improve profitability, a regional strategy, originating in the federal state of North Rhine-Westphalia, included the bundling of hedgerows through an online registration system '*Wallhecken information system* (WALLIS)', which is an aid to deciding maintenance measures for these hedgerows and the allocation of buyers for wood used as biomass feedstock (Bröckling and Meyer, 2010).

The assessment of hedgerow condition suggested that managing and maintaining the structure is necessary in improving the provision of multiple ecosystem services. These ecosystem services were not accounted for in the assessment of profitability. Translating the benefits of hedgerow management and maintenance to protect soil, water and biodiversity in monetary terms may shift the profitability of hedgerow management significantly. According to Graves et al. (2015), only 20% of the estimated annual costs of soil degradation in England and Wales were associated with loss of the ecosystem services linked to agricultural production. In Germany, comprehensive studies calculating the costs of soil erosion and water contamination caused by agriculture are needed for the decision-making processes. Nevertheless, the utilisation of hedgerow biomass contributes to hedgerow restoration and maintenance, which in turn improves the condition of ecosystem services and enhances sustainability in agricultural landscapes.

8 Summary and future trends

This chapter has attempted to highlight the importance of hedgerow management in improving the multifunctionality of agricultural landscapes in terms of biomass production, wind and water protection, habitat provision and landscape aesthetics. A classification of hedgerows was developed based on characteristics such as type, structure and degree of naturalness. Hedgerows with a closed structure and M or NN vegetation could be associated with a good condition of ecosystem services, especially these related to regulation and, to some extent, to production functions. Hedgerows with SG and NN vegetation were associated with a good condition of ecological and amenity functions. A poor condition of multiple ecosystem services was suggested in hedgerows of all types having BG and containing E vegetation.

The investigation of existing hedgerows in the Brandenburg study area revealed a prevalence of tree-dominated hedgerows and a predominantly poor condition of the ecosystem services of wind and water protection. This was mainly associated with a lack of management in these hedgerows. Developing a strategy for hedgerow restoration and management in agricultural landscapes can simultaneously supply biomass and enhance the sustainability of land use. The hedgerow types TH and PTH were associated with the highest biomass potential as their average marketable tree volume was higher than that of forests in Germany. However, it should be remembered that hedgerows are small-scale elements which may be widely spread in the landscape, with consequent high management and maintenance costs.

As a consequence of high labour costs and relatively low current prices for wood, the results of the economic evaluation of the two example scenarios, applied to selected hedgerows in the study area, suggested a profit loss. Selling trunk wood was shown to be the better alternative when compared with producing biomass feedstock. In order for the production function of hedgerows to become profitable, the development of regional concepts for innovative wood utilisation and higher prices for biomass products are crucial. However, the utilisation of hedgerow biomass is required for the hedgerow restoration and maintenance which improves the condition of multiple ecosystem services and benefits which are provided to society. In the long term, estimating the costs of soil erosion, water pollution and biodiversity loss in monetary terms is necessary for understanding the total economic value of hedgerow management.

9 Acknowledgements

The AUFWERTEN project (Reference No. 033L129AN) was funded by the Federal Ministry of Education and Research (BMBF).

10 Where to look for further information

Selected sources of detailed information for hedgerow development and distribution in Europe as well as guidelines for hedgerow management are listed below:

- The book *Europe's Field Boundaries* by G. Müller (http://www.wallhecke.de/index.html)
- The UK partnership Hedgelink that brings stakeholders interested in hedgerows together (http://www.hedgelink.org.uk/index.php)

- The internet platform for supporting nature by the Swiss association Biodivers (https://www.biodivers.ch/de/index.php/Hecke).

In addition, the following educational materials and sources of extensive information about recent advances in the field of agroforestry and hedgerow management are recommended:

- Innovation leaflets by AGFORWARD - AGroFORestry that Will Advance Rural Development (http://agforward.eu/index.php/en/Innovation-leafl ets.html)
- Innovation Group AUFWERTEN - Agroforestry for Environmental Services, Energy Production and Added Value (http://agroforst-info.de/)
- Agroforst-App (*German language*) for Android and iOS (http://agroforst-info.de/app/).

11 References

Ackermann, I., Baals, C., Hundsdorfer, M., Kraut, D., Rothenburger, W. and Sauer, N. 2005. *Datensammlung Landschaftspflege: Daten zur Kalkulation von Arbeitszeit und Maschinenkosten*. KTBL - Kuratorium für Technik und Bauwesen in der Landwirtschaft e.V., Darmstadt.

Augenstein, I. 2002. *Die Ästhetik der Landschaft - ein Bewertungsverfahren für die planerische Umweltvorsorge*. Weißensee-Verlag, Berlin.

Axe, M. S., Grange, I. D. and Conway, J. S. 2017. Carbon storage in hedge biomass–a case study of actively managed hedges in England. *Agriculture, Ecosystems and Environment* 250, 81-8. doi:10.1016/j.agee.2017.08.008.

Bach, M. 2000. Fließgewässer XIII-7.15.1 Gewässerrandstreifen - Aufgaben und Pflege. In: Konold, W., Böcker, R. and Hampicke, U. (Eds), *Handbuch Naturschutz und Landschaftspflege*. Ecomed-Verlag, Landsberg.

Baudry, J., Bunce, R. G. H. and Burel, F. 2000. Hedgerows: an international perspective on their origin, function and management. *Journal of Environmental Management* 60(1), 7-22. doi:10.1006/jema.2000.0358.

Best, L. B., Freemark, K. E., Dinsmore, J. J. and Camp, M. 1995. A review and synthesis of habitat use by breeding birds in agricultural landscapes of Iowa. *American Midland Naturalist* 134, 1-29.

Bioland, KÖN, Bio Austria and FiB. 2011. Hecken planen, pflanzen, pflegen - Eine praktische Anleitung für Landwirte. Available at: http://m.bioland.de/fileadmin/dateien/HP_Dokumente/Verlag/Hecken.pdf (Accessed on 25 April 2018).

Bluemlein, B. 2009. Land care in Germany. Available at: https://www.lpv.de/fileadmin/user_upload/data_files/Publikationen/Englische_Dokumente/Landcare_Germany_070312.pdf (Accessed on 25 April 2018).

BMVEL. 2001. *Gute fachliche Praxis zur Vorsorge gegenüber Bodenschadverdichtungen und Bodenerosion*. Bundesministerium für Verbraucherschutz Ernährung und Landwirtschaft, Bonn.

Böhm, C., Quinkenstein, A. and Freese, D. 2011. Yield prediction of young black locust (*Robinia pseudoacacia* L.) plantations for woody biomass production using allometric relations. *Annals of Forest Research* 54, 215–27.

Böhm, C., Kanzler, M. and Freese, D. 2014. Wind speed reductions as influenced by woody hedgerows grown for biomass in short rotation alley cropping systems in Germany. *Agroforestry Systems* 88(4), 579–91. doi:10.1007/s10457-014-9700-y.

Borin, M., Passoni, M., Thiene, M. and Tempesta, T. 2010. Multiple functions of buffer strips in farming areas. *European Journal of Agronomy* 32(1), 103–11. doi:10.1016/j.eja.2009.05.003.

Brandle, J. R., Hodges, L. and Zhou, X. H. 2004. Windbreaks in North American agricultural systems. *Agroforestry Systems* 61–62, 65–78.

Brandle, J., Hogdes, L., James, J. and Sudmeyer, R. A. 2009. Windbreak practices. In: Garrett, H. E. (Ed.), *North American Agroforestry: an Integrated Science and Practice* (2nd edn.). America Society of Agronomy, Inc, Madison.

Bright, P. W. 1998. Behaviour of specialist species in habitat corridors: arboreal dormice avoid corridor gaps. *Animal Behaviour* 56(6), 1485–90. doi:10.1006/anbe.1998.0921.

Bröckling, F. and Meyer, A. 2010. Heckenmanagement im Münsterland – die Renaissance eines alten Kulturlandschaftselementes. *Westfalen Regional*. Available at: https://www.lwl.org/westfalen-regional-download/PDF/110n_Heckenmanagement.pdf (Accessed on 29 January 2019).

Brosowski, A., Thrän, D., Mantau, U., Mahro, B., Erdmann, G., Adler, P., Stinner, W., Reinhold, G., Hering, T. and Blanke, C. 2016. A review of biomass potential and current utilisation e status quo for 93 biogenic wastes and residues in Germany. *Biomass and Bioenergy* 95, 257–72. doi:10.1016/j.biombioe.2016.10.017.

Burel, F. 1992. Effect of landscape structure and dynamics on species-diversity in hedgerow networks. *Landscape Ecology* 6(3), 161–74. doi:10.1007/BF00130028.

Carroll, Z. L., Bird, S. B., Emmett, B. A., Reynolds, B. and Sinclair, F. L. 2004. Can tree shelterbelts on agricultural land reduce flood risk? *Soil Use and Management* 20(3), 357–9. doi:10.1079/SUM2004266.

Christen, B. and Dalgaard, T. 2013. Buffers for biomass production in temperate European agriculture: a review and synthesis on function, ecosystem services and implementation. *Biomass and Bioenergy* 55, 53–67. doi:10.1016/j.biombioe.2012.09.053.

Cleugh, H. A. 1998. Effects of windbreaks on airflow, microclimates and crop yields. *Agroforestry Systems* 41(1), 55–84. doi:10.1023/A:1006019805109.

Collinge, S. K. 1996. Ecological consequences of habitat fragmentation: implications for landscape architecture and planning. *Landscape and Urban Planning* 36(1), 59–77. doi:10.1016/S0169-2046(96)00341-6.

Conrady, D., Kathke, S., Johst, A. and Rösch, W. 2014. Energieholz und Biodiversität: Die Nutzung von Energieholz als Ansatz zur Erhaltung und Entwicklung national bedeutsamer Lebensräume. *Nationalpark-Jahrbuch Unteres Odertal* 7, 94–100.

Croxton, P. J., Franssen, W., Myhill, D. G. and Sparks, T. H. 2004. The restoration of neglected hedges: a comparison of management treatments. *Biological Conservation* 117(1), 19–23. doi:10.1016/S0006-3207(03)00258-1.

DEFRA. 2007. *Hedgerow Survey Handbook – a Standard Procedure for Local Surveys in the UK*. Department for Environment, Food & Rural Affairs, London.

Deumlich, D., Kiesel, J., Thiere, J., Reuter, H. I., Völker, L. and Funk, R. 2006. Application of the SIte COmparison Method (SICOM) to assess the potential erosion risk - a basis for the evaluation of spatial equivalence of agri-environmental measures. *CATENA* 68(2-3), 141-52. doi:10.1016/j.catena.2006.04.003.

Dondina, O., Kataoka, L., Orioli, V. and Bani, L. 2016. How to manage hedgerows as effective ecological corridors for mammals: a two-species approach. *Agriculture, Ecosystems and Environment* 231, 283-90. doi:10.1016/j.agee.2016.07.005.

DVL. 2000. *Hinweise zur Biotop- und Landschaftspflege - 'Flurgehölze'*. Deutscher Verband für Landschaftspflege e.V., Lychen (Brandenburg).

DVL. 2006. *Landschaftselemente in der Agrarstruktur - Entstehung, Neuanlage und Erhalt. DVL-Schriftenreihe 'Landschaft als Lebensraum'.* Deutscher Verband für Landschaftspflege e.V., Ansbach.

Eichhorn, M. P., Paris, P., Herzog, F., Incoll, L. D., Liagre, F., Mantzanas, K., Mayus, M., Moreno, G., Papanastasis, V. P., Pilbeam, D. J., et al. 2006. Silvoarable systems in Europe - past, present and future prospects. *Agroforestry Systems* 67(1), 29-50. doi:10.1007/s10457-005-1111-7.

F.-O. Lürssen Baumschulen GmbH & Co. KG. 2017. *LÜRSSEN-Katalog 2017-2018.*

Forman, R. T. T. and Baudry, J. 1984. Hedgerows and hedgerow networks in landscape ecology. *Environmental Management* 8(6), 495-510. doi:10.1007/BF01871575.

Garbaczewska, S., Cieniecka-Rosłonkiewicz, A., Michalczyk, A. and Kielczewska, A. 2014. The use of components of the extract from the leaves of walnut *Juglans regia* L against the fungus *Ascosphaera apis*. *Chemik* 68, 117-22.

Gelling, M., Macdonald, D. W. and Mathews, F. 2007. Are hedgerows the route to increased farmland small mammal density? Use of hedgerows in British pastoral habitats. *Landscape Ecology* 22(7), 1019-32. doi:10.1007/s10980-007-9088-4.

Graves, A. R., Morris, J., Deeks, L. K., Rickson, R. J., Kibblewhite, M. G., Harris, J. A., Farewell, T. S. and Truckle, I. 2015. The total costs of soil degradation in England and Wales. *Ecological Economics* 119, 399-413. doi:10.1016/j.ecolecon.2015.07.026.

Handke, K. 1999. Fließgewässer XIII-7.15. Gräben. In: Konold, W. and Böcker, R. and Hampicke, U. (Eds), *Handbuch Naturschutz und Landschaftspflege*. Landsberg.

Herdt, T., Friske, V., Nickel, E., Barz, H.-P., Hämmerle, K., Schaber-Schoor, G., Ostermann, R. and Striebel, J. 2007. *Gehölze an Fließgewässern*. Landesanstalt für Umwelt, Messungen und Naturschutz Baden-Württemberg (LUBW), Rheinstetten.

Herzog, F. 2000. The importance of perennial trees for the balance of northern European agricultural landscapes. *Unasylva* 200(51), 42-8.

Hinsley, S. A. and Bellamy, P. E. 2000. The influence of hedge structure, management and landscape context on the value of hedgerows to birds: a review. *Journal of Environmental Management* 60(1), 33-49. doi:10.1006/jema.2000.0360.

HM Government. 1997. The Hedgerow Regulations, SI 1997/1160. Available at: http://www.legislation.gov.uk/uksi/1997/1160/contents/made (Accessed on 21 January 2019).

Hübner, R. 2016. Landscape metrics calculation as a tool to optimize the landscape aesthetics of new agroforestry systems? In: *3rd European Agroforestry Conference*. Montpellier, France.

Hübner, R., Härtl, J., Zehlius-Eckert, W. and Pukall, K. 2016. Agroforst-Standardtypen und deren Wahrnehmung - erste Ergebnisse. In: Böhm, C. (Ed.), *Bäume in der Land(wirt)schaft - von der Theorie in die Praxis*. Brandenburgische Technische Universität Cottbus-Senftenberg, IKMZ - Universitätsbibliothek.

Hübner, R., Zehlius-Eckert, W. and Augenstein, I. 2018. Auswirkungen von Agroforstsystemen auf das Landschaftsbild: qualitative Expertenbewertung der landschaftlichen Veränderungen der Praxisversuchsflächen im Forschungsprojektes 'AUFWERTEN'. *6. Forum Agroforstsysteme 'Brücken bilden'*, Göttingen.

Huwer, A. and Wittig, R. 2012. Changes in the species composition of hedgerows in the Westphalian Basin over a thirty-five-year period. *Tuexenia* 32, 31-53.

Jacobs, J. H., Clark, S. J., Denholm, I., Goulson, D., Stoate, C. and Osborne, J. L. 2009. Pollination biology of fruit-bearing hedgerow plants and the role of flower-visiting insects in fruit-set. *Annals of Botany* 104(7), 1397-404. doi:10.1093/aob/mcp236.

Jose, S., Gillespie, A. R. and Pallardy, S. G. 2004. Interspecific interactions in temperate agroforestry. *Agroforestry Systems* 61-2, 237-55.

Joswig, W., Beiersdorf, H. and Ullmann, S. 2012. *Kostendatei für Maßnahmen des Naturschutzes und der Landschaftspflege - Vollversion überarbeitet 2010/2011*. Bayerisches Landesamt für Umwelt, Augsburg.

Knauer, N. 1993. *Ökologie und Landwirtschaft: Situation Konflikte Lösungen*. Ulmer, Stuttgart.

Knickel, K., Janßen, B., Schramek, J. and Käppel, K. 2001. Naturschutz und Landwirtschaft: Entwicklung eines Kriterienkataloges zur Bewertung der 'Guten fachlichen Praxis' aus naturschutzfachlicher Sicht. *Angewandte Landschaftsökologie* 41.

Kramer, H. and Akça, A. 2008. *Leitfaden zur Waldmesslehre*. J. D. Sauerländer's Verlag, Frankfurt am Main.

Kurz, P., Machatschek, M. and Iglhauser, B. 2001. *Hecken: Geschichte und Ökologie - Anlage, Erhaltung & Nutzung*. Graz.

Landkreis Elbe-Elster. 2013. Verordnung des Landkreises Elbe-Elster zum Schutz von Bäumen und Hecken (Gehölzschutzverordnung - GehölzSchVO EE) vom 12. Februar 2013.

Larcher, F. and Baudry, J. 2013. Landscape grammar: a method to analyse and design hedgerows and networks. *Agroforestry Systems* 87(1), 181-92. doi:10.1007/s10457-012-9534-4.

Lee, K. H., Isenhart, T. M. and Schultz, R. C. 2003. Sediment and nutrient removal in an established multi-species riparian buffer. *Journal of Soil and Water Conservation* 58, 1-8.

LfL. 2005. Hecken, Feldgehölze und Feldraine in der landwirtschaftlichen Flur. LfL-Information (11th edn.). Bayerische Landesanstalt für Landwirtschaft, Freising.

Litza, K. and Diekmann, M. 2017. Resurveying hedgerows in Northern Germany: plant community shifts over the past 50 years. *Biological Conservation* 206, 226-35. doi:10.1016/j.biocon.2016.12.003.

Lutze, G., Schultz, A., Wuntke, B., Voss, M., Kiesel, J. and Wenkel, K.-O. 2010. Brutvogelausstattung der Agrarlandschaften Brandenburgs - zwischen Artenreichtum und Artenarmut. Waldökologie, Landschaftsforschung und Naturschutz. *Waldökologie, Landschaftsforschung und Naturschutz* 9, 79-93.

McCollin, D., Jackson, J. I., Bunce, R. G. H., Barr, C. J. and Stuart, R. 2000. Hedgerows as habitat for woodland plants. *Journal of Environmental Management* 60(1), 77-90. doi:10.1006/jema.2000.0363.

Morandin, L. A. and Kremen, C. 2013. Bee preference for native versus exotic plants in restored agricultural hedgerows. *Restoration Ecology* 21(1), 26-32. doi:10.1111/j.1526-100X.2012.00876.x.

Mosquera-Losada, M. R., Santiago Freijanes, J. J. and Pisanelli, A. 2016. Extent and success of current policy measures to promote agroforestry across Europe. Deliverable 8.23 for EU FP7 Research Project: AGFORWARD 613520 (8 December 2016), 95p.

Müller, G. 2013. *Europas Feldeinfriedungen - Wallhecken (Knicks), Hecken, Feldmauern (Steinwälle), Trockenstrauchhecken, Biegehecken, Flechthecken, Flechtzäune und traditionelle Holzzäune*. Neuer Kunstverl, Stuttgart.

Nerlich, K., Graeff-Hönninger, S. and Claupein, W. 2013. Agroforestry in Europe: a review of the disappearance of traditional systems and development of modern agroforestry practices, with emphasis on experiences in Germany. *Agroforestry Systems* 87(5), 1211-. doi:10.1007/s10457-013-9618-9.

Nuberg, I. K. 1998. Effect of shelter on temperate crops: a review to define research for Australian conditions. *Agroforestry Systems* 41(1), 3–34. doi:10.1023/A: 1006071821948.

Pavlidis, G. and Tsihrintzis, V. A. 2018. Environmental benefits and control of pollution to surface water and groundwater by agroforestry systems: a review. *Water Resources Management* 32(1), 1–29. doi:10.1007/s11269-017-1805-4.

Pe'er, G., Zinngrebe, Y., Hauck, J., Schindler, S., Dittrich, A., Zingg, S., Tscharntke, T., Oppermann, R., Sutcliffe, L. M. E., Sirami, C., et al. 2017. Adding some green to the greening: improving the EU's ecological focus areas for biodiversity and farmers. *Conservation Letters* 10(5), 517–30. doi:10.1111/conl.12333.

Peters, W., Schicketanz, S., Kinast, P. and Plöchl, M. 2014. *Landschaftspflegematerial im Land Brandenburg - Potenzialermittlung und Möglichkeiten der energetischen Verwertung - Endbericht*. Berlin.

Plieninger, T. and Schaich, H. 2014. Socialist and postsocialist land-use legacies determine farm woodland composition and structure: lessons from Eastern Germany. *European Journal of Forest Research* 133(4), 597–610. doi:10.1007/s10342-014-0788-4.

Reif, A. and Achtziger, R. 2000. Biotoptypen XI-2.2: Gebüsche, Hecken, Waldmäntel, Feldgehölze (Strauchformationen). In: Konold, W., Böcker, R. and Hampicke, U. (Eds), *Handbuch Naturschutz und Landschaftspflege*. Landsberg.

Reif, A. and Walentowski, H. 2008. The assessment of naturalness and its role for nature conservation and forestry in Europe. *Waldökologie, Landschaftsforschung und Naturschutz* 6, 63–76.

Reppin, N. and Augenstein, I. 2018. Zur Gestaltung von Agrarholzflächen unter landschaftsästhetischen Gesichtspunkten. In: Veste, M. and Böhm, C. (Eds), *Agrarholz: Schnellwachsende Bäume in der Landwirtschaft - Biologie - Ökologie - Management*. Springer Spektrum, Berlin.

Roßkamp, T. 1999. *Die Vegetation der Feld- und Wallhecken in Niedersachsen - Gebüsch- und Saumgesellschaften der Hecken sowie Trockenrasengesellschaften der gehölzfreien Wälle*. Martina-Galunder-Verlag, Wiehl.

Romer, T., Kaufman, T. and Kammermeier, M. 2016. Ökonomische Wertigkeit von Flurgehölzen und Kurzumtriebsplantagen in der Agrarlandschaft. B.Sc.-Project, Technical University of Munich.

Ryszkowski, L. and Kedziora, A. 2007. Modification of water flows and nitrogen fluxes by shelterbelts. *Ecological Engineering* 29(4), 388–400. doi:10.1016/j. ecoleng.2006.09.023.

Sang, N., Miller, D. and Ode, Å. 2008. Landscape metrics and visual topology in the analysis of landscape preference. *Environment and Planning B: Planning and Design* 35(3), 504–20. doi:10.1068/b33049.

Santiago-Freijanes, J. J., Rigueiro-Rodríguez, A., Aldrey, J. A., Moreno, G., den Herder, M., Burgess, P. and Mosquera-Losada, M. R. 2018. Understanding agroforestry practices in Europe through landscape features policy promotion. *Agroforestry Systems* 92(4), 1105–15. doi:10.1007/s10457-018-0212-z.

Sauerbrei, R., Aue, B., Krippes, C., Diehl, E. and Wolters, V. 2017. Bioenergy and biodiversity: intensified biomass extraction from hedges impairs habitat conditions for birds. *Journal of Environmental Management* 187, 311–9. doi:10.1016/j.jenvman.2016.11.052.

Saunders, D. A., Hobbs, R. J. and Margules, C. R. 1991. Biological consequences of ecosystem fragmentation – a review. *Conservation Biology* 5(1), 18–32. doi:10.1111/j.1523-1739.1991.tb00384.x.

Schleyer, C. and Plieninger, T. 2011. Obstacles and options for the design and implementation of payment schemes for ecosystem services provided through farm trees in Saxony, Germany. *Environmental Conservation* 38(4), 454–63. doi:10.1017/S0376892911000361.

Schmidt, C. 2011. Zur ökonomischen Bewertung von Agroforstsystemen. PhD Thesis, Justus-Liebig-Universität Gießen.

Schmucki, R., De Blois, S., Bouchard, A. and Domon, G. 2002. Spatial and temporal dynamics of hedgerows in three agricultural landscapes of southern Quebec, Canada. *Environmental Management* 30(5), 651–64. doi:10.1007/s00267-002-2704-9.

Schober, R. 1995. *Ertragstafeln wichtiger Baumarten bei verschiedener Durchforstung*. J. D. Sauerländer Verlag, Frankfurt am Main.

Schultz, R. C., Isenhart, T. M., Simpkins, W. W. and Colletti, J. P. 2004. Riparian forest buffers in agroecosystems – lessons learned from the Bear Creek Watershed, central Iowa, USA. *Agroforestry Systems* 61–62(1–3), 35–50. doi:10.1023/B:AGFO.0000028988.67721.4d.

Schulze, E. D. 1982. *Symposium Hecken und Flurgehölze – Struktur, Funktion und Bewertung. Laufener Seminarbeiträge*. Akademie für Naturschutz und Landschaftspflege, Bayreuth.

SIAS. 2017. Gemeinsames Tarifregister Berlin und Brandenburg – Kurzübersichten über tarifliche Arbeitsbedingungen in verschiedenen Branchen in den Ländern Berlin und Brandenburg – Stand: Juni 2017. Senatsverwaltung für Integration Arbeit und Soziales.

Smith, J., Pearce, B. D. and Wolfe, M. S. 2012. A European perspective for developing modern multifunctional agroforestry systems for sustainable intensification. *Renewable Agriculture and Food Systems* 27(4), 323–32. doi:10.1017/S1742170511000597.

Staley, J. T., Bullock, J. M., Baldock, K. C. R., Redhead, J. W., Hooftman, D. A. P., Button, N. and Pywell, R. F. 2013. Changes in hedgerow floral diversity over 70 years in an English rural landscape, and the impacts of management. *Biological Conservation* 167, 97–105. doi:10.1016/j.biocon.2013.07.033.

Staley, J. T., Amy, S. R., Adams, N. P., Chapman, R. E., Peyton, J. M. and Pywell, R. F. 2015. Re-structuring hedges: rejuvenation management can improve the long term quality of hedgerow habitats for wildlife in the UK. *Biological Conservation* 186, 187–96. doi:10.1016/j.biocon.2015.03.002.

Thünen Institut 2012. Dritte Bundeswaldinventur – Ergebnisdatenbank. Available at: https://bwi.info (Accessed on 12 December 2017).

Torita, H. and Satou, H. 2007. Relationship between shelterbelt structure and mean wind reduction. *Agricultural and Forest Meteorology* 145(3-4), 186-94. doi:10.1016/j.agrformet.2007.04.018.

Tsonkova, P., Mirck, J., Böhm, C. and Fütz, B. 2018a. Addressing farmer-perceptions and legal constraints to promote agroforestry in Germany. *Agroforestry Systems* 92(4), 1091-103. doi:10.1007/s10457-018-0228-4.

Tsonkova, P., Böhm, C. and Hübner, R. 2018b. The biomass potential of existing linear woody-features in the agricultural landscape. In: *4th European Agroforestry Conference - Agroforestry as Sustainable Land Use*, Nijmegen, 27-30 May 2018, Agroforestry and multiple products value chain, pp. 527-31.

Weber, H. E. 2003. *Gebüsche, Hecken, Krautsäume*. Stuttgart.

Wehling, S. and Diekmann, M.2009. Hedgerows as an environment for forest plants: a comparative case study of five species. *Plant Ecology* 204(1), 11-20. doi:10.1007/s11258-008-9560-5.

Westaway, S., Chambers, M., Crossland, M., Wolton, R. and Smith, J. 2016. Managing traditional hedges for biofuel. In: *12th European International Farming Systems Association Symposium 'Social and Technological Transformation of Farming Systems: Diverging and Converging Pathways'*, Harper Adams, UK.

Wiegmann, K., Heintzmann, A., Peters, W., Scheuermann, A., Seidenberger, T. and Thoss, C. 2007. Bioenergie und Naturschutz: Sind Synergien durch die Energienutzung von Landschaftspflegeresten möglich? Endbericht an das Bundesministerium für Umwelt, Naturschutz und Reaktorsicherheit, Darmstadt.

Wolton, R. J., Morris, R. K. A., Pollard, K. and Dover, J. W. 2013. Understanding the combined biodiversity benefits of the component features of hedges. Report of Defra project BD5214, 57p.

Wolton, R. J., Pollard, K. A., Goodwin, A. and Norton, L. 2014. Regulatory services delivered by hedges: the evidence base. Report of Defra project LM0106, 99p.

Zianis, D., Muukkonen, P., Mäkipää, R. and Mencuccini, M. 2005. Biomass and stem volume equations for tree species in Europe. *Silva Fennica Monographs*. The Finnish Society of Forest Science, The Finnish Forest Research Institute.

Chapter 4

Temperate alley cropping systems

Diomy S. Zamora, University of Minnesota, USA; Samuel C. Allen, New Mexico State University, USA; Kent G. Apostol, Independent Researcher and Editor, USA; Shibu Jose, University of Missouri, USA; and Gary Wyatt, University of Minnesota, USA

1 Introduction

The intensive agricultural production systems in industrialized nations, such as the United States, have undergone modernization over the past century, which has subsequently led to a remarkable increase in crop yields. The implementation of modern agricultural practices has largely excluded trees from the rural landscape causing negative environmental impacts such as soil fertility depletion, soil erosion, non-point-source pollution and biodiversity losses. Recent studies, however, have shown that the integration of trees in the agricultural landscape using alley cropping may mitigate these negative impacts (Jose et al. 2000a,b, 2012; Thevathasan and Gordon 2004; Palma et al. 2007; Zamora et al. 2009a). Alley cropping, an agroforestry practice, involves the planting of row crops or forage in alleys formed by widely spaced single or multiple rows of trees or shrubs (Garrett and Buck 1997; Gillespie et al. 2000; University of Missouri Center for Agroforestry (UMCA) 2015). In temperate settings, the tree-row and alley arrangement allows the use of standard farm equipment and reduces the need for manual labour, in

http://dx.doi.org/10.19103/AS.2018.0041.04

contrast with the configurations of most alley cropping systems in the humid tropical regions.

Alley cropping - also known as silvoarable in Europe - has recently been recognized and adopted in temperate climates (Eichhorn et al. 2006; Regueiro et al. 2009), while having been a common practice in the tropics for many years. In North America, alley cropping is most popular in the southern and Midwestern United States (US) where high-value hardwoods are used to create alleys of various widths that support conventional row, forage or horticultural crops (Fig. 1). Such practices are developed with both production and conservation (protection) benefits. In this region of the US, cotton (*Gossypium hirsutum* L.), peanut (*Arachis hypogea* L.), maize (*Zea mays* L.), soybean (*Glycine max* L. (Merr)), wheat (*Triticum* spp.) and oats (*Avena* spp.) are important crops for alley cropping that are normally combined with trees such as black walnut

Figure 1 Examples of temperate alley cropping systems: (a) black walnut intercropped with soybean; (b) pear intercropped with lettuce; (c) poplar intercropped with wheat; (d) pecan intercropped with forages; and (e) slash pine intercropped with cotton. Photo source: (a) and (e), University of Missouri Center for Agroforestry. (d), Shibu Jose, University of Missouri. (b) and (c), National Agroforestry Center, Lincoln, Nebraska.

(*Juglans nigra* L.), pecan (*Carya illinoinensis* K. Koch) (Allen et al. 2004a,b; Wanvestraut et al. 2004; Zamora et al. 2006, 2007) and some pine trees (*Pinus* spp.) (Zamora et al. 2009a,b; Hagan et al. 2009). However, the relatively large initial investment and long time needed to maturity for trees and shrubs is a substantial economic hurdle to adoption (Dyack et al. 1999), although leveraging multispecies systems (Malézieux et al. 2009) and high-value tree crops (Molnar et al. 2013) could lessen this burden, provided that farmers are aware of this option.

2 Potential of alley cropping

With 2.2 million agricultural farms in the US covering 373 million ha (USDA 2018), for example, alley cropping has the greatest potential to be implemented in temperate regions among the various agroforestry practices such as riparian buffers, windbreaks, forest farming and silvopasture. However, its adoption is potentially hampered by apprehensions of landowners, whose cultural practices do not typically employ tree-crop systems, and who are unfamiliar with the management skills of such complex systems (Zinkhan and Mercer 1997). In the tropics, alley cropping is often employed to address soil erosion issues, as farming often takes place in highly erodible areas.

While limited data exist that provide reliable estimates of the hectares established in alley cropping in the US, an estimate of land suitable for alley cropping is possible. In the Midwestern US, there are more than 8.1 million ha of non-federal croplands that are characterized as highly erodible soil, with an erodibility index (EI) ≥ 8 (USDA NRCS 2007). The EI provides a numerical expression of the potential for a soil to erode considering its physical and chemical properties and the climatic conditions where it is located; the higher the index, the greater the investment needed to maintain the sustainability of the soil resource base if intensively cropped. Soils with a low EI (<5) are only slightly susceptible to erosion and farmers wishing to plant row crops have a wider range of alternatives on these lands. An EI from 8 to 15 indicates that cropland bare of cover could erode at least eight or more times the tolerance value while an EI of 15 or higher suggests that erosion could occur at 15 or more times the tolerance value (USDA NRCS 2007). Certain areas of the Midwest and other regions have soils that are typically susceptible to damage, and in many cases these soils are cultivated in a manner that leads to the loss of excessive quantities of topsoil. Thus, the potential for alley cropping on highly erodible cropland is significant, given that it can protect fragile soils through a network of fine roots and biomass produced by trees and companion crops.

Furthermore, there are many hectares of land that are under orchard management in the US, and could be adapted or converted to alley cropping practices especially during the establishment phase. Intercropping is a traditional practice that has been used in the establishment of fruit and nut trees for hundreds of years in both tropical and temperate areas. In the tropics, intercropping is commonly used to reforest marginal lands. In North America, intercropping with row crops has been shown to be feasible and a viable option for income diversification. For example, the use of a number of native tree species in pecan production has been examined in the Southeast (Gordon and Williams 1991). Similarly, approximately 7 million ha of pastureland have high potential for use as cropland and could be alley cropped. When alley cropping is properly planned and managed, it can provide landowners excellent market opportunities for their wood products (Smith et al. 2013). By selecting intercropping species for which known viable markets exist, landowners could improve the financial conditions of their farm operations. However, the success of alley cropping depends in many ways on its design.

3 Design considerations

Agroforestry, which is perceived as a practice that can provide optimum production while maintaining ecological sustainability, can enhance many of the biophysical cornerstones of ecologically sound agricultural production (Gordon et al. 1997). Temperate alley cropping has two distinct system components: trees or shrubs, and the companion crop(s). The design of alley cropping practice is flexible to reflect landowner needs and site potential. However, there are physical interactions (e.g. above ground competition for light, and below ground competition for water and nutrients) between the rows of woody species and the companion crop that should be understood and reflected in plans for alley cropping practice design (Jose et al. 2004). Proper tree/shrub companion crop selection, tree/shrub planting design, layout and orientations, and available farm area are among the major considerations of an adequate alley cropping design.

3.1 Tree or shrub selection

The tree or shrub selection in alley cropping is the most critical aspect in designing an alley cropping system, as this determines the long-term sustainability and also profitability of the practice. Many tree species of high economic value exist that could be used in alley cropping. Species selection, however, must be based on matching the species both to the site and to the companion crop (Udawatta et al. 2005; Garrett et al. 2009; Jose et al. 2009; Jose 2011), which can be a challenging task. A tree or shrub species must be selected that best accommodates the sunlight requirements of a specific crop. The tree species should also satisfy the criteria established by the landowner, which would include providing income to the family, creating a microenvironment suitable for the companion crop, protecting the site and having wildlife benefits. Various species may be identified that are potentially capable of addressing these production and protection functions; however, these functions can only be fully achieved when species are compatible with the specific site's climate and soil conditions. Table 1 shows tree and shrub species suitable for alley cropping systems in a temperate region, particularly in North America.

Table 1 Common tree and shrub species suitable for temperate alley cropping systems

Species	Site requirements
Apple (*Malus* spp.)	Tolerate a wide range of soil types but prefer well-drained, sandy loam to sandy clay loams. Some rootstocks are more tolerant of heavier, wetter soils. An ideal pH would be around 6.5, but a pH of 5.5-7.5 is acceptable.
Black walnut (*Juglans nigra* L.)	Require fertile, deep soil with good drainage and pH of 6.0-7.5. Sensitive to spring frost injury, so avoid low-lying frost pockets if nuts are desired.
Pecan (*Carya illinoinensis* K. Koch)	Grow in full-sun environment, and prefer acidic, alkaline, loamy, moist, rich, sandy, silty loam, well-drained, wet and clay soils.
Chinese chestnut (*Castanea mollissima* Blume)	Require well-drained loamy to sandy loam soils as they are susceptible to root rot on wet, heavy soil; pH requirement of 5.5-6.5.
Hazelnut (*Corylus avellana* L.; *C. americana* L.)	Withstand wetter areas than chestnuts or walnuts, but still prefer well-drained soil with a pH of 5.5-7.5.
Aronia (*Aronia melanocarpa* Michx)	Tolerate a wide range of soil conditions but prefer well-drained soils with slightly acid to neutral pH. Full sun is required for optimal production.
Black currant (*Ribes* spp.)	Grow well in any soil but avoid heavy clay soils or standing water. Ideal pH is 5.5-6.5. Self-pollinating, but better crops occur with cross-pollination.
Saskatoon, juneberry, serviceberry (*Amelanchier* spp.)	Adaptable to a variety of soil types but do best on gently sloping hills with good air and soil drainage and a pH of 6.0-8.0.
American persimmon (*Diospyros virginiana* L.)	Adaptable to a variety of soils, from wet bottomland to thin ridge tops. Young trees can handle partial shade, but full sun is recommended for fruit production.
Honey locust (*Gleditsia triacanthos* L.)	Thrive in a wide range of soils and climate zones. They have wide adaptability and tolerance to drought, compaction and pollution.
Elderberry (*Sambucus canadensis* L.)	Adapted to many growing conditions, and grow well on well-drained loamy soil. Avoid planting in dry sites.

Adapted and modified from Wilson (2007).

3.2 Tree planting design, layout and orientation

Plantings can consist of a single tree species or a number of species. Similarly, single tree rows or multiple rows may be used, depending on the objectives of the farmers.

3.2.1 Single species versus mixed species

The alley cropping planting design is flexible and can be done to achieve the landowner's conservation and production goals. The establishment of a pure orchard or tree plantation and converting it to an alley cropping may be best for some farmers; however, the mixing of tree species offers advantages to other farmers. The planting of two or more woody (tree) species in alley cropping with similar growth patterns and site requirements may provide greater economic and ecological diversity and create a more stable stand. There are factors to consider when deciding how many rows to establish and the arrangements of trees within the row depending on the desired benefits. These factors include (1) annual crop produced and the area removed from production by tree/shrubs rows; (2) desired tree/shrub crops and management needed to enhance production, including minimum widths needed for farm machinery operation; (3) erosion concerns that multiple rows and combinations of trees/shrubs/grasses can better address; and (4) wildlife habitats created through multiple rows of combined trees/shrubs/grasses (UMCA 2015).

3.2.2 Single- versus multiple-row sets of trees

Conventional wisdom might suggest that single-row tree patterns are better suited than multiple-row patterns for alley cropping. However, alley cropping is unique because unlike conventional forestry, two or more crops are grown in an intimate mixture on the same piece of land. Because of the nature of interactions (either facilitative or competitive) between system components (trees/shrubs and crops) and the goal of optimizing economic gain, a mix of trees (Table 1) and companion crops may be created to provide the highest economic potential to the landowner. The selection of a single or multiple rows of trees in planting design and practice depends on the ultimate product that the landowner will derive from the trees (Garrett et al. 2009). In some plantings, this may be single-row patterns; in others, it will be a multiple of rows. In the US, the selection of single- or multiple-row sets of trees is dictated in part by the federal government's cost-share programme(s) that a farmer may apply to in order to establish the alley cropping practice. It is noteworthy that some tree species do not adapt well to single-row configurations under the open conditions, especially when the ultimate goal is to have high-quality timber. Under such conditions, a triple-row configuration using a 'trainer' species in the outside rows that does not show a strong light response (e.g. pine trees) might be more economically viable.

Planting multiple rows consisting of more than one species or mixing species within a row must be carefully planned, as the juvenile growth patterns of different species can vary significantly. Multiple tree rows have, in some studies, also been shown to be better than single rows in optimizing wood and forage production, which suggests multiple rows might also be better with some other companion crops. Lewis et al. (1985) in their study involving assessment of the effects of single-row versus double-row configurations of slash pine (*Pinus elliotti* Engelm) on forage yield showed that the latter configurations were in several ways superior to single rows.

3.2.3 Between- (and within-) row spacing and tree orientation

The spacing between individual trees within rows and between rows, and the orientation of trees in the alleys, is a critical consideration in designing an alley cropping system, as this will directly impact aboveground competition for light. In meeting the needs of light-demanding species such as conventional row crops, spacing between tree rows can be wide while the within-row spacing is narrow. An East–West orientation of tree rows will maximize the sunlight received by an alley crop. Many alley crop varieties and species are shade intolerant (i.e. row crops, forages, small berry crops etc.), therefore, alleyways must be sufficiently wide enough to accommodate their light requirements. Choices related to between-row spacing are critical in alley cropping practices and will directly influence the system's success. Between-row spacing of trees depends on a number of management decisions including the emphasis placed on wood production versus some other tree-related crops (Garrett et al. 2009). For example, if the emphasis is on wood production, between-row spacing will be less than if nut production is emphasized. The spacing in alley cropping practice is also determined by the duration of the cropping regime selected. Studies in Missouri (Garrett and Harper 1999) and Florida (Wanvestraut et al. 2004) have shown that a single row spaced 12 m apart can likely only accommodate light and moisture needs of maize and cotton, respectively, within 10 years of establishment, when lateral roots are not pruned to minimize competition between tree and crops.

The spacing between trees in alley cropping practices will also vary depending on the environmental protection goals of the farmer. In the tropics, alley cropping practice is employed in many cases to control erosion problems, while optimizing the use of the land is of great importance among farmers in the temperate region. However, as earlier noted, severe soil erosion is also rampant in agricultural production systems in the US; thus, trees should be closely spaced to provide an immediate effect, or a shade-tolerant shrub with a shallow and highly diffused root system should be strategically combined with the trees.

3.3 Available farm machinery and operating requirements

Often, farmers in temperate regions are hesitant to adopt alley cropping because of farm equipment compatibility issues with the practice. Tree/shrub spacing should be adjusted based on multiples of the widest farm equipment to be used in the alley. Alley widths should be planned such that full or multiple passes of the equipment are utilized.

3.4 Selecting companion crops

Companion crops are planted in the alleys between tree rows. Corn, soybean, wheat, barley (*Hordeum vulgare* L.), oats, pumpkins (*Cucurbita* spp.), lettuce (*Lactuca sativa* L.), peas, among others have demonstrated success in alley cropping. The choice of companion crop with the trees will vary depending on the types of trees selected and the crop(s) desired by the farmer. There are three major groups of crops which can be grown in alley cropping. These include row/cereal and forage crops, fruits and specialty crops, and biomass-producing crops. Initially the growing environment in the alley will be favourable to row crops requiring full sun (corn, soybean, wheat) or forages. As trees grow taller and develop larger crowns and root systems, they will exert greater influence on the growing environment in the alley with increased shade, water and nutrient competition. While availability of sunlight is a major factor that determines how well row crops or forages perform in the alley, water and nutrient competition is even more significant. The tree canopy density will be partially determined by the spacing of trees within a row and the width between tree rows, and this spacing will also influence below ground competition.

4 Functions/benefits of alley cropping

The benefits of the alley cropping practice can be broadly classified as achieving financial productivity (i.e. production goals) and sustaining environmental integrity (i.e. protection goals).

4.1 Economic benefits

4.1.1 Diversification of farm products and supplemental income

Alley cropping diversifies farm enterprises by providing short-term cash flow from annual crops while also providing medium- to long-term products from the woody components. Timber and non-timber products may contribute to income generation from the farm. Intercropping with row crops has been shown to be feasible in the US. The combination of trees and annual crops creates dynamic agroecosystems that when properly designed, may increase and diversify farm income. For example, in the US, studied systems have grown maize along with black walnut (Jose et al. 2000b; Gillespie et al. 2000), cotton along with pecan (Allen et al. 2004a,b; Wanvestraut et al. 2004; Zamora et al. 2008), and with slash pine (Zamora et al. 2009b), and ornamental shrubs with longleaf pine (*P. palustris*) (Hagan et al. 2009), while wheat is grown under rows of hybrid poplar (*Populus* spp.) in Ontario, Canada. Alley cropping also offers an opportunity to produce annual or perennial biomass crops for biofuel (Holzmueller and Jose 2012). In addition to the potential for producing nuts, berries and fruits, well-managed timber can provide a long-term investment. While opportunities abound in terms of crop diversification, it must be noted that caution must be observed when considering large-scale adoption of diverse cropping systems, as economic path dependency associated with intensive agricultural production (such as in the US Corn Belt) can work against the incorporation of diverse crops, as described by Roesch-McNally et al. (2018). They conclude that this issue needs to be further addressed by adjusting policy and economic incentives that foster diverse systems and reduce climate risk.

4.1.2 A source of food supply

The role of agroforestry in food security is now getting recognized. Food security entails having access to sufficient, safe and nutritious food to meet the dietary needs of an individual (Dawson et al. 2013). In the tropics (e.g. Philippines), the rapid decline in forest area can be attributed to the large and rapid conversion of the uplands into permanent annual cropping to meet the food requirements of a rapidly expanding population (Magcale-Macandog et al. 2010). This activity causes environmental problems such as accelerated soil erosion. However, implementing agroforestry such as alley cropping can address this problem. Planting fruit-bearing trees or shrubs integrated with annual crops (e.g. corn, soybean, wheat etc.) can serve as an immediate source of food for the farmer. Alley cropping is considered as a sustainable agricultural practice because it addresses the establishment of a resilient food system, which focuses on the pivotal role of enhancing food production in fighting food insecurity and achieving a certain degree of food supply.

The role of agroforestry, alley cropping in particular, in food security cannot be underestimated. Garrity (2004) outlined various food security issues that can be addressed by adopting agroforestry practices. These include eradication of hunger through the establishment of basic production systems in marginal lands based on agroforestry methods of soil fertility and land regeneration; poverty reduction among the rural poor through market-driven, locally led tree cultivation systems that generate income and build assets that increase purchasing power; and improvement in health and nutrition of rural communities from products of agroforestry systems. In general, food from trees in agroforestry systems is of particular importance to farmers in developing areas and may contribute up to 25–50% of their annual food requirements (FAO 2013).

4.2 Ecological benefits

4.2.1 Soil erosion control and water quality improvement

Soil erosion is a serious concern in agricultural production systems, particularly in areas with an EI greater than eight (UMCA 2015). Croplands lose soil at an average rate of 15 tons ha^{-1} $year^{-1}$ with over 90% of the land losing at a rate above the sustainable level (Pimentel et al. 1995). Intensive agricultural production in highly erodible sites could result in erosion at a level above the tolerance (USDA NRCS 2007). Properly designed alley cropping practices can reduce erosion and sediment transport through the trapping capacity of the forested strip or in combination with ground covers. The practices can regulate erosion by developing a network of surface roots. Wanvestraut et al. (2004) and Zamora et al. (2007) in their separate assessment of root dynamics of pecan in alley cropping in Florida found that fibrous root systems of pecan concentrate in the upper 30 cm of the soil profile. These researchers concluded that such root concentration in the soil profile could play a major role in reducing soil erosion.

The excessive application of nitrogen (N) fertilizer in commercial agriculture and forestry can result in leaching of nitrate (NO_3-N) into surface and sub-surface drainage water. Ground water contamination through N leaching is a long-term environmental problem resulting from intensive agricultural production (Bonilla et al. 1999; Ng et al. 2000; Allen et al. 2004a,b; Jose et al. 2009). Over-application of N increases the production costs of farmers by millions of dollars every year (Marshall and Bennett 1998). The negative effects of nitrate leaching on rivers, lakes and residential wells are of increasing public and scientific concern. Thus, the effects of trees in alley cropping systems are of interest environmentally because the deep roots of trees may serve as a 'safety net' for capturing the N that is leached below the roots of the crops, and thus may help in improving nutrient use efficiency and in mitigating groundwater contamination (Rowe et al. 1999; Allen et al. 2004a,b; Zamora et al. 2009a). There is direct and indirect evidence of the safety-net hypothesis in the literature. A study in Sweden by Browaldh (1995) on the effects of tree harvesting on natural N levels in the soil showed that natural N levels increased in the vicinity of harvested sites due to the lack of N uptake by tree roots of harvested sites. In other studies, isotopic tracers such as ^{15}N-enriched fertilizer have been used to trace movement of applied N in alley cropping systems (Rowe et al. 1999; Jose et al. 2000b; Allen et al. 2004a,b; Zamora et al. 2009a). For instance, Allen et al. (2004b) found that pecan tree roots played a significant role in alleviating groundwater nitrate leaching through their safety-net role in a pecan-based alley cropping system. Similarly, Zamora et al. (2009a) observed the same pattern in a loblolly pine alley cropping system intercropped with cotton.

4.2.2 Pest and disease protection

Exploiting the principles of biological control of crop pests is an important management strategy for farmers. Information on the population dynamics of insects, pests and their natural enemies is essential for utilizing habitat management as a tool for biological control of pests. Mixed systems such as alley cropping with diverse crops are known to encounter less pest problems than pure crops (Girma et al. 2000). Alley cropping systems differ widely in their plant diversity and are more diverse than monoculture crops. According to Stamps and Linit (1998), agroforestry in general is potentially useful for reducing pest problems because tree-crop combinations provide greater niche diversity and complexity than annual crops alone. This effect may be explained in one or more of the following ways: (1) wide spacing of host plants of the intercropping scheme may make the plants more difficult for herbivores to find; (2) one plant species may serve as a trap crop to prevent herbivores from finding the other crops; (3) one plant species may serve as a repellent to the pest; (4) one plant species may serve to disrupt the ability of the pest to efficiently attack its intended host; and (5) the intercropping situation may attract more predators and parasites than monocultures, thus reducing pest density through predation and parasitism (Root 1973; Vandermeer 1989; Jose et al. 2004, 2009).

The effects of multi-species systems in controlling pests and diseases have been demonstrated in many studies. Altieri (1983) found that intercropped maize in Colombia recorded 25% fewer leafhoppers (*Empoasca kraemeri*), 45% fewer beetles (*Diabrotica balteata*) and 25% less fall army worms (*Spodoptera frugiperda*) than sole-cropped maize. Studies with pecan, for example, have looked at the influence of ground covers on arthropod densities in tree-crop systems. Bugg et al. (1991) observed that cover crops (e.g. annual legumes and grasses) sustained lady beetles (Coleoptera: Coccinellidae) and other arthropods that may be useful for biological control of pests in pecan. Eckberg et al. (2015) also observed that alley cropping involving willow and soybean provided host conditions for aphid predators on soybean. Stamps et al. (2002) observed similar results in a black walnut-based alley cropping system in Missouri in the Central US. In another study of alley cropping trials with peas (*Pisum sativum* L.) and four tree species (*Juglans*, *Platanus*, *Fraxinus* and *Prunus* spp.), Peng et al. (1993) found an increase in insect diversity and improved natural enemy abundance compared to monoculture peas. The greater diversity of birds in alley cropping systems (Gillespie et al. 1995; Gibbs et al. 2016) could also provide beneficial services for pest reduction. In essence, the diversity of species in alley cropping systems reduces damage from insect pests by reducing crop visibility, diluting pest hosts due to plant diversity, interfering with pest movement and creating habitat more favourable to beneficial insects.

4.2.3 Wildlife habitat enhancement

Intensive agricultural production and continued fragmentation of forest cover has been shown to have detrimental effects on wildlife populations (Wilcove et al. 1998; MacLeod et al. 2012; Vasileiadis et al. 2013; Gibbs et al. 2016). Wildlife species living within agroecosystems, including avian species, are especially impacted by conventional monocrop agriculture, which provides negligible nesting, foraging and refuge habitat (MacLeod et al. 2012), and has been linked to biodiversity losses (Pekin and Pijanowski 2012). Many solutions aimed at mitigating the consequences of unsustainable agriculture and subsequent biodiversity loss are suggested in the literature. One such strategy is the implementation of agroforestry practices in the landscape as a means of creating more heterogeneous and complex ecosystem structures than are found in conventional agricultural systems (Bakermans et al. 2012; Gibbs et al. 2016).

Alley cropping can play a vital role in providing wildlife habitat and landscape improvement. Incorporating the practice into an agricultural landscape increases the habitat diversity for wildlife (Garrett et al. 2009). This increased diversity improves habitat conditions and can increase wildlife use. Several studies have shown the positive impact of linear plantings on diversity and density of bird species. Yahner (1982) found that increased vertical complexity of the woody vegetation is correlated with increased bird numbers, and must be used in designing alley cropping systems. Increased vertical complexity of managed grassland in North Dakota, USA, has also been shown to increase bird usage (Renken 1983). The corridors created in alley cropping could provide more woody cover and foster greater total numbers of birds and more bird species (Garrett et al. 2009; Gibbs et al. 2016). Another advantage of linear plantings such as those in alley cropping is found in their edge effects. Because of their extensive edge-to-volume ratios, alley cropping can serve as a reservoir for many kinds of insects that attract bird species. Strong correlations between the amount of edge and bird species and bird abundance have been observed (Best et al. 1990).

Alley cropping may be viewed as a complex series of tree–crop interactions guided by utilization of light, water, soil and nutrients. Therefore, an understanding of the biophysical processes and mechanisms involved in the utilization of these resources is essential for the development of ecologically sound agroforestry systems. The following section discusses key issues, limitations and challenges of implementing alley cropping systems.

5 Competition for growth resources

The incorporation of multiple species in single ecological systems such as alley cropping brings about a unique set of ecological interactions among different species (Jose et al. 2004). By combining multiple resources in the same system, potential exists to increase nutrient use efficiency, control sub-surface water levels, improve soil and water quality, provide favourable habitats for plants, insects, or animal species and create a more sustainable agricultural systems (Garret and McGraw 2009; Garrity 2004; Garret et al. 2009; Jose et al. 2009) as a result of their component interactions. According to Nair (1993), interaction refers to the influence that one or more components of a system has on the performance of both another component of the system and the overall system itself. If two species are competing for the same resources, and do so equally, both species will likely exhibit lower productivity when compared to their potential for independent growth. However, alley cropping systems can be designed so that the physiological needs for a particular resource are spatially or temporally different for the individual species growing in the system. As such, then the system would lead to a greater productivity than the cumulative production of individual system components when grown separately on equal land area. As an association of plant communities, alley cropping is deliberately designed to optimize the use of spatial, temporal and physical resources by maximizing positive interactions (facilitation) and minimizing the negative ones (competition) between trees and crops (Jose et al. 2000b; Zamora et al. 2006). These interactions are outlined below.

5.1 Above ground competition for light

The productivity of alley cropping is dependent on the crop-tree interactions. Understanding of the biophysical processes and mechanisms of resource capture such as above ground competition for light in alley cropping is necessary. Light availability and its effect on above ground production are equally important to agricultural system sustainability. Trees or shrubs in alley cropping can increase the amount of shading that plant species, primarily those in the understorey, experience compared to growing in a monoculture. When plant growth and productivity are not limited by water or nutrients, they can be altered by the amount of radiant energy intercepted by the foliage (Monteith et al. 1991).

Numerous studies have examined shading and its effect on crop growth (Monteith et al. 1991; Gillespie et al. 2000; Reynolds et al. 2007), and those studies indicate that light availability caused by shading is a major factor influencing production (e.g. reduction in crop yield). Reynolds et al. (2007), in their study on alley cropping in Ontario, Canada, showed that low levels of photosynthetically active radiation (PAR), 400-700 nm wavelength, resulting from overhead shading by hybrid poplar (*Populus* spp.) and silver maple (*Acer saccharinum* L.) significantly reduced maize and soybeans yields. Similar results have also been reported in a temperate silvopastoral system, although pastures may be somewhat less prone to reduced productivity due to greater overall surface area (Pardini et al. 2008). Lin et al. (1999) observed a significant decrease in mean dry weight of warm season grasses under silvopasture in Missouri, USA. However, cotton production under pecan alley cropping system in Florida demonstrated that light availability was not a major factor influencing cotton yield (Zamora et al. 2008). See Section 6 'Evaluating system performance' for details of the impacts of shading on cotton production.

The physiological response of plants to shading depends on the pathway used by crops to fix carbon (C_3 or C_4). In C_3 plants, as PAR increases from complete shade to approximately 25-50% of full sun, there is a corresponding increase in photosynthetic rate (P_{net}) (Fig. 2); however, as more light becomes available, P_{net} does not increase, but rather levels off despite the additional increase in PAR. In contrast, in C_4 plants, P_{net} does not level off as PAR increases in full sunlight, but rather continues to increase with increasing PAR (Fig. 2). In accordance with the theory that C_3 plants would not have reduced growth growing under shaded conditions, Zamora et al.

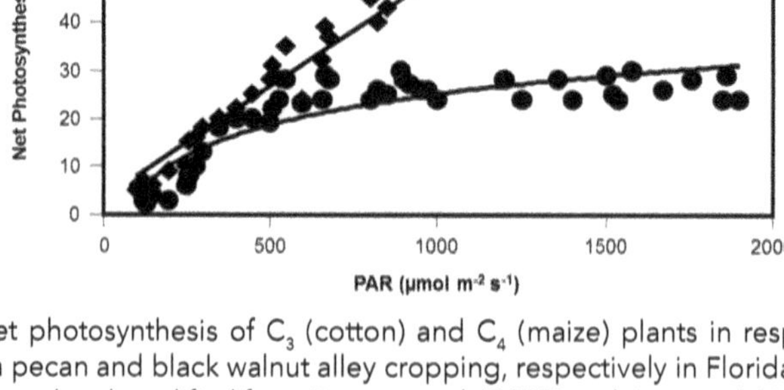

Figure 2 Net photosynthesis of C_3 (cotton) and C_4 (maize) plants in response to PAR growing in a pecan and black walnut alley cropping, respectively in Florida and Indiana. Source: Adapted and modified from Zamora et al. (2008) and Jose et al. (2009).

(2008) concluded that light was not a major factor influencing production of cotton, a C_3 plant species growing under moderate shading in a temperate pecan alley cropping system described above in Florida, USA. Contrary to an anticipated yield decrease in maize, a C_4 species, in response to shading, Gillespie et al. (2000) reported that there are no effects of shading in both a black walnut - maize and red oak (*Quercus rubra* L.) - and maize alley cropping system in Indiana, USA, which was not expected given the known strong positive correlation between PAR and P_{net} in C_4 plants. Although Gillespie et al. (2000) found that the edge rows received lower PAR compared to the middle rows, particularly in the red oak-maize system because of the higher canopy leaf area, once competition for water and nutrients was removed through trenching and polyethylene root barriers, there was no indication of yield reduction because of reduced PAR, leading the authors to conclude that light was not a limiting factor of these systems, but results from other studies may differ on this point.

5.2 Below ground competition for water and nutrients

Minimizing resource competition between trees and crops while maximizing the use of available resources is central to improving yield and overall productivity in alley cropping and similar agroforestry systems. The component species in alley cropping systems share the same biophysical resources and draw from a common resource pool.

5.2.1 Competition for water

Researchers in the temperate zone, humid tropics and semi-arid tropics alike have identified competition for water as the major limiting factor influencing productivity in alley cropping systems (e.g. Lehmann et al. 1998; Govindajaran et al. 1996; Jose et al. 2000b; Wanvestraut et al. 2004). In other studies, although certain alley cropping configurations generated benefits, they were overshadowed by production losses due to competition for water. For example, Govindajaran et al. (1996) in an alley cropping stand of maize and leucaena (*Leucaena leucocephala* Lam) found that nitrogen requirement for maize was met through the presence of leucaena but maize yield decreased 39-49% due to water competition. In a temperate alley cropping system involving hardwoods (oak and black walnut) and maize, Gillespie et al. (2000) reported crop yield reductions of up to 40% due to severe competition for water when trees were 11 years old.

Drought stress has been found to affect crop plants. Multiple studies have reported large reductions in crop plant height (NeSmith and Ritchie 1992; Wanvestraut et al. 2004) and leaf area (NeSmith and Ritchie 1992; Jose et al.

2000b; Wanvestraut et al. 2004; Zamora et al. 2008), and yield (Gillespie et al. 2000; Wanventraut et al. 2004; Zamora et al. 2007) when water availability is limited. In a silvopastoral study in the Canterbury region of New Zealand, Yunusa et al. (2005) reported a slightly lower water potential in radiata pine (*Pinus radiata* D. Don) trees planted in an alfalfa (*Medicago sativa* L.) pasture compared to trees growing in a vegetation-free environment. In that study, competition for water led to a decrease in photosynthetic rates in radiata pine compared to trees growing in the alfalfa-free treatment.

Competition for water in alley cropping systems becomes increasingly intense when water availability decreases throughout the soil profile (Miller and Pallardy 2001; Jose et al. 2009). Factors that limit plant growth and productivity due to competition for water include drought, soil water-holding capacity and irrigation. Competition for water in alley cropping can be minimal given adequate levels of either or both precipitation and irrigation. Furthermore, competition for water can also be minimized by proper tree-crop selection and combinations, as some deep-rooted trees have hydraulic lift capabilities. Hydraulic lift is the process in which some deep-rooted plants take in water from lower soil layers and release that water into upper, drier soil layers. This phenomenon has been reported to be an appreciable water source for neighbouring plants (Corak et al. 1987; Caldwell and Richards 1989). Trees in alley cropping systems can benefit associated crop plants through the mechanism of hydraulic lift, thus providing more water for surrounding vegetation during dry periods (van Noordwijk et al. 1996; Burgess et al. 1998; Ong et al. 1999; Jose et al. 2009). Recent evidence shows that this phenomenon can help promote greater plant growth, and could have important implications for net primary productivity, as well as ecosystem nutrient cycling and water balance (Horton and Hart 1998).

5.2.2 Competition for nutrients

Below ground competition for nutrients is most likely to occur when two or more species have developed a specialized root system that directs them to explore the same soil strata for resources (van Noordwijk et al. 1996). Thus, the extent of competition in a mixed system such as alley cropping will depend on nutrient availability, root architecture, rooting depth and proximity to competing roots, and temporal demands (Jose et al. 2000b; Allen et al. 2004a,b). As in the case of conventional agriculture, nutrients can often be limiting in agroforestry systems. Nitrogen (N) is generally the most limiting soil nutrient in temperate alley cropping systems due to various reasons. First, N is lost via various biogeochemical mechanisms such as volatilization, denitrification and leaching. N is also lost when crop biomass is removed from the field following harvest. In addition, plants of the same species and growth stage can compete heavily for N when zones of depletion in the soil overlap with neighbouring plants. In alley cropping systems, competition for nutrients can be even more intense, as most tree species have the bulk of their fine, feeder roots in the top 30 cm layer of the soil. The root systems of all component species in three temperate alley cropping systems consisting of maize and black walnut or red oak (Jose et al. 2000b), and pecan-cotton (Zamora et al. 2007) were found to most heavily occupy the top 30 cm of soil and to decrease in density with depth. Consequently, yield of maize was reduced by 35% and 33% in the black walnut and red oak systems (Jose et al. 2000b), respectively, and yield of cotton was reduced by 38% in the pecan system (Zamora et al. 2008).

Although it is difficult to separate the below ground competition for water from that for nutrients, it is widely recognized that crop production in agroforestry systems in semi-arid regions is likely limited by competition for water. When considering alley cropping systems in temperate regions, it is not clear as to which competitive vector will limit productivity, although competition for water has been reported (Jose et al. 2000b; Wanvestraut et al. 2004). Research on competition for nutrients in temperate agroforestry is rather limited (Jose et al. 2000b; Miller and Pallardy 2001; Allen et al. 2003; Zamora et al. 2009a). Many of the reported agroforestry systems were fertilized at the conventional agronomic level as needed for the crop component. Jose et al. (2000) reported that competition for nitrogen fertilizer was minimal in their black walnut-maize alley cropping system, since nutrient acquisition was not simultaneous among the system's components. Water availability, however, was a factor in nutrient competition, as competition for water by tree roots was responsible for the biomass reduction in intercropped maize resulting in decreased efficiency of fertilizer use (Jose et al. 2000b). The same trend was also reported by Zamora et al. (2009) in the production of cotton in a loblolly pine alley cropping system in Northwestern Florida. Under resource competition, plants tend to inhibit further growth. For example, in that study, cotton plants intercropped with loblolly pine showed reduced growth, and so were unable to utilize the available fertilizer efficiently. This phenomenon was also observed for maize growing in a black walnut system in Missouri (Jose et al. 2000b) and with cotton planted under pecan (Allen et al. 2004b) due to competition for water.

5.2.3 Allelopathy

The mixture of multiple species in an alley cropping system may not only increase competition for growth resources such as water, light and nutrients but may bring about other negative effects such as allelopathy. Allelopathy is a biological phenomenon by which one plant may produce and release some chemical compounds that inhibit the growth of nearby plants. These chemicals are released into the soil by root exudation and aboveground litterfall.

Allelochemicals are known to affect germination, growth, development, distribution and reproduction of a number of plant species (Inderjit and Malik 2002; Jose and Gillespie 1998a,b; Jose et al. 2009). Several examples of allelopathy have been shown in temperate agroforestry systems (Jose and Gillespie 1998b; Thevathasan et al. 1998; Jose and Holzmueller 2008) such as those based on black walnut or pecan. These authors reported declines in yields of maize, cotton and peanut when these companion crops were exposed to juglone in hydroponic cultures. However, the effects of allelopathy can be minimized (at least in research plots) by separating the roots of the tree from the crop through the use of polyethylene barrier treatment (Jose and Gillespie 1998a,b). Jose et al. (2000) demonstrated that by using this barrier approach, maize crop yield became similar to that of monoculture maize. Jose and Gillespie (1998a,b) showed that the barrier also eliminated detectable levels of juglone in the soil in a black walnut alley cropping system. Perhaps the most effective approach to address allelopathic effects is proper species combinations when considering an alley cropping system as a production system.

5.3 Design complexities

While alley cropping is practised in temperate regions it is not widely acknowledged or adopted because of its apparent complexities. In their survey to identify landowners' and extension professionals' perceptions of agroforestry in the Southeast US, Workman et al. (2003) identified the lack of skills and machinery compatibility with the alley cropping operations as major factors in the slow adoption of practices. As with other forms of multi-cropping systems, alley cropping requires more intensive technical management skills and marketing knowledge. Alley cropping requires a more intensive management system including specialized equipment for the tree management and additional managerial skills and training to manage multiple crops at a given site. Removal of land from annual crop production is another barrier to alley cropping adoption because the practice may not provide a financial return from the trees for several years. Further, trees in alley cropping may be an obstacle to crop cultivation if not carefully planned and designed. Alley cropping also requires a marketing infrastructure for the tree products that may not be present in the local area. As earlier discussed, the competition for growth resources between the trees and crops in alley cropping systems may limit productivity; thus, proper design of the practice may address or minimize the impacts of these issues.

6 Evaluating system performance: the case of the pecan-cotton alley cropping system

Realization of the potential benefits of alley cropping depends on the production, protection (environmental benefits) and adoptability of the practice. To shed light on these three criteria (production, protection and adoptability), it is useful to examine them in the context of a specific documented system as a case study. Hence, this section will examine a pecan-cotton alley cropping system in Northwestern Florida, USA.

6.1 The system

A 47-year-old pecan orchard was converted into an alley cropping system in 2001 (Fig. 3). The pecan trees were planted at a uniform spacing of 18.3 m and the orchard remained under grass cover for 29 years until it was converted into an alley cropping practice. The rows of trees were oriented in a north-south direction. Alleys measured 55 m in length and 18.3 m in width, with a practical cultivable width of 16.2 m. Before planting the cotton as the intercropped species, each alley was divided into two plots to assess the impacts of competition for water and nutrients on cotton production: barrier and non-barrier plots. The barrier plots were subjected to a root pruning treatment, in which a trenching machine was used to dig a 0.2 m wide and 1.2 m deep trench along both sides of the plot at a distance of 1.5 m from the trees to separate root systems of pecan and cotton. A double layer of 0.15-mm thick polyethylene sheeting was used to line the ditch prior to mechanical backfilling. The barrier plots thus served as the tree root exclusion treatment, preventing

Figure 3 Pecan-cotton alley cropping system in Northwestern Florida, USA.

interaction of tree and cotton roots, while the non-barrier plots, which did not receive this treatment, served as the tree-crop competition treatment. Sixteen rows of cotton were planted in each alley. Measurements were made on rows 1, 4, 8, 9 and 16. The barrier and non-barrier treatment received standard fertilizer and pesticide application for cotton, and no irrigation was applied.

6.2 Production (yield)

Lint yield of cotton in the barrier treatment (677 kg ha^{-1}) was higher than that of the non-barrier treatment (502 kg ha^{-1}) but was not different from the monoculture stand (634 kg ha^{-1}). Within the non-barrier treatment, the highest yield was observed in the centre of the alley (row 8), with row 1 (at the tree-crop interface) yielding less than the centre row (P = 0.0388 for row 8; P = 0.0216 for row 9). The presence of the barrier had the greatest (positive) impact on plants in the two edge rows (1 and 16) of the barrier treatment (Allen et al. 2004a,b; Wanvestraut et al. 2004; Zamora et al. 2006, 2007, 2008). When high yield of the edge rows is removed (as they theoretically had no crop competitor on one side), the barrier treatment yield (556 kg ha^{-1}) was not different from the monoculture cotton yield (634 kg ha^{-1}). The vegetative development of cotton was strongly enhanced by the presence of the barrier treatment, which outperformed the non-barrier plants. Leaf area index in the barrier treatment was higher than that of the non-barrier plants. Although the absorbed PAR was 42% lower for the barrier plants compared to the monoculture plants, lint yield was similar for both treatments. This indicates that light was not a major factor influencing production in that system, but below ground competition for water and nutrients was present as evidenced by the significant reduction of yield in non-barrier treatment. Cotton tolerated moderate shading and provided acceptable yield when below ground competition was alleviated (Zamora et al. 2008).

6.3 Environmental impacts

Alley cropping is viewed as a possible strategy to clean up agricultural contaminants in groundwater. In conventional agricultural systems, less than half of the applied N and phosphorus fertilizer is taken up by the crops. Consequently, excess fertilizer is washed away from agricultural fields via surface run-off or leached into sub-surface water supply, thereby contaminating water sources and decreasing water quality (Cassman 1999; Jose 2009). Trees or shrubs with deep root systems can theoretically improve water quality by serving as a 'safety net' whereby excess nutrients are captured by tree roots and then recycled back into the system through root turnover and litter fall, increasing the nutrient use efficiency of the system (Allen et al. 2004a).

In the pecan-cotton alley cropping system, NO_3-N concentrations varied between the barrier and non-barrier treatments. At 0.3 and 0.9 m depths, NO_3-N levels in barrier treatments were recorded to be 14.36 and 10.82 mg L^{-1}, respectively, representing a 24.7% decrease with depth. Mean concentrations for non-barrier at corresponding depths were 10.16 and 7.15 mg L^{-1}, respectively, representing a 29.6% decrease. Furthermore, average NO_3-N leaching was higher in the barrier treatment (3.40 kg ha^{-1} $month^{-1}$) than in the non-barrier treatment (1.28 kg ha^{-1} $month^{-1}$). During the whole growing season, the amount of NO_3-N at 0.3 m depth in the barrier treatment was 121.94 kg ha^{-1}, which was significantly higher than in the non-barrier treatment (63.83 kg ha^{-1}), a decrease of 47.7% (Allen et al. 2004a,b). A similar trend occurred at 0.9 m depth, where cumulative NO_3-N leaching in barrier (45.56 kg ha^{-1}) was higher than in non-barrier (13.05 kg ha^{-1}), a 71.4% decrease. Therefore, these numbers indicate that the competitive presence of trees can be utilized to reduce N leaching in alley cropping systems.

Allen et al. (2004a) further observed competition for N in the pecan-cotton alley cropping system, where cotton plants in the barrier treatment had a 59% higher aboveground biomass compared to plants in the non-barrier treatment. Although a companion study indicated that competition for water was also a factor (Wanvestraut et al. 2004), the authors hypothesized that because pecan trees leaf out earlier in the spring and have a high nutrient demand early in the growing season, the soil N depleted before cotton plants were established later in the growing season. Therefore, in this particular system, cotton plants were more likely to rely on supplemental N to fulfil plant needs. In addition, Allen et al. (2005) showed that N mineralization rates differed between barrier and non-barrier treatments in this system. Higher rates of N mineralization were observed in the non-barrier treatment (26.05 mg kg^{-1} $month^{-1}$) compared to rates observed in the barrier treatment (19.78 mg kg^{-1} $month^{-1}$), indicating that competitive interactions for water and N in the non-barrier treatment may have led to decreased ability of cotton plants to uptake N (Allen et al. 2005; Jose et al. 2009).

Furthermore, integrating trees in the agricultural landscape using alley cropping can enhance soil physical, chemical and biological properties by adding significant amounts of above and below ground organic matter and thus releasing and recycling nutrients within the systems. Trees can increase nutrient inputs by retrieving nitrogen from deep soil horizons. Lee and Jose (2003a,b) reported a higher soil organic matter and microbial biomass level in the previously described pecan-cotton system compared to monocropped cotton. Generally, nutrients are lost in monoculture row crop systems due to total or substantial harvest of biomass that prevents the build-up of organic matter in the soil. Agroforestry's role in enhancing and maintaining long-term soil productivity and sustainability has been well documented (Nair 1993;

Young 1997; Schroth and Sinclair 2003). The incorporation of trees or shrubs in agroforestry systems can increase the amount of carbon sequestered compared to a monoculture field of crops (Sharrow and Ismail 2004; Kirby and Potvin 2007; Jose 2009). Carbon sequestered in agroforestry systems could be sold in carbon credit markets where such opportunities exist; so this is an example of how they may be capable of providing other long-term economic returns to farmers.

6.4 Adoptability

Alley cropping can contribute to achieving various environmental impacts. However, the success of agroforestry adoption in the landscape depends on how the practice can address many of the socioeconomic conditions of the farmers including enhancement of revenue, labour and machinery needs, time and effort to do it, among others. Although its potential as a sustainable agricultural practice exists, farmers are hesitant to adopt it due to lack of skills and scientific knowledge to implement it such as in the case for pecan-based alley cropping. To increase adoption of this practice, the system must be properly designed to address the economic bottom line of the farmers, and the adoption must conform to the farmer's available resources (e.g. machinery, manpower etc.). Furthermore, there is a need to understand factors influencing adoption behaviour, identifying impacts of land- and tree-tenure, developing methods for integrating socioeconomics and biophysical approaches to increase adoption. Understanding or improving agroforestry cost-benefit and impact analyses including incorporating of risk and uncertainty, estimating and incorporating non-market goods and services produced by agroforestry systems, multi-input and output production function analysis are keys to adoption of alley cropping systems.

Zinkhan and Mercer (1997), in their survey of land-use professionals in the Southern US and related literature review, found that silvopastoral systems were the most common type of agroforestry in the region, which is also similar in Europe (Mosquera-Losada et al. 2016). In this context, the most frequently mentioned benefits associated with these systems included the potential for increased economic returns, diversification and the enhancement of the timing of cash flows. However, more study is needed on the verification of economic benefits from various temperate agroforestry systems, including ground-level agricultural economics studies. These are by nature very situation-specific and include various economic inputs and outputs that must be weighed against monocrop production benefits and farmer economic priorities.

Lastly, it should be noted that, in today's highly competitive and market-driven economy, caution must be observed when considering large-scale adoption of diverse cropping systems, as economic path dependency associated with intensive agricultural production (such as in the US Corn Belt) can work against the incorporation of diverse crops, as described by Roesch-McNally et al. (2018). They conclude that this issue needs to be further addressed by adjusting policy and economic incentives that foster diverse systems and reduce climate risk.

7 Future trends and conclusion

Considered as a holy grail in agriculture, properly designed alley cropping has shown promise to achieve the cornerstones of sustainable agriculture in temperate regions compared to conventional agriculture. Alley cropping can promote economic productivity while enhancing ecological integrity. As a perennial system, enhancing the positive interactions between the tree or shrub and crop while minimizing the negative ones can enhance economic production. Even though crop yield may be decreased when compared to monoculture systems, by incorporating multiple species in a single system there is a possibility to increase overall biomass production and economic value. Environmentally, the trees in alley cropping could play the safety-net role, which shows the potential improvement of water quality by reducing surface and sub-surface contamination.

Although the adoption of the practice is still slow, the growing intensity of land-use practices, the increasing public interest in achieving ecosystem services (clean water, wildlife habitat, carbon sequestration, among others) all serve to encourage farmers to evaluate the potential of alley cropping in their farm operation to meet their land management needs. Alley cropping can advance temperate zone stewardship by (1) converting degraded or marginal land into being productive estate, (2) protecting sensitive lands and (3) enhancing, diversifying and promoting farm production systems.

Temperate alley cropping research is still limited, and several information gaps still exist that need immediate attention by the scientific community. There is a need to quantify the interactive effects of trees and crops at landscape level and at different geographic scales. Furthermore, appropriate design of alley cropping systems at various levels and scales needs to be explored. Likewise, new technologies stemming from GPS data, precision agriculture, online weather databases and Landsat-type satellite imagery need to be incorporated into land-use decision making. Also, appropriate incentives and policies need to be pursued in order to create an environment where alternative cropping systems are financially viable and geared towards long-term climate change response. In sum, while there are many challenges to the adoption of alley cropping and other alternative systems, there are also many opportunities, as more and more landowners and consumers recognize the importance of achieving mutual sustainability in the agricultural and environmental landscapes.

8 Where to look for further information

National Agroforestry Center: https://www.fs.usda.gov/nac/practices/alleycropping.shtml

The USDA National Agroforestry Center (NAC) accelerates the application of agroforestry through a national network of partners. NAC conducts research, develops technologies and tools, coordinates demonstrations and training, and provides useful information to natural resource professionals on agroforestry.

The Center for Agroforestry at the University of Missouri: http://www.centerforagroforestry.org, Alley Cropping: http://www.centerforagroforestry.org/practices/ac.php

The Center for Agroforestry at the University of Missouri, established in 1998, is one the world's leading centres contributing to the science underlying agroforestry. The faculty at the Center has a number of members from various disciplines doing research and outreach activities on agroforestry around the United States.

AGFORWARD: http://www.agforward.eu/index.php/en/silvoarable-agroforestry-in-the-uk.html

AGFORWARD (AGroFORestry that Will Advance Rural Development) is a four-year research project funded by the EU's Seventh Framework Program for Research and Technological Development. The project builds on existing agroforestry experiments, current on-farm agroforestry trials and previous research projects such as 'Silvoarable Agroforestry For Europe (SAFE)'. The European Agroforestry Federation is a partner. The start of the project coincides with the launch of EU Rural Development Regulations that can support the establishment of agroforestry systems.

Association for Temperate Agroforestry (AFTA): http://www.aftaweb.org/

AFTA is a group of natural resource professionals, researchers and farmers whose mission is to promote the wider adoption of agroforestry by landowners in temperate regions of North America. With members in the United States, Canada and overseas, AFTA pursues its mission through activities such as networking, information exchange, public education and policy development.

Savanna Institute: http://www.savannainstitute.org/

The Savanna Institute is a nonprofit organization working to lay the groundwork for widespread adoption of agroforestry in the Midwest US. The Institute works in collaboration with farmers and scientists to develop perennial food and fodder crops within multifunctional polyculture systems grounded in ecology and inspired by the savanna biome. It strategically enacts this mission via research, education and outreach.

9 References

Allen, S., Jose, S., Nair, P. K. R. and Brecke, B. J. 2004a. Competition for ^{15}N labeled nitrogen in a pecan-cotton alley cropping system in the southern United States. *Plant and Soil* 263:151–64.

Allen, S., Jose, S., Nair, P. K. R., Brecke, B. J. and Nkedi-Kizza, P. 2004b. Safety net role of tree roots: Experimental evidence from an alley cropping system. *Forest Ecology and Management* 192:395–407.

Allen, S., Jose, S., Nair, P. K. R., Brecke, B. J., Nair, V. D., Graetz, D. and Ramsey, C. L. 2005. Nitrogen mineralization in a pecan (*Carya illinoensis* K. Koch)-cotton (*Gossypium hirsutum* L.) alley cropping system in the southern United States. *Biology and Fertility of Soils* 41:28–37.

Altieri, M. A. 1983. *Agroecology: The Scientific Basis of Alternative Agriculture*. Golden Horn Press, Berkeley, CA.

Bakermans, M., Rodewald, A., Vitz, A. and Rengifo, C. 2012. Migratory bird use of shade coffee: The role of structural and floristic features. *Agroforestry Systems* 85:85–94.

Best, L. B., Whitmore, R. C. and Booth, G. M. 1990. Use of cornfields by birds during the breeding season. The importance of edge habitat. *American Midland Naturalist* 123:84–99.

Bonilla, C. A., Munoz, J. F. and Vauclin, M. 1999. Opus simulation of water dynamics and nitrate transport in a field plot. *Ecological Modeling* 122:69–80.

Browaldh, M. 1995. The influence of trees on N dynamics in an agrisilvicultural system in Sweden. *Agroforestry Systems* 30:131–8.

Bugg, R. L., Sarrantonio, M., Dutcher, J. D. and Phatak, S. C. 1991. Understory cover crops in pecan orchards: Possible management systems. *American Journal of Alternative Agriculture* 6:50–62.

Burgess, S. O., Adams, M. A., Turner, N. C. and Ong, C. K. 1998. The redistribution of soil water by tree roots systems. *Oecologia* 115:306–11.

Caldwell, M. M. and Richards, J. H. 1989. Hydraulic lift: Water efflux from upper roots improves effectiveness of water uptake by deep roots. *Oecologia* 79:1–5.

Cassman, K. G.1999. Ecological intensification of cereal production systems: Yield potential, soil quality, and precision agriculture. *Proceedings of the National Academy of Sciences* 96:5952–9.

Corak, S. J., Blevins, D. G. and Pallardy, S. G. 1987. Water transfer in an alfalfa-maize association: Survival of maize during drought. *Plant Physiology* 84:582–6.

Dawson, I. K., Place, F., Torquebiau, E., Malézieux, E., Iiyama, M., Sileshi, G. W., Kehlenbeck, K., Masters, E., McMullin, E. and Jamnadass, R. 2013. Agroforestry, food and nutritional security. *FAO*. http://www.fao.org/forestry/37082-04957fe26afbc90d1e9c0356c48185295.pdf (Last accessed 22 January 2018).

Dyack, B., Rollins, K. and Gordon, A. M. 1999. An economic analysis of market and non-market benefits of a temperate intercropping system in southern Ontario, Canada. *Agroforestry Systems* 44:197–214.

Eckberg, J., Johnson, G., Heimpel, G., Sheaffer, C., Peterson, J., Plecas, M., Kaser, J. and Wyse, D. 2015. Integrative cropping systems to enhance biological control and increase soybean yield. *University of Minnesota - Monsanto Fellows Meeting*, 29–30 January.

Eichhorn, M. P., Paris, P., Herzog, F., Incoll, L. D., Liagre, F., Mantzanas, K., Mayus, M., Moreno, G., Papanastasis, V. P., Pilbeam, D. J., Pisanelli, A. and Dupraz, C. 2006. Silvoarable systems in Europe – past, present and future prospects. *Agroforestry Systems* 67:29-50.

FAO. 2013. Advancing agroforestry on the policy agenda: A guide for decision-makers. Agroforestry Working Paper No.1. FAO, Rome.

Garrett, H. E. and Buck, L. E. 1997. Agroforestry practice and policy in the United States of America. *Forest Ecology and Management* 91:5-15.

Garrett, H. E. and Harper, L. S. 1999. The science and practice of black walnut agroforestry in Missouri, USA: A temperate zone assessment. *In* Buck, L. E., Lassoie, J. P. and Fernandes, E. C. M. (Eds), *Agroforestry in Sustainable Agriculture Systems*. CRC Press, New York, NY, pp. 97–110.

Garrett, H. E., McGraw, R. L. and Walter, W. D. 2009. Alley cropping practices. *In* Garrett, H. E. (Ed.), *North American Agroforestry: An Integrated Science and Practice*, 2nd Edition. American Society of Agronomy, Madison, WI.

Garrity, D. 2004. Agroforestry and the achievement of the millennium development goals. *Agroforestry Systems* 61:5–17.

Gibbs, S., Koblents, H., Coleman, B., Gordon, A., Thevathasan, N. and Williams, P. 2016. Avian diversity in a temperate tree-based intercropping from inception to now. *Agroforestry Systems* 90:905–16.

Gillespie, A. R., Miller, B. K. and Johnson, K. D. 1995. Effects of ground cover on tree survival and growth in filter strips of the Cornbelt Region of the midwestern US. *Agriculture, Ecosystems and Environment* 53:263-70.

Gillespie, A. R., Jose, S., Mengel, D. B., Hoover, W. L., Pope, P. E., Seifert, J. R., Biehle, D. J., Stall, T. and Benjamin, T. 2000. Defining competition vectors in a temperate alleycropping system in the midwestern USA: I. Production physiology. *Agroforestry Systems* 48:25-40.

Girma, H., Rao, M. R. and Sithanantham, S. 2000. Insect pests and beneficial arthropod populations under different hedgerow intercropping in semiarid Kenya. *Agroforestry Systems* 50:279–92.

Gordon, A. M. and Williams, P. A. 1991. Intercropping valuable hardwood tree species and agricultural crops in southern Ontario. *The Forestry Chronicle* 67:200–8.

Gordon, A. M., Newman, S. M. and Williams, P. A. 1997. *Temperate Agroforestry System: An Overview*. CAB International, Wallingford, UK, pp. 1-6.

Govindarajan, M., Rao, M. R., Mathuva, M. N. and Nair, P. K. R. 1996. Soil-water and root dynamics under hedgerow intercropping in semiarid Kenya. *Agronomy Journal* 88:513–20.

Hagan, D., Jose, S., Thetford, M. and Bohn, K. 2009. Production physiology of three native shrubs interplanted in a young longleaf pine plantation. *Agroforestry Systems* 76:283–94

Holzmueller, E. J. and Jose, S. 2012. Bioenergy crops in agroforestry systems: Potential for the U.S. North Central Region. *Agroforestry Systems* 85:305–14

Horton, J. L. and Hart, S. C. 1998. Hydraulic lift: A potentially important ecosystem process. *Trends in Ecological Evolution* 13:232-5.

Inderjit, and Malik, A. U. 2002. *Chemical Ecology of Plants: Allelopathy in Aquatic and Terrestrial Ecosystems*. Birkhauser-Verlag, Berlin, Germany, 272pp.

Jose, S. 2009. Agroforestry for ecosystem services and environmental benefits: An overview. *Agroforestry Systems* 76:1-10.

Jose, S. 2011. Managing native and non-native plant species in agroforestry. *Agroforestry Systems* 83:101-5.

Jose, S. and Gillespie, A. R. 1998a. Allelopathy in black walnut (*Juglans nigra* L.) alley cropping: I. Spatio-temporal variation in soil juglone in a black walnut-corn (*Zea mays* L.) alley cropping system. *Plant and Soil* 203:191-7.

Jose, S. and Gillespie, A. R. 1998b. Allelopathy in black walnut (*Juglans nigra* L.) alley cropping. II. Effects of juglone on hydroponically grown corn (*Zea mays* L.) and soybean (*Glycine max* (L.) Merr.) growth and physiology. *Plant and Soil* 203:199-205.

Jose, S. and Holzmueller, E. 2008. Black walnut allelopathy: Implications for intercropping. *In* Zeng, R. S., Mallik, A. U. and Luo, S. M. (Eds), *Allelopathy in Sustainable Agriculture and Forestry*. Springer, New York, NY.

Jose, S., Gillespie, A. R., Seifert, J. R. and Biehle, D. J. 2000a. Defining competition vectors in a temperate alleycropping system in the midwestern USA: II. Competition for water. *Agroforestry Systems* 48:41-59.

Jose, S., Gillespie, A. R., Seifert, J. E. and Pope, P. E. 2000b. Defining competition vectors in a temperate alleycropping system in the midwestern USA: III. Competition for nitrogen and litter decomposition dynamics. *Agroforestry Systems* 48:61-77.

Jose, S., Gillespie, A. R. and Pallardy, S. G. 2004. Interspecific interactions in temperate agroforestry. *Agroforestry Systems* 61:237-55.

Jose, S., Holzmueller, E. J. and Gillespie, A. R. 2009. Tree-crop interactions in temperate agroforestry. *In* Garrett, H. E. (Ed.), *North American Agroforestry: An Integrated Science and Practice*, 2nd Edition. American Society of Agronomy, Madison, WI. Chapter 4.

Jose, S., Gold, M. A. and Garrett, H. E. 2012. The future of temperate agroforestry in the United States. *In* Garrity, D. and Nair, P. K. R. (Eds), *Agroforestry: The Way Forward*. Springer Science, the Netherlands.

Kirby, K. R. and Potvin, C. 2007. Variation in carbon storage among tree species: Implications for the management of a small-scale carbon sink project. *Forest Ecology and Management* 246:208-21.

Lee, K. H. and Jose, S. 2003a. Soil respiration, fine root production and microbial biomass in cottonwood and loblolly pine plantations along a soil nitrogen gradient. *Forest Ecology and Management* 185:263-73.

Lee, K. H. and Jose, S. 2003b. Soil respiration and microbial biomass in a pecan-cotton alley cropping system in southern USA. *Agroforestry Systems* 58:45-54.

Lehmann, J., Peter, I., Steglich, C., Gebauer, G., Huwe, B. and Zech, W. 1998. Belowground interactions in dryland agroforestry. *Forest Ecology and Management* 111:157-69.

Lewis, C. E., Tanner, G. W. and Terry, W. S. 1985. Double vs. single-row pine plantations for wood and forage production. *Southern Journal of Applied Forestry* 9:55-61.

Lin, C. H., McGraw, R. L., George, M. F. and Garrett, H. E. 1999. Shade effects on forage crops with potential in temperate agroforestry practices. *Agroforestry Systems* 44:109-19.

MacLeod, C., Blackwell, G. and Benge, J. 2012. Reduced pesticide toxicity and increased woody vegetation cover account for enhanced native bird densities in organic orchards: Orchard management and native birds. *Journal of Applied Ecology* 49:652-60.

Magcale-Macandog, D. B. M., Ranola, F. M., Ranola, R. and Vidal, P. A. B. 2010. Enhancing the food security of upland farming households through agroforestry in Claveria, Misamis Oriental Philippines. *Agroforestry Systems* 79:327-42.

Malézieux, E., Crozat, Y., Dupraz, C., Laurans, M., Makowski, D., Ozier-Lafontaine, H., Rapidel, B., de Tourdonnet, S. and Valantin-Morison, M. 2009. Mixing plant species in cropping systems: Concepts, tools and models. A review. *Agronomy for Sustainable Development* 29:43–62.

Marshall, M. G. and Bennett, C. F. (Eds). 1998. *Outcomes of Nitrogen Fertilizer Management Programs, Vol. 3. National Extension Targeted Water Quality Program, 1992–1995*. Texas A&M University, State College, TX, Cooperating with Cooperative State Research, Education, and Extension Service (CSREES), USDA.

Miller, A. W. and Pallardy, S. G. 2001. Resource competition across the crop-tree interface in a maize-silver maple temperate alley cropping stand in Missouri. *Agroforestry Systems* 53:247–59.

Molnar, T., Kahn, P., Ford, T., Funk, C. and Funk, C. 2013. Tree crops, a permanent agriculture: concepts from the past for a sustainable future. *Resources* 2, 457–88.

Monteith, J. L., Ong, C. K. and Corlett, J. E. 1991. Microclimatic interactions in agroforestry. *Forest Ecology and Management* 45:31–44.

Mosquera-Losada, M. R., Santiago Freijanes, J. J., Pisanelli, A., Rois, M., Smith, J., den Herder, M., Moreno, G., Malignier, N., Mirazo, J. R., Lamersdorf, N., Ferreiro Domínguez, N., Balaguer, F., Pantera, A., Rigueiro-Rodríguez, A., Gonzalez-Hernández, P., Fernández-Lorenzo, J. L., Romero-Franco, R., Chalmin, A., Garcia de Jalon, S., Garnett, K., Graves, A. and Burgess, P. J. 2016. Extent and success of current policy measures to promote agroforestry across Europe. Deliverable 8.23 for EU FP7 Research Project: AGFORWARD 613520, 95pp.

Nair, P. K. R. 1993. *An Introduction to Agroforestry*. Kluwer Academic Publishers, Dordrecht, the Netherlands, pp. 499.

NeSmith, D. S. and Ritchie, J. T. 1992. Short- and long-term responses of corn to pre-anthesis soil water deficit. *Agronomy Journal* 84:107–13.

Ng, H. Y. F., Drury, C. F., Serem, V. K., Tan, C. S. and Gaynor, J. D. 2000. Modeling and testing of the effect of tillage, cropping and water management practices on nitrate leaching in clay loam soil. *Agricultural Water Management* 43:111–31.

Ong, C. K., Deans, J. D., Wilson, J., Mutua, J., Khan, A. A. H. and Lawson, E. M. 1999. Exploring belowground complementarity in agroforestry using sap flow and root fractal techniques. *Agroforestry Systems* 44:87–103.

Palma, J., Graves, A. R., Burgess, P. J., van der Werf, W. and Herzog, F. 2007. Integrating environmental and economic performance to assess modern silvoarable agroforestry in Europe. *Ecological Economics* 63:759–67.

Pardini, A., Longhi, F. and Natali, F. 2008. Pastoral systems and agro-tourism in marginal areas of Central Italy. *Options Méditerranéennes* 79:97–102.

Pekin, B. and Pijanowski, B. 2012. Global land use intensity and the endangerment status of mammal species. *Diversity and Distribution* 18:909–18.

Peng, K., Incoll, R. D., Sutton, L. L., Wright, S. and Chadwick, C. 1993. Diversity of airborne arthropods in a silvoarable agroforestry system. *Journal of Applied Ecology* 30 551–60.

Pimentel, D. Harvey, C. Resosudarmo, P., Sinclair, K., Kurz, D., McNair, M., Crist, S., Shpritz, L., Fitton, R., Saffouri, R. and Blair, R. 1995. Environmental and economic costs of soil erosion and conservation benefits. *Science* 267:1117–23.

Regueiro, R. A., McAdam, J. and Mosquera-Lozada, M. R. 2009. *Agroforestry in Europe: Current Status and Future Prospects*. Springer Science Business Media B.V., Dordrecht, the Netherlands, pp. 295–320.

Renken, R. B. 1983. Breeding bird communities and bird habitat associations on North Dakota waterfowl production areas of three habitat types. Unpublished MS Thesis, Iowa State University, Ames, IA.

Reynolds, P. E., Simpson, J. A., Thevathasan, N and Gordon, A. M. 2007. Effects of tree competition on corn and soybean photosynthesis, growth, and yield in a temperate tree-based agroforestry intercropping system in southern Ontario, Canada. *Ecological Engineering* 29:362–71.

Roesch-McNally, G. E., Arbuckle, J. G. and Tyndall, J. C. 2018. Barriers to implementing climate resilient agricultural strategies: The case of crop diversification in the U.S. Corn Belt. *Global Environmental Change* 48:206–15.

Root, R. 1973. Organization of a plant-arthropod association in simple and diverse habitats: The fauna of collards (*Brassica oleracea*). *Ecological Monographs* 43:95–124.

Rowe, E. C., Hairiah, K, Giller, K. E., van Noordwijk, M. and Cadisch, G. 1999. Testing the safety-net role of hedgerow tree roots by ^{15}N placement at different soil depths. *Agroforestry Systems* 43:81–93.

Schroth, G and Sinclair, F. 2003. *Trees Crops and Soil Fertility: Concepts and Research Methods*. CABI, Wallingford, UK, p. 464.

Sharrow, S. H. and Ismail, S. 2004. Carbon and nitrogen storage in agroforests, tree plantations, and pastures in western Oregon, USA. *Agroforestry Systems* 60:123–30.

Smith, J., Pearce, B. D. and Wolfe, M. 2013. Reconciling productivity with protection of the environment: Is temperate agroforestry the answer? *Renewable Agriculture and Food Systems* 28(1):80–92.

Stamps, W. T. and Linit, M. J. 1998. Plant diversity and arthropod communities: Implications for temperate agroforestry. *Agroforestry Systems* 39:73–89.

Stamps, W. T., Wood, T. W., Linit, M. G. J. and Garrett, H. E. 2002. Arthropod diversity in alley cropped black walnut (*Juglans nigra* L.) stands in eastern Missouri, USA. *Agroforestry Systems* 56:167–75.

Thevathasan, N. V. and Gordon, A. M. 2004. Ecology of tree intercropping systems in north temperate region: Experience from southern Ontario, Canada. *Agroforestry Systems* 61:257–68.

Thevathasan, N. V., Gordon, A. M. and Voroney, R. P. 1998. Juglone (5-hydroxy-1,4 napthoquinone) and soil nitrogen transformation interactions under a walnut plantation in southern Ontario, Canada. *Agroforestry Systems* 44:51–162.

Udawatta, R. P., Nygren, P. and Garett, H. E. 2005. Growth of three oak species during establishment of an agroforestry practice for watershed protection. *Canadian Journal of Forestry Research* 35:602–9.

University of Missouri Center for Agroforestry (UMCA). 2015. Alley cropping. *In Training Manual for Applied Agroforestry Practices*. Columbia, MO. http://www.centerforagroforestry.org/pubs/training/index.php

USDA. 2018. Agricultural census. https://www.agcensus.usda.gov/ (Last accessed 22 January 2018).

USDA NRCS. 2007. Soil erosion on cropland. https://www.nrcs.usda.gov/wps/portal/nrcs/detail/national/home/?cid=stelprdb1041887 (Last accessed 22 January 2018).

Vandermeer, J. 1989. *The Ecology of Intercropping*. Cambridge University Press, Cambridge, UK, 249pp.

van Noordwijk, M., Lawson, G., Soumaré, A., Groot, J. J. R. and Hairiah, K. 1996. Root distribution of trees and crops: Competition and/or complementarity. *In* Ong,

C. K. and Huxley, P. (Eds), *Tree-Crop Interactions: A Physiological Approach*. CAB International, Wallingford, UK, pp. 319–64.

Vasileiadis, V., Moonen, A., Sattin, M., Otto, S., Pons, X., Kudsl, P., Veres, A., Dorner, Z., van der Weide, R., Marraccini, E., Pelzer, E., Angevin, F. and Kiss, J. 2013. Sustainability of European maize-based cropping systems: Economic, environmental and social assessment of current and proposed innovative IPM-based systems. *European Journal of Agronomy* 48:1–11.

Wanvestraut, R., Jose, S., Nair, P. K. R. and Brecke, B. J. 2004. Competition for water in a pecan-cotton alley cropping system in the southern United States. *Agroforestry Systems* 60:167–79.

Wilcove, D., Rothstein, D., Dubow, J., Phillips, A. and Losos, E. 1998. Quantifying threats to imperiled species in the United States. *Bioscience* 48:607–15.

Wilson, M. 2007. Perennial pathways: Planning and establishment practices for edible agroforestry. Master of Science Thesis, University of Illinois, Urbana, IL.

Workman, S. W., Bannister, M. E. and Nair, P. K. R. 2003. Agroforestry potential in the southeastern United States: Perceptions of landowners and extension professionals. *Agroforestry Systems* 59:73–83.

Yahner, R. 1982. Avian use of vertical strata and plantings in farmstead shelterbelts. *Journal of Wildlife Management* 46:50–60.

Young, A. 1997. *Agroforestry for Soil Management*, 2nd ed. CAB International, Wallingford, UK, p. 307.

Yunusa, I. A. M., Thomson, S. E., Pollock, K. P., Youwei, L and Mead, D. J. 2005. Water potential and gas exchange did not reflect performance of *Pinus radiata* D. Don in an agroforestry system under conditions of soil-water deficit in a temperate environment. *Plant and Soil* 275:195–206.

Zamora, D., Jose, S. and Nair, P. K. R. 2006. Interspecific interaction in a pecan-cotton alleycropping system in the southern United States: The production physiology. *Canadian Journal of Botany* 84:1686–94.

Zamora, D., Jose, S. and Nair, P. K. R. 2007. Morphological plasticity of cotton roots in response to interspecific competition with pecan in an alleycropping system in the southern United States. *Agroforestry Systems* 69:107–16.

Zamora, D. S., Jose, S., Nair, P. K. R., Jones, J. W., Brecke, B. J. and Ramsey, C. L. 2008. Interspecific competition in a pecan-cotton alleycropping system in the southern United States: Is light the limiting factor?. *In* Jose, S. and Gordon, A. M. (Eds), *Toward Agroforestry Design: An Ecological Approach*. Springer Science, the Netherlands, pp. 81–95.

Zamora, D., Jose, S. and Napolitano, K. 2009a. Competition for ^{15}N labeled nitrogen in a loblolly pine-cotton alleycropping system in the southern United States. *Agriculture, Ecosystems and Environment* 131:40–50.

Zamora, D., Jose, S., Jones, J. W. and Cropper, W. P. 2009b. Modeling cotton production in a pecan alleycropping system using CROPGRO. *Agroforestry Systems* 76:423–35.

Zinkhan, F. C. and Mercer, D. E. 1997. An assessment of agroforestry systems in southern USA. *Agroforestry Systems* 35:303–21.

Chapter 5

Agroforestry practices: silvopastoralism

Gerardo Moreno and Victor Rolo, INDEHESA, Institute for Silvopastoralism Research, University of Extremadura, Spain

1 Introduction

Silvopastoral systems (SPs) intentionally combine livestock with trees or other woody perennials, either native or planted, and forages on the same unit of land such that mutual benefits to each component arise (Jose et al., 2017). Among the several categories of agroforestry systems found worldwide, SPs are presumably the most common and extended types (Sharrow, 1999; Rigueiro-Rodríguez et al., 2009; Orefice et al., 2016; Jose et al., 2017). Worldwide, SPs include scattered trees in pasturelands, managed plant succession, live fences, windbreaks, fodder tree banks, cut-and-carry systems, tree plantations with livestock grazing, pastures between tree alleys and various kinds of intensive management practices (Mosquera-Losada et al., 2009; Murgueitio et al., 2016; Fig. 1). So, SPs range from densely forested landscapes (grazed forest) to very open wood pastures (savannoid landscapes with scattered trees). Although in most of them understorey herbaceous layer is the main forage resource, in dry regions trees and shrubs can become the main forage resource (Papanastasis et al., 2008). Nair (2012) classified SPs broadly into two major forms: grazing systems and tree-fodder systems. In grazing systems, cattle are allowed to graze on pasture under widely spaced or scattered trees, whereas in the tree-fodder systems the animals are stall-fed with fodder from trees or shrubs grown in blocks on farms. Frequently, livestock farms combine SPs with arable

http://dx.doi.org/10.19103/AS.2018.0041.05

Figure 1 Images of different types of SPs across the world. Images 1 and 2 show natural savannas, grazed by wild animals in Africa and by livestock in California (USA). Images 3 and 4 show anthropic savannas with scattered oak trees after the clearance of the native forests, in Spain and Iran, respectively; besides grass, in both the trees are important to feed the livestock, with acorns in Spain and with branches in Iran. Image 5 shows an SP based on scattered native shrub species in very arid regions where trees hardly can grow. Images 6 and 7 show traditional live fences that surround paddocks in Europe; these trees are still pollarded (Image 7) to feed livestock and to have firewood. Image 8 shows a timber plantation managed with cattle in the United States. Images 9 and 10 show fruit orchards grazed in Europe, the first is the traditional German Streuobstwiesen and the second an innovative poultry farm in the Netherlands. Images 11 and 12 show innovative silvopastoral systems framed with alleys of Tagasaste and grasses (Image 11 in Australia) and multiple shrubs and trees species (Image 12 in Colombia).

land providing fodder crops to be used as supplement for livestock during less favoured seasons (Ovalle et al., 1990; Moreno and Pulido, 2009). In any case, as Jose et al. (2017) state, while combinations of trees/shrubs, grasses and livestock may exist, if the four 'I' requirements (intentional, integration, interactions and intensive management) are not met, they do not qualify as SPs. In accordance with these authors, extensive shrublands (e.g. European heathlands, García et al., 2013) and forests (Vera, 2000) grazed by livestock are not included in this chapter.

SPs provide multiple woody and non-woody plant products, high-quality food resources, livestock and game products, biofuel, as well as recreational or cultural services through multiple activities conducted with a comparatively low environmental impact. Although the use of this biomass as timber or energy (firewood and charcoal) has almost disappeared in Northern countries, this is still quite important in Asia, Africa and South America. Even in Europe, SP can be a promising solution within the EU target of renewable energy. New opportunities are also open for high-quality timber for which new plantations are being managed as SPs. Other products have been traditionally produced in SPs, such as fruits, cork, honey, mushrooms and medicinal plants. SPs also supply important ecosystem goods and services such as biodiversity maintenance, carbon sequestration, soil fertilization, microclimate amelioration, shelter for livestock, control against erosion and desertification, and increased local water balance (Manning et al., 2006; Wallace, 2007; Bugalho et al., 2011; Moreno et al., 2018). This agroforestry practice is lately gaining prominence as a sustainable and climate-resilient livestock production system. In this context, silvopastoralism provides an effective synergy between climate change adaptation and mitigation, by directly contributing to sequestration of greenhouse gases while at the same time buffering livelihoods against variability/recurrent drought as trees are less affected by variability in climate than grass pastures (Murdiyarso et al., 2005; Nair, 2012; Montagnini et al., 2013; Balehegn, 2017). Given that SPs can provide, when compared with widely used livestock production systems, efficient feed conversion, higher biodiversity and better animal welfare, SPs could replace existing systems in many parts of the world and should be further developed (Broom et al., 2013).

SPs are mostly restricted to marginal areas where orography, low soil fertility and/or climate constraints have hampered the development of intensive agriculture (Plieninger et al., 2015). A major challenge that SPs will face in the coming decades is the need to provide sustainable pasture yields while conserving ecosystem services. However, many traditional SPs are facing several threats that compromise their long-term persistence because changes/loss of multiple management practices that concern both the grazing schemes and forestry practices such as tree logging, coppicing, pollarding and shrub clearing (Plieninger et al., 2015). Unfortunately, most of the traditional practices have largely vanished in more developed countries: firewood has been replaced by fuel energies, fodder browse by commercial concentrates and forages, herders by fences, transhumance by continuous grazing. This is leading SPs to two divergent trends named intensification of land use in some areas, and land abandonment-encroaching in others (Pinto-Correia, 2000; Papanastasis, 2004; Bugalho et al., 2011; Plieninger and Bieling, 2012).

In many cases, woody vegetation has been barely managed and/or management did not follow ecological and economic objectives. Areas of woody vegetation are being degraded by loss of trees and soil fertility (Plieninger and Wilbrand, 2001; Rigueiro-Rodríguez et al., 2009). This progressive degradation and deforestation of SPs has been reported for Mediterranean wood pastures (Moreno and Pulido, 2009) and other semiarid (Yadav et al., 1997) and tropical (Angelsen and Kaimowitz, 2004) regions. On the contrary, and especially in mountain areas, land abandonment results in woody encroachment of SPs with important changes on their functioning and productivity (Asner et al., 2004; Mazzoleni et al., 2004; Archer, 2010). Ultimately, while the production of high-quality food (meat, milk, cheese) together with the recreational and cultural services are important drivers of the economic revitalization of SPs, there is an existing true challenge of integrating the duality of intensification (e.g. pasture improvement that could perform against tree regeneration) with the periodical woody encroachment, which can guarantee soil and tree regeneration and thus, the sustainability of the system.

1.1 Objectives

This review compiles and summarizes the following topics concerning the SPs of the world:

- Extension and geographical distribution;
- Commercial goods provided by SPs;
- Importance of the trees for the livestock production, analysed at three different levels:
  - the net effect of trees on the understorey pasture production,
  - the importance of trees and shrubs as forage, and
  - the potential benefit of the shading/sheltering on livestock production;
- Importance of SPs for biodiversity and regulating services highlighting the role played by trees for carbon sequestration and water quality;
- Main threats and challenges that currently SPs have to face across the world, followed by the presentation of innovative designs and management practices proposed and tested to overcome these challenges;
- Tentative agenda for future research to reinforce the role of the SP in the achievement of the millennium goals.

2 Silvopastoral systems (SPs)

Worldwide, the integration of livestock with forest lands is predominantly not done according to silvopastoral principles. Still, integration of livestock with trees represents a dominant land use system in both the tropical and temperate regions of the world alike (Jose et al., 2017). Livestock ranching can be viewed as another way of deforestation, because browsing and trampling frequently cancels opportunities for tree regeneration (Pulido et al., 2010) and because periodical fires have been a commonly associated management practice (Grove and Rackham, 2001). Indeed, SPs are frequently diffuse transitions from agriculture to forest lands (Silva-Pando et al., 2002a; Cubbage et al., 2012; Hartel and Plieninger, 2014). Nevertheless, in certain regions trees have been intentionally maintained and scattered trees are still a prominent element of many human-modified landscapes around the world (Gibbons et al., 2008). Unfortunately, there is still a lack of reliable maps showing distribution or any official statistics of their extension.

Zomer et al. (2014) using remote sensing data estimated that in 2010 around 43% of agricultural lands in the world had at least 10% of tree cover (1 billion ha), which is reported as a rough estimation of the agroforestry area. Unfortunately, the spatial resolution used (1 km^2) overestimates the extension of agroforestry. In their estimation any mosaic of treeless farm with interspaced woodlots results in an agroforestry pixel. For instance, for Europe den Herder et al. (2017) estimated 15.4 million ha of agroforestry (and this is already an overestimation of the deliberate agroforestry practised by farmers), while Zomer's estimation amounts to 110 million ha. In spite of the overestimation, Zomer's data inform us that the distribution of tree cover on agricultural land broadly followed bioclimatic zones – in particular aridity. High tree cover (>45%) was found in the humid regions such as Southeast Asia, Central America, Eastern South America and central and coastal West Africa. Tree cover was moderate (10–30%) in the majority of agricultural areas in South Asia, subhumid Africa, Central and Western Europe, Amazonian South America and Midwestern North America. At the other extreme are agricultural areas with low (<10%) tree cover such as Eastern China, Northwest India and the Punjab, Western Asia, the southern border of the Sahara, the northern prairies of North America and Southwest Australia (Zomer et al., 2016; Fig. 2). Unfortunately, Zomer's estimation does not discern between silvopastures and other agroforestry systems (e.g. silvoarable). Nair (2012) gives a rough estimation of 450 million ha of silvopasture in the world, mostly semiarid and subhumid lands in Africa, India and the Americas. Nevertheless, only for a few regions, sound estimation of coverage of different types of silvopastures are available. The following examples cover these concepts.

Wood pastures as land use are now historical or neglected in most parts of Europe, but still recognizable and widespread as encroached scrub or woodland formations. Exceptions are reindeer pastoral woodland with birch and pine in subarctic Europe (Jernsletten and Klokov, 2002), mountain summer pastures that extend into montane woodlands, and pastoral woodlands and scrublands in parts of Eastern Europe, the Mediterranean and the Balkans, where wood pastures to some extent retain their traditional usage (Bergmeier et al., 2010). Recently, Plieninger et al. (2015) estimated that SPs cover 20 million ha in Europe. Excluding ungrazed wood pastures, den Herder et al. (2017) reduced this estimation to ~15.1 million ha. These amount roughly to the

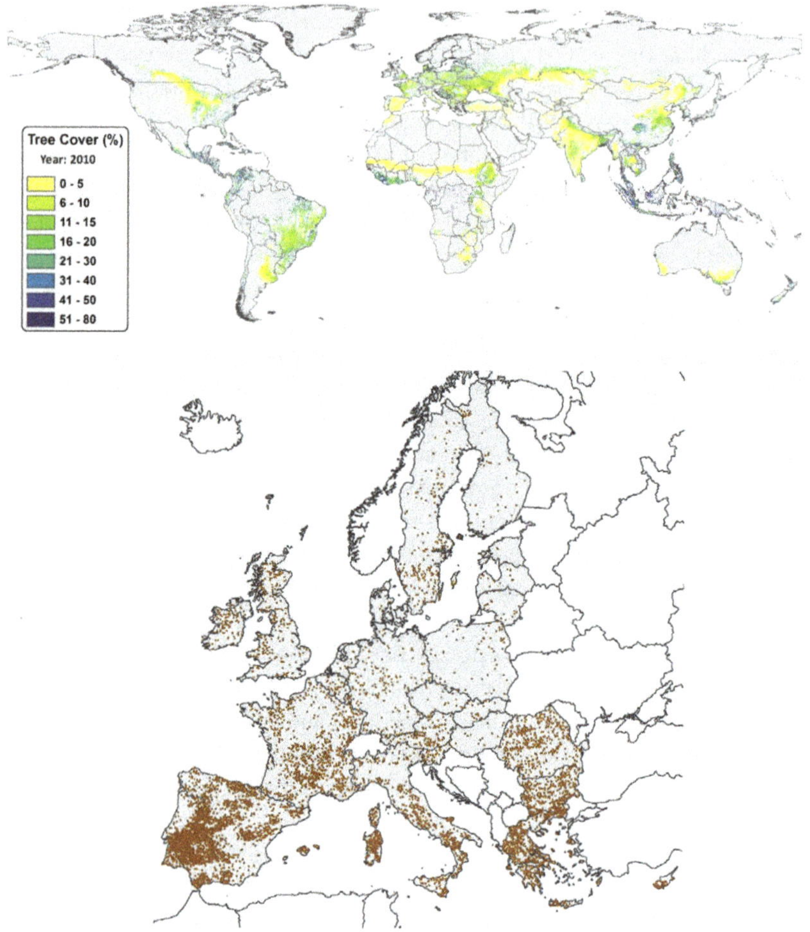

Figure 2 World map of agroforestry, defined by Zommer et al. (2016; above) by the presence of at least 10% of the tree cover in agricultural lands (Zommer et al., 2016; above). This map includes silvopastoral and silvoarable systems but also permanent crops (e.g. olive groves, fruit orchards, nut trees). European map of silvopastoral systems defined by the presence of the tree and grass covers and evidences of grazing by den Herder et al. (2017; below).

35% of the actually grazed lands in Europe (Fig. 2). The largest extent of SPs are found in the Mediterranean countries, are outstanding the well-known dehesas and montados in the Iberian Peninsula. Wood pastures with ancient trees are still present and in use in Eastern Europe (Romania and Hungary; ~15 000 ha; Moreno et al., 2018). Porqueddu et al. (2016) also reported the importance of SPs in other Mediterranean regions, such as 'WANA' region (Western Asia and Northern Africa), although no figures about the extension of SPs in these regions were reported.

Although SPs are the most common type of agroforestry across South America (Coelho, 2017; Murgueitio et al., 2011), no data about the extension are available. Espinales represents the most widespread agroforestry system of the unirrigated portions of the Mediterranean climate region of central Chile (about 2 million ha; Ovalle et al., 1990). In Argentina, while SP implementation using exotic tree species planted in natural grasslands (or with exotic pasture species) had reached ~140 000 ha by 2012, native forests under silvopastoral use is ~6.8 million ha with different degrees of management strategies (Peri et al., 2016a). In the United States alone, there are about 70 million ha of land where forests are accessed by livestock (Jose et al., 2017), but only 810 ha were adapted from 2008 to 2012 with the US assistance programme (Jose et al., 2018).

In South East Asia, up to 210 million ha under permanent tree crops and forests are used for animal production (Devendra and Ibrahim, 2004). In China, silvopastoralism is common both in semiarid regions of northwestern China and subtropical of southern China (Zou and Sandford, 1990). However, only 0.39 out of 45.2 million ha of agroforestry in China is practised as SPs (Huang et al., 1997). By contrast, in India, up to 70 million ha out of the 120 million ha devoted to grazing can be defined as SP (Faroda, 1997). Indeed, in these countries, only 6 million ha out of 67 million ha under forest are real forests, as most of them are opened/degraded by the pastoral use (Saxena and Singh, 1997). Many other examples document the importance of SPs across the world, although they do not provide information of the surface occupied by SPs (e.g. Donaghy et al., 2010 for central Queensland in Australia; Benavides et al., 2009 in New Zealand; Franzel et al., 2014 in Africa).

Most SPs in Africa and other developing regions of the world involve extensive open grazing by free-roaming animals under scattered natural stands of trees and shrubs mostly in semiarid to arid areas, as in the parklands of sub-Saharan Africa. More intensive grazing systems of silvopasture are practised in Latin America and Europe where animals are penned in barbwired parcels and grazing is regulated (Nair, 2012).

2.1 Guiding case study: Iberian Dehesa

The Mediterranean wooded pasturelands known as 'dehesa' in Spain and 'montado' in Portugal are agroforestry systems of high natural and cultural value that cover around 3.5 million ha of the southwestern Iberian Peninsula, where they are the main land use systems (Moreno et al., 2018) and form one of the largest agroforestry system in Europe (den Herder et al., 2017).

They are agrosilvopastoral systems originating from the clearing of evergreen woodlands where trees, native grasses, crops and livestock interact positively under specific management practices. The scattered tree layer is dominated by Mediterranean evergreen oaks (30–50 trees ha^{-1}, principal species *Quercus ilex* and *Q. suber*). Mature trees are regularly pruned with the aim of maximizing acorn production and providing browse in summer and winter, when the herbage production is low, and of providing wood fuel for household use and sale. Trees also provide shelter from heat in summer, prevent soil erosion and desertification, enhance the vegetation and structural

Figure 3 Images of the Iberian dehesa, probably the most studied SPs in the world. Image 1 shows the typical extension of dehesa landscapes. Images 2 and 3 show that the grass understorey grow better beneath than out of the canopies in winter and early summer, respectively. Images 4, 5 and 6 show pigs, sheep and cattle, the most common livestock species grazing in the dehesa. The last images also show trees pruned to feed cattle with browse and to provide firewood and charcoal.

complexity of the ecosystem, provide habitat and resources for many species, and are an important food resource for livestock, especially for pigs (Fig. 3).

The importance of dehesas rests on both environmental and socio-economic values. First, dehesas play a prominent role in the economy of rural areas in southwestern Spain where high-quality food derived from livestock is produced (Campos et al., 2013), notably the production of Iberian pigs that consume large amounts of acorns for fattening. In addition, dehesas have been valued at an international policy-making level for their biodiversity, aesthetic qualities and potential for tourism and recreation. Dehesas are included in the EU habitat directive as habitat with community-wide interest (Díaz et al., 2013).

Nevertheless, over the last few decades, dehesas and montados are facing several threats (Plieninger et al., 2015). Increased mechanization and increased stocking rates, together with the oversimplification of the management practices (notably a lack of livestock herding), have enlarged at least three sources of environmental degradation: (1) soil erosion rates due to changes in vegetation, soil properties and hydrological processes (Schnabel et al., 2013); (2) over-aged oak stands due to a prolonged lack of regeneration (Plieninger et al., 2010); and (3) loss of diversity at various spatial scales (Díaz et al., 2013). In this context, the sustainability of the dehesa system has been seriously questioned (Moreno and Pulido, 2009).

3 Importance of trees for livestock production

Although SPs are multifunctional systems, and provide a series of products, they are mainly used for livestock rearing. In this context, the importance of tree canopy for livestock needs to be analysed as a forage source, but also by the effects on pasture understorey and the shading and sheltering capacity.

3.1 Tree effects on pasture production

Between trees and pasture, competition for soil resources (nutrients and water) occurs, and trees reduce the light and water reaching the soil. However, they may smooth microclimate (reducing wind speed and evapotranspiration, and alleviating extreme temperatures) (Silva-Pando et al., 2002b; Karki and Goodman, 2015; Jose et al., 2017), and improve soil properties (reinforced organic matter, nutrient availability, water infiltration and holding capacity) (Young, 1997; Sinclair and Joshi, 2000; Jose, 2005). Karki and Goodman (2015) reported lower average values for all the measured microclimatic variables in SPs compared to open pasture (e.g. average air and soil (5 cm) temperatures in the silvopasture were 2.3°C and 2.1°C lower, on average). Jose et al. (2017) reported that the potential evapotranspiration was reduced by 70% under the canopy and the grass remained greener for a longer period of time. Cubera and

Moreno (2007) reported that soil bulk density and water-holding capacity was reduced and increased, respectively, by roughly 10% under trees compared to the open pastures.

The net effects of trees on pasture production are strongly context-dependent, but they are overall neutral to positive for deciduous tree species, and neutral to negative for evergreen ones, as recent meta-analysis (Rivest et al., 2013; Mazía et al., 2016), a comparative review (Benavides et al., 2009) and several empirical studies reveal (Table 1). Rozados-Lorenzo et al. (2007a) reported that pasture production was lower under conifer trees (*Pseudotsuga menziesii*, *Pinus pinaster*, *Pinus radiata*) than under broadleaved trees (*Betula alba*, *Quercus rubra*, *Castanea sativa*). In other studies, deciduous broadleaves show a rather neutral effect (e.g. Teklehaimanot et al., 2002; Sigurdsson et al., 2005), and evergreen broadleaves show divergent results in function on soil and climate (e.g. *Quercus ilex*; Gea-Izquierdo et al., 2009). Since evergreen trees intercept light and rainfall, and use soil resources, all year round, they leave less resources for pasture growth than deciduous trees (Moreno et al., 2013).

There is still, nevertheless, uncertainty under which ecological conditions and plant traits, the tree impacts on pasture understorey change from net competitive to positive (Rivest et al., 2013; Blaser et al., 2013). Forage understorey plants reach a light saturation point at approximately 50% and 85% of full sunlight (cool-season (C3) and warm-season (C4), respectively) (Gardner et al., 2017; Jose et al., 2017). C3 will not have their growth or yield adversely affected by certain degrees of tree shade, while with C4 grasses, tree shade could affect more negatively pasture production (Lin et al., 1999; Broom et al., 2013). Gea-Izquierdo et al. (2009) reported a positive effect of oak canopy on dehesa pasture yield in average climatic years, but the interaction changed with increasing abiotic water stress. In that study, during a dry year, the higher fertility beneath the canopy could not be used for plant growth because of the lack of moisture and the effect of the oak canopy was neutral. The decreased positive effect of trees with aridity in Spanish dehesas indicates that competition for soil water is an outstanding factor in the balance of positive and negative effects of trees on pasture. This kind of exception to the soil gradient hypothesis is common in Mediterranean ecosystems (Moreno, 2008; Maestre et al., 2006), especially when the abiotic stress gradient is driven by a resource such as soil water in arid and semiarid ecosystems.

For livestock, the extension of the active growing period of pasture can be more important than the overall pasture yield (Moreno et al., 2013). A well-designed SP can potentially extend the grazing season period on both ends of the growing season (Kallenbach, 2009; Jose et al., 2017; Fig. 3). Forage quality can also be affected by the presence of tree canopy. For instance, Kallenbach et al. (2006) found that the crude protein content was increased but not the

Table 1 Compilation of cases that report the net effect of trees on pasture understorey production

Reference	Location	Tree group	Herb layer	Tree effect
Abdallah et al. (2008)	Tunisia	N_2-fixing	Natural	Positive to neutral
Abule et al. (2005)	Ethiopia	N_2-fixing	Natural	Positive to neutral
Ash (1998)	Australia	Evergreen	Natural	Neutral
Belsky et al. (1993)	Kenya	N_2-fixing/evergreen	Natural	Positive
Cubera et al. (2009)	Portugal	Evergreen	Natural	Negative
de Mendonca et al. (2017)	Brazil	N_2-fixing	Improved	Negative
Durr and Rangel (2002)	Australia	N_2-fixing	Natural	Positive
Douglas et al. (2006)	New Zealand	Deciduous	Natural/improved	Negative
East and Felker (1993)	USA	N_2-fixing	Natural	Positive
Gea-Izquierdo et al. (2009)	Spain	Evergreen	Natural	Positive
Guevara-Escobar et al. (2007)	New Zealand	Deciduous	Natural	Negative
Jackson and Ash (1998)	Australia	Evergreen (*Eucalyptus*)	Natural	Negative to neutral
Kallenbach et al. (2006)	USA	Mix forest	Sown	Negative
Moreno (2008)	Spain	Evergreen	Natural	Positive
Nordenstahl et al. (2011)	Argentina	Deciduous	Improved	Negative
Paciullo et al. (2010)	Brazil	N_2-fixing/evergreen	Improved	Negative
Rolo et al. (2012)	Spain	Evergreen	Natural	Negative
Rozados-Lorenzo et al. (2007a,b)	Spain	Coniferous	Natural	Negative
Rusch et al. (2014)	Nicaragua	Deciduous	Natural	Negative
Sánchez-Jardón et al. (2010)	Chile	Deciduous	Natural	Negative
Seddaiu et al. (2018)	Italy	Evergreen	Natural	Negative
Serrano et al. (2018)	Portugal	Evergreen	Natural	Negative
Silva-Pando et al. (2002b)	Spain	Coniferous	Improved	Negative
Singh et al. (2008)	India	N_2-fixing/evergreen	Natural	Negative

Only studies that measured herbaceous above-ground biomass beneath the canopy of isolated trees and away from their influence (i.e. open areas) are reported. When a study included different temporal measurements of herbaceous biomass, only the result at the peak biomass is reported. Note that the compilation is not exhaustive.

acid detergent and neutral detergent fibres, in the vicinity of pine and walnut trees in the United States. Rolo et al. (2015) reported the increase of nutrient content (N, P, K and CA) of pasture beneath trees of the Iberian dehesas. Frost et al. (1991), East and Felker (1993) and Jackson and Ash (1998) reported similar results for oaks in California, honey mesquites in Texas and *Eucalyptus* in Australia, respectively.

There is a need for widening the choice of available grass and legume species and cultivars best adapted to shade and specifically selected for SPs. The genetic base of currently sown pastures is very narrow: more than three quarters of the grass cultivars registered in the European Union are of just six species, and *Lolium perenne* and *L. multiflorum* account for more than 80% of the forage grass seed sold in the EU (Batello et al., 2008). They were all selected under full sunlight conditions.

3.2 Trees as fodder resource

Fodder trees and shrubs (FTS) are valued because of their capacity to be a source of forage for ruminants especially during seasons where other plant species with higher nutritional value are not present (González-Hernández and Silva-Pando, 1999; Vandermeulen et al., 2018). Besides, many FTS are rich in protein, energy, vitamins and minerals and frequently are recommended as alternatives to commercial concentrates (Balehegn, 2017; Table 2). Traditionally, trees were pruned under different schemes to feed livestock in SPs (Figs. 3 and 4). In some SPs, trees constitute the main source of fodder as in the Iranian Galazarini (Valipour et al., 2014), where every 3 years all the branches are pruned by the end of the summer and stored in the canopy of the trees, to feed livestock later in winter (Fig. 4e).

Traditional FTS are nowadays rarely managed in temperate countries, and pruning practices to feed animals are now marginal, although the use of woody species is being promoted again on grasslands (some examples are given below) through environment-friendly policies due to the ecosystem services they provide such as carbon sequestration, control of soil erosion, limitation of airborne pollutants and biodiversity conservation (Vandermeulen et al., 2018; Hernandez-Morcillo et al., 2018). By contrast, indigenous FTS are still quite important in tropical countries, where they are also important complements of the agro-biodiversity (Le Houérou, 2000). For instance, over 40 000 ha have been planted in northern Chile, mainly in the Coquimbo Region to supplement the shortage of fodder in summer (Perret and Mora, 2001, in Peri et al., 2016a). More than 200 000 smallholders plant trees to feed dairy cows in the highlands of Eastern Africa (Franzel et al., 2014). Temperate species such as *Salix* spp. and *Populus* spp. are fed to sheep and cattle in fodder blocks or by pruning trees in New Zealand, *Leucaena leucocepahala* and *Desmanthus* spp. browsing is common in Australia, *Fraxinus* spp. or *Corylus avellana* in hedgerows supply forage to livestock in Belgium (Vandermeulen et al., 2018), *Fraxinus excelsior, Alnus glutinosa* and *Salix viminalis* in the Netherlands (Luske and van Eekeren, 2018), and oaks are periodically pruned to feed cattle in winter in the Iberian Peninsula (Moreno et al., 2018). Recently, tree legumes have been claimed as an underexploited resource in warm climate SPs (Dubeux Junior et al., 2017). For Iberian dehesas, fodder banks with leguminous shrubs (*Medicago arborea, Ceratonia siliquoa* (carob), *Acacia* sp.), but also with other protein-rich shrub species such as mulberry (*Morus* spp.) and *Atriplex* sp., are currently viewed as a promising way to overcome seasonal shortages of forage (Moreno et al., 2018).

Although trees can reduce the pasture biomass production, the total metabolizable energy produced at plot scale can increase in the presence of trees because of the pastoral value of fruit (e.g. acorns), tree leaves and shrubs (López-Díaz et al., 2015). This is often true especially in the case of deciduous tree species, as their leaves on the ground can be a significant nutritional resource. For example, red alder leaves can provide a significant source of digestible protein during fall in SPs of Northwestern United States, when availability of nutrients in most forage species is generally declining as both green tree leaves and senescent contain a significant amount of protein (González-Hernández et al., 2000). Unlike many of the 'improved' introduced FTS species, indigenous ones are well adapted to local conditions and are multipurpose and provide benefits and services such as food, fibre, shade, soil improvement and conservation, timber, firewood and live fences (Balehegn, 2017). As this author points out, indigenous FTS are preferred over the improved forage plants for being of low cost, available and accessible to local communities, adaptable to local conditions and requiring little or no management input. So, re-incorporating indigenous FTS into SPs will not only help to mitigate the livestock feed problem, but also contribute to environmental rehabilitation, stability and improve livelihoods (Murgueitio et al., 2011). Key challenges constraining the uptake of fodder trees include limited species appropriate to different agroecological zones, shortages in seed and that farmers lack knowledge and skills needed to grow them (Franzel et al., 2014). Nowadays, ensiling and pelleting techniques are being developed as a way to store browse forage. As the renewed interest in using shrubs and trees to feed ruminants is growing recently, especially in temperate regions, additional research about the optimal introduction of this resource within systems is needed (Vandermeulen et al., 2018).

Nutraceutical effects of browse on rumen digestion and parasite control have also been documented across different plant species and regions (Osoro et al., 2017; Vandermeulen et al., 2018). Moreno-Gonzalo et al. (2012) found in Spain that supplementation with tannin-rich shrubs in grazing goats significantly reduced the level of gastrointestinal nematode egg excretion and that the

Table 2 Changes in forage availability and quality in a monoculture of a pasture plant (*Cynodon plectostachyus*) and in an SP combining this pasture plant with a leguminous shrub (*Leucaena leucocephala*)

Variable (per year)	Monoculture of *Cynodon plectostachyus*	SP of *Leucaena leucocephala* (10 000 ha^{-1}) with *Cynodon plectostachyus*	Difference (%)
Biomass (Mg ha^{-1})	23.2	29.9	+29
Crude protein (Mg ha^{-1})	2.5	4.1	+64
Metabolizable energy (Mcal ha^{-1})	56.9	70.2	+23
Calcium (kg ha^{-1})	83.2	142.3	+71
Phosphorus (kg ha^{-1})	74.0	88.8	+20

Source: Adapted from Broom et al. (2013).

amount of tannins consumed by the goats did not seem to be associated with anti-nutritional effects which eventually resulted in a better animal performance in the animals incorporating these shrubs in their diet. Also Galina-Hidalgo

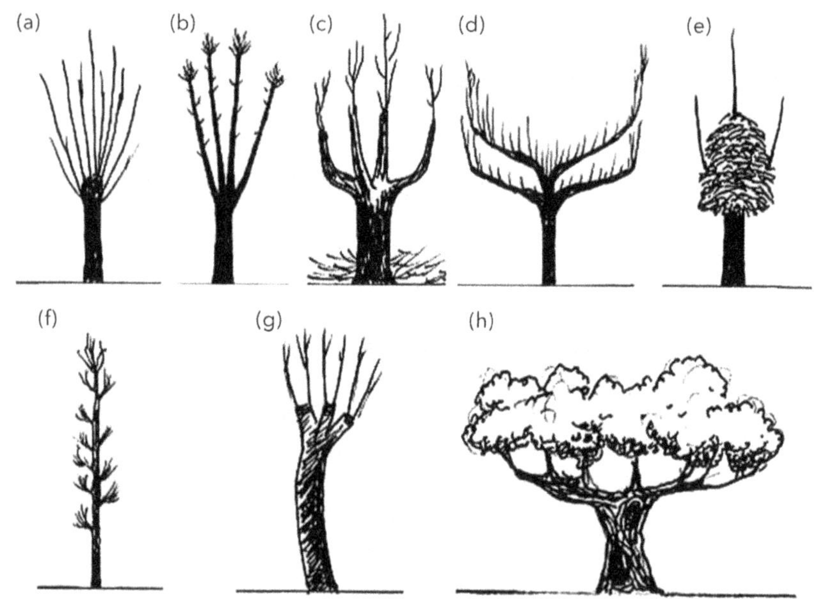

Figure 4 Traditional management of fodder trees in Greece: (a) pollarded tree of any species, (b, c) pollarded oak trees, (d) pollarded mulberry tree, (e) pollarded tree for storage of fodder, (f) shredded oak tree, (g) grafted tree for fruit production and (h) lopped olive tree for olives and fodder production. Source: Adapted from Papanastasis et al. (2009).

et al. (2016) working with tannin-rich shrubs in Brazil also concluded that the mixed SP is a promising system controlling parasitic nematode infection in bovines. Despite their beneficial traits, these phenolic compounds can also act as nutritional constraints in various degrees depending on the browsing animal species (Starkey et al., 1999), as those browsers adapted to rich tannin diets (goat and deer for example) would have advantages over grazers (Osoro et al., 2017). A wide range of studies have shown that tannins are abundant in many shrubs and trees potentially present in livestock diets (González-Hernández et al., 2003), and therefore both features enhance the importance of ligneous plant species present in SPs. Other potential benefits of fodder trees are summarized in Fig. 5.

3.3 Shading and sheltering effect

One rarely explored advantage of wood pastures is the reduction of fodder needs of livestock thanks to tree shelter under unfavourable climatic conditions (heat or cold stress; Higgins and Dodd, 1989; Oosterbaan and Kuiters, 2009; Garrett et al., 2000). Animals, which are unable to shelter from the direct sun during the heat of the day (or chilly wind), will have to expend energy to deal with their discomfort and/or reduce their feeding activity. Their productivity will decline in proportion to the time spent under these unfavourable conditions. Increased livestock production by shading and/or sheltering is often promoted in SPs as one of the benefits of having trees. Indeed, in some SPs trees are mostly maintained/planted for livestock shading/sheltering (Orefice et al., 2017). However, there is very little published research available to either verify or disprove this widely held belief (Sharrow, 2000).

Ruminants' sweating mechanisms are poor and they rely on respiration to cool themselves (De Rensis and Scaramuzzi, 2003). Cattle without shade have a physiological (i.e. higher respiration rate) and behavioural (i.e. less active) stress response to heat that negatively affects productivity (Jose et al., 2017). Schütz et al. (2010) reported that respiration rate was higher when dairy cows had less shade available (62, 57 and 51 breaths min^{-1} for no shade, 2.4 and 9.6 m^2 shade/cow, respectively). Roman-Ponce et al. (1977) also reported improved respiration ($54 < 82$ breaths min^{-1}) and rectal temperatures ($38.9°C < 39.4°C$) from providing shade structure to dairy cows, with positive consequences for reproduction and lactation (16.6 and 15.0 kg day^{-1} for shade and no shade). Betteridge et al. (2012) monitored the body temperature and quantified grazing time of cows in paddocks with and without space-planted willows that provide shade in summer. He found that shade reduced the temperature of cow backs from 40.8°C to 36.8°C, and shaded cows grazed for 30–40 min longer than 'no shade' cows (7% increase in grazing time). Mitlohner et al. (2001) found that cattle provided with shade reached their target bodyweight 20 days earlier than those not afforded shade. Kallenbach (2009) found that by integrating silvopasture into traditional pasture systems, cow–calf producers could lower winter feeding costs, decrease calving problems and produce heavier calves.

All together, the net balance of tree effects on pasture understorey, the forage value of trees and the metabolic energy saved by tree shading and sheltering determine that the presence of trees in the SPs has profound consequences for the productivity and management of these agropastoral systems. However, there remains a need to develop systematic and scientific comparisons of the foraging potential of SPs against open pastures under different edapho-climatic conditions and vegetation structures. Nonetheless, some nice examples of the advantages of silvopastures vs monopastures have been experimentally documented. For instance, Paciullo et al. (2011) showed that the annual body weight gain was higher in SPs (*Brachiaria decumbens* combined with four tree species planted in Brazil) than in the respective monopastures, and they concluded that the SP was more efficient for rearing dairy heifers based on the higher body weight gain per heifer and per area. Yamamoto et al. (2007) showed similar values for dual-purpose cattle production in Nicaragua by maintaining useful naturally regenerated trees in grazing areas with relatively moderate tree density (20%) and low stocking rates and limited amounts of supplemental fodder.

4 Trees for biodiversity and ecosystem services

There is a general acknowledgement of the important contribution of the tree component in agroforestry systems to multiple ecosystem services such as C sequestration, greenhouse gas mitigation, improvement of water quality, control of soil erosion, providing shade, windbreaks and habitat for pollinators. Tree layer is also known to support biological N_2-fixation, nutrient cycling, soil fertility and soil health, photosynthesis and primary productivity, and provide cultural ecosystem services and biodiversity conservation. However, the full potential of SPs is still to be acknowledged, and more examples are still needed. Here we show some examples well documented in the literature (biodiversity, water quality, carbon sequestration and GHG emission), mostly for European cases, and intend to identify some gaps in knowledge.

4.1 Carbon sequestration

During the last 20 years, there has been an increasing appreciation of the extent of the negative externalities associated with livestock production, either as grazed ruminants or animals such as pigs and poultry (Burgess and Morris, 2009). The key negative externalities include methane production by ruminants, ammonia and nitrous oxide production by all forms of livestock production

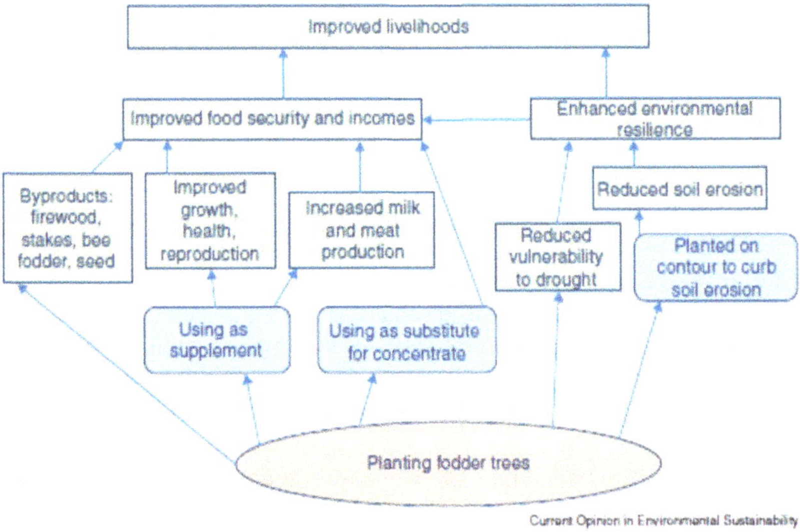

Figure 5 Principal ways that fodder trees contribute to improved food security, incomes and livelihoods. Source: Adapted from Franzel et al. (2014).

and emissions related to feed production. Recently it was estimated that Europe's meat and dairy consumption was responsible for 14% of the total CO_2 emissions in the EU (Weidema et al., 2008), and livestock production is seen as a key driver of global land use changes, with resulting impacts on climate change and biodiversity. The integration of trees and shrubs with livestock is considered a farming practice that could store carbon as an offset for methane and nitrous oxide production (Hristov et al., 2013; Broom et al., 2013).

The magnitude of C sequestration and climate change mitigation in SPs will vary depending on the nature and level of management of the systems, with rather low levels in the extensive system to relatively high levels in the intensively managed systems (Nair, 2012). The trees in each system enable the storage of greater amounts of above-ground carbon than the comparable agricultural system. Below ground the effect of trees in pasture on carbon sequestration is mixed. Upson et al. (2016) reported that trees on grazing land may reduce carbon in the first 0–10 cm of soil, whereas topsoil organic carbon content under old hedgerows can be up to 2.5 times higher than that in the adjacent crop field (Walter, 2011). Despite some exceptions, the contribution of trees to the build-up of the soil carbon pool has been repeatedly reported (Lorenz and Lal, 2014) and comparison among treeless pastures and SPs show that the C pool is significantly higher in the latter (Nair, 2012). Howlett et al. (2011a,b) reported good examples for Atlantic and Mediterranean silvopastoralism where soil C was up to 30% higher in wood pastures compared to neighbouring treeless pastures, and besides C was stored on average deeper and thereby for longer in the wood pastures. The importance of SPs for storing soil carbon has been also noted for other hedgerows and wood pastures in Europe (e.g. Howlett et al., 2011a,b; Seddaiu et al., 2013; Francaviglia et al., 2014; Zianis et al., 2017). Udawatta and Jose (2011) concluded that SPs had the greatest potential among all agroforestry practices to sequester C in the United States (6.1 Mg C ha^{-1} $year^{-1}$). Verchot et al. (2005) estimated that the potential for carbon sequestration on SPs across different tropical regions ranged from roughly 20 to 200 Mg C ha^{-1} (50 years). Grazing has been also identified as an efficient practice to reduce the wildfire risk, given that grazed grasslands, wood pastures and other rangelands are less prone to burn than forest and shrublands under seasonal dry climate (Etienne et al., 1996; Pardini, 2002; Rigueiro-Rodríguez et al., 2009; Casals et al., 2009; Ruiz-Mirazo et al., 2011; Franca et al., 2012). Hence, carbon stored in the grasslands, especially if they are regularly grazed, can retain carbon far longer than woody ecosystems in wildfire-prone regions.

The increasing intensification of many SPs, especially those in arid regions, can produce a worrying loss of soil quality (Coelho et al., 2004), and new management practices that favour the increment of soil organic matter are now needed. Three kinds of management practices have been shown as useful for this purpose: ramial wood (Dodelin et al., 2007), biomass charcoal (Kimetu et al., 2008) and legume-rich forages (Teixeira et al., 2015). Long-term effects of improved pasture establishment in Portuguese montados showed higher organic C than those under unmanaged pasture, and the effects were stronger in the presence of oak trees (Gómez-Rey et al., 2012). The opportunity to produce (or use the excess of) wood biomass in SPs as a local or in-farm source of energy/carbon has gained recent interest. Incorporating life cycle assessments that compute C cycle and overall C sequestration/emission for agrarian systems and products is also needed, and presumably would give a new momentum to SPs (Upson and Burgess, 2012). Also, the use of woody plants as tannin-rich forages is gaining interest as tannins (especially condensed tannins) decreased methane production (Piluzza et al., 2014).

4.2 SPs to improve water quality

Water quality degradation results from an inadequate use of fertilizer that promotes the loss of nutrients, unrestrained farm effluents (e.g. livestock wastes), the use of pesticides such as sheep-dipping chemicals, and bacterial and protozoan infections that come from livestock manures and slurries. There has been a general uncoupling of nutrient cycles, with frequent problems of unacceptable high nutrient deposition or uneven spatial distribution (Hooda et al., 2000). The increased use of all these practices since the second half of the twentieth century, together with the degradation of native forest to create agricultural lands, particularly in the tropics, has led to a global degradation of water quality (Novotny, 1999). Maintaining, therefore, low diffuse pollution levels from agricultural practices is crucial for keeping an adequate quality of water. Nevertheless, actions taken to reduce diffuse pollution to water impose important costs to livestock farms (Fezzi et al., 2008). For example the societal costs of 0–20 and 30–70 kg N ha^{-1} $year^{-1}$ from extensive and intensive grassland respectively have been reported to be 100 and 400 € ha^{-1} $year^{-1}$ (Osterburg et al., 2007; Matzdorf et al., 2010).

Planting trees in grasslands has been shown by a number of studies to be beneficial for the preservation of water quality (Jose, 2005). Different studies have shown the capacity of deep rooting trees to capture nutrients from deep layer that are no longer utilizable by herbaceous plants, and thus reduce nutrient leaching in SPs (e.g. Bambo et al. (2009) and López-Díaz et al. (2011) for N; Nair et al. (2007) and Blazier et al. (2008) for P). In soil nitrate leaching, comparing different land use systems in north Florida, Bambo et al. (2009) reported that silvopastures were the most efficient system in reducing nitrate leaching at 1.2 m depth as compared with forest and grasslands. Similarly, SPs have a lower chance of P losses than in treeless areas (Michel et al., 2007). However, the potential of silvopastures to reduced faecal coliforms has been shown to be limited (Boyer and Neel, 2010). The increased proliferation of tree roots at deeper soil layers is commonly argued as being the main mechanism by which silvopastures are efficient systems to reduce N leaching (López-Díaz et al., 2011). A decreased nitrate leaching was also demonstrated in France by increasing the hedgerow/tree density (Benhamou et al., 2013) and planting energy trees in free-range pig farms (Jørgensen et al., 2018). Recent research showed integrating pigs in a willow/miscanthus plantation results in a much smaller risk for N leaching than typically seen on grassland with the same stocking rate (Sørensen, 2010) and that the pigs only caused limited damage to the trees (Horsted et al., 2012). The efficiency of trees to control the wash-out of veterinary antibiotics in silvopastures has been also reported, and explained by the increased enzymatic activity stimulated by the tree rhizosphere (Chu et al., 2010; Lin et al., 2010; Jose et al., 2017). The optimal tree cover needed for an efficient control of chemical leaching is however unknown.

The complementary use of soil resources between herbaceous species, that use the uppermost soil layer, and the trees, that use the deeper soil layers, promote a higher utilization of nutrients in the whole soil profile (Rolo and Moreno, 2012). The increased usage of deep water promoted by trees in SPs has other beneficial effects. For instance, Alconada Magliano et al. (2009) reported that SPs can help to avoid the environmental deterioration resulting from the effect of floods. Moreover, the combination of grazing and improved pastures, such as legume sowing could be an efficient means to minimize negative interactions of understorey vegetation, to limit mineral fertilization usage and, at the same time, avoid nitrate losses (López-Díaz et al., 2019).

On the other hand, shade and shelter may result in increased localized pollution (critical source areas) where overland flow carries nutrients and pollutants to waterways and lakes (Betteridge et al., 2010). These authors reported that 50% of urination events were returned to between 6% and 10% of the paddock areas (i.e. campsites) in steep hill country. While shade-providing trees may create critical source areas, where they are spaced across the slopes of hill country pastures, these trees will attract cows away from stream banks and/or riparian zones. Schütz et al. (2010) found that dairy cows spent more time around the water trough when little or no shade was provided (11%, 5% and 2% of observations within 4.5 m of water trough for no shade, 2.4 and 9.6 m^2 shade/cow). Thus, the presence of scattered trees also contributes to better manure spread in the paddocks (Bestman et al., 2014; Jørgensen et al., 2018). There is still much need of experimental studies to provide an increased understanding of the adequate management of tree cover and grazing intensity that could help to improve the beneficial effect of SPs on water quality.

4.3 SPs promote biodiversity conservation

Agricultural intensification has led to a widespread decline in farmland biodiversity across many different taxa (Benton et al., 2003), but farmland also hosts many species that depend on appropriate agricultural management (Kleijn and Sutherland, 2003). The positive role of SPs on biodiversity conservation has been highlighted in the literature (García et al., 2013; Torralba et al., 2016; Peri et al., 2016b). SPs are characterized by having a high heterogeneity both in space and time. This heterogeneity provides for multiple resource gradients at different scales, from local to landscape, that is key to maintaining high diversity values across trophic levels. A first source of heterogeneity is promoted by the presence of trees at the local scale. Trees actively modify the availability of resources (e.g. nutrient, water and light) with marked consequences in the composition of the herbaceous communities of underlying grasslands (Abdallah and Chaieb, 2012). When grasslands are transformed into SPs by the plantation of trees, there is a surge in biodiversity levels because, for example, trees attract a number of invertebrates which constitute the food supply of birds (Mcadam et al., 2007). Indeed, SPs can be a suitable habitat for forest birds as compared to treeless habitats (McDermott and Rodewald, 2014). This cascading effect is particularly important when ancient trees are present. Large trees are keystone elements because, among other things, they harbour multiple niche opportunities for different taxa (Manning et al., 2006). In ancient wood pastures of central Romania, the presence of large trees has been identified as the most important factor for the occurrence of the endangered tree pipit (*Anthus trivialis*) (Moga et al., 2009). Large trees usually bear cavities which are used as nesting sites for bird species that cannot create their own breeding cavities (Robles et al., 2011). Ramírez-Hernández et al. (2014) reported a positive effect of old mature trees on the diversity of saproxylic species in Iberian dehesas.

The presence of large trees is a key feature of ancient wood pastures, particularly those located in the old world (Plieninger et al., 2015). In these traditionally managed SPs, the interplay among large trees, treeless areas and marginal low-quality habitats determine to a great extent the occurrence of high biodiversity values. Santos-Reis and Correia (1999) found 264 fungi, 75 bryophytes, 304 vascular plants and 121 vertebrate species in a 220 ha Portuguese wood pasture. In a boreal wood pasture, even rocks and woody debris exposed to the effect of grazers, or lack of it, are a suitable habitat for the establishment of rare non-vascular species, such as certain bryophytes (Oldén and Halme, 2016). In Iberian dehesas, high dissimilarity values in species occurrence among habitats support the hypothesis of this 'habitat-mix' effect (Díaz et al., 2013; Moreno et al., 2016). Moreover, red listed species such as Iberian lynx, Iberian imperial eagle, black vulture and the black stork have their core distributional range in Iberian dehesas and montados (Díaz et al., 2013). Despite the anthropogenic origin of the Iberian wood pastures, their unique features have allowed them to be included under the protection of the European Habitat Directive (Díaz and Pulido, 2009). Díaz (2008) and Marañón et al. (2009) compiled a wide number of studies that reported a noticeably high diversity of vascular plants, birds, mammals, lizards and butterflies for Iberian dehesas compared to other adjacent land uses.

In SPs where a traditional structural complexity is concomitant, such as European wood pastures, tree gradients are also an important determinant of biodiversity levels. Low to medium tree densities have been shown to maximize biodiversity values (e.g. 30–60 trees ha^{-1} for Iberian dehesas; Díaz et al., 2013) due to the addition of forest generalist and open field species. Bergmeier et al. (2010) reported high richness of vascular plants, birds, snails and beetles for different European SPs. Hungarian and Romanian wood pastures are especially rich in fungi, epiphytes, macroinvertebrates, bats and birds. In an ancient 133 ha oak wood pasture in Romania, there were 476 species of vascular plants, 121 species of macromycetes, 281 species of Lepidoptera, 40 species of xylophagous beetles, 27 species of nesting birds and 38 species of mammals (Hartel et al., 2013). An increase in invertebrate species and numbers was observed when moving from open grassland to agroforestry conditions for carabid beetles in Northern Ireland (Cuthbertson and McAdam, 1996) and for four arthropod groups in Scotland (Dennis et al., 1996). Rolo et al. (2016) reported that the number of herbaceous species peaked at intermediate values of woody cover in Iberian dehesas (Fig. 6). In Swedish wood pastures, a tree density of around 50 trees ha^{-1} maximized bird diversity (Jakobsson and Lindborg, 2017). Nevertheless, this pattern is not unique to ancient wood pastures. For instance, in grazed forests of Chile's Aysén region, low to medium tree density (~40 trees ha^{-1}) allowed a better preservation of native species than open areas (Sánchez-Jardón et al., 2010). Gillet et al. (1999) determined an optimum of plant species richness at 30% of tree cover for SPs of the Swiss Jura mountains. This pattern accords with the landscape complexity hypothesis that suggests landscapes with intermediate complexity have higher conservation potential than clearer (treeless) or more complex landscapes (Tscharntke et al., 2012).

The clearance of native forest usually results in the loss of biodiversity. This is especially relevant in tropical areas where the pressure to create new agricultural areas has resulted in the loss of large areas of native forest in recent decades. Nevertheless, in this degraded scenario, SPs can help to halt the loss of biodiversity in a number of ways. On the one hand, keeping large trees embedded within the pasture matrix can act as keystone elements and help to conserve higher biodiversity levels than treeless areas. On the other hand, degraded treeless areas that are the result of the combined effect of the loss of the original forest and the intensification of livestock grazing can be restored by planting low to intermediate densities of trees (Villanueva-Partida et al., 2016). Besides tree-based gradients that increase the spatial heterogeneity at plot scale (Moreno et al., 2013), trees are essential sources of food and refuge for many species. Indeed, they have a disproportionate value for different taxa, as reported by Fischer et al. (2010) for birds and bats in an Australian livestock grazing landscape. Compared to treeless sites, bird richness doubled with the presence of the first tree; bat richness tripled with the presence of 3–5 trees (2-ha sites; n = 108 sites of 33 farms); and bat activity increased by a factor of 100 with the presence of 3–5 trees.

Similarly to tree density gradients, intensification gradients, mainly due to the effect of grazers, can have important consequences in the overall conservation potential of SPs. Grazers are an additional source of spatial heterogeneity that affects grassland dynamics by introducing a mechanical disturbance, changing seed dispersal patterns and altering soil fertility (Pittarello et al., 2016; Probo et al., 2016). Grazing intensification affects the functional diversity of the herbaceous communities with important consequences for nutrient cycling and the provision of other ecosystem services (Chillo et al., 2018). At the other extreme, gradual abandonment of grazing in SPs has been linked to marked consequences on grassland biodiversity and soil properties (Peco et al., 2006). The potential of SPs to help in the conservation of biodiversity in agricultural lands will depend ultimately in the proper management of its inherent complexity (Silva-Pando et al., 2002a). Managing the complex interaction among its tree, grassland and animal component is crucial to maintaining and promoting the conservation potential of SPs. Promoting the presence of large trees while keeping a mix of habitats of intermediate complexity seems to be a feasible option to reconcile agricultural production and biodiversity conservation in SPs.

In a recent study conducted in Iberian dehesas, Moreno et al. (2016) reported higher species richness for vascular plants (primary producers), bees (pollinators), spiders (depredators) and earthworms (decomposers) in wood pastures than in adjacent monopastures. Moreover, there was very high β diversity (spatial heterogeneity) among different types of pasture within farms. While monopasture plots shared a high number of species among them, plots with wood pastures exhibited very different species assemblages. Finally, although wood pastures and monopastures displayed high numbers of species (alpha diversity), the fine mosaic of habitats within farms, including some marginal habitats and linear features (common in many SPs) played a very significant role in terms of species richness at farm-level and landscape-level (γ diversity). Indeed, habitat heterogeneity at multiple spatial scales has been revealed as key for biodiversity conservation (Benton et al., 2003; Concepción et al., 2012).

5 Designing and managing silvopastures

Modern intensive, but low-input, SPs (ISPs) that are seen today throughout the world are much simplified from their natural analogues (Jose et al., 2017; Murgueitio et al., 2016). In the tropics, ISPs are a form of agroforestry for animal production that integrates fodder shrubs planted at high densities (more than 10000 plants ha^{-1}), intercropped with improved, highly productive pastures and timber trees planted in east-west lines to minimize shading, all combined in a system that can be directly grazed by livestock (Murgueitio and Solorio, 2008). In Northern Australia, *Leucaena* hedgerows are sown with grasses to form highly productive grass-legume grazing systems for cattle (Dalzell et al., 2006; Shelton, 2009). Approximately 150000 ha of these sustainable *Leucaena* pastures existed by 2006, some of which have remained productive over the past 30 years.

Under conditions of low population density, farmers appear more ready to adopt passive systems of protecting useful species found in their fields and forests than to engage in active afforestation programmes (Jose et al., 2017). Traditionally, extensive SPs result from a simplification, in structure and species richness, of native forests, and are attained by tree clearance, elimination of shrubs and favouring grasses by means of grazing and occasionally forage sown. Farmer interest to expand the agricultural land base by transforming forests into silvopastures still persists in many regions (e.g. Northeast United States; Orefice et al., 2016). These authors showed that silvopasture in that region outperformed open pasture and thinned forest (19-6 m^2 h^{-1} of basal

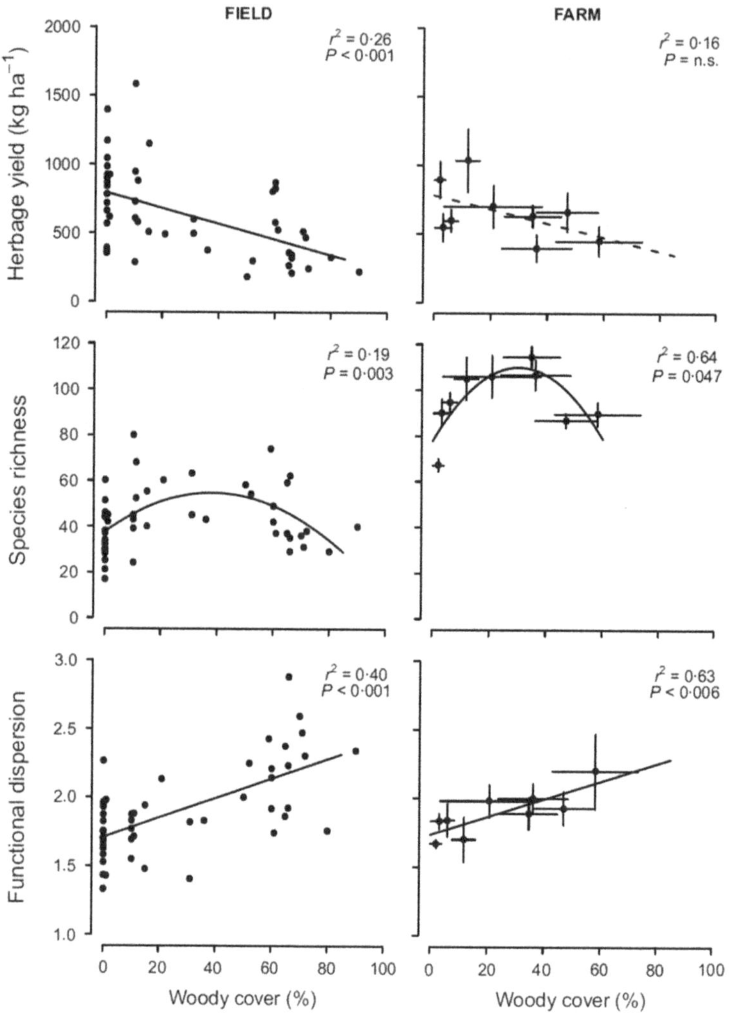

Figure 6 Relationship between herbage yield (kg ha^{-1}), species richness and functional dispersion against woody cover (% of shrubs and trees) at the field (left column) and farm (right column) scale. Error bars at the farm scale depict standard errors. Dashed lines indicate non-significant relationships between variables. Source: Adapted from Rolo et al. (2016).

area) treatments in terms of both internal rate of return and net present value per ha.

These semi-natural SPs are maintained by a balance between divergent ecological processes such as grazing pressure and tree regeneration (Buttler et al., 2009). The coexistence of patches of pastures, woodlands and/or isolated trees is therefore a result of an unstable equilibrium, which can lead either to closed forests or to open pastures. By increasing grazing pressure, a decline in palatable perennial herbaceous species and a lack of tree regeneration is sometimes followed by a complete disappearance of vegetation and subsequent soil erosion (Chaideftou et al., 2011; Moreno and Pulido, 2009). Among the components of wood pasture management, limiting the grazing pressure, choosing the grazing regime and allowing for time and space gaps between grazing activities are relevant practices for ensuring tree regeneration while halting the encroachment of dense shrub cover (Van Uytvanck et al., 2008; Plieninger et al., 2015).

Although in the last two decades an intense debate has been developed on the sustainability of SPs in view of the lack of tree regeneration and stagnation of tree population (Smit et al., 2005; McEvoy et al., 2006; Mayer et al., 2009; Moreno and Pulido, 2009; Plieninger et al., 2010; Fischer et al., 2009; Pinto-Correia and Godinho, 2013; Peri et al., 2016a), it seems that they are resilient enough to recover to a natural stand structure after being set aside for several years (Smit et al., 2008; Pulido et al., 2010; Bergmeier et al., 2010; Plieninger et al., 2015). Nevertheless, cases of irreversible degradation have also been reported. For instance, Acácio et al. (2007) showed that in cork oak dehesas densely encroached after years of overgrazing, cultivation and/or fires, shrub understorey affects the production, dispersal and germination of acorns and the survival of the seedlings, which in turn results in an arrested succession to productive wood pastures. Matching appropriate timing of grazing and forage availability is crucial to mitigating the detrimental impact of livestock on soil properties (Jose et al., 2017). When not adequately practised in terms of timing and stocking rate, continuous, or unregulated, grazing practices increase compaction, run-off and decreases soil moisture available for growing plants (Coelho et al., 2004). By contrast, under rotational grazing, soil quality and plant diversity remain higher and trees can perform better (Sharrow, 2007; Lehmkuhler et al., 1999; Southorn and Cattle, 2004; Pittarello et al., 2018), although this depends on the instantaneous stocking rate and the type of animal/s chosen in the SPs (Briske et al., 2008). These authors report some cases where the advantages of rotational grazing were not confirmed experimentally. Besides, continuous grazing can be more efficient to reduce combustible biomass (thus preventing fire; Silva-Pando and González-Hernández, 1992) and to increase spatial heterogeneity (Perotti et al., 2018).

Seasonal (Carmona et al., 2013) and multi-year grazing exclusions (Smit et al., 2008) have been proposed to ensure oak regeneration in Mediterranean wood pastures. Tree regeneration in wood pastures should be an integrated part of regular management practices instead of sporadic pulses of artificial afforestation. A rotational stage of grazing exclusion per plot could be included in the farm management plan to guarantee the continuous renewal of the tree cover. Nevertheless, some assistance could be needed to make grazing compatible with tree regeneration, especially in areas with very low tree density (seeding and planting) or very low-growing species (sheltering young trees).

More recently, adapted grazing in multi-paddocks, with instantaneous high stocking rates and long periods of pasture recovery, has been proposed as a grazing scheme to recover soils and pasture quality of extensive livestock farms (Briske et al., 2008, 2011; Teague et al., 2011, 2013). Also, the use of new GPS-based technologies to ease herding and to improve pasture use is gaining momentum (Brunberg et al., 2015; Choi et al., 2016; Bengaly et al., 2016; Campbell et al., 2018; Marini et al., 2018).

6 Summary

Trees can favour pasture production through the improvement of soil physical and chemical fertility, but trees and pasture also compete for soil water and light. A good knowledge of the effects of mature trees on forage production and the effects of understorey structure on tree production (e.g. acorn production) is critical for ranchers to determine the optimal design and management of the system (tree density, thinning and pruning, canopy form and size, fertilization, pasture improvement, encroachment control, grazing regime, periodical cropping etc.). Similarly, the effects of understorey structure on tree recruitment must be taken into account to ensure the stability (long persistence) of the system (e.g. tree regeneration). Pasture production, tree growth and/or fruit production and tree regeneration need to be analysed taking into account the complexity of the overstorey-understorey. Trees and shrubs in SPs are a source of forage resources that help to overcome seasons with limited grass understorey growth. Browsing trees and shrubs can reduce the risk of intestinal diseases and infections. Shading and sheltering by trees and shrubs increase animal welfare and could reduce significantly the metabolic requirements of livestock, thus increasing farm productivity.

Spatial heterogeneity in SPs operates from the ubiquitous fine-scale gradients caused by scattered canopies to land use and land cover heterogeneity that often define extensive silvopastoral farms and landscapes. In addition, trees are sources of refuge and fodder for many species. All these features have the potential to create ecological niches for a wide range of organisms, and SPs are frequently more biodiverse than comparable forests and open pastures in farmlands. Carbon sequestration in SPs can offset sufficiently the emission of other greenhouse gases associated with livestock rearing. Trees can reduce the nutrient leaching and odours in more intensive/fertilized silvopastures.

SPs worldwide are managed for livestock production in varied forms, although reliable statistics and/or maps are not available yet. Nevertheless, in many cases SPs are not deliberately designed and managed but are a consequence of the extensive grazing of forest and rangelands.

Silvopastoralism in some cases is producing a progressive deforestation of former forest, but in cases of abandonment of grazing practices is reversing many SPs to forests. Indeed, although SP is the more extended and varied agroforestry category, its importance for livestock production is decreasing. There is a need to reconcile grazing with the long persistence of tree cover and diversity and to adapt management practices to new socio-economic context and functions expected from traditional SPs. Besides, novel intensive SPs based on protein-rich trees and shrubs combined with a highly productive grass understorey are gaining momentum in the context of climate change and the desirable fodder autonomy of the farms.

7 Future trends in research

There is an urgent need for promotion of SP practices across the pastoral areas that will advance sustainable rural development, that is innovative practices to assure the ecological persistence of SPs, improved competitiveness, and social and environmental enhancement. Explicit long-term strategies should be designed to promote management practices that ensure SP conservation and implantation of recent innovations. However, in order to convince landowners, better knowledge is still needed. For instance, studies focusing on the conditions under which the net balance of trees is positive (facilitation) or negative (competition) for pasture understorey are still needed. Similarly, the optimal tree species and density in SPs under different uses and ecological constraints (namely water shortage) has not been adequately addressed yet. The analysis of consequences and opportunities of woody encroachment of extensive pastoral systems and landscapes also deserve more attention. SP encroachment could be favourable for regeneration, but it is doubtful whether shrub encroachment would keep stands functioning (e.g. hydric and nutritional tree status, biodiversity) and profitability (e.g. livestock carrying capacity).

From the lessons learned across three decades of scientific studies, we estimate as a priority the following:

1 The elaboration of a comprehensive and categorized map, and associated database of pastoral systems.
2 Compilation of traditional and modern management practices as adaptive solutions to environmental and socio-economical contexts.
3 Identification, selection and introduction of tree, shrub and annual pasture plants of species suitable for different pedo-climatic conditions and grazing characteristics.
4 A selection of programmes to develop specific legume forage species suitable for silvopastoral purposes, where herbaceous plants have to cope with shade and competition for resources imposed by trees.

5 Design new intensive SPs or woody forage banks to take advantage of the nutraceutical value of the browse and to gain farm fodder autonomy, especially in regions with long seasonal shortages of green herbaceous forage.
6 Studies focusing on the conditions under which net balance of trees is positive (facilitation) or negative (competition) for pasture understorey, which surprisingly is still lacking.
7 Development of appropriate silvicultural management practices to make trees more productive and resilient to grazing pressure and climate change (e.g. determine the optimum tree density).
8 The analysis of consequences and opportunities of woody encroachment of extensive pastoral systems (as fodder and nursery plants) also deserves more attention.
9 The scientific evaluation of environmental services, such as reducing forest fires, C storage reinforcements, control of water loss and quality, and biodiversity preservation, under different environmental and management contexts of the SP.
10 Identification of synergies and trade-offs occurring among products, and among livestock production, ecosystem services and biodiversity.
11 Diversification and increment of forage offerings (mitigating seasonal shortages) and other marketable products of SPs.
12 Economical evaluation of SP including environmental goods and services (green accounting).
13 The popularization of the links between silvopastoral practices with the production of high-quality products and the provision of public environmental services.
14 And finally, there is a need for an increased understanding of landholders' decision-making processes, and the drivers that mostly affect their choices, so that better targeted public interventions can be developed.

8 Where to look for further information

Blackshaw, J.K. and Blackshaw, A.W. (1994). Heat stress in cattle and the effect of shade on production and behaviour: A review. *Australian Journal of Experimental Agriculture*, 34(2), pp. 285–95.

Broom D.M., Galindo F.A., Murgueitio E. (2013). Sustainable, efficient livestock production with high biodiversity and good welfare for animals. *Proc R Soc B*, 280: 20132025.

Campos, P., Ovando, P., Mesa, B., and Oviedo, J.L. (2018). Environmental income of livestock grazing on privately-owned silvopastoral farms in Andalusia, Spain. *Land Degradation & Development*, 29(2), pp. 250–61.

Cubbage, F., Balmelli, G., Bussoni, A., Noellemeyer, E., Pachas, A. N., Fassola, H., Colcombet, L., Rossner, B., Frey, G., Dube, F., Lopes de Silva, M., Stevenson, H.,

Hamilton, J. (2012). Comparing silvopastoral systems and prospects in eight regions of the world. *Agroforestry Systems*, 86(3), pp. 303–14.
Hartel, T. and Plieninger, T. (2014). *European wood-pastures in transition: A social-ecological approach*. London: Routledge. p. 303.
Peri, P.L., Dube, F. and Varella, A. (Eds). (2016). *Silvopastoral Systems in Southern South America*. Advances in Agroforestry Book-Series 11. Springer.
Rotherham, I.D. (Ed.). (2013). *Trees, Forested Landscapes and Grazing Animals: A European Perspective on Woodlands and Grazed Treescapes*. London: Routledge. p. 411.
Vandermeulen, S., Ramírez-Restrepo, C.A., Beckers, Y., Claessens, H. and Bindelle, J. (2018). Agroforestry for ruminants: A review of trees and shrubs as fodder in silvopastoral temperate and tropical production systems. *Animal Production Science*, 58(5), pp. 767–77.

9 References

Abdallah, F. and Chaieb, M. 2012. The influence of trees on nutrients, water, light availability and understorey vegetation in an arid environment. *Applied Vegetation Science* 15(4), 501–12. doi:10.1111/j.1654-109X.2012.01201.x.
Abdallah, F., Noumi, Z., Touzard, B., Belgacem, A. O., Neffati, M. and Chaieb, M. 2008. The influence of *Acacia tortilis* (Forssk.) subsp. raddiana (Savi) and livestock grazing on grass species composition, yield and soil nutrients in arid environments of South Tunisia. *Flora - Morphology, Distribution, Functional Ecology of Plants* 203(2), 116–25. doi:10.1016/j.flora.2007.02.002.
Abule, E., Smit, G. N. and Snyman, H. A. 2005. The influence of woody plants and livestock grazing on grass species composition, yield and soil nutrients in the Middle Awash Valley of Ethiopia. *Journal of Arid Environments* 60(2), 343–58. doi:10.1016/j.jaridenv.2004.04.006.
Acácio, V., Holmgren, M., Jansen, P. A. and Schrotter, O. 2007. Multiple recruitment limitation causes arrested succession in Mediterranean cork oak systems. *Ecosystems* 10(7), 1220–30. doi:10.1007/s10021-007-9089-9.
Alconada Magliano, M. M., Bussoni, A., Rosa, R. and Rivera, J. J. C. 2009. Bio-drainage as control to water excess in the Pampa Arenosa, Buenos Aires, Argentina. *Investigaciones Geograficas* 68, 50–72.
Angelsen, A. and Kaimowitz, D. 2004. *Is Agroforestry Likely to Reduce Deforestation? Agroforestry and Biodiversity Conservation in Tropical Landscapes*. Washington DC: Island Press, pp. 87–106.
Archer, S. R. 2010. Rangeland conservation and shrub encroachment: New perspectives on an old problem. In: Toit, J. T., Kock, R. and Deutsch, J. (Eds), *Wild Rangelands: Conserving Wildlife While Maintaining Livestock in Semi-Arid Ecosystems*. New York, NY: John Wiley & Sons, pp. 53–97.
Ash, A. J. 1998. Tree-grass relationships in open eucalypt woodlands of northeastern Australia: Influence of trees on pasture productivity, forage quality and species distribution. *Agroforestry Systems* 40, 159–76.
Asner, G. P., Elmore, A. J., Olander, L. P., Martin, R. E. and Harris, A. T. 2004. Grazing systems, ecosystem responses, and global change. *Annual Review of Environment and Resources* 29(1), 261–99. doi:10.1146/annurev.energy.29.062403.102142.

Balehegn, M. 2017. Silvopasture using indigenous fodder trees and shrubs: The underexploited synergy between climate change adaptation and mitigation in the livestock sector. In: Leal Filho, W., Simane, B., Kalangu, J., Wuta, M., Munishi, P. and Musiyiwa, K. (Eds), *Climate Change Adaptation in Africa*. Bangkok, Thailand: Springer International Publishing, pp. 493–510.

Bambo, S. K., Nowak, J., Blount, A. R., Long, A. J. and Osiecka, A. 2009. Soil nitrate leaching in silvopastures compared with open pasture and pine plantation. *Journal of Environmental Quality* 38(5), 1870–7. doi:10.2134/jeq2007.0634.

Batello, C., Mannetje, L., Martinez, A. and Suttie, J. 2008. *Plant Genetic Resources of Forage Crops, Pasture and Rangelands: Thematic Background Study*. Rome, Italy: FAO, pp. 5–7.

Belsky, A. J., Mwonga, S. M., Amundson, R. G., Duxbury, J. M. and Ali, A. R. 1993. Comparative effects of isolated trees on their undercanopy environments in high- and low-rainfall savannas. *The Journal of Applied Ecology* 30(1), 143–55. doi:10.2307/2404278.

Benavides, R., Douglas, G. B. and Osoro, K. 2009. Silvopastoralism in New Zealand: Review of effects of evergreen and deciduous trees on pasture dynamics. *Agroforestry Systems* 76(2), 327–50. doi:10.1007/s10457-008-9186-6.

Bengaly, K., Traoré, S., Cibils, A. F., Moses, J. and Nyamurekung'e, S. 2016. Rainy season herding patterns of agro-pastoral livestock smallholders in southwestern Mali: A preliminary GPS-based assessment. In: Iwaasa, A., Lardner, H. B., Schellenberg, M., Willms, W. and Larson, K. (Eds), *10th International Rangeland Congress*, 16–22 July 2016, Saskatoon, Canada, p. 1127.

Benhamou, C., Salmon-Monviola, J., Durand, P., Grimaldi, C. and Merot, P. 2013. Modeling the interaction between fields and a surrounding hedgerow network and its impact on water and nitrogen flows of a small watershed. *Agricultural Water Management* 121, 62–72. doi:10.1016/j.agwat.2013.01.004.

Benton, T. G., Vickery, J. A. and Wilson, J. D. 2003. Farmland biodiversity: Is habitat heterogeneity the key? *Trends in Ecology and Evolution* 18(4), 182–8. doi:10.1016/S0169-5347(03)00011-9.

Bergmeier, E., Petermann, J. and Schroder, E. 2010. Geobotanical survey of woodpasture habitats in Europe: Diversity, threats and conservation. *Biodiversity and Conservation* 19(11), 2995–3014. doi:10.1007/s10531-010-9872-3.

Bestman, M., van Eekeren, N., Luske, B., Vonk, M., Anssems, E., Boosten, M. and van Bree, M. 2014. Introducing trees in dairy and poultry farms – Experiences dairy and poultry farmers' networks in The Netherlands. Paper at: IFOAM Organic World Congress, Istanbul, Turkey, 13–15 October 2014.

Betteridge, K., Hoogendoorn, C., Costall, D., Carter, M. and Griffiths, W. 2010. Sensors for detecting and logging spatial distribution of urine from grazing cows and ewes. *Computers and Electronics in Agriculture* 73(1), 66–73. doi:10.1016/j.compag.2010.04.005.

Betteridge, K., Costall, D., Martin, S., Reidy, B., Stead, A. and Millner, I. 2012. Impact of shade trees on Angus cow behaviour and physiology in summer dry hill country: Grazing activity, skin temperature and nutrient transfer issues. In: *Advanced Nutrient Management: Gains from the Past – Goals for the Future*. Occasional Report No. 25. Palmerston North, New Zealand: Fertilizer and Lime Research Centre, Massey University, 10p. Available at: http://flrc.massey.ac.nz/publications.html.

Blaser, W. J., Sitters, J., Hart, S. P., Edwards, P. J. and Olde Venterink, H. 2013. Facilitative or competitive effects of woody plants on understorey vegetation depend on N-fixation, canopy shape and rainfall. *Journal of Ecology* 101(6), 1598–603. doi:10.1111/1365-2745.12142.

Blazier, M. A., Gaston, L. A., Clason, T. R., Farrish, K. W., Oswald, B. P. and Evans, H. A. 2008. Nutrient dynamics and tree growth of silvopastoral systems: Impact of poultry litter. *Journal of Environmental Quality* 37(4), 1546–58. doi:10.2134/jeq2007.0343.

Boyer, D. G. and Neel, J. P. S. 2010. Nitrate and fecal coliform concentration differences at the soil/bedrock interface in Appalachian silvopasture, pasture, and forest. *Agroforestry Systems* 79(1), 89–96. doi:10.1007/s10457-009-9272-4.

Briske, D. D., Derner, J. D., Brown, J. R., Fuhlendorf, S. D., Teague, W. R., Havstad, K. M., Gill, R. L., Ash, A. J. and Willms, W. D. 2008. Rotational grazing on rangelands: Reconciliation of perception and experimental evidence. *Rangeland Ecology and Management* 61(1), 3–17. doi:10.2111/06-159R.1.

Briske, D. D., Sayre, N. F., Huntsinger, L., Fernandez-Gimenez, M., Budd, B. and Derner, J. D. 2011. Origin, persistence, and resolution of the rotational grazing debate: Integrating human dimensions into rangeland research. *Rangeland Ecology and Management* 64(4), 325–34. doi:10.2111/REM-D-10-00084.1.

Broom, D. M., Galindo, F. A. and Murgueitio, E. 2013. Sustainable, efficient livestock production with high biodiversity and good welfare for animals. *Proceedings. Biological Sciences* 280(1771), 20132025. doi:10.1098/rspb.2013.2025.

Brunberg, E. I., Bøe, K. E. and Sørheim, K. M. 2015. Testing a new virtual fencing system on sheep. *Acta Agriculturae Scandinavica, Section A – Animal Science* 65(3–4), 168–75. doi:10.1080/09064702.2015.1128478.

Bugalho, M. N., Caldeira, M. C., Pereira, J. S., Aronson, J. and Pausas, J. G. 2011. Mediterranean cork oak savannas require human use to sustain biodiversity and ecosystem services. *Frontiers in Ecology and the Environment* 9(5), 278–86. doi:10.1890/100084.

Burgess, P. J. and Morris, J. 2009. Agricultural technology and land use futures: The UK case. *Land Use Policy* 26, S222–9. doi:10.1016/j.landusepol.2009.08.029.

Buttler, A., Kohler, F. and Gillet, F. 2009. The Swiss mountain wooded pastures: Patterns and processes. In: Rigueiro-Rodríguez, A., McAdam, J. and Mosquera-Losada, M. R. (Eds), *Agroforestry in Europe: Current Status and Future Prospects*. Dordrecht, The Netherlands: Springer Science & Business Media, pp. 377–96.

Campbell, D. L. M., Lea, J. M., Haynes, S. J., Farrer, W. J., Leigh-Lancaster, C. J. and Lee, C. 2018. Virtual fencing of cattle using an automated collar in a feed attractant trial. *Applied Animal Behaviour Science* 200, 71–7. doi:10.1016/j.applanim.2017.12.002.

Campos, P., Huntsinger, L., Oviedo, J. L., Starrs, P. F., Diaz, M., Standiford, R. B. and Montero, G. (Eds). 2013. *Mediterranean Oak Woodland Working Landscapes: Dehesas of Spain and Ranchlands of California*. Series: Landscape Series, vol 16. New York, NY: Springer, p. 508.

Carmona, C. P., Azcárate, F. M., Oteros-Rozas, E., González, J. A. and Peco, B. 2013. Assessing the effects of seasonal grazing on holm oak regeneration: Implications for the conservation of Mediterranean dehesas. *Biological Conservation* 159, 240–7. doi:10.1016/j.biocon.2012.11.015.

Casals, P., Baiges, T., Bota, G., Chocarro, C., de Bello, F., Fanlo, R., Sebastià, M. T. and Taull, M. 2009. Silvopastoral systems in the Northeastern Iberian Peninsula: A multifunctional perspective. In: Rigueiro-Rodríguez, A., McAdam, J. and Mosquera-Losada, M. R.

(Eds), *Agroforestry in Europe: Current Status and Future Prospects*. Dordrecht, The Netherlands: Springer Science & Business Media, pp. 161-81.

Chaideftou, E., Thanos, C. A., Bergmeier, E., Kallimanis, A. S. and Dimopoulos, P. 2011. The herb layer restoration potential of the soil seed bank in an overgrazed oak forest. *Journal of Biological Research* 15, 47-57.

Chillo, V., Vázquez, D. P., Amoroso, M. M. and Bennett, E. M. 2018. Land-use intensity indirectly affects ecosystem services mainly through plant functional identity in a temperate forest. *Functional Ecology* 32(5), 1390-9. doi:10.1111/1365-2435.13064.

Choi, S. G., Chimeddorj, G., Altankhuyag, B. and Dunkhorol, S. 2016. Design and implementation of a GPS-enabled mobile wireless sensor network for livestock herd tracking in Mongolian nomadic herding. *11th International Forum on Strategic Technology (IFOST)*, Novosibirsk, 2016, pp. 423-7.

Chu, B., Goyne, K. W., Anderson, S. H., Lin, C. H. and Udawatta, R. P. 2010. Veterinary antibiotic sorption to agroforestry buffer, grass buffer, and cropland soils. *Agroforestry Systems* 79(1), 67-80. doi:10.1007/s10457-009-9273-3.

Coelho, G. C. 2017. Ecosystem services in Brazilian's southern agroforestry systems. *Tropical and Subtropical Agroecosystems* 20(3), 475-92.

Coelho, C. O. A., Ferreira, A. J. D., Laouina, A., Hamza, A., Chaker, M., Naafa, R., Regaya, K., Boulet, A. K., Keizer, J. J. and Carvalho, T. M. M. 2004. Changes in land use and land management practices affecting land degradation within forest and grazing ecosystems in the western Mediterranean. In: Schnabel, S. and Ferreira, A. (Eds), *Sustainability of Agrosilvopastoral Systems*. Advances in GeoEcology, vol. 37. Reiskirchen, Germany: Catena Verlag, pp. 137-53.

Concepción, E. D., Díaz, M., Kleijn, D., Báldi, A., Batâry, P., Clough, Y., Gabriel, D., Herzog, F., Holzschuh, A., Knop, E., et al. 2012. Interactive effects of landscape context constrain the effectiveness of local agrienvironmental management. *Journal of Applied Ecology* 49, 695-705.

Cubbage, F., Balmelli, G., Bussoni, A., Noellemeyer, E., Pachas, A. N., Fassola, H., Colcombet, L., Rossner, B., Frey, G., Dube, F., et al. 2012. Comparing silvopastoral systems and prospects in eight regions of the world. *Agroforestry Systems* 86(3), 303-14. doi:10.1007/s10457-012-9482-z.

Cubera, E. and Moreno, G. 2007. Effect of single Quercus ilex trees upon spatial and seasonal changes in soil water content in dehesas of central western Spain. *Annals of Forest Science* 64(3), 355-64. doi:10.1051/forest:2007012.

Cubera, E., Nunes, J. M., Madeira, M. and Gazarini, L. 2009. Influence of Quercus ilex trees on herbaceous production and nutrient concentrations in southern Portugal. *Journal of Plant Nutrition and Soil Science* 172(4), 565-71. doi:10.1002/jpln.200800191.

Cuthbertson, A. and McAdam, J. 1996. The effect of tree density and species on carabid beetles in a range of pasture-tree agroforestry systems on a lowland site. *Agroforestry Forum* 7, 17-20.

Dalzell, S. A., Shelton, H. M., Mullen, B. F., Larsen, P. H. and McLaughlin, K. G. 2006. *Leucaena, a Guide to Establishment and Management*. Sydney, Australia: Meat and Livestock Australia Ltd., p. 70.

de Mendonca, R. M. A., Lana, Â. M. Q., Lana, R. M. Q., Filho, J. P. L., Toth, G. and do Carmo, T.D. 2017. Meteorological, light and grass characteristics under trees in a Silvopasture in the Brazilian cerrado. *Australian Journal of Crop Science* 11(2), 177-83. doi:10.21475/ajcs.17.11.02.p224.

De Rensis, F. and Scaramuzzi, R. J. 2003. Heat stress and seasonal effects on reproduction in the dairy cow: A review. *Theriogenology* 60(6), 1139-51. doi:10.1016/S0093-691X(03)00126-2.

den Herder, M., Moreno, G., Mosquera-Losada, R. M., Palma, J. H. N., Sidiropoulou, A., Freijanes, J. J. S., Crous-Durán, J., Paulo, J. A., Tomé, M., Pantera, A., et al. 2017. Current extent and stratification of agroforestry in the European Union. *Agriculture, Ecosystems and Environment* 241, 121-32. doi:10.1016/j.agee.2017.03.005.

Dennis, P., Shellard, L. J. F. and Agnew, R. D. M. 1996. Shifts in arthropod species assemblages in relation to silvopastoral establishment in upland pastures. *Agroforestry Forum* 7(3), 14-7.

Devendra, C. and Ibrahim, M. 2004. Silvopastoral systems as a strategic for diversification and productivity enhancement from Livestock in the tropics. In: Mannetje, L., Ramírez, M., Ibrahim, C., Sandoval, N., Ojeda, N. and Ku, J. (Eds), *The Importance of Silvopastoral Systems in Rural Livelihoods to Provide Ecosystem Services. Proceedings of the Second International Symposium of Silvopastoral Systems*. Mérida, Yucatán, México, pp. 10-24.

Díaz, M. 2008. Biodiversity in the dehesa. In: Mosquera-Losada, M. R., Fernández, J.L. and Rigueiro, A. (Eds), *Agroforestry Systems as a Technique for Sustainable Territorial Management*. Madrid: Programme Azahar, Ministerio de Asuntos Exteriores-AECID, pp. 209-26.

Díaz, M. and Pulido, F. J. 2009. 6310 Dehesas perennifolias de *Quercus* spp. In: *Bases ecológicas preliminares para la conservación de los tipos de hábitat de interés comunitario en España*. Madrid: Ministerio de Medio Ambiente, y Medio Rural y Marino, p. 69.

Díaz, M., Tietje, W. D. and Barrett, R. H. 2013. Effects of management on biological diversity and endangered species. In: Campos, P., Huntsinger, L., Oviedo, J. L., Starrs, P. F., Diaz, M., Standiford, R. B. and Montero, G. (Eds), *Mediterranean Oak Woodland Working Landscapes*. Series: Landscape Series, vol 16. New York, NY: Springer, pp. 213-43.

Dodelin, B., Eynard-Machet, R. and Athanaze, P. J. A. (Eds). 2007. Les rémanents en foresterie et agriculture - Les branches: Matériau d'Avenir. *Proceedings of the International Conference*, Lyon, France, 1-2 February 2007. Hermès-Lavoisier, Paris: Tec & Doc, p. 384.

Donaghy, P., Bray, S., Gowen, R., Rolfe, J., Stephens, M., Hoffmann, M. and Stunzer, A. 2010. The bioeconomic potential for agroforestry in Australia's northern grazing systems. *Small-Scale Forestry* 9(4), 463-84. doi:10.1007/s11842-010-9126-y.

Douglas, G. B., Walcroft, A. S., Hurst, S. E., Potter, J. F., Foote, A. G., Fung, L. E., Edwards, W. R. N. and Van Den Dijssel, C. 2006. Interactions between widely spaced young poplars (*Populus* spp.) and introduced pasture mixtures. *Agroforestry Systems* 66, 165-78.

Dubeux Junior, J. C. B., Muir, J. P., Apolinário, V. XdO., Nair, P. K. R., Lira, MdA. and Sollenberger, L. E. 2017. Tree legumes: An underexploited resource in warm-climate silvopastures. *Revista Brasileira de Zootecnia* 46(8), 689-703. doi:10.1590/s1806-92902017000800010.

Durr, P. A. and Rangel, J. 2002. Enhanced forage production under Samanea saman in a subhumid tropical grassland. *Agroforestry Systems* 54(2), 99-102. doi:10.1023/A:1015070203455.

East, R. M. and Felker, P. 1993. Forage production and quality of 4 perennial grasses grown under and outside canopies of mature *Prosopis glandulosa* Torr. var. *glandulosa* (mesquite). *Agroforestry Systems* 22(2), 91–110. doi:10.1007/BF00705139.

Etienne, M., Derzko, M. and Rigolot, E. 1996. Browse impact in silvopastoral systems participating in fire prevention in the French Mediterranean region. In: Etienne, M. (Ed.), *Western European Silvopastoral Systems*. Montpellier, France: INRA, pp. 93–102.

Faroda, A. S. 1997. Silvi-pastoral approach for arid and semi-arid areas: Issues and strategies. In: Yadav, M. S., Singh, Manjit, Sharma, S. K., Tewari, I. C. and Burman, U. (Eds), *Silvipastoral Systems in Arid and Semiarid Ecosystems*. Jodhpur, India: Central Arid Zone Research Institute, Indian Council of Agricultural Research, pp. 1-10.

Fezzi, C., Rigby, D., Bateman, I. J., Hadley, D. and Posen, P. 2008. Estimating the range of economic impacts on farms of nutrient leaching reduction policies. *Agricultural Economics* 39(2), 197–205. doi:10.1111/j.1574-0862.2008.00323.x.

Fischer, J., Stott, J., Zerger, A., Warren, G., Sherren, K. and Forrester, R. I. 2009. Reversing a tree regeneration crisis in an endangered ecoregion. *Proceedings of the National Academy of Sciences* 106(25), 10386–91. doi:10.1073/pnas.0900110106.

Fischer, J., Stott, J. and Law, B. S. 2010. The disproportionate value of paddock trees. *Biological Conservation* 143(6), 1564–7. doi:10.1016/j.biocon.2010.03.030.

Franca, A., Sanna, F., Nieddu, S., Re, G. A., Pintus, G. V., Ventura, A., Duce, P., Salis, M. and Arca, B. 2012. Effects of grazing on the traits of a potential fire in a Sardinian wooded pasture. *Options méditerranéennes. Série A: Séminaires méditerranéens, CIHEAM, Centre international de hautes études agronomiques méditerranéennes* 102, 307–11.

Francaviglia, R., Benedetti, A., Doro, L., Madrau, S. and Ledda, L. 2014. Influence of land use on soil quality and stratification ratios under agro-silvo-pastoral Mediterranean management systems. *Agriculture, Ecosystems and Environment* 183, 86–92. doi:10.1016/j.agee.2013.10.026.

Franzel, S., Carsan, S., Lukuyu, B., Sinja, J. and Wambugu, C. 2014. Fodder trees for improving livestock productivity and smallholder livelihoods in Africa. *Current Opinion in Environmental Sustainability* 6, 98–103. doi:10.1016/j.cosust.2013.11.008.

Frost, W. E., McDougald, N. K. and Demment, M. W. 1991. Blue oak canopy effect on seasonal forage production and quality. In: Standiford, R. B. (Tech. Coord.), *Proceedings of the Symposium on Oak Woodlands and Hardwood Rangeland Management*, 31 October – 2 November 1990, Davis, CA. Gen. Tech. Rep. PSW-GTR-126. Berkeley, CA: Pacific Southwest Research Station, Forest Service, U.S. Department of Agriculture, pp. 307–11.

Galina-Hidalgo, M. A., Pineda, L. J., Olvera, E., Ortiz, R. M., Cuellar, A. and Galinado, E. 2016. Effect of tannins-rich plants to control gastrointestinal nematodes in Zebu cows grazing in a sub-tropical sylvopastoral system. *Archivos Latinoamericanos de Producción Animal* 24(4), 241–51.

García, R. R., Fraser, M. D., Celaya, R., Ferreira, L. M. M., García, U. and Osoro, K. 2013. Grazing land management and biodiversity in the Atlantic European heathlands: A review. *Agroforestry Systems* 87(1), 19–43. doi:10.1007/s10457-012-9519-3.

Gardner, F. P., Pearce, B. B. and Mitchell, R. L. 2017. *Physiology of Crop Plants* (2nd edn.). Jodhpur, India: Scientific Publishers, p. 327.

Garrett, H. E., Rietveld, W. J. and Fisher, R. F. (Eds). 2000. *North American Agroforestry: An Integrated Science and Practice*. Madison, WI: Agronomy Society of America, 402p.

Gea-Izquierdo, G., Montero, G. and Cañellas, I. 2009. Changes in limiting resources determine spatiotemporal variability in tree-grass interactions. *Agroforestry Systems* 76(2), 375-87. doi:10.1007/s10457-009-9211-4.

Gibbons, P., Lindenmayer, D. B., Fischer, J., Manning, A. D., Weinberg, A., Seddon, J., Ray, P. and Barrett, G. 2008. The future of scattered trees in agricultural landscapes. *Conservation Biology: The Journal of the Society for Conservation Biology* 22(5), 1309-19. doi:10.1111/j.1523-1739.2008.00997.x.

Gillet, F., Murisier, B., Buttler, A., Gallandat, J. D. and Gobat, J. M. 1999. Influence of tree cover on the diversity of herbaceous communities in subalpine wooded pastures. *Applied Vegetation Science* 2(1), 47-54. doi:10.2307/1478880.

Gómez-Rey, M. X., Garcês, A. and Madeira, M. 2012. Soil organic-C accumulation and N availability under improved pastures established in Mediterranean oak woodlands. *Soil Use and Management* 28(4), 497-507. doi:10.1111/j.1475-2743.2012.00428.x.

González-Hernández, M. P. and Silva-Pando, F. J. 1999. Nutritional attributes of understory plants known as components of deer diets. *Journal of Range Management* 52(2), 132-8. doi:10.2307/4003506.

González-Hernández, M. P., Starkey, E. E. and Karchesy, J. 2000. Seasonal variation in concentrations of fiber, crude protein, and phenolic compounds in leaves of red alder (Alnus rubra): Nutritional implications for cervids. *Journal of Chemical Ecology* 26(1), 293-301. doi:10.1023/A:1005462100010.

González-Hernández, M. P., Karchesy, J. and Starkey, E. E. 2003. Research observation: Hydrolyzable and condensed tannins in plants of northwest Spain forests. *Journal of Range Management* 56, 461-5.

Grove, A. T. and Rackham, O. 2001. *The Nature of Mediterranean Europe: An Ecological History*. New Haven, CT: Yale University Press, p. 384.

Guevara-Escobar, A., Kemp, P. D., Mackay, A. D. and Hodgson, J. 2007. Pasture production and composition under poplar in a hill environment in New Zealand. *Agroforestry Systems* 69(3), 199-213. doi:10.1007/s10457-007-9038-9.

Hartel, T. and Plieninger, T. (Eds). 2014. *European Wood-Pastures in Transition: A Social-Ecological Approach*. London and New York: Routledge.

Hartel, T., Dorresteijn, I., Klein, C., Máthé, O., Moga, C. I., Öllerer, K., Roellig, M., von Wehrden, H. and Fischer, J. 2013. Wood-pastures in a traditional rural region of Eastern Europe: Characteristics, management and status. *Biological Conservation* 166, 267-75. doi:10.1016/j.biocon.2013.06.020.

Hernandez-Morcillo, M., Burgess, P., Mirck, J., Pantera, A. and Plieninger, T. 2018. Scanning agroforestry-based solutions for climate change mitigation and adaptation in Europe. *Environmental Science and Policy*, 80, 44-52. doi:10.1016/j.envsci.2017.11.013.

Higgins, K. P. and Dodd, V. A. 1989. A model of the bioclimatic value of shelter to beef cattle. *Journal of Agricultural Engineering Research* 42(3), 149-64. doi:10.1016/0021-8634(89)90047-4.

Hooda, P. S., Edwards, A. C., Anderson, H. A. and Miller, A. 2000. A review of water quality concerns in livestock farming areas. *Science of the Total Environment* 250(1-3), 143-67. doi:10.1016/S0048-9697(00)00373-9.

Horsted, K., Kongsted, A. G., Jorgensen, U. and Sorensen, J. 2012. Combined production of free-range pigs and energy crops - Animal behaviour and crop damages. *Livestock Science* 150(1-3), 200-8. doi:10.1016/j.livsci.2012.09.006.

Howlett, D. S., Mosquera-Losada, M. R., Nair, P. K. R., Nair, V. D. and Rigueiro-Rodríguez, A. 2011a. Soil carbon storage in silvopastoral systems and a treeless pasture in

Northwestern Spain. *Journal of Environmental Quality* 40(3), 825–32. doi:10.2134/jeq2010.0145.

Howlett, D. S., Moreno, G., Mosquera Losada, M. R., Nair, P. K. R. and Nair, V. D. 2011b. Soil carbon storage as influenced by tree cover in the Dehesa cork oak silvopasture of central-western Spain. *Journal of Environmental Monitoring: JEM* 13(7), 1897–904. doi:10.1039/c1em10059a.

Hristov, A. N., Oh, J., Lee, C., Meinen, R., Montes, F., Ott, T., Firkins, J., Rotz, A., Dell, C., Adesogan, A., et al. 2013. Mitigation of greenhouse gas emissions in livestock production – A review of technical options for non-CO_2 emissions. FAO Animal Production and Health Paper No. 177. Rome, Italy: FAO, p. 231.

Huang, W., Kanninen, M., Xu, Q. and Huang, B. 1997. Agroforestry in China: Present state and future potential. *Ambio – A Journal of the Human Environment* 26(6), 394–8.

Jackson, J. and Ash, A. J. 1998. Tree-grass relationships in open eucalypt woodlands of northeastern Australia: Influence of trees on pasture productivity, forage quality and species distribution. *Agroforestry Systems* 40(2), 159–76. doi:10.1023/A:1006067110870.

Jakobsson, S. and Lindborg, R. 2017. The importance of trees for woody pasture bird diversity and effects of the European Union's tree density policy. *Journal of Applied Ecology* 54(6), 1638–47. doi:10.1111/1365-2664.12871.

Jernsletten, J. L. L. and Klokov, K. B. 2002. *Sustainable Reindeer Husbandry*. Tromsø, Norway: Centre for Saami Studies University of Tromsø, p. 157.

Jørgensen, U., Thuesen, J., Eriksen, J., Horsted, K., Hermansen, J. E., Kristensen, K. and Kongsted, A. G. 2018. Nitrogen distribution as affected by stocking density in a combined production system of energy crops and free-range pigs. *Agroforestry Systems* 92(4), 987–99. doi:10.1007/s10457-018-0200-3.

Jose, S. 2005. Ecological interactions in silvopastoral systems: A synthesis. *Annals of Arid Zone* 44, 327–36.

Jose, S., Walter, D. and Kumar, B. M. 2017. Ecological considerations in sustainable silvopasture design and management. *Agroforestry Systems*. doi:10.1007/s10457-016-0065-2.

Jose, S., Gold, M. A. and Garrett, H. 2018. Temperate agroforestry in the United States: Current trends and future directions. In: Gordon, A. M., Newman, S. M. and Coleman, B. (Eds), *Temperate Agroforestry Systems* (2nd edn.). Wallingford, UK: CABI Publishing, pp. 50–71.

Kallenbach, R. L. 2009. Integrating silvopastures into current forage-livestock systems. In: *Agroforestry Comes of Age: Putting Science into Practice. Proceedings of the 11th North American Agroforestry Conference*, Columbia, Missouri, USA, 31 May – 3 June, pp. 455–61.

Kallenbach, R. L., Kerley, M. S. and Bishop-Hurley, G. J. 2006. Cumulative forage production, forage quality and livestock performance from an annual ryegrass and cereal rye mixture in a pine-walnut silvopasture. *Agroforestry Systems* 66(1), 43–53. doi:10.1007/s10457-005-6640-6.

Karki, U. and Goodman, M. S. 2015. Microclimatic differences between mature loblolly-pine silvopasture and open-pasture. *Agroforestry Systems* 89(2), 319–25. doi:10.1007/s10457-014-9768-4.

Kimetu, J. M., Lehmann, J., Ngoze, S. O., Mugendi, D. N., Kinyangi, J. M., Riha, S., Verchot, L., Recha, J. W. and Pell, A. N. 2008. Reversibility of soil productivity decline with organic matter of differing quality along a degradation gradient. *Ecosystems* 11(5), 726–39. doi:10.1007/s10021-008-9154-z.

Kleijn, D. and Sutherland, W. J. 2003. How effective are European agri-environment schemes in conserving and promoting biodiversity? *Journal of Applied Ecology* 40(6), 947–69. doi:10.1111/j.1365-2664.2003.00868.x.

Le Houérou, H. N. 2000. Restoration and rehabilitation of arid and semiarid Mediterranean ecosystems in North Africa and West Asia: A review. *Arid Soil Research and Rehabilitation* 14(1), 3–14. doi:10.1080/089030600263139.

Lehmkuhler, J. W., Kerley, M. S., Garrett, H. E., Cutter, B. E. and McGraw, R. L. 1999. Comparison of continuous and rotational silvopastoral systems for established walnut plantations in southwest Missouri, USA. *Agroforestry Systems* 44, 267–79.

Lin, C. H., McGraw, R. L., George, M. F. and Garrett, H. E. 1999. Shade effects on forage crops with potential in temperate agroforestry practices. *Agroforestry Systems* 44, 109–19.

Lin, C. H., Goyne, K. W., Kremer, R. J., Lerch, R. N. and Garrett, H. E. 2010. Dissipation of sulfamethazine and tetracycline in the root zone of grass and tree species. *Journal of Environmental Quality* 39(4), 1269–78. doi:10.2134/jeq2009.0346.

López-Díaz, M. L., Rolo, V. and Moreno, G. 2011. Trees' role in nitrogen leaching after organic and mineral fertilization: A greenhouse experiment. *Journal of Environmental Quality* 40(3), 853–9. doi:10.2134/jeq2010.0165.

López-Díaz, M. L., Rolo, V., Benítez, R. and Moreno, G. 2015. Shrub encroachment of Iberian dehesas: Implications on total forage productivity. *Agroforestry Systems* 89(4), 587–98. doi:10.1007/s10457-015-9793-y.

López-Díaz, M. L., Benítez, R., Rolo, V. and Moreno, G. 2019. Managing high quality timber plantations as silvopastoral systems: Tree growth, soil water dynamics and nitrate leaching risk. *Soil Tillage Research* (under review).

Lorenz, K. and Lal, R. 2014. Soil organic carbon sequestration in agroforestry systems: A review. *Agronomy for Sustainable Development* 34(2), 443–54. doi:10.1007/s13593-014-0212-y.

Luske, B. and van Eekeren, N. 2018. Nutritional potential of fodder trees on clay and sandy soils. *Agroforestry Systems* 92(4), 975–86. doi:10.1007/s10457-017-0180-8.

Maestre, F. T., Valladares, F. and Reynolds, J. F. 2006. The stress-gradient hypothesis does not fit all relationships between plant-plant interactions and abiotic stress: Further insights from arid environments. *Journal of Ecology* 94(1), 17–22. doi:10.1111/j.1365-2745.2005.01089.x.

Manning, A. D., Fischer, J. and Lindenmayer, D. B. 2006. Scattered trees are keystone structures - Implications for conservation. *Biological Conservation* 132(3), 311–21. doi:10.1016/j.biocon.2006.04.023.

Marañón, T., Pugnaire, F. I. and Callaway, R. M. 2009. Mediterranean-climate oak savannas: The interplay between abiotic environment and species interactions. *Web Ecology* 9(1), 30–43. doi:10.5194/we-9-30-2009.

Marini, D., Llewellyn, R., Belson, S. and Lee, C. 2018. Controlling within-field sheep movement using virtual fencing. *Animals: An Open Access Journal from MDPI* 8(3), 31:doi:10.3390/ani8030031.

Matzdorf, B., Reutter, M. and Hübner, C. 2010. Gutachten-Vorstudie Bewertung der Ökosystemdienstleistungen von HNVGrünland (High Nature Value Grassland). Leibniz-Zentrum für Agrarlandschaftsforschung (ZALF) e.V. Müncheberg, p. 71.

Mayer, R., Kaufmann, R., Vorhauser, K. and Erschbamer, B. 2009. Effects of grazing exclusion on species composition in high-altitude grasslands of the Central Alps. *Basic and Applied Ecology* 10(5), 447–55. doi:10.1016/j.baae.2008.10.004.

Mazía, N., Moyano, J., Perez, L., Aguiar, S., Garibaldi, L. A. and Schlichter, T. 2016. The sign and magnitude of tree-grass interaction along a global environmental gradient. *Global Ecology and Biogeography* 25(12), 1510–9. doi:10.1111/geb.12518.

Mazzoleni, S., di Pasquale, G., Mulligan, M., di Martino, P. and Rego, F. 2004. *Recent Dynamics of the Mediterranean Vegetation and Landscape*. Chichester, England: Wiley Online Library, p. 305.

Mcadam, J. H., Sibbald, A. R., Teklehaimanot, Z. and Eason, W. R. 2007. Developing silvopastoral systems and their effects on diversity of fauna. *Agroforestry Systems* 70(1), 81–9. doi:10.1007/s10457-007-9047-8.

McDermott, M. E. and Rodewald, A. D. 2014. Conservation value of silvopastures to Neotropical migrants in Andean forest flocks. *Biological Conservation* 175, 140–7. doi:10.1016/j.biocon.2014.04.027.

McEvoy, P. M., McAdam, J. H., Mosquera-Losada, M. R. and Rigueiro-Rodríguez, A. 2006. Tree regeneration and sapling damage of pedunculate oak *Quercus robur* in grazed forest in Galicia, NW Spain: A comparison of continuous and rotational grazing systems. *Agroforestry Systems* 66(2), 85–92. doi:10.1007/s10457-005-2916-0.

Michel, G. -A., Nair, V. D. and Nair, P. K. R. 2007. Silvopasture for reducing phosphorus loss from subtropical sandy soils. *Plant and Soil* 297(1–2), 267–76. doi:10.1007/s11104-007-9352-z.

Mitlohner, F. M., Morrow, J. L., Dailey, J. W., Wilson, S. C., Galyean, M. L., Miller, M. F. and McGlone, J. J. 2001. Shade and water misting effects on behaviour, physiology, performance, and carcass traits of heat-stressed feedlot cattle. *Journal of Animal Science* 79(9), 2327–35. doi:10.2527/2001.7992327x.

Moga, C. I., Hartel, T. and Öllerer, K. 2009. Ancient oak wood-pasture as a habitat for the endangered tree pipit Anthus trivialis. *Biologia* 64(5), 1011–5. doi:10.2478/s11756-009-0167-7.

Montagnini, F., Ibrahim, I. and Murgueitio, E. 2013. Silvopastoral systems and climate change mitigation in Latin America. *Bois et Forêts des Tropiques* 316(2), 3–16.

Moreno, G. 2008. Response of understorey forage to multiple tree effects in Iberian dehesas. *Agriculture, Ecosystems and Environment* 123(1–3), 239–44. doi:10.1016/j.agee.2007.04.006.

Moreno, G. and Pulido, F. 2009. The functioning, management and persistence of Dehesas. In: Rigueiro-Rodriguez, A., McAdam, J. and Mosquera-Losada, M. R. (Eds), *Agroforestry in Europe, Current Status and Future Prospects*. Advances in Agroforestry. Heidelberg: Springer, pp. 127–60.

Moreno, G., Bartolome, J. W., Gea-Izquierdo, G. and Cañellas, I. 2013. Overstory-understorey relationships. In: Campos, P., Huntsinger, L., Oviedo, J. L., Starrs, P. F., Díaz, M., Standiford, R. B. and Montero, G. (Eds), *Mediterranean Oak Woodland Working Landscapes*. Landscape Series, 16. Dordrecht, The Netherlands: Springer, pp. 145–79.

Moreno, G., Gonzalez-Bornay, G., Pulido, F., Lopez-Diaz, M. L., Bertomeu, M., Juárez, E. and Diaz, M. 2016. Exploring the causes of high biodiversity of Iberian dehesas: The importance of wood pastures and marginal habitats. *Agroforestry Systems* 90(1), 87–105. doi:10.1007/s10457-015-9817-7.

Moreno, G., Aviron, S., Berg, S., Crous-Duran, J., Franca, A., de Jalón, S. G., Hartel, T., Mirck, J., Pantera, A., Palma, J. H. N., et al. 2018. Agroforestry systems of high nature and cultural value in Europe: Provision of commercial goods and other ecosystem services. *Agroforestry Systems* 92(4), 877-91. doi:10.1007/s10457-017-0126-1.

Moreno-Gonzalo, J., Ferre, I., Celaya, R., Frutos, P., Ferreira, L. M. M., Hervás, G., Gracia, U., Ortega-Mora, L. M. and Osoro, K. 2012. Potential use of heather to control gastrointestinal nematodes in goats. *Small Ruminant Research* 103(1), 60-8. doi:10.1016/j.smallrumres.2011.10.019.

Mosquera-Losada, M. R., McAdam, J., Romero-Franco, R., Santiago-Freijanes, J. J. and Riguero-Rodríquez, A. 2009. Definitions and components of agroforestry practices in Europe. In: Rigueiro-Rodríguez, A., McAdam, J. and Mosquera-Losado, M. (Eds), *Agroforestry in Europe: Current Status and Future Prospects*. Dordrecht, the Netherlands: Springer Science + Business Media B.V., pp. 3-19.

Murdiyarso, D., Robledo, C., Brown, S., Coto, O., Drexhage, J., Forner, C., Kanninen, M., Lipper, L., North, N. and Rondón, M. 2005. Linkages between mitigation and adaptation in land-use change and forestry activities. In: Robledo, C., Kanninen, M. and Pedroni, L. (Eds), *Tropical Forests and Adaptation to Climate Change: In Search of Synergies*. Bogo, Indonesia: CIFOR, pp. 122-53.

Murgueitio, E. and Solorio, B. 2008. El sistema silvopastoril intensivo, un modelo exitoso para la competitividad ganadera en Colombia y México. *V Congreso Latinoamericano de Agroforestería para la Producción Pecuaria Sostenible (Proceedings)*. Maracay, Venezuela: Universidad Rómulo Gallegos, Universidad Central de Venezuela, Universidad de Zulia (electronic publication).

Murgueitio, E., Calle, Z., Uribe, F., Calle, A. and Solorio, B. 2011. Native trees and shrubs for the productive rehabilitation of tropical cattle ranching lands. *Forest Ecology and Management* 261(10), 1654-63. doi:10.1016/j.foreco.2010.09.027.

Murgueitio, E., Galindo, W., Chará, J. and Uribe, F. 2016. *Establecimiento y manejo de sistemas silvopastoriles intensivos con leucaena*. Cali, Colombia: Editorial CIPAV, p. 220.

Nair, P. K. R. 2012. Climate change mitigation: A low-hanging fruit of agroforestry. In: Nair, P. K. R. and Garrity, D. (Eds), *Agroforestry - The Future of Global Land Use*. Advances in Agroforestry 9. Dordrecht, The Netherlands: Springer, pp. 31-67.

Nair, V. D., Nair, P. K. R., Kalmbacher, R. S. and Ezenwa, I. V. 2007. Reducing nutrient loss from farms through silvopastoral practices in coarse-textured soils of Florida, USA. *Ecological Engineering* 29, 192-9.

Nordenstahl, M., Gundel, P. E., Clavijo, M. P. and Jobbágy, E. G. 2011. Forage production in natural and afforested grasslands of the Pampas: Ecological complementarity and management opportunities. *Agroforestry Systems* 83(2), 201-11. doi:10.1007/s10457-011-9383-6.

Novotny, V. 1999. Diffuse pollution from agriculture - A worldwide outlook. *Water Science and Technology* 39(3), 1-13. doi:10.2166/wst.1999.0124.

Oldén, A. and Halme, P. 2016. Grazers increase β-diversity of vascular plants and bryophytes in wood-pastures. *Journal of Vegetation Science* 27(6), 1084-93. doi:10.1111/jvs.12436.

Oosterbaan, A. and Kuiters, A. T. 2009. Agroforestry in the Netherlands. In: Rigueiro-Rodríguez, A., McAdam, J. and Mosquera-Losado, M. (Eds), *Agroforestry in Europe: Current Status and Future Prospects*. Dordrecht, The Netherlands: Springer Science + Business Media B.V., pp. 331-41.

Orefice, J., Smith, R. G., Carroll, J., Asbjornsen, H. and Howard, T. 2016. Forage productivity and profitability in newly-established open pasture, silvopasture, and thinned forest production systems. *Agroforestry Systems*. doi:10.1007/s10457-016-0052-7.

Orefice, J., Carroll, J., Conroy, D. and Ketner, L. 2017. Silvopasture practices and perspectives in the Northeastern United States. *Agroforestry Systems* 91(1), 149–60. doi:10.1007/s10457-016-9916-0.

Osoro, K., Ferreira, L. M. M., García, U., Martínez, A. and Celaya, R. 2017. Forage intake, digestibility and performance of cattle, horses, sheep and goats grazing together on an improved heathland. *Animal Production Science* 57(1), 102–9. doi:10.1071/AN15153.

Osterburg, B., Rühling, I., Runge, T., Schmidt, T. G., Seidel, K., Antony, F., Gödecke, B. and Witt-Altfelder, P. 2007. Kosteneffiziente Maßnahmenkombinationen nach Wasserrahmenrichtlinie zur Nitratreduktion in der Landwirtschaft. Maßnahmen zur Reduzierung von Stickstoffeinträgen in die Gewässer-eine wasserschutzorientierte Landwirtschaft zur Umsetzung der Wasserrahmenrichtlinie. *Landbauforschung Völkenrode, Sonderheft* 307, 156.

Ovalle, C., Aronson, J., Del Pozo, A. and Avendano, J. 1990. The espinal: Agroforestry systems of the Mediterranean–type climate region of Chile. *Agroforestry Systems* 10(3), 213–39. doi:10.1007/BF00122913.

Paciullo, D. S. C., de Castro, C. R. T., Gomide, C. A. M., Fernandes, P. B., da Rocha, W. S. D., Müller, M. D. and Rossiello, R. O. P. 2010. Soil bulk density and biomass partitioning of Brachiaria decumbens in a silvopastoral system. *Scientia Agricola* 67, 598–603.

Paciullo, D. S. C., de Castro, C. R. T., de Miranda Gomide, C. A., Maurício, R. M., Pires, M. F. Á., Müller, M. D. and Xavier, D. F. 2011. Performance of dairy heifers in a silvopastoral system. *Livestock Science* 141(2–3), 166–72. doi:10.1016/j.livsci.2011.05.012.

Papanastasis, V. P. 2004. Vegetation degradation and land use changes in agrosilvopastoral systems. In: Schnabel, S. and Ferreira, A. (Eds), *Sustainability of Agrosilvopastoral Systems*. Advances in GeoEcology, vol. 37. Reiskirchen, Germany: Catena Verlag, pp. 1–12.

Papanastasis, V. P., Yiakoulaki, M. D., Decandia, M. and Dini-Papanastasi, O. 2008. Integrating woody species into livestock feeding in the Mediterranean areas of Europe. *Animal Feed Science and Technology* 140(1–2), 1–17. doi:10.1016/j.anifeedsci.2007.03.012.

Papanastasis, V. P., Mantzanas, K., Dini-Papanastasi, O. and Ispikoudis, I. 2009. Traditional agroforestry systems and their evolution in Greece. In: Riguero-Rodriguez, A., Mosquera-Losada, M. R. and McAdam, J. (Eds), *Agroforestry Systems in Europe: Current Status and Future Prospects*. Advances in Agroforestry Series. The Netherlands: Springer, pp. 89–109.

Pardini, A. 2002. Mediterranean pastoral systems and the threat of globalization. *Options Méditerranéennes* 62, 155–68.

Peco, B., Sánchez, A. M. and Azcárate, F. M. 2006. Abandonment in grazing systems: Consequences for vegetation and soil. *Agriculture, Ecosystems and Environment* 113(1–4), 284–94. doi:10.1016/j.agee.2005.09.017.

Peri, P. L., Dube, F. and Varella, A. C. 2016a. Silvopastoral systems in the subtropical and temperate zones of South America: An overview. In: Peri, P. L., Dube, F. and Varella, A. (Eds), *Silvopastoral Systems in Southern South America*. Advances in Agroforestry, Book-Series 11. Switzerland: Springer International Publishing, pp. 1–8.

Peri, P. L., Bahamonde, H. A., Lencinas, M. V., Gargaglione, V., Soler, R., Ormaechea, S. and Pastur, G. M. 2016b. A review of silvopastoral systems in native forests of Nothofagus antarctica in southern Patagonia, Argentina. *Agroforestry Systems* 90(6), 933–60. doi:10.1007/s10457-016-9890-6.

Perotti, E., Probo, M., Pittarello, M., Lonati, M. and Lombardi, G. 2018. A 5-year rotational grazing changes the botanical composition of sub-alpine and alpine grasslands. *Applied Vegetation Science* 21(4), 647–57. doi:10.1111/avsc.12389.

Pieninger, T. and Wilbrand, C. 2001. Land use, biodiversity conservation, and rural development in the dehesas of Cuatro Lugares, Spain. *Agroforestry Systems* 51(1), 23–34. doi:10.1023/A:1006462104555.

Piluzza, G., Sulas, L. and Bullitta, S. 2014. Tannins in forage plants and their role in animal husbandry and environmental sustainability: A review. *Grass and Forage Science* 69(1), 32–48. doi:10.1111/gfs.12053.

Pinto-Correia, T. 2000. Future development in Portuguese rural areas: How to manage agricultural support for landscape conservation? *Landscape and Urban Planning* 50(1–3), 95–106. doi:10.1016/S0169-2046(00)00082-7.

Pinto-Correia, T. and Godinho, S. 2013. Changing agriculture - Changing landscape: What is going on in the high valued Montado landscapes of Southern Portugal? In: Ortiz-Miranda, D., Moragues-Faus, A. and Arnalte-Alegre, E. (Eds), *Agriculture in Mediterranean Europe: Between Old and New Paradigms*. Research in Rural Sociology and Development, vol. 19. Bingley, UK: Emerald Group Publishing Limited, pp. 75–90.

Pittarello, M., Probo, M., Lonati, M. and Lombardi, G. 2016. Restoration of sub-alpine shrub-encroached grasslands through pastoral practices: Effects on vegetation structure and botanical composition. *Applied Vegetation Science* 19(3), 381–90. doi:10.1111/avsc.12222.

Pittarello, M., Lonati, M., Gorlier, A., Perotti, E., Probo, M. and Lombardi, G. 2018. Plant diversity and pastoral value in alpine pastures are maximized at different nutrient indicator values. *Ecological Indicators* 85, 518–24. doi:10.1016/j.ecolind.2017.10.064.

Plieninger, T. and Bieling, C. (Eds). 2012. *Resilience and the Cultural Landscape: Understanding and Managing Change in Human-Shaped Environments*. Cambridge, UK: Cambridge University Press, p. 347.

Plieninger, T., Rolo, V. and Moreno, G. 2010. Large-scale patterns of Quercus ilex, Quercus suber, and Quercus pyrenaica regeneration in Central-Western Spain. *Ecosystems* 13(5), 644–60. doi:10.1007/s10021-010-9345-2.

Plieninger, T., Hartel, T., Martín-López, B., Beaufoy, G., Bergmeier, E., Kirby, K., Montero, M. J., Moreno, G., Oteros-Rozas, E. and Van Uytvanck, J. 2015. Wood-pastures of Europe: Geographic coverage, social-ecological values, conservation management, and policy implications. *Biological Conservation* 190, 70–9. doi:10.1016/j.biocon.2015.05.014.

Porqueddu, C., Ates, S., Louhaichi, M., Kyriazopoulos, A. P., Moreno, G., del Pozo, A., Ovalle, C., Ewing, M. A. and Nichols, P. G. H. 2016. Grasslands in 'Old World' and 'New World' Mediterranean-climate zones: Past trends, current status and future research priorities. *Grass and Forage Science* 71(1), 1–35. doi:10.1111/gfs.12212.

Probo, M., Pittarello, M., Lonati, M. and Lombardi, G. 2016. Targeted grazing for the restoration of sub-alpine shrub-encroached grasslands. *Italian Journal of Agronomy* 11(4), 268–72. doi:10.4081/ija.2016.775.

Pulido, F., García, E., Obrador, J. J. and Moreno, G. 2010. Multiple pathways for tree regeneration in anthropogenic savannas: Incorporating biotic and abiotic drivers into management schemes. *Journal of Applied Ecology* 47(6), 1272–81. doi:10.1111/j.1365-2664.2010.01865.x.

Ramírez-Hernández, A., Micó, E., Marcos-García, M. Á., Brustel, H. and Galante, E. 2014. The 'dehesa', a key ecosystem in maintaining the diversity of Mediterranean saproxylic insects (Coleoptera and Diptera: Syrphidae). *Biodiversity and Conservation* 23(8), 2069–86. doi:10.1007/s10531-014-0705-7.

Rigueiro-Rodríguez, A., Fernández-Núñez, E., González-Hernández, P., McAdam, J. H. and Mosquera-Losada, M. R. 2009. Agroforestry systems in Europe: Productive, ecological and social perspectives. In: Rigueiro-Rodríguez, A., McAdam, J. and Mosquera-Losada, M. R. (Eds), *Agroforestry in Europe: Current Status and Future Prospects*. Dordrecht, The Netherlands: Springer Science & Business Media, pp. 43–65.

Rivest, D., Paquette, A., Moreno, G. and Messier, C. 2013. A metaanalysis reveals mostly neutral influence of scattered trees on pasture yield along with some contrasted effects depending on functional groups and rainfall conditions. *Agriculture, Ecosystems and Environment* 165, 74–9. doi:10.1016/j.agee.2012.12.010.

Robles, H., Ciudad, C. and Matthysen, E. 2011. Tree-cavity occurrence, cavity occupation and reproductive performance of secondary cavity-nesting birds in oak forests: The role of traditional management practices. *Forest Ecology and Management* 261(8), 1428–35. doi:10.1016/j.foreco.2011.01.029.

Rolo, V. and Moreno, G. 2012. Interspecific competition induces asymmetrical rooting profile adjustments in shrub-encroached open oak woodlands. *Trees-Structure and Function* 26(3), 997–1006. doi:10.1007/s00468-012-0677-8.

Rolo, V., López-Díaz, M. L. and Moreno, G. 2012. Shrubs affect soil nutrients availability with contrasting consequences for pasture understory and tree overstory production and nutrient status in Mediterranean grazed open woodlands. *Nutrient Cycling in Agroecosystems* 93(1), 89–102. doi:10.1007/s10705-012-9502-4.

Rolo, V., Rivest, D., López-Díaz, M. L. and Moreno, G. 2015. Microhabitat effects on herbaceous nutrient concentrations at the community and species level in Mediterranean open woodlands: The role of species composition. *Grass and Forage Science* 70(2), 219–28. doi:10.1111/gfs.12110.

Rolo, V., Rivest, D., Lorente, M., Kattge, J. and Moreno, G. 2016. Taxonomic and functional diversity in Mediterranean pastures: Insights on the biodiversity-productivity trade-off. *Journal of Applied Ecology* 53(5), 1575–84. doi:10.1111/1365-2664.12685.

Roman-Ponce, H., Thatcher, W. W., Buffington, D. E., Wilcox, C. J. and Van Horn, H. H. 1977. Physiological and production responses of dairy cattle to a shade structure in a subtropical environment. *Journal of Dairy Science* 60(3), 424–30. doi:10.3168/jds.S0022-0302(77)83882-4.

Rozados-Lorenzo, M. J., González-Hernández, M. P. and Silva-Pando, F. J. 2007a. Pasture production under different tree species and densities in an Atlantic silvopastoral system. *Agroforestry Systems* 70(1), 53–62. doi:10.1007/s10457-007-9032-2.

Rozados-Lorenzo, M. J., Garcia De Yzaguirre, A. and Silva Pando, F. J. 2007b. Comparison of selected ecotypes and cultivars of cocksfoot (*Dactylis glomerata* L.) under pinewoods in Spain. *Journal of Sustainable Agriculture* 29(2), 5–18. doi:10.1300/J064v29n02_03.

Ruiz-Mirazo, J., Robles, A. B. and Gónzalez-Rebollar, J. L. 2011. Two-year evaluation of fuelbreaks grazed by livestock in the wildfire prevention program in Andalusia (Spain). *Agriculture, Ecosystems and Environment* 141(1-2), 13-22. doi:10.1016/j.agee.2011.02.002.

Rusch, G. M., Zapata, P. C., Casanoves, F., Casals, P., Ibrahim, M. and DeClerck, F. 2014. Determinants of grassland primary production in seasonally-dry silvopastoral systems in Central America. *Agroforestry Systems* 88(3), 517-26. doi:10.1007/s10457-014-9711-8.

Sánchez-Jardón, L., Acosta, B., del Pozo, A., Casado, M. A., Ovalle, C., Elizalde, H. F., Hepp, C. and de Miguel, J. M. 2010. Grassland productivity and diversity on a tree cover gradient in *Nothofagus pumilio* in NW Patagonia. *Agriculture, Ecosystems and Environment* 137(1-2), 213-8. doi:10.1016/j.agee.2010.02.006.

Santos-Reis, M. and Correia, A. I. 1999. *Caracterização da flora e da fauna do montado da Herdade da Ribeira Abaixo (Grândola-Baixo Alentejo)*. Lisboa: Centro de Biologia Ambiental.

Saxena, S. K. and Singh, S. 1997. Rehabilitation of degraded grazing lands. In: Yadav, M. S., Singh, Manjit, Sharma, S. K., Tewari, I. C. and Burman, U. (Eds), *Silvipastoral Systems in Arid and Semiarid Ecosystems*. Jodhpur, India: Central Arid Zone Research Institute, Indian Council of Agricultural Research, pp. 123-36.

Schnabel, S., Dahlgren, R. A. and Moreno, G. 2013. Soil and water dynamics. In: Campos, P., Huntsinger, L., Oviedo, J. L., Starrs, P. F., Diaz, M., Standiford, R. B. and Montero, G. (Eds), *Mediterranean Oak Woodland Working Landscapes. Dehesas of Spain and Ranchlands of California*. Series: Landscape Series, vol. 16. Dordrecht, The Netherlands: Springer Science Business Media, pp. 91-121.

Schütz, K. E., Rogers, A. R., Poulouin, Y. A., Cox, N. R. and Tucker, C. B. 2010. The amount of shade influences the behavior and physiology of dairy cattle. *Journal of Dairy Science* 93(1), 125-33. doi:10.3168/jds.2009-2416.

Seddaiu, G., Porcu, G., Ledda, L., Roggero, P. P., Agnelli, A. and Corti, G. 2013. Soil organic matter content and composition as influenced by soil management in a semi-arid Mediterranean agro-silvo-pastoral system. *Agriculture, Ecosystems and Environment* 167, 1-11. doi:10.1016/j.agee.2013.01.002.

Seddaiu, G., Bagella, S., Pulina, A., Cappai, C., Salis, L., Rossetti, I., Lai, R. and Roggero, P. P. 2018. Mediterranean cork oak wooded grasslands: Synergies and trade-offs between plant diversity, pasture production and soil carbon. *Agroforestry Systems* 92(4), 893-908. doi:10.1007/s10457-018-0225-7.

Serrano, J., Shahidian, S., da Silva, J. M. and de Carvalho, M. 2018. A holistic approach to the evaluation of the montado ecosystem using proximal sensors. *Sensors* 18(2), 570. doi:10.3390/s18020570.

Sharrow, S. H. 1999. Silvopastoralism: Competition and facilitation between trees, livestock, and improved grass-clover pastures on temperate rainfed lands. In: Buck, L. E., Lassoie, J. and Fernandez, E. C. M. (Eds), *Agroforestry in Sustainable Agricultural Systems*. Boca Raton, FL: CRC Press, pp. 111-30.

Sharrow, S. H. 2000. Trees in pastures: Do cattle benefit from shade. *Temperate Agroforester* 8(3), 3-9.

Sharrow, S. H. 2007. Soil compaction by grazing livestock in silvopastures as evidenced by changes in soil physical properties. *Agroforestry Systems* 71(3), 215-23. doi:10.1007/s10457-007-9083-4.

Shelton, M. 2009. Feeding Leucaena to ruminant livestock: The Australian experience. In: *Proceedings II Congreso sobre Sistemas Silvopastoriles Intensivos, en camino hacia núcleos de ganadería y bosques*. Morelia, Michoacán: Fundación Produce Michoacán (electronic document).

Sigurdsson, B. D., Magnusson, B., Elmarsdottir, A. and Bjarnadottir, B. 2005. Biomass and composition of understory vegetation and the forest floor carbon stock across Siberian larch and mountain birch chronosequences in Iceland. *Annals of Forest Science* 62(8), 881-8. doi:10.1051/forest:2005079.

Silva-Pando, F. J. and González Hernández, M. P. 1992. Agroforestry helps prevent forest fires. *Agroforestry Today* 4(4), 7-8.

Silva-Pando, F. J., Lorenzo, M. J. R. and Hernández, M. P. G. 2002a. Grasslands and scrublands in the northwest of the Iberian Peninsula: Silvopastoral systems and nature conservation. In: Redecker, B., Härdtle, W., Finck, P., Riecken, U. and Schröder, E. (Eds), *Pasture Landscapes and Nature Conservation*. Berlin, Heidelberg: Springer, pp. 271-83.

Silva-Pando, F. J., González-Hernández, M. P. and Rozados-Lorenzo, M. J. 2002b. Pasture production in a silvopastoral system in relation with microclimate variables in the Atlantic coast of Spain. *Agroforestry Systems* 56, 203-11.

Sinclair, F. L. and Joshi, L. 2000. Taking local knowledge about trees seriously. In: Lawrence, A. (Ed.), *Forestry Forest Users and Research: New Ways of Learning. European Tropical Research Network (ETFRN)*. ETFRN Publication Series. Wageningen, The Netherlands: European Tropical Forest Research Network, pp. 45-58.

Singh, G., Rathod, T. R., Mutha, S., Upadhyaya, S. and Bala, N. 2008. Impact of different tree species canopy on diversity and productivity of understorey vegetation in Indian desert. *Tropical Ecology* 49, 13-23.

Smit, C., Béguin, D., Butler, A. and Müller-Schärer, H. 2005. Facilitation of tree regeneration in pasture woodlands. In: Mosquera-Losada, M. R., McAdam, J. and Rigueiro-Rodríguez, A. (Eds), *Silvopastoralism and Sustainable Land Management*. Wallingford, UK: CABI Publishing, p. 202.

Smit, C., den Ouden, J. and Díaz, M. 2008. Facilitation of *Quercus ilex* recruitment by shrubs in Mediterranean open woodlands. *Journal of Vegetation Science* 19(2), 193-200. doi:10.3170/2007-8-18352.

Sørensen, J. 2010. Nitrogen distribution and potential nitrate leaching in a combined production system of energy crops and free range pigs. Master of Science Thesis. Denmark: Aarhus University, p. 59.

Southorn, N. and Cattle, S. 2004. The dynamics of soil quality in livestock grazing systems. *Third Australian New Zealand Soils Conference*. SuperSoil 2004, Australia: University of Sydney.

Starkey, E. E., Happe, P. J., Gonzalez-Hernandez, M. P., Lange, K. M. and Karchesy, J. J. 1999. Tannins as nutritional constraints for elk and deer of the coastal Pacific Northwest. In: Gross G. G., Hemingway R. W., Yoshida T. and Branham, S. J. (Eds), *Plant Polyphenols 2: Chemistry, Biology, Pharmacology, Ecology*. Boston, MA: Springer, pp. 897-908.

Teague, W. R., Dowhower, S. L., Baker, S. A., Haile, N., DeLaune, P. B. and Conover, D. M. 2011. Grazing management impacts on vegetation, soil biota and soil chemical, physical and hydrological properties in tall grass prairie. *Agriculture, Ecosystems and Environment* 141(3-4), 310-22. doi:10.1016/j.agee.2011.03.009.

Teague, R., Provenza, F., Kreuter, U., Steffens, T. and Barnes, M. 2013. Multi-paddock grazing on rangelands: Why the perceptual dichotomy between research results and rancher experience? *Journal of Environmental Management* 128, 699-717. doi:10.1016/j.jenvman.2013.05.064.

Teixeira, R. F. M., Proença, V., Crespo, D., Valada, T. and Domingos, T. 2015. A conceptual framework for the analysis of engineered biodiverse pastures. *Ecological Engineering* 77, 85-97. doi:10.1016/j.ecoleng.2015.01.002.

Teklehaimanot, Z., Jones, M. and Sinclair, F. L. 2002. Tree and livestock productivity in relation to tree planting configuration in a silvopastoral system in North Wales, UK. *Agroforestry Systems* 56(1), 47-55. doi:10.1023/A:1021131026092.

Torralba, M., Fagerholm, N., Burgess, P. J., Moreno, G. and Plieninger, T. 2016. Do European agroforestry systems enhance biodiversity and ecosystem services? A meta-analysis. *Agriculture, Ecosystems & Environment* 230, 150-61.

Tscharntke, T., Tylianakis, J. M., Rand, T. A., Didham, R. K., Fahrig, L., Batáry, P., Bengtsson, J., Clough, Y., Crist, T. O., Dormann, C. F., et al. 2012. Landscape moderation of biodiversity patterns and processes - Eight hypotheses. *Biological Reviews of the Cambridge Philosophical Society* 87(3), 661-85. doi:10.1111/j.1469-185X.2011.00216.x.

Udawatta, R. P. and Jose, S. 2011. Carbon sequestration potential of agroforestry practices in temperate North America. In: Kumar, B. M. and Nair, P. K. R. (Eds), *Carbon Sequestration Potential of Agroforestry Systems*. Dordrecht, The Netherlands: Springer Science, pp. 17-42.

Upson, M. and Burgess, P. J. 2012. Comparison of carbon storage by agroforestry and agriculture. *UK Farm Woodland Forum Annual Meeting*. Bangor University.

Upson, M. A., Burgess, P. J. and Morison, J. I. L. 2016. Soil carbon changes after establishing woodland and agroforestry trees in a grazed pasture. *Geoderma* 283, 10-20. doi:10.1016/j.geoderma.2016.07.002.

Valipour, A., Plieninger, T., Shakeri, Z., Ghazanfari, H., Namiranian, M. and Lexer, M. J. 2014. Traditional silvopastoral management and its effects on forest stand structure in northern Zagros, Iran. *Forest Ecology and Management* 327, 221-30. doi:10.1016/j.foreco.2014.05.004.

Van Uytvanck, J., Decleer, K. and Hoffmann, M. 2008. Establishment patterns of woody species in low intensity-grazed pastures after the cessation of intensive agricultural use. *Forest Ecology and Management* 256(1-2), 106-13. doi:10.1016/j.foreco.2008.04.008.

Vandermeulen, S., Ramírez-Restrepo, C. A., Beckers, Y., Claessens, H. and Bindelle, J. 2018. Agroforestry for ruminants: A review of trees and shrubs as fodder in silvopastoral temperate and tropical production systems. *Animal Production Science* 58(5), 767-77. doi:10.1071/AN16434.

Vera, F. W. M. 2000. *Grazing Ecology and Forest History*. Wallingford, UK: CABI Publishing.

Verchot, L. V., Mackensen, J., Kandji, S., van Noordwijk, M., Tomich, T., Ong, C., Albrecht, A., Bantilan, C., Anupama, K. V. and Palm, C. 2005. Opportunities for linking adaptation and mitigation in agroforestry systems. In: Robledo, C., Kanninen, M. and Pedroni, L. (Eds), *Tropical Forests and Adaptation to Climate Change: In Search of Synergies*. Bogo, Indoneia: CIFOR, pp. 103-21.

Villanueva-Partida, C., Casanova-Lugo, F., Villanueva-López, G., González-Valdivia, N., Oros-Ortega, I. and Díaz-Echeverría, V. 2016. Influence of the density of scattered trees in pastures on the structure and species composition of tree and grass cover in southern Tabasco, Mexico. *Agriculture, Ecosystems and Environment* 232, 1-8.

Wallace, K. J. 2007. Classification of ecosystem services: Problems and solutions. *Biological Conservation* 139(3-4), 235-46. doi:10.1016/j.biocon.2007.07.015.

Walter, D. 2011. Silvopasture's effect on growth and development of white and black oaks in an intensively managed upland Central Hardwood Forest. Ph.D. dissertation. University of Missouri-Columbia.

Weidema, B. P., Wesnas, M., Hermansen, J., Kristensen, T., Halberg, N., Eder, P. and Delgado, L. 2008. Environmental improvement potentials of meat and dairy products. JCR Scientific and Technical Reports, EUR 23491 EN – 2008, p. 190.

Yadav, M. S., Singh, M., Sharma, S. K., Tewari, J. C. and Burman, U. 1997. *Silvipastoral Systems in Arid and Semi-Arid Ecosystems*. Jodhpur, India: Central Arid Zone Researh Institute, Indian Council of Agricultural Research, p. 472.

Yamamoto, W., Ap Dewi, I. A. and Ibrahim, M. 2007. Effects of silvopastoral areas on milk production at dual-purpose cattle farms at the semi-humid old agricultural frontier in central Nicaragua. *Agricultural Systems* 94(2), 368–75. doi:10.1016/j.agsy.2006.10.011.

Young, A. 1997. *Agroforestry for Soil Management*. Wallingford, UK: CAB International, p. 320.

Zianis, D., Pantera, A., Papadopoulos, A. and Mosquera-Losada, M. R. 2017. Bayesian and classical biomass allometries for open grown valonian oaks (*Q. ithaburensis* subs. *macrolepis* L.) in a silvopastoral system. *Agroforestry Systems*. doi:10.1007/s10457-016-0060-7.

Zomer, R. J., Trabucco, A., Coe, R., Place, F., Van Noordwijk, M. and Xu, J. C. 2014. Trees on farms: An update and reanalysis of agroforestry's global extent and socio-ecological characteristics. Working Paper 179. Bogor, Indonesia: World Agroforestry Centre (ICRAF) Southeast Asia Regional Program 54.

Zomer, R. J., Neufeldt, H., Xu, J., Ahrends, A., Bossio, D., Trabucco, A., van Noordwijk, M. and Wang, M. 2016. Global tree cover and biomass carbon on agricultural land: The contribution of agroforestry to global and national carbon budgets. *Scientific Reports* 6, 29987.

Zou, X. and Sanford, R. L. 1990. Agroforestry systems in China: A survey and classification. *Agroforestry Systems* 11(1), 85–94. doi:10.1007/BF00122813.

Chapter 6

Agroforestry practices: forest farming

Kenneth Mudge, Cornell University, USA

1 Introduction: setting the scene

How might we visualize forest farming? People who visit a place where forest farming is practised may be calmed by its shady surroundings on a hot summer day, and intrigued by mushrooms bursting from the log seeded by the farmer, but also by wild mycorrhizal mushrooms springing from the ground that are nutritionally collaborating with their nearby tree. They are sure to notice the gentle arc of a 6-foot-tall stalk of black cohosh (*Actaea racemosa*) flowers, as well as the arc of a black raspberry's (*Rubus occidentalis*) primocane with its tip touching the ground where roots just naturally reach out to the soil despite your attempt to train it to a wire. In autumn the palm-like leaves of wild-simulated ginseng arising on a single stalk might remind one of a mini green medusa with

http://dx.doi.org/10.19103/AS.2018.0041.06

three or four 5-fingered hands trying to keep a cluster of red berries from falling to the ground. Goldenseal's (*Hydrastis canadensis*) delicate white flowers, which, turn into a cluster of bright red berries like ginseng berries growing out of a maple leaf.

A majestic black walnut (*Juglans nigra*) by the creek is truly a multipurpose tree. It is renowned for its valuable timber that is used for veneers, gun stock and other fine woodworking. Its nuts are much loved by squirrels. They are tasty, but seriously difficult to crack, and the outer green husk quickly deteriorates to black goo that stains ones hands, shirt and even hair, which is why some folks like to use it to dye their hair brown. Black walnut casts less shade on the forest floor than most other deciduous trees in the woods like maples, oaks, hickory, ash and so on. This makes the environs beneath its canopy well suited for two valued fruit species in particular that tolerate partial shade - black raspberry and pawpaw (*Asimina trilobum*). Black raspberry is a forest delight. It does well along forest edges or in full sun. The partial sun of the forest edge is much like the partial shade beneath a black walnut tree. Pawpaw is the other lover of partial shade. In the spring, the pawpaw's flowers are dark red, like venous blood which lures the kind of pollinator insects that are attracted to fresh meat. In the fall, the globoid green fruit is like a ridiculous chubby dill pickle that is filled with a sweet custard, as the skin turns yellow, fading to black.

2 What is forest farming?

Forest farming involves deliberately cultivating high-value non-timber forest crops beneath the canopy of an existing actively managed forest, with an emphasis on ecological and economic sustainability (Mudge, 2010). Grace described forest farming as a place 'where non-timber forest crops are cultivated in the unique microclimate created by a forest canopy'. Depending on the site and circumstances, forest farming may also include:

- manipulation of the light environment by pruning or tree removal to meet the needs of specific crops;
- management for non-timber forest products (NTFP) as well as for timber;
- management to restore or maintain a healthy forest ecosystem;
- intentional cultivation of crops at either hobby or commercial scale;
- gathering of wild NTFPs (wildcrafting) in addition to cultivation.

These specialist crops grown under the forest canopy are known as non-timber forest products (NTFP). NTFPs suitable for temperate zone forest farming include food, medicinals, nursery crops and specialty wood-based and floral products. Examples of each of these categories are given in Table 1.

Table 1 Categories and examples of non-timber forest products suitable for temperate zone forest farming

Category	NTFP
Food	Mushrooms (oyster, shiitake, lion's mane, nameko, wine-cap stropharia)
	Ramps
	Tree fruit (pawpaw, persimmon)
	Berries (raspberry, blackberry, goji berry, currants, elderberry) Nuts (walnut, hickory, chestnut, hazelnut)
	Syrups (maple, birch, walnut)
Medicinals	Ginseng
	Goldenseal
	Virginia snakeroot
	Blue cohosh
	Black cohosh Bloodroot Fairy wand
Nursery stock for transplant	Landscape perennials (hosta, ferns, orchids, black cohosh)
	Small fruits (raspberry, currant, blackberry)
Specialty wood-based and floral products	Wreaths, walking sticks, carving stock, pine cones, floral greens

Source: adapted from Chamberlain et al. (2009), Mudge and Gabriel (2015), Alexander and McLain (2001).

The forest canopy provides multiple ecosystem services, which include protecting and regenerating soils, restoring aquifers, controlling floods and preventing drought, as well as supporting biodiversity (Sholto and Hart, 1985). In a forest farming system shade is paramount, but shade provided by the forest canopy is both enabling on the one hand and limiting on the other. Shade or partial shade enables the cultivation of the shade-adapted NTFPs typically found in a forest farm such as ginseng, mushrooms, ramps and so on. Crops that require or thrive under full sun are not necessarily excluded from the forest farm, provided they are located along the perimeter of the forest, or locations where partial shade exists, although yields may be lower than they would be under full sun.

The effect of forest shade on microclimate is more than just its effect on the level of photosynthetically active radiation available for photosynthesis, but also other ecosystem services, including reduction in air temperature which reduces transpiration of understorey plants, and reduced soil temperature (Scott and Sullivan, 2007) which reduces evaporation of soil moisture. A leafy canopy may reduce wind turbulence and increase atmospheric relative humidity, both of which further contribute to minimizing water stress of understorey vegetation.

The more valuable shade-tolerant NTFPs (e.g. ginseng, mushrooms) could be and sometimes are grown in the open by constructing artificial shade. As will be discussed below, ginseng grown under artificial shade is not forest farming, nor is it any other agroforestry practice. The key to cultivating ginseng as a forest farming crop is to find a location suitable for ginseng cultivation (natural shade, slope, aspect, soil). Site selection for American ginseng cultivation is discussed below.

A key reason for the interest in forest farming is the amount and type of forest available in countries such as the United States. Despite a trend towards urbanization, much of the eastern United States remains forested, and in fact for many eastern states the percentage of forested land has been steadily increasing since the early twentieth century as marginal agricultural land was being abandoned and open fields were reverting to secondary forest. In the eastern United States, where forest cover is greatest, per cent forest cover ranges from about 60% to 80%. Thirteen per cent of the United Kingdom is forested, which is approaching levels in 1750 AD (Forestry Statistics, 2017). Forestation of Europe as a whole is 43% (Eurostat Statistics Explained, 2017), with the greatest forest cover in Russia at 81%.

Most forested land in the United States and elsewhere is largely or entirely unmanaged, resulting in tree density, high enough that the quality of timbers is compromised. Private forest owners in search of income often resort to high grading which means harvesting the only best trees and leaving the rest of lesser size and quality. As is the case with agroforestry in general, forest farming is usually practised on agricultural marginal land that is not well suited for conventional agriculture. Poor soils (rocks, fertility, drainage) and topography (steep slopes) account for much of the forested land and hence, what is available for forest farming.

It should be noted that the agroforestry practice of silvopasture is similar to forest farming in that it takes place under an established forest canopy. Instead of plants and fungi, the NTFPs in a silvopasture system are the 'crops that walk', that is livestock. Other agroforestry practices including alley cropping, wind breaks, riparian buffers or forest gardens usually start 'from scratch' and usually require at least several years lead time until newly planted trees or shrubs have grown enough to become functional as a source of fruits, nuts or services like shade, wind abatement or stream bank stabilization. On the other hand, forest farming 'hits the ground running' because the shade, a key element in forest farming, is already in place.

There is a tendency of academics, government scientists and others who write about agroforestry to emphasize the differences among various agroforestry practices, but to many small-scale farmers with woodlots or other private forest or woodlot owners, their land is likely to be heterogeneous in terms of terrain, soils, access to water and so on, such that no single agroforestry practice is appropriate for the entire farm, so that more than one agroforestry practice may be operating simultaneously on the same farm.

It is also important to highlight that a forest farm is a place not only for growing non-timber forest crops but also a place where productive conservation is practised. In addition to cultivation of NTFP and conservation of forest resources and ecosystems, another goal, explicit or not, is public education; a place where people come to relax and to learn about productive conservation, that is forest farming. Most will expect to see the rock stars of forest farming - ginseng (*Panax quinquefolius*) and mushrooms - but in a well-designed forest farm they should also come away with an appreciation of the balance between cultivation of NTFP and a naturalistically managed forest ecosystem that is like the one that supports these plants and many others in the wild. Forest farming can be an opportunity for ecotourism such as in 'Old MacDonald's Farm' (traditional nursery rhyme), whether it is for fee or for free.

3 Estimating yields in forest farming

In the case of a forest farm managed for multiple NTFPs, it is important to note that yield of each crop on the site may not be optimized, but rather that the total yield (and monetary value) may be greater than the sum of all of the crops as if grown individually on separate parcels of land. A standard approach for quantifying the relative yields of intercropping versus sole cropping is land equivalent ratio (LER), which is widely used to estimate the efficiency of multicropping of two or more species compared to monocropping of the same two species. LER is defined as:

intercrop1/pure crop1 + intercrop2/pure crop2 + etc. = LER (Kantor, 1994)

LER is more frequently applied to agronomic or horticultural crops such as orchard fruit trees interplanted with berries, arranged in orderly rows. It is more difficult to apply LER to a forest farming setting where (deliberately or not) irregular placement of trees or other NTFPs is likely to occur. For this reason, LER has not been reported for NTFPs in a forest farming situation. *In lieu* of that, Table 2 shows a hypothetical approximation of LER applied to intercropping of black walnut and pawpaw versus sole crop of each in an orchard setting.

Pawpaw is resistant to both walnut toxicity and deer foraging. Since the black walnut canopy in the overstorey receives the same amount of light regardless of pawpaw in the midstorey, its yield would be largely unaffected by the pawpaw. Pawpaw, on the other hand, performs best in full sunlight, so its yield would be reduced under the walnut canopy. Based on realistic but hypothetical yields for each crop grown together (multicrop) compared to reported (not hypothetical) yields of each grown separately, Table 2 shows that

Table 2 Land equivalent ratio (LER) in an orchard setting, calculated using reported yield for sole crops of walnut (Harper, 1998) and of pawpaw (Cantaluppi, 2016), and expected (reduced) yields of each of these crops grown in a double intercropped system, with pawpaw under a canopy of black walnut

Crop	Yield of intercropped crop (lb/acre)	Yield of sole crop	Partial LER
Black walnut	900	1000	0.9
Pawpaw	1320	1650	0.8
Total LER			1.7

LER = BW intercrop/BW sole + PP intercrop/PP sole crop.

total LER is 1.7. LER >1 indicates that the total yield for the multicrop is greater than the sum of the crops grown separately.

Determining LER for some other crop combinations in a forest farming setting would be less revealing or more difficult to interpret. For example, consider a sugar maple (*Acer saccharum*) overstorey with a shiitake mushroom (*Lentinula edodes*) and/or an American ginseng understorey. Sugar maple yield (litres of syrup) would not be affected by either understorey crop, but neither the mushrooms nor the ginseng could survive in full sun without the sugar maple overstorey, so 'sole crop' yield used to calculate LER could not be applied in this situation.

Nonetheless there are important positive interactions between the sugar maple and ginseng. Based on the work by Beyfuss described in Davis and Persons (2014), he found that ginseng has a high requirement for calcium as indicated by its positive response to the addition of gypsum ($CaSO_4$). Sugar maple leaves have a high calcium content (Mitchell, 1937) so the ginseng would benefit by annual leaf fall and decay of calcium-rich sugar maple leaves. Research to develop novel approaches to quantify the efficiency of intercropping in forest farming systems would be an useful endeavour.

4 Site selection and characteristics

Throughout the history of agriculture, site modification has been a key element in successful crop cultivation. Ploughing, fertilization, irrigation, mechanization and other interventions have contributed to the reliability and sufficiency of food production. In the case of forest farming, however, opportunities for similar site modification are greatly restricted and impractical. Site modification is limited to a few site characteristics such as fencing to control deer, and possibly access to water by diverting a stream, or thinning or pruning of trees to increase available light levels, but site modification through fertilization, irrigation or mechanization is not practical in most cases. Usually the forest farmer must choose the crops to fit the site rather than modify the site to suit the crop. Without the appropriate level of shade, aspect, soil, slope and drainage to begin with, the probability of success is reduced.

Broadly speaking, there are two ways to select a site for forest farming. The first assumes that the forest farmer starts out with a fixed geographic location and no other choice of location is available. In this case it is more important to ask what crops are appropriate for that site. On the other hand, prospective forest farmers may (and often do) start with the question 'How do I grow ginseng on my site?' If siting of the forest farm is limited to a particular piece of land with only southern exposure, ginseng is probably not a viable option unless steps are taken to improve the site (see Section 12). Without any appropriate aspect, such as lighting, slope and drainage, successful cultivation of ginseng is unlikely. Instead, a better approach would be to select NTFPs that are appropriate to existing site conditions, such as brambles (berries in the genus *Rubus*) on a relatively warm, semi-shaded southern exposure. On the other hand, if there is some flexibility as to site selection and if growing ginseng is an overriding goal, then it would be prudent to select a site with a northeast-facing slope and a high-degree canopy cover. Site selection is more than just the appropriate biophysical characteristics. The forest owner's goals and aspirations are just as, if not more important.

Keep in mind that site conditions are often heterogeneous over an area designated as a forest farm. For example, at the five-acre MacDaniels Nut Grove at Cornell University there is a south-facing slope curving around to a west-facing slope, partially enclosing flat land bisected by a stream with well-drained alluvial soil on the south side and a poorly drained clay soil on the other (Fig. 1).

When selecting an appropriate site for locating a forest farm and non-timber forest crops that are suitable for that site, there are several site characteristics that should be taken into consideration. These are discussed further in subsequent sections.

4.1 Plant hardiness zones

Plant hardiness zones are intended to predict which plants are most likely to thrive at a given location based on winter hardiness. The original plant hardiness zone map, developed by the US Department of Agriculture, which has been refined recently (http://planthardiness.ars.usda.gov/PHZMWeb/), is based on average annual winter minimum temperature. The same criterion has been applied to the development of hardiness zone maps throughout much of the world (http://www.edible-landscape-design.com/plant-hardiness-zone.html #region). While minimum winter temperature is an important predictor of plant survival, the USDA map does not take into consideration other regional limiting factors. On the other hand, the Canadian hardiness zone maps developed by Natural Resources Canada (http://planthardiness.gc.ca/?m=15&lang=en) are based not only on minimum winter temperature but also other limiting factors, including number of frost-free days, rainfall, maximum temperature, snow depth and wind gust.

Figure 1 MacDaniels Nut Grove at Cornell University. Looking South, across the stream, at 'Walnut Island' which includes, in addition to black walnut trees, pawpaw, black raspberries, wine-cap stropharia (*Stropharia rugoso-annulata*) mushrooms and stream bank willows (*Salix matsudana*).

4.2 Light

In forest farming, the degree of light transmission/shading by a tree canopy is obviously a critical factor for the survival of shade-loving plants or plants that require or will tolerate moderate shade. With respect to shade in a forest farming situation, all tree species are not equal. Canham et al. (1994) ranked light transmission of the leafy canopy for six forest tree species:

> Hemlock (*Tsuga canadensis*) (1% full sun) < beech (*Fagus grandidentata*) < sugar maple < red maple (*Acer rubrum*) < red oak (*Quercus rubra*) < ash (*Fraxinus* sp.) (6% full sun).

Two species that cast low-to-moderate shade include black walnut (Pavlis, 2017) and honey locust (*Gleditsia triacanthos*) (Nowak, 1996) which cast less shade than most other temperate canopy tree species.

There are several methods for determining light transmission through a leafy canopy. These include hemispherical photography, densitometer and ocular estimation. Hemispherical photography is the most precise but requires a camera, a wide-angle lens (Fig. 2) and appropriate software (Chianucci

Figure 2 Hemispheric photo of the Bavarian Forest in 2011. Permission by Wegman, through Wikimedia Commons. Photo taken with a wide-angle lens. Software is then required to estimate per cent canopy cover.

and Cutini, 2012). A densitometer is an inexpensive handheld instrument with intermediate accuracy (Lemmon, 1957), and the third method, ocular estimation, is the simplest but least accurate. It involves merely standing under a leafy canopy to estimate per cent light transmission compared to a printed template. Ocular estimation requires no instrumentation and minimal training other than a template published by Barnhardt and Swiecki (1991).

Keep in mind that a deciduous canopy can cast considerable shade (reduced light transmission) during the leafy growing season but light transmission through defoliated tree canopy will be higher during the dormant season. For most NTFPs, this seasonal variation in lighting will be inconsequential because crops like ginseng will have gone dormant (died back to ground level) in the fall, along with defoliated overstorey trees, but in the case of mushrooms cultivated on logs, winter sun can result in significant drying of the logs which may reduce their productive life. Evergreen conifers on the other hand cast shade throughout the year, with hemlock casting the greatest amount, making it ideal for a mushroom laying yard.

Vertical stratification is another characteristic of the light environment in forest farming system, which renders the NTFP growing space three dimensional:

- the upper storey (canopy) of tall trees;
- a midstorey of smaller trees and shrubs (e.g. brambles, pawpaw);
- an understorey of (mostly) herbaceous plants and fungi (e.g. ginseng, mushrooms).

Some plants can grow across these storeys such as vertical vines. However, for temperate forest farming, there are few shade-tolerant vines that produce a marketable NTFP other than the hardy kiwi (*Actinidia chinensis*), but which performs best under full sun.

4.3 Soil and water

Soil is an important consideration in any agricultural operation including forest farming. Soil type (order/series) for a given location and appropriate land uses for different soils within the United States can be accessed from the NRCS Web Soil Survey (https://websoilsurvey.sc.egov.usda.gov/App/HomePage.htm), and in Europe from the European Soil Data Center (https://esdac.jrc.ec.europa.eu/resource-type/maps). These are broad (low-resolution) classifications that may be of limited value for a small and somewhat heterogeneous forest site. Some soil tests can be conducted on site. The characteristics which are likely to be most useful and most predictive of successful cultivation of NTFPs are soil texture which is an indicator of drainage, soil pH and soil organic matter content.

Water is a resource that may be overly abundant during some seasons and in some locations, but at other times and other places it may be quite limiting to the growth of NTFPs. Not much can be done about the latter unless a well, spring or stream is on the property; however, delivering water from any of these sources to plants in need of irrigation may be a nontrivial matter. Irrigation in the forest from an off-site source is not usually an option unless one has an extremely long hose to provide plumbed water from the home or other source. Water harvesting and soil conservation measures can be handled through 'passive' water management. A system of swales and berms is an effective way to slow down the rate of water flowing downhill and thereby minimize erosion. A swale is simply a ditch that is dug on a contour along a hillside. The soil removed to create the ditch is piled on the downhill side of the ditch, creating a berm. As rapid run-off during a rain event, that might otherwise cause erosion, is intercepted and slowed down by flowing into a swale, some of that impounded water is absorbed by the berm. NTFPs planted on the berm are passively irrigated, using water that might otherwise create erosion and be lost to the system as it rapidly runs off the hillside (Mudge and Gabriel, 2015). Another approach to passive water management that takes advantage of seasonal variation in rainfall is the use of roof catchment, which is designed to collect water off the roof of a building or other raised structure and channel it through a gutter and pipe that directs it to a cistern or directly to plants, for use during periods when rainfall is not reliably available (Daily and Wilkins, 2012).

4.4 Slope and aspect

Slope refers to the angle of a hill from the horizontal, expressed in degrees or per cent. Per cent slope is calculated from the vertical rise (elevation) over the run (horizontal distance). Hence, a 45° slope = 100% slope. In the case of most non-timber forest crops, the steeper the slope, the more inconvenient in terms of access for planting, maintenance or harvesting. As described above, steep slopes are subject to erosion during heavy rainfall and soil moisture deficit during drought. Mitigation of erosion may require the labour-intensive installation of swales to make the hillside productive. On the other hand, ginseng or other NTFPs that do not tolerate standing water benefit by run-off down a moderately steep slope.

Aspect refers to the compass direction that a slope is facing which affects the temperature and soil moisture content. A south- or southwest-facing slope, which intercepts afternoon sun, tends to be warmer and dryer than a north- or northeast-facing slope.

5 Growing mushrooms

As has been noted above, two leading NTFPs are shiitake mushrooms and ginseng. Shiitake mushrooms are a companion crop that are particularly well suited to complement ginseng cultivation. Rather than the 8–10 years needed for ginseng cultivation from seed to harvest, mushrooms typically produce a crop within a year of log inoculation, and a well-managed log can produce up to two or more crops in a single growing season. Furthermore, a log can continue to be productive for several years. The value of shiitake mushroom, depending on the market, is typically in the vicinity of US$10 or more per pound fresh weight. Hence, shiitake mushroom can begin to generate significant income much sooner than ginseng and continue to produce for several years, in contrast to ginseng which is gone after a single harvest almost a decade after planting.

It is important to understand that shiitake is not the only mushroom species that can be cultivated in a forest farming context, but it is a 'best bet', as will be discussed below. Others include oyster (*Pleurotus* sp.), lion's mane (*Hericium* sp.), wine cap (*Stropharia rugoso-annulata*) (Fig. 3) and nameko (*Pholiota microspora*) (not shown). All of these are saprophytic basidiomycetes that feed on dead organic matter. In the forest farm, they grow almost exclusively on wood substrate. Oyster mushrooms are less fastidious in terms of their substrate preference. Many high-cellulosic substrates such as straw, paper, coffee grounds and sawdust will do, but none of these are particularly well suited to forest farming.

Several well-known forest mushrooms that are considered choice edibles by collectors of wild mushrooms and have never been successfully cultivated under forest farming conditions include porcini (*Boletus edulis*), chanterelles (*Cantharellus cibarius*) and matsutake (*Tricholoma matsutake*). These form obligate, mutualistic symbiotic associations, called ectomycorrhizae, with the roots of certain living trees. Cultivating mushrooms from these two-part mycorrhizal symbioses have proven to be extremely challenging, and in most cases cultivation was never successful. Honey mushrooms (*Armillaria mellea*) and maitake (*Grifola frondosa*) are not mycorrhizal, but they have nonetheless not been successfully cultivated by forest farmers. Forest cultivation of these recalcitrant mushroom species could be quite useful especially considering growing public demand for non-traditional (white button) mushrooms (*Agaricus bisporus*).

Truffles are an exception. They have been successfully cultivated for a long time in Europe. In North America truffle cultivation has proved to be challenging. Truffles are highly valuable ectomycorrhizal ascomycetes. Perigord truffle (*Tuber*

Figure 3 Mushroom species suitable for forest farm. Clockwise from top left, shiitake, oyster, wine-cap stropharia and lion's mane. (Nameko not shown).

melanosporum) is grown in France, and the most valuable is the white truffle (*T. magnatum*) grown in Italy. Both are extremely difficult to cultivate, hence extremely valuable (>>US$1000/lb fresh weight) and attempts to grow them in North America have met with limited success. As a result of these astronomical prices, there is considerable public interest in truffle farming, but it can be reliably said that in North America, sellers of truffle-inoculated tree seedlings have made considerably more money than anyone that has actually grown truffles from inoculated seedlings to maturity and sold them. Successful truffle cultivation is more of an orchard practice rather than a forest farming practice.

Shiitake mushrooms are native to East Asia. They are the third most commonly cultivated mushroom in the world after button (*Agaricus*) and oyster mushrooms. Forest cultivation of shiitake is limited to the more humid (eastern) parts of North America, as well as parts of Mexico, Central America, Brazil, the United Kingdom and countries in Europe. Commercial production of shiitake outside of Japan was not undertaken until the early 1980s in North America, in response to a seminal publication by Leatham (1982).

Shiitake cultivation can occur either indoors (not forest farming) or outdoors (forest farming). The former is the source of most commercial production, but it requires more investment in infrastructure to provide the necessary climate control. Consequently, production costs (lighting, humidity and temperature control) are higher for indoor compared to outdoor production. Commercial forest cultivation on hardwood logs ranges from several hundred logs to several thousands. Log-grown outdoor cultivated shiitake mushrooms are considered by many consumers to be a superior product compared to indoor factory-grown shiitake, as indicated by the fact that consumers are willing to spend one and a half time to twice or more for a pound of log-grown shiitake (US$10 at a minimum) than for factory-grown shiitake (~US$7–8) (Mudge and Matthews, 2011). Despite the price differential, there is considerable public demand for log-grown shiitake.

A wide range of deciduous hardwood tree species are suitable for shiitake production including the oaks (*Quercus* sp.), American beech and other substrate species including hard maple (*Acer saccharum*), ironwood (*Carpinus betulaefolia*), hop hornbeam (*Ostrya carpinifolia*) and tulip poplar (*Liriodendron tulipifera*) especially in the south. Red maple and poplar (*Populus* sp.) are considered by growers to be relatively poor substrates. Oak is usually considered the 'gold standard' (best). Beech is a good substrate when managed correctly, but it has not been frequently used because of the bark's tendency to peel off which accelerates loss of moisture. Conifers, black locust and ash should be avoided.

Why do different substrate tree species perform better than others? Our experiment shown in Table 3 indicates that beech significantly outperformed red oak, aspen and red maple in terms of mushroom production over a 4-year

Table 3 Effect of substrate tree species on initial (2006) % log moisture content (LMC) and final (2009) LMC, and on cumulative shiitake mushroom production (g dry wt./log)

Tree species	Initial LMC (%) (2006)	Final LMC (%) (2009)	Mushroom (g dry wt./log)
Red maple	43	40	39b
American beech	41	39	74a
Big tooth aspen	47	41	5c
Red oak	38	42	42b

Source: adapt from Mudge et al. (2011).

period. Red maple and red oak were not significantly different, but poplar was a very poor substrate. This experiment was designed to test the hypothesis that mushroom yield for a given tree species is inversely related to moisture loss from substrate logs, that is tree species that retain more moisture should have higher mushroom yield. However, we did not observe the expected correlation between log moisture content (LMC) and mushroom production. At the end of the experiment, LMC of all the four tree species was about the same, whereas there were large differences in mushroom yield among species (Mudge et al., 2011; Mudge and Gabriel, 2015). Ranking of mushroom production among substrate tree species in this experiment (beech > red oak > red maple > aspen) differs from results often reported by commercial growers that red oak is the 'gold standard' (i.e. outperforms other species). This is probably due to site-to-site environmental effects, that is differential effects of temperature, rainfall and so forth on mushroom production from different tree species. It may also be the case that effects of environmental conditions on tree growth, nutrient content, age and so on may account for difference among tree species on mushroom production. Here, we recommend that a beginning grower test several locally available substrate tree species rather than rely entirely on 'conventional wisdom'.

6 Shiitake mushroom cultivation

Live trees are usually cut while dormant (winter or early spring). A 2- to 3-week 'rest' between tree cutting and inoculation is recommended. Holes are drilled into the 3-foot-long logs (bolts) and spawn is inserted into the holes, and each hole is covered with wax (Fig. 4). Spawn is a mixture of substrate (sawdust, wooden dowels or grain) colonized with fungal mycelium (not spores) under sterile laboratory conditions. The period of time during which logs, inoculated with spawn, are colonized by the fungus, known as spawn run, ranges from as little as 8 to 18 months, with 1 year being typical. This takes place in the laying yard (Fig. 5) which must be located under a well-shaded tree canopy. The most important management consideration during the spawn run and afterwards is to avoid drying of the log. The most important way to achieve that is to keep the

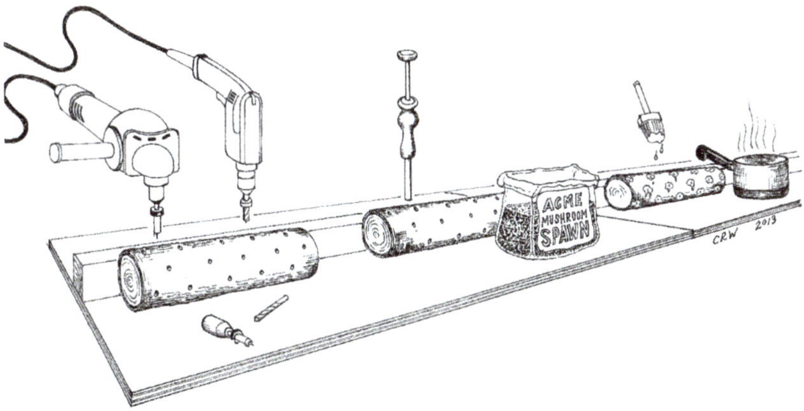

Figure 4 Steps involved in inoculation of a substrate log with shiitake spawn, from left to right. Drilling can be either performed with a shop drill (~2500 rpm) or preferably with a modified angle grinder (10 000 rpm) since it drills holes three times faster. The bag contains sawdust spawn (fungal mycelium) and the inoculation tool, shown at the same log, injects a measured amount of spawn into each hole. Holes are waxed to avoid drying. Note that the term 'spawn' refers to the vegetative mycelium of the fungus, not to spores.

Figure 5 Laying yard where inoculated logs are colonized with the fungus (spawn run), for up to about 1 year. Logs arranged in crib stacks, shown below, or other configurations. Note heavily shading hemlock canopy fulfils the absolute requirement for shade to avoid moisture loss from logs.

logs shaded at all times. Logs are sometimes irrigated, during hot dry weather, by soaking for about 2 h or by sprinkling for several hours to maintain high LMC.

One important marketing advantage with shiitake compared to other forest-cultivatable fungi (oyster, lion's mane, wine cap, nameko) is that shiitake can be forced to fruit after spawn run 'on demand' by soaking the log in water (e.g. cattle trough) for 12–24 h. In a few days 'pins' appear (early stage mushroom primordia) and about a week later mushrooms are ready to be harvested (Fig. 6). The same log can be force fruited again after an 8-week resting period, for a total of at least two flushes each growing season, annually for 3 or 4 years. An average forcing from a single log is typically about half-pound fresh weight of mushrooms. Hence over a 3-year life, a typical log, harvested twice per year, can produce approximately 3-4 lb of mushrooms, which, if sold at US$10/lb, generate (conservatively) US$30–40 income.

For shiitake and some other mushrooms, there are a number of commercially available strains (mycelial clones) of spawn that can be used to facilitate scheduling of fruiting, in order to match production to demand. Three temperature-related categories of spawn strains are available that differ in the temperature range at which they fruit. These are CW, cool weather (45–70°F); WR, wide range (55–75°F); and WW, warm weather (50–85°F). By using a combination of CW and WR or WW strains, a grower can extend the growing season by many weeks since the CW strains fruit early in the cool spring while WR and WW strains have not begun to fruit. The same applies to CW strains in the cool fall, which fruit for some weeks after WW and WR strains are no longer fruiting. Furthermore,

Figure 6 Shiitake mushrooms growing on a red oak log.

CW strains do not require the cold water forcing that is necessary to get WR and WW strains to fruit during their optimum fruiting temperature.

Other edible fungi that can be cultivated as an NTFP for forest farming include oyster, lion's mane, stropharia and nameko. Of these, only oyster is well known to the public, but, in our experience, oyster mushrooms are not particularly well suited for outdoor cultivation. The market demand for log-grown lion's mane, stropharia and nameko is far less than that for shiitake. Most people have never heard of them. It also has to do both with reliability and productivity of the cultivation process. None of these other specialty forest mushrooms can be reliably forced by soaking as is the case for shiitake so that their fruiting is unpredictable, depending more on natural variation in temperature, rainfall and humidity than a grower's schedule. Although oyster mushrooms may be highly productive indoors on cellulose-rich substrates like straw, under more climate-controlled conditions, oyster mushrooms grown on bolts (logs 4–6" diameter × 36" long), under a forest canopy, are not as productive as shiitake. Similarly, lion's mane is less productive than shiitake when grown on bolts, but our research at Cornell has shown that over several years, lion's mane grown on totems produced about the same yield (without forcing) as typical shiitake production over a similar period of time (Grace and Mudge, 2015). This conclusion arose from an experimental comparison of several wild strains of lion's mane that had been collected locally (Central New York). A project by the European Network for Rural Development is undertaking a comparison of wild strains of several local mushrooms in an attempt to develop new income stream for Finnish forest owners (EAFRD, 2014–20).

Although a great deal is known about forest cultivation of shiitake mushrooms, most modern practices are based on grower findings under non-experimental conditions (lacking controls, unreplicated and large inter-farm variation). One exception is a series of experiments at the University of Missouri Center for Agroforestry where Bruhn et al. (2009) have contributed considerably to our understanding of the effects of cultural conditions on shiitake cultivation. Research on the effects of shiitake strain and substrate tree species by Sabota (1996) found relatively little difference among strains and more pronounced differences among substrate tree species. As described above (Table 3), Mudge et al. (2011) reported on the relationship between LMC and shiitake production on four substrate tree species, and found that mushroom production was not correlated with LMC, regardless of substrate tree species. These publications are rather few and far between. The field would benefit greatly by research directed towards how forest-site conditions (soil, exposure, rainfall, season, tree age) influence the growth of different tree species and in turn their efficacy as substrates for shiitake cultivation. Site to site variation in external conditions could help explain why ranking of different substrate tree species often varies among commercial growers. Even less research has been conducted on outdoor cultivation of the other four candidate mushroom species described above (oyster, lion's mane, nameko and wine-cap stropharia). Grace and Mudge (2015) compared the performance of different wild-collected strains of lion's mane. Bruhn et al. (2010) compared woodchip versus straw substrates for outdoor cultivation of wine-cap stropharia. There are some mushroom species that are being grown commercially indoors but have not been successfully grown outdoors including maitake, sulphur shelf, bluets and others. Research could help growers introduce these 'new' mushroom species for commercial outdoor cultivation. Farm crop diversity is generally considered a good thing.

An excellent practical resource (pamphlet) on shiitake cultivation is *Growing Shiitake Mushrooms in a Continental Climate* (Kozak and Krawczyk, 1993; currently out of print). Their website contains a wealth of practical information about mushroom cultivation (fieldforest.net).

7 Growing woodland medicinal plants

There are many forest herbs marketed or personally wildcrafted (collected from the wild) for health-related uses, but only a few of these are actually profitable when deliberately cultivated under forest shade. One reason for this is that the herbal products industry tends to pay the same prices for either wild-collected or forest-cultivated material for most woodland medicinals, with the exception of woods-cultivated American ginseng. Hence, there is little incentive for cultivation of minor medicinals that could be more easily collected from the wild. This may change as certain medicinal species become increasingly scarce because of over harvesting from the wild, or from habitat destruction and so on.

Although the term woodland medicinal is used to describe plants that, broadly speaking, are used to promote health, in the United States they are not legally classified as pharmaceutical medicines but rather they are classified as dietary supplements which are not required to undergo the rigorous testing required to meet the 'safe and effective' standard applied to pharmaceuticals.

In the European Union, woodland medicinals are classified as 'traditional herbal medicinal products'. As in the United States, proof of clinical effectiveness is not required but they must present an acceptable level of safety. These regulations in the United States, and in Europe, are worth noting since they affect marketing and sale of wildcrafted or cultivated woodland medicinals. The relative lack of regulatory scrutiny is a boon to marketing of woodland medicinals. On the other hand, acquiring proof of safety and efficacy is expensive. Many in the pharmaceutical and medical professions contend that this lack of rigorous testing may put consumers at risk (Mayo Clinic Staff, 2017).

Cultivation of woodland medicinals fall into two categories. First ginseng. Second, everything else. American ginseng is by far the most valuable but not necessarily the best choice of NTFP for a beginning forest farmer. All other

Figure 7 Second only to American ginseng, fairy wand is a valuable (US$105/dry lb) woodland medicinal herb.

woodland medicinals are far less valuable than ginseng, but some are still worth considering for forest farming, including goldenseal and fairy wand.

Aside from ginseng, there is very little deliberate forest cultivation of other woodland medicinals, for which wildcrafting supplies nearly all of the demand from the herbal products industry. Without a premium paid for cultivation, there is little incentive to undertake the time, expense and effort involved in cultivation of these less-valuable forest herbs. There are exceptions however. Based on economic modelling involving the woods cultivation of nine medicinal herbs including ginseng, goldenseal, bloodroot (*Sanguinaria canadensis*), black cohosh, blue cohosh (*Caulophyllum thalictroides*), fairy wand (Fig. 7), trillium (*Trillium erectum*), spikenard (*Aralia racemosa*) and mayapple (*Podophyllum peltatum*), Burkhart and Jacobson (2009) concluded that only ginseng and to a much lesser (marginal) extent goldenseal could be forest farmed profitably. Davis and Persons (2014) were more optimistic about the potential profitability of goldenseal. In a hypothetical enterprise budget, using highest reported yield, 0.1 acre of woods-cultivated goldenseal generated US$8219 after 4 years. This hypothetical profit could encourage others to attempt to cultivate goldenseal as a forest farming crop. These projections suggest that profitable forest cultivation of goldenseal is not out of the question but should be approached with caution.

8 Ginseng cultivation

Forest owners are frequently attracted to forest farming because of the eye-popping value of wild-collected or wild-simulated ginseng which approaches US$1000/lb dry weight. Even if a given prospective site happens to be well suited for ginseng, a beginning forest farmer would be well advised to consider growing another NTFP in addition to ginseng for crop diversity and 'insurance' against crop failure. In the case of ginseng the risks of poaching and predation stand out especially since eight or more years must elapse before harvest.

It should be noted that American ginseng (*Panax quinquefolius*) (Fig. 8) is an entirely different species from Asian ginseng (*Panax ginseng*). Cultivation, collection and commerce of ginseng in North America is almost entirely American ginseng, most of which is exported to China. Asian ginseng is rarely cultivated in North America, but it is exported from China to North America for beverages, energy boosters and other questionable formulations. Both species are used in traditional Chinese medicine. Reference to ginseng in this chapter refers entirely to American ginseng.

American ginseng is by far the most valuable forest herb, followed distantly by goldenseal and fairy wand (Burkhart and Jacobson, 2009). The price paid for ginseng depends on several factors, as discussed below. Retail prices for goldenseal is US$25/lb and for fairy wand (*Chamaelirium luteum*) US$105/lb.

There are four different methods for cultivating or collecting American ginseng, ranked here from most intensive in terms of labour (and other factors) to least intensive. They are:

1 Field cultivation (artificial shade)
2 Woods cultivation
3 Wild simulated
4 Wildcrafted (collected from the wild)

8.1 Field cultivation

Field cultivation (FC) under artificial shade does not qualify as forest farming per se because it does not involve trees for shade. Nonetheless, FC is worth mentioning in this chapter since it is the main method of production worldwide for this valuable crop. The particulars of FC are useful to understand the inverse relationship between intensity of cultivation (labour and infrastructure) and the value of the crop as shown in Table 4. FC takes this shade-loving forest herb (ginseng) out of the woods and grows it under artificial shade, that is, in some cases, acres in extent. The necessary infrastructure and maintenance of shade structures is decidedly more costly and labour intensive than natural shade from a forest canopy. Construction of artificial shade structures for FC of American ginseng, as is practised mainly in parts of Wisconsin and Ontario, costs US$15 000–35 000/acre (Eric Burkhart, *pers. comm.*). Ironically, intensively cultivated ginseng under artificial shade is worth far less per pound than WC, WS or wildcrafted ginseng (Table 4). Cultivation of ginseng under artificial shade is as horticulturally intensive as practically any high-value intensively

Figure 8 Morphology of American ginseng. Ginseng is a slow-growing, shade-requiring forest herb. It is mainly the root and rhizome (neck) that are commercially valuable, but there is a market for seeds as well, which growers use to propagate new plants. Permission: Carl Whittaker.

managed field-grown crop such as cut flowers, wine from grapes or specialty nursery crops. Field-cultivated ginseng's high yield per acre can be attributed primarily to its high planting density (Table 4). Such densely planted ginseng, only a few inches between plants, is prone to fungal diseases, especially *Phytophthora* and *Alternaria* blight, which necessitates liberal application of fungicides. Another factor that contributes is the high maintenance cost, but

Table 4 Inputs (expenses) and outputs (income) for cultivating American ginseng by three different production systems on 0.1 acre, and wildcrafting

	Seeds/sq. ft.	No. plants/ sq. ft.	Years to harvest	Value (US$/lb)	Yield (US$)	Labour (US$)	Gross profit (US$)	Total expenses (US$)	Net profit (US$)
Field cultivated (artificial shade)	20	15	3–4	70	200	6000	14000	7415	6585
Woods cultivated	11	8–10	6–8	300	60	5850	18000	7051	10949
Wild simulated	4–5	1–2	9–10	900	16	1170	14400	2221	11779
Wildcrafted		<1	10+	1000	na	na	na	na	na

With respect to wildcrafting, the size (acreage) depends on the collector, location and distribution of wild plants so direct comparison to the three cultivation systems is not applicable (na).
Value (US$/lb. dry wt.) refers to the amount paid by a buyer to a grower or collector (retail).
Beyfuss is a recognized authority on collecting, cultivating and marketing ginseng in the northeast (Beyfuss, 2017).
Source: adapted from Davis and Persons (2014) and Beyfuss (2017).

another detrimental factor, not reflected in Table 4, is replant disease. Ginseng performs poorly if it is planted back on a site where it was recently grown. This has been attributed, at least in part, to pathogen build-up in soils (Westerveld, 2014). To avoid this problem the site must be fumigated before replanting or, a costly new shade structure must be constructed on a new site. For this and other reasons, field-grown ginseng gets poor marks for sustainability.

8.2 Woods-cultivated ginseng

The woods-cultivated method of producing ginseng, unlike FC ginseng, qualifies as forest farming because it depends on natural shade provided by the tree canopy. The process begins with constructing raised beds. Planting density is less than FC. Management after seeding may include weeding, disease and pest control, and even control of mega pests by the use of fencing to exclude deer and poachers. A crop of roots can be harvested after 6–8 years. The price paid to the grower is intermediate between FC and WS (Table 4).

8.3 Wild simulated

Careful site selection is, without a doubt, the most important consideration for success with the WS production method. As the name WS implies, the point is to find a forest site that either does grow natural ginseng, has in the past or could do so. The task of site selection has been greatly facilitated by Bob Beyfuss, retired Cornell University Cooperative Extension agent (Greene County). He is a ginseng specialist who has published a site evaluation tool called the Visual Site Index (Beyfuss, 2001). It allows a landowner to rank a prospective WS ginseng site based on several criteria including presence or absence of wild ginseng, canopy tree species, herbaceous indicator plants, soil, slope and security (Beyfuss, 2017) Once a suitable site has been selected, leaves are raked off, seeds are scattered and leaf mulch reapplied. That's it. Seeds are left to germinate and grow with minimal or no further input, before harvest 8–10 years hence. Assuming that the site has been well chosen and disturbance kept to a minimum the harvested roots should be essentially identical in appearance to wild roots and the value paid to the grower should be nearly the same. Although maintenance labour is practically nil, the risk of poaching and/or deer browse can be significant, especially over such a long period of time.

8.4 Wildcrafted (collected from the wild)

Wildcrafting refers to collecting ginseng (or other NTFP) from the wild. It is a traditional practice that has gone on, especially in the Appalachian Mountains, ever since the early to mid-eighteenth century. Wildcrafting is still a significant source of seasonal income for ginseng collectors ('sang hunters') in the eastern and especially southeastern Appalachian mountain region of the United States, but collection of wild ginseng is illegal throughout Canada due to perceived over collection. Today, in the United States, wildcrafting is regulated at the state and federal levels in order to implement US commitments to the CITES treaty which lists ginseng in Appendix II. State regulations define when, where and at what age wild ginseng may be collected. Collection is not permitted at all on most federal or state lands. Despite these restrictions, avid 'sang hunters' (wildcrafters) scour the eastern forests, in the fall, in response to considerable demand for the herb by Asian buyers.

How do these different methods of cultivation compare? Sustainability is an important characteristic of WS ginseng compared to FC and WC ginseng. When FC and WC ginseng is mature the entire block (raised bed) is harvested at once. Unlike a perennial crop like asparagus which regrows yearly from a crown, in the case of ginseng the root and rhizome are completely harvested so there is nothing to regrow from. Furthermore the site cannot be reseeded and used again because of replant disease, discussed earlier. Replant disease appears not to be a problem with WS ginseng because of the wide spacing, making plant-to-plant disease transmission unlikely.

In the context of crop production, sustainability means stable long-term production with minimal external inputs (fertilizer, pesticide, machinery) and minimal harm to the environment. Neither FC nor WC ginseng production rank high on a sustainability index based on these criteria. Neither can be planted twice on the same site. FC requires machinery for shade construction and ploughing and harvest. Both rely on chemical pest control. On the other hand none of these drawbacks apply to WS ginseng production. Little if any site preparation is required, and disease pressure is minimal. However, the most important contribution to sustainability of WS ginseng production is the plant's natural life cycle (Fig. 9). It begins with the grower scattering seed on a suitable site, mulching and then walking away for years at a time. Seedlings arising from the initial seeding (first generation) typically flowers for the first time during the fourth year, and continues to flower every year thereafter. Second-generation seedling begins flowering when they are 4 years old, and by that time the original first-generation plants are nearing harvest at year 9–10. A new generation of seedlings begins every year after the fourth year. Harvest of older plants can proceed every year without diminishing the size of the overall population.

This scenario is idealized, and does not take into consideration possibility of predation, poaching, annual variation in seed set, disease or other stochastic factors. This sustainable WS production cycle has been ongoing at Cornell University's Arnot Forest for 15 years and the population is not in decline. It is the gift that keeps on giving with no end in sight.

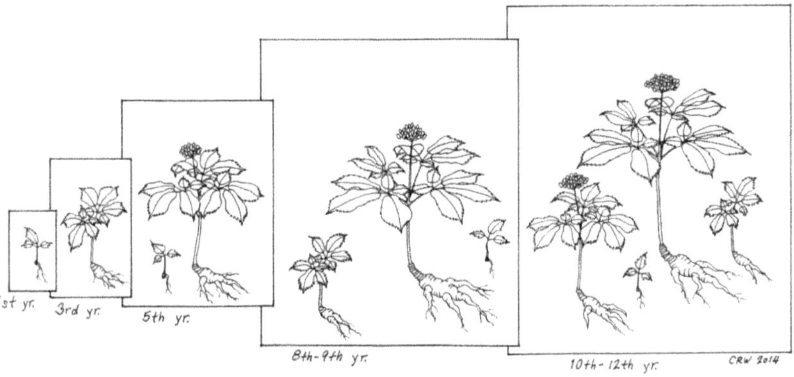

Figure 9 Sustainability of wild-simulated ginseng. The first generation grows from seed planted by the forest farmer (far left). It produces seeds which grow into seedlings. When these are 4 years old they produce seed which grow into the second generation which repeats every year until first-generation plants are 10 years old at which time they are harvested, and younger plants grow to harvestable size each successive year, indefinitely. Permission: Carl Whittaker.

For American ginseng, its value (price in US$/lb) paid by a buyer to a grower (or wildcrafter) depends on how it is grown, as described above. Table 4 compares the value and profitability for the four ginseng production methods, listed in order of intensity of cultivation from least to most: wildcrafted < WS < WC < FC. WS and WC qualify as forest farming (WC and WS), but FC does not. Unlike most crops the value of American ginseng is based not only on yield (weight) but also on resemblance of a cultivated root to a wildcrafted root (shape, texture, colour). These somewhat subjective criteria are considered of great importance to Asian buyers for export, mostly to China. In fact, in Chinese (Mandarin) 'ginseng' is known as Ren Shen which translates as *man root*. The more person-like the shape of the root (arms, legs etc.), the more Asian buyers will pay for it (Fig. 10).

While retail price to the grower or wild collector is clearly inversely related to the intensity of cultivation, the other considerations that tend to compensate for the price differential is the number of years to harvest (growth rate) and planting density (yield), both of which increases with intensity of cultivation. Table 4 also shows the effect of method on overall profitability (income − expense = profit). There is an increased yield for FC ginseng (200 lb/0.1 acre) compared to WC (60 lb/0.1 acre) or WS ginseng (16 lb/0.1 acre). On the other hand, net profit for FC is lower (US$6585/0.1 acre) than the less intensively managed WC (US$10 949) and the much less intensively cultivated WS (US$11 779). If wild ginseng were abundant and easy to collect, the price would not be so astronomical, and the Asian demand would be readily met, but that is not the case. In fact, collection of wild ginseng does not nearly satisfy the seemingly

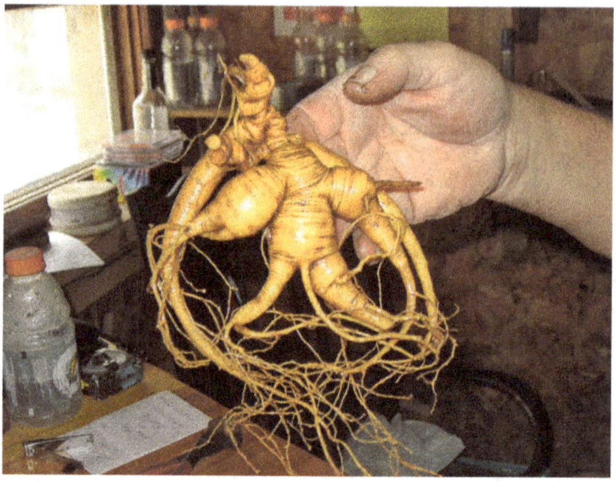

Figure 10 American ginseng. The Chinese word for ginseng is Ren Shen which translates as *man root*. The more a ginseng root looks like a person, the greater its value. This exceptionally large cultivated root is at least 30 years old. This root is extremely valuable because of its age and its man-like appearance.

inexhaustible Asian demand, leaving considerable opportunity for forest cultivation of this valuable North American crop.

9 Growing fruits, nut trees and shrubs

The shady forest floor is not a good place for growing typical garden vegetables like carrots, lettuce, radish, sweet corn and so on. However, there are two relatively high-value vegetables that can be cultivated under forest conditions. These include ramps (*Allium tricoccum*) (also known as wild leek) (Section 13) and ostrich fern (*Matteuccia struthiopteris*). Ramps produce small onion-like bulbs similar in size and shape to shallots. Ramps are widely wildcrafted, but few forest farmers are deliberately cultivating them. Ferns are not usually considered a vegetable, but in early spring while fern fronds are emerging from the ground they are rolled into a tight 'fiddlehead' which is an edible delicacy. They can be obtained by wildcrafting or forest cultivation.

With few exceptions, fruit- and nut-bearing trees and shrubs that are cultivated agriculturally are not well adapted to shady conditions found under a mature forest canopy, but some may be acceptably, but not optimally productive under partial shade. Other NTFPs like ginseng or log-grown mushrooms are not only tolerant of shade but absolutely require deep shade in the understorey to survive. Eastern black walnut, shagbark hickory (*Carya ovata*) happen to be mature canopy nut trees not well suited to growing under shade.

Black walnut (*Juglans nigra*) is a multipurpose tree that produces not only edible nuts but also can be tapped for sap that can be boiled down to produce tasty black walnut syrup. Its foliage casts moderate shade which could be beneficial for understorey species like brambles that are not typically found in shady situations, but may perform adequately there. Black walnut is also a very valuable timber species.

On the down side, black walnut is associated with a syndrome known as walnut toxicity which can kill susceptible plants. It is caused by an allopathic chemical known as juglone, which is present in all parts of the plant, but is only of concern when the roots of black walnut come in contact with or close proximity to the roots of another plant of a susceptible species. It is widely acknowledged that not all plant species are susceptible to walnut toxicity. In fact, they are more resistant than susceptible species, and some of these are ideal candidates for a forest farming situation. The Morton Arboretum lists 44 species of trees and 35 species of shrubs that are resistant to black walnut toxicity (Morton Arboretum), among which several that are well suited for forest farming including black raspberry, currant (*Ribes nigrum*), elderberry (*Sambucus* sp.), persimmon (*Diospyros virginiana*) and pawpaw. Furthermore, weed pressure from juglone-susceptible species may be reduced by this natural 'herbicide' (juglone). Scott and Sullivan (2007) suggested that walnut toxicity should be 'destigmatized' because the advantages of black walnut often outweigh the disadvantage associated with walnut toxicity.

In addition to black walnut, there are other temperate nut tree species that might be considered for a forest farming system since they produce a potentially marketable nut crop and when mature they contribute to the overstorey canopy. These include shagbark hickory and shellbark hickory (*C. laciniosa*), and pecan (*C. illinoensis*) in southern latitudes. Along with black walnut, they are all in the Juglandaceae family. There are improved clonal (grafted) selections of all of them, but pecan is the only one for which there is a well-established market. Today, the popularity of hickory nuts is largely limited to amateur nut growers. The Northern Nut Growers Association (NNGA, http://www.nutgrowing.org/) and statewide nut growers associations are a valuable source of information and inspiration about these and other temperate nut tree species.

Attempts to introduce nut trees or any mature canopy species into a forest farming system, from seedlings or grafted saplings, are unlikely to succeed because these nut trees normally grow under high light conditions. Even under optimum conditions (full sun), black walnut and hickories require about 40 years to maturity. Under a shady canopy, they grow slowly compared to the same species or other hardwoods planted in full sun or on the periphery of a forest farm. About 12 years ago, at the MacDaniels Nut Grove (Cornell University, Forest Farming Education and Research Center) (see Section 12), we grafted scions of the shagbark hickory cultivar Yoder #1, onto 3" diameter seedling understocks (inlay bark graft, Fig. 14), beneath a mature canopy of black walnut and shagbark hickory. The graft unions formed a functional graft union satisfactorily in the first year but in the 12 years since they have grown only as would be expected in 3 or 4 years from the same grafted tree grown in full sun. After 12 years, they have not begun to produce nuts even yet (2017). Under full sun conditions, they would be expected to produce nuts as early as 6–8 years. One approach to get a selected nut cultivar into production sooner than a nursery seedling is to 'top work' (graft) a cultivar (scion) onto a well-established tree.

Another problem with large canopy nut trees is that harvesting nuts can be a challenge if most of them are out of reach. Waiting until they drop on their own can be problematic if squirrels are in direct competition, but during mast years (heavy crop of nuts once in several years) this is less of a problem because there are plenty of nuts for squirrels and man to share. In commercial-scale nut orchards, a tractor-driven tree shaker is often the harvesting method of choice, but this would not be practical on most forest farms because of limited access. Another nut harvesting strategy that works on a small scale (several trees) is the use of an arborist's sling shot to shoot a small-weighted bag attached to a thin line over a nut-bearing branch. Then, holding both ends of the line, the branch is vigorously shaken as the nuts drop to the ground.

For optimum (commercial) production, most fruit-bearing shrubs (e.g. brambles - *Rubus* species etc.) are grown in full sun. This is also the case with the pawpaw tree. Nonetheless, in a multispecies forest farm, they may yield adequately under partial shade. Most temperate nut trees (walnut, hickory, Chinese chestnut (*Castanea mollissima*) will not produce nuts at an acceptable level under shade, but one nut-bearing shrub that may be an exception is the hazelnut. Research by Hampson et al. (1996) indicates that under 30% shade, bush hazelnut (*Corylus* sp.) will yield 70% of the crop it would produce under full sun (Fig. 11).

Small fruit breeders are focused on optimizing production under full sun. Nevertheless, several fruit crops that are normally grown in full sun but

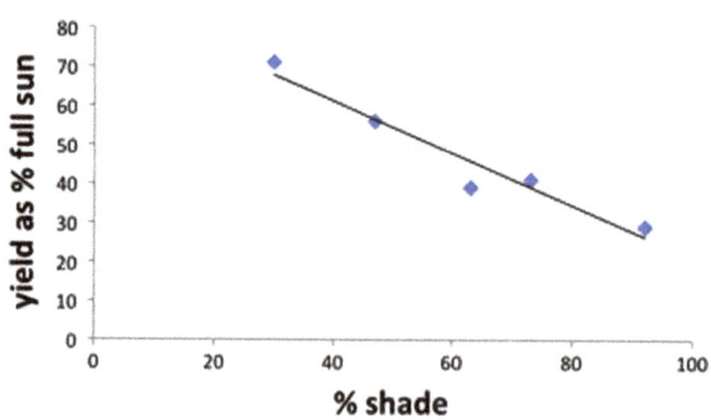

Figure 11 Hazelnut production vs. shade level. Redrawn from Hampson et al. (1996).

might be regarded as acceptable under partial shade include but are not limited to blackberry, raspberry (*Rubus idaeus*), hardy kiwi (*Actinidia chinensis*), chokeberry (*Aronia arbutifolia*), currant and gooseberry (*Ribes* sp.) (Nelson, 2009). With the exception of Gallagher et al. (2015), there has been no research intended to quantify yields under less than full sun. However, in a forest farming context, Gallagher et al. (2015) investigated the shade tolerance of blackberry (*Rubus fruticosus*) under a range of shade levels and reported that 'Illini Hardy' blackberry may benefit from partial shading at 70% of full sunlight.

Another fruit species that can produce a reasonable yield in the shade is the midstorey tree, pawpaw (15 to 20+ feet tall) (Mudge and Gabriel 2015) (Fig. 12), which is a North American fruit tree native to eastern and southern United States.

Pawpaw is in the Anonaceae family, along with custard apple and other tropical species. It is the largest of Native American fruits. It shares the common name pawpaw with the tropical fruit tree otherwise referred to as papaya (*Carica papaya*), but the two are not related. Although its natural habitat is shady forest, nevertheless temperate pawpaw produces greater yield in full sunlight. It reproduces naturally, both by seed and by natural suckers, the latter from shoot buds arising on roots. In a decade or so after planting, suckers from a single individual pawpaw tree can proliferate into a pawpaw patch that is a small grove of trees that are genetically identical to the original tree, or only to the rootstock, in the case of a grafted tree. Pawpaw is predominantly self-incompatible. Pollen from another genetically different tree is required for fertilization and fruiting. Researchers at Kentucky State University and University

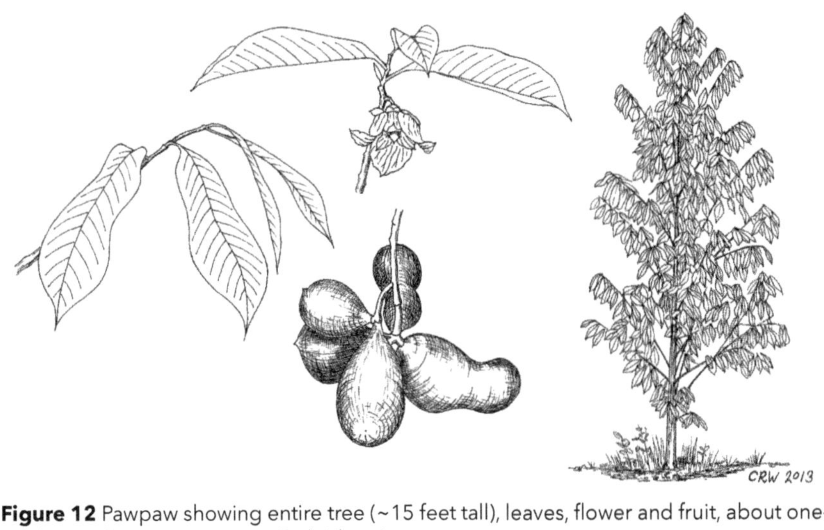

Figure 12 Pawpaw showing entire tree (~15 feet tall), leaves, flower and fruit, about one-tenth actual size. Permission: Carl Whittaker.

of Kentucky (Pomper, 2009), as well as amateur breeders, have introduced improved clonal selections (cultivars). Cultivars have been selected for fruit size, taste and flesh-to-seed ratio, resistance to pests and diseases and overall productivity. For optimal productivity and trueness to type, grafted cultivars are recommended over seedlings. Bud grafting onto a seedling rootstock is the only practical method for clonal propagation of cultivars.

10 Cultivating tree syrups

The most valuable of any NTFP that is well suited for forest farming is maple syrup. For 2017, the total crop value in the United States was approximately US$149 million gallons at US$35/gallon (National Agricultural Statistical Service, 2017), and considerably higher for Canada. Tree syrup is rarely mentioned as a forest farming crop, and many maple producers wouldn't even recognize the name 'forest farming'. Nonetheless, it can be highly profitable. Maple syrup is most commonly associated with the sugar maple tree, although red maple can also be tapped. Sap collection for syrup from other deciduous trees common to the northeastern forest include black walnut and birch (*Betula* sp.).

The 'sugaring' process begins with healthy mature sugar maple or other suitable trees. In early spring, holes are drilled through the bark into the sapwood, then spouts are inserted to channel the rising sap into a collection vessel (bucket). In larger-scale operations, PVC tubing that is under a slight vacuum increases the sap flow from multiple trees into a common collection tank. Necessary conditions that favour sap flow from maple trees are freezing nights followed by warmer days. Typical home-scale sugar bush will have 5–50 taps. Commercial-scale operations will tap 500 trees at the low end, and some larger-scale operations will have thousands of trees (Mudge and Gabriel, 2015). Tapping the same trees for many years does no harm to the tree if it is done conservatively. The process of turning sap into syrup is energy intensive and time consuming. This is accomplished entirely or in part by boiling to evaporate off most of the water in sap, which begins at 2% sugar, into syrup which is 67% sugar. Evaporators are fuelled by wood, propane or oil. The ratio of sap to syrup is about 40 to 1 (40 gallons of sap makes 1 gallon of syrup). Larger-scale operations begin the process by using reverse osmosis to partially concentrate the sugar content of the sap considerably, before using an evaporator to drive off the remainder of the water. This saves time and fuel. To bypass this energy-intensive process, some producers are selling fresh sap (2% sugar) as a beverage or for cooking, and so forth. From this prospective, the tree can be seen as a pump which brings water up through the soil and passes it along to the consumer. For black walnut and birch, the same basic approach is used for collecting sap and converting it into syrup. Black walnut produces sap of about the same initial sap-sugar concentration (2–4%), but it produces a lesser volume of syrup compared to sugar maple over an equivalent period of time

(Mudge and Gabriel, 2015). As mentioned above, American ginseng is well suited for growing under the canopy of sugar maple. It is no coincidence that wild ginseng is often found beneath sugar maple trees at a suitable forest site.

11 Nursery crops in forest farming

Fruits and nuts, medicinals, mushrooms, ornamentals and tree sap are the main categories of NTFPs considered so far in this chapter, but an additional category, nursery crops, represents a promising opportunity for the forest farmer. Broadly defined, a nursery is a protected area set aside for propagating planting stock intended to be used in situ (on the forest farm), or sold-off site. A forest nursery can include medicinals, fruits, nuts crops already discussed, with the exception of mushrooms. The 'nursery' for forest-cultivated mushrooms is usually a commercial laboratory in which spawn (mycelium) is produced under aseptic conditions, rather than an outdoor nursery.

11.1 Propagation

Propagation is where the forest nursery begins. There are two broad categories of propagation that may be considered. These are sexual and vegetative propagation. Sexual propagation, of course, involves seed. Vegetative propagation, otherwise known as asexual or clonal propagation, involves propagation by cuttings, division, grafting or layering. The choice of propagation method is not arbitrary. Things that factor into a choice of propagation method (seed or vegetative) are economics, genetic considerations, seasonal considerations, the end use for the propagated plant and ease or difficulty of seed or vegetative propagation.

First, seed propagation, because it is usually the easiest. Most plant species that a forest farmer is likely to deal with naturally reproduce from seed. In some ways a seed is the ideal propagule. It is small, transports easily, stores easily and what could be simpler than sowing a seed? One problem with seed propagation is lack of uniformity. Seedlings, even from a single parent plant, or even from a single fruit, do not always come 'true-to-type' (genetically identical to parent plant). This seedling variation may be negligible for many plants like American ginseng or in the case of the rootstock of grafted trees like pawpaw. There are clonal rootstocks for fruit trees like apple, but not for any of the grafted trees likely to be used in forest farming.

Although many crops raised from seed are genetically different from the parent, the differences are minor and have no detrimental effect on their performance of seedlings in terms of yield, and so forth. American ginseng is a good example. This is a good thing because, for all intents and purposes, ginseng is nearly impossible to propagate vegetatively.

Seed propagation is generally easier than vegetative propagation in terms of skill, labour, infrastructure, transport and storage. Nonetheless, propagation from seed requires more involvement than just placing the seed in the soil. This is because most temperate woody or herbaceous perennials have some sort of seed dormancy, which requires either stratification or scarification to overcome. Stratification to break embryo dormancy involves moist chilling for 30–90 days (species specific) at about 5°C. Seeds that require a cold stratification include pawpaw, ginseng and goldenseal. On the other hand, scarification is required to break seed dormancy of other species. As the name implies scarification involves breaking (scarifying) an impermeable seed coat by abrasion with a file, sandpaper or even soaking in concentrated acid. In the case of natural scarification, passage of a seed through the acidic gut of a mammal or freeze-thaw cycles are often involved. Weakening the seed coat allows oxygen and water to gain entry to the seed embryo. Species that require scarification include honey locust, red bud (*Cercis canadensis*) and Kentucky coffee tree (*Gymnocladus dioicus*), as well as most other woody legumes. Recommendations for overcoming seed dormancy of tree species found in North America can be found in the US Forest Service online publication *The Woody Plant Seed Manual* (https://www.fs.fed.us/rm/pubs_series/wo/wo_ah727.pdf). Or the more recent print version *Seeds of Woody Plants in the United States* (Young and Young, 1992).

Vegetative propagation is used when propagation of a variety from seed would not produce an acceptable true-to-type seedling. This is true for many varieties of woody and herbaceous perennials that have been selected for some horticultural trait not present in the wild type, including cultivated varieties (cultivars) of nut trees, pawpaw, willow and so on. Many vegatatively propagated cultivars are protected by the US or other plant patents which legally precludes others from vegetative propagation of the cultivar.

In addition to berry crops (raspberry, black raspberry, elderberry, currant, gooseberry, goji berry etc.) and fruit trees like pawpaw must be vegetatively propagated in order to produce true-to-type offspring. Appropriate techniques include layering (blackberry, black raspberry (*Rubus occidentalis*), cuttings (currant, gooseberry, goji berry and elderberry), division (red raspberry) and grafting (pawpaw, walnut and hickory). Several of these can be, and often are, propagated from seed, but the offspring are not identical to the parent. Another reason some species are vegetatively propagated is simply because that is easier than propagation from seed (e.g. willow, brambles).

11.1.1 Cutting

Cutting propagation is the method of choice for willow, elderberry, currant and gooseberry, especially since they are extremely easy to propagate from cuttings and difficult to propagate from seed. In the case of willow, large, leafless stake cuttings (truncheons), up to an inch or more in diameter and 3 feet long can simply be stuck or pounded into the ground where they will strike roots and grow on. A tree or shrub rooted from a large cutting will produce a functional plant-bank stabilization in the case of willow, or fruit production in the case of elderberry - years sooner than a seedling of the same species. When a cutting is severed from its parent plant, it cannot readily take up water since it has no root system, and with the exception of a few species that can tolerate this water shortage (willow truncheons), cuttings of most species would dry out and die before striking roots. Minimizing water loss from the cutting is necessary for survival and rooting. This can be accomplished by increasing the humidity by putting the cutting on a flat object (tray) containing suitable rooting medium (sand, peat etc.) into a polyethylene 'tent' (Fig. 13) (Mudge and Gabriel, 2015). It is absolutely necessary that a simple humidity chamber be kept out of the sun to avoid overheating.

11.1.2 Layering

Layering is similar to rooting of cuttings except that a stem is induced to form roots before it is detached from the parent plant. There are several kinds of layering (e.g. tip layering, air layering) but all share this basic trait. For some species, like black raspberry, layering is successful whereas cuttings are not. Black raspberry propagates by natural layering with very little intervention by

Figure 13 Cuttings rooting under a poly tent.

the forest farmer. During the course of a growing season, the tip of its arching primocanes (current seasons shoot growth arising from ground level) come in contact with soil. Here the tip roots into the soil, so that, the following spring, the newly rooted shoot (rooted layer) will have formed a new vegetative bud. This rooted layer can be cut away from the original primocane, and the new plant can be dug and transplanted to a new location or into a container to be grown in the nursery before sale/outplanting.

11.1.3 Grafting

Grafting is somewhat different from other vegetative propagation techniques. It involves bringing together cut surfaces of two different plants so they will fuse together and grow as one plant (Mudge and Gabriel, 2015). Shaping the cuts so the stock and scion (upper and lower portions respectively of the grafted plant) will fit together tightly (the 'carpentry' of grafting), applying pressure to hold them together and assuring that stock and scion are sufficiently closely related (genetic compatibility) are the essential requirements of successful grafting. Additionally, the appropriate season for grafting differs among species and grafting methods, but most grafting is done when the stock and scion are dormant (to conserve moisture). The scion is selected from a tree or shrub with desirable characteristics. The scion, before and after it is grafted onto a rootstock, is genetically identical to the scion donor plant ('mother') as is any other grafted plant obtained from scions taken from the original 'mother' plant. Collectively all those genetically identical plants and their vegetatively propagated descendants constitute a clone. Note: the rootstocks of these grafted plants, usually grown from seed, are not genetically identical to the original clone but that is usually irrelevant. On the other hand, if a plant can be clonally propagated from cuttings or layerings there is usually no point in grafting, at least for species, for which genetically improved clonal rootstocks have not been selected. An inlay bark graft of shagbark hickory used to graft a small diameter scion onto a large diameter rootstock is illustrated in Fig. 14.

11.1.4 Division

Division is a vegetative propagation technique that involves splitting a multi-stemmed plant with a large knife or other tool so that each piece retains a functionally complete plant, including part of its original stems and roots. Each division can be transplanted separately to function as two or more genetically identical plants (clones). Division is often the method of choice for herbaceous perennials, especially those used as ornamentals such as ferns, hosta and iris. Ramps, an edible bulb, can also be propagated by division.

Figure 14 Inlay bark graft, of shagbark hickory. This technique is used to graft a leafless scion onto a large diameter (2-4" diameter) understock, in the early spring before leaf out.

11.2 Growing on

Propagation is the first step in forest nursery production, but oftentimes the new plant does not leave the nursery immediately after propagation. Whereas as successfully propagated plants may be outplanted or sold, others may be grown in the nursery for some period before they are large enough for outplanting.

Growing plants in the nursery after propagation will involve one of three different production systems - field production, container production or pot-in-pot (PIP) container production. They may be grown directly in-ground, commonly known as field production. Field nursery production is typically used for trees that will remain in the ground for several years to attain a desirable size. One major disadvantage of field production is the considerable labour involved in digging up a field-grown tree for transplanting to its final location, by a process known as ball and burlap. In addition to the labour involved, another disadvantage is the inevitable damage to the root system when the plant is dug for transport to another location.

Container production, on the other hand, involves growing a propagated plant in a pot (plastic nursery container, tin can or paper pot), which can be easily carried off site with relatively little disturbance to the root system. Plants in containers are more likely to require irrigation during dry periods. Most container nursery production at large-scale commercial nurseries is above-grown, that is the irrigated containers are sitting directly on the surface (soil, gravel, plastic).

In the north temperate zone, winter injury or death caused by low temperatures is a major disadvantage for above-ground container production. To begin with, soil temperature during winter is usually several degrees warmer than air temperature because it is not exposed to the cold circulating air and is protected by the insulating environment of the soil (Mathers, 2003). On the other hand, the temperature of the above-ground root system of a container grown plant is essentially the same as the air temperature. Above-ground cold hardiness of roots is usually several degrees lower than that of the shoots. Shoots can tolerate much lower temperatures than roots. Since below-ground soil temperature is higher than air temperature, the roots of field-grown plants are more likely to survive during periods of low temperature. This is clearly an advantage of field production compared to above-ground container production in areas with low winter temperatures.

Considering the previously stated disadvantages of field production (labour, root injury), fortunately there is a third alternative to field or container production. PIP container production has the 'best of both worlds' in a sense. It involves placing a 'socket pot' below the ground and then lowering a plant pot into the socket pot, as shown in Fig. 15. This involves the extra labour of digging a hole and placing an empty 'socket' pot into the hole. When the plant pot is placed into the socket pot, its roots (but not the shoot) are entirely below the ground. The roots are never exposed to low air temperature, but rather they are at the somewhat higher temperature and insulated by the surrounding soil. PIP solves another problem common to above-ground container production - wind.

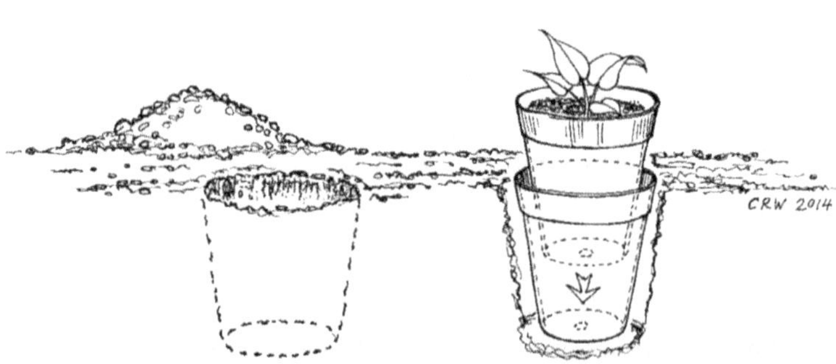

Figure 15 Pot-in-pot nursery container production. Socket pot and nested plant pot are below ground, which insulates the root system from the colder air temperature. Permission: Carl Whittaker.

Above-ground pots, especially with tall plants in them, are easily blown over by wind. Pots anchored below the ground (PIP) are prevented from blowing over.

12 Case study: overcoming limitations at the MacDaniels Nut Grove Forest Farming Research and Education Center

In 1926, Lawrence MacDaniels was hired as an assistant professor in the Department of Pomology at Cornell University in Ithaca, NY. His main responsibility, commodity wise, was apples, but his passion was temperate nut trees. Shortly after his arrival at Cornell he undertook the establishment of a temperate nut tree collection (grove) for evaluation of new cultivars, and to determine the suitability of the Ithaca area for commercial nut production. He began by clearing a 5-acre section of an oak hickory forest. What happened next is not well documented, but we know that he planted walnut and hickory seedlings and encouraged the natural recruitment of nut tree seedlings and for the next 20 or so years he cloned exceptional trees donated by farmers and other sources by grafting scions (pencil-sized twigs) from those trees onto the seedling rootstocks. Hence, these grafted trees on the Ithaca site were genetically identical to the original trees from the farmer's site. These grafted trees included shagbark hickory, shagbark × shellbark hybrid hickories, eastern black walnut and Japanese walnut. He also planted Chinese chestnuts, pecans and filberts along the northern perimeter of the woodlot. For the next 30 years, he evaluated the performance of these cultivars. Eventually he concluded that Ithaca was not well suited for commercial nut production due to insufficiently long growing seedling. His active involvement in the site declined by about the mid-1960s, more than a decade after he retired from Cornell.

Today, many of the grafted nut trees remain at the MacDaniels Nut Grove and constitute most of the mature canopy at our forest farming activities there today. Unfortunately, we do not have a complete historical record of what cultivars he planted, when he planted them and how they performed over time. Over the next approximately 40 years, labels were lost and no map of the site was ever found, so most of the grafted trees today are unidentified. Over the decades the site reverted to secondary forest so we were able to recognize that the site was deliberately planted to clonal nut trees. It was the obvious graft unions on these 70+-year-old tree that gave it away, along with other forest crops he had planted including pawpaw and persimmon. We rediscovered the site in 2002, and with the help of students began renovating it to develop a multispecies forest farm, and called it the MacDaniels Nut Grove (Fig. 1). The work began by removing the invasive honeysuckle, privet and multiflora rose, as well as selected individuals of oak and maple on the site. To improve access, students constructed a stone stairway from the eastern boundary of the site, down a steep southwest-facing hillside. This slope curved around to the north (south facing) creating a flat pocket. To the south a small stream flowing east to west bisects the site and on the south side of the stream is an alluvial flood plain populated by mature black walnut trees along with pawpaws and brambles planted about 12 years ago.

All the major categories of non-timber forest crops, including medicinals, mushrooms, small and tree fruits, have been planted at the site since 2002. Mushrooms are being grown in a laying yard along the stream bank at the base of a steep west-facing hillside. Students have participated in inoculating logs with shiitake mushroom spawn, and lion's mane, oyster, nameko on logs have been grown, as well as stropharia on wood chip beds. One limitation was inadequate space for cultivating log-grown shiitake mushrooms. To overcome this limitation, we created usable space for shiitake logs by resorting to the Japanese hillside stacking method (Mudge and Gabriel, 2015) which has rarely been used outside of Japan. It enables logs to be arranged in long rows in a stair step fashion climbing directly up a steep hill. It allowed us to more than double usable space for cultivating mushrooms.

Another critical element for forest mushroom cultivation on logs is avoidance of moisture loss, and the primary way to avoid that is to site the laying yard under deep shade, ideally of evergreen tree species which cast sufficient shade all-year round, in order to avoid both summer and winter sun. Hemlock, such as we have at another laying yard at Cornell's Arnot Forest, is ideal since it casts the deepest shade of any canopy tree in the northeastern forest, accounting for the frequent observation that there is very little understorey vegetation below hemlock. At the MacDaniels Nut Grove site, however, the forest is entirely deciduous - there are no conifers, so light exposure in the winter is less than ideal. Our approach to overcoming that limitation was to cover mushroom log stacks with 70% greenhouse shade cloth. The shade cloth was simply laid directly on top of the log stack rather than constructing a raised frame work.

Our third limitation involved cultivation of the important woodland medicinal, American ginseng. As discussed above, recommendations for site selection for American ginseng put considerable emphasis on a north- or northeast-facing slope because it is cooler and moister than southern exposure. At the MacDaniels Nut Grove there is no north or north east facing slope available. Furthermore, the soil on our south-facing slope was a poorly drained, dense clay loam. With these two strikes against us, did that mean that we should have given up on growing ginseng? If obtaining a commercially viable yield of ginseng was an important goal, perhaps we should not have bothered to try. However, in keeping with the educational use of our forest farm (MacDaniels Nut Grove Forest Farming Research and Education Center), we felt it was important to expose students to this very important NTFP,

American ginseng. So what we did to overcome the limitation of our poorly drained south-facing slope was to construct raised beds, framed by 1 × 10″ boards, and filled with soil enriched with peat moss, to increase water-holding capacity and drainage. It worked! Ginseng has been growing in these beds for years, and even dropping seed which has yielded multiple generations of plants.

Of course, white-tailed deer do forage on ginseng, even if it is not their favourite food. They also forage rather indiscriminately on young trees, shrubs and herbs, so deer presented another limitation not just with respect to ginseng but to other NTFPs on our site. As for the ginseng, the solution was mechanical - we simply covered the raised bed with wire mesh. Another solution at another location in our forest farm was to choose deer-resistant plants. Some critics might scoff at this, arguing that deer eat anything. But that's not entirely true. Over, across the stream, at Walnut Island we faced the double limitation not only of deer foraging, but also the effect of juglone on susceptible species. With no assurance that our choice of species would solve either limitation, we proceeded to plant black and purple raspberries, as well as pawpaw trees. As it has turned out, both the brambles and the pawpaw are thriving beneath the shade of black walnut to the consternation of the deer.

13 Case study: from ramps to riches

Yes, the title is a bit of an exaggeration. Nonetheless, cultivation of ramps is an emerging income opportunity for forest farming. Ramps are also called wild leeks. Botanically, they are classified as *Allium tricoccum*, a close relative of onions, garlic and scallions. Wild ramps occur in hardwood forests throughout eastern North America, from South Carolina to Canada. They inhabit moist soils that are often saturated or nearly so in the early spring. They are found beneath deciduous hardwoods like oak, beech, maple, but not beneath evergreen trees. Like many forest-dwelling spring wildflower like cohosh, trillium, bloodroot and others, ramps are spring ephemerals. They thrive photosynthetically during the brief period of when the overstorey is leafless and the light reaching the understorey is relatively high. Shortly after canopy closure, the leaves of the ramps die back leaving little or no sign that they were there at all, until leafless flower stocks emerge in mid-summer. By fall an inflorescence sheds 20 or so hard black seed about the size of pepper corns.

For generations, ramps have been considered a spring staple among Appalachian mountain folk, and in recent years they have become a seasonal delicacy in some upscale restaurants. They are used in soups, stews, fried with eggs and as least as in many different ways as onions. The fresh leaves are a particularly tasty treat. They have a distinctive garlic/peppery taste which makes them a special addition to salads. Traditionally ramps have been collected throughout the Appalachian Mountains for generations as the first leafy vegetable after a long winter without greens in the diet. Lore has it that school-aged children were especially fond of ramps, not only because they taste so good, but also because their breath would take on such a garlicky odour that their teachers would send them home for the day (of play).

To this day, 'ramp festivals' are held throughout Appalachia, some attracting hundreds if not thousands of visitors. The ramps gathered for a single such occasion could make a huge dent in even the largest of wild populations or many smaller populations. Overharvesting in some areas has led to concerns about the long-term sustainability of wildcrafting of this forest vegetable. In parts of Canada, for example, wild collection of ramps is limited to personal use only.

Recently, in the latter days of April, Jim Chamberlin, John Munsell and I visited a ramp patch in the woods in upstate NY, where a healthy population was thriving beneath a canopy of beech, birch and sugar maple. It was early in the morning and the sun was streaming through the leafless tree branches at a low angle, lighting up the clumps of ramps so their leaves fairly glowed. Other herbs of the forest floor were just beginning to emerge, like blue cohosh and toothwort and trillium. The biggest ramps were clustered around a narrow stream where the soil was nearly saturated. From there a bank covered with ramps rose steeply to a place where the forest floor levelled out about 20 feet above the stream. This particular ramp patch was about one-tenth acre in extent. A couple of determined collectors could collect just about all the ramps here in less than a day, and sell them for a healthy profit. The rate of population increase of ramps is exceedingly slow, so sustainability of wild harvest in a situation like that is probably non-existent.

13.1 Growth habit

Ramps have 6–8" long, somewhat floppy leaves, that are an inch and a half or so at their widest point. The similarity between the leafy foliage of ramps and of lily of the valley has been noted more than once. Exercise caution because the latter is poisonous. Lily of the valley is an ornamental garden perennial, not a forest native, but from time to time perennials like this 'escaped' garden and become naturalized in the forest.

Ramp plants do not usually grow alone. They commonly grow in clumps of 10–20 individuals with their leaves overlapping and their roots intertwined. A clump like this may have begun as a single or several seeds which flowered after several years and set seed that fell directly beside the original seedling(s), and germinated 18 months later. Ramps multiply asexually as well, by bulb splitting (think daffodil) and by sprouting along a short chubby rhizome.

13.2 Cultivation

The population studies described above make a strong case that wildcrafting of ramps should be discouraged or curtailed in some instances because harvesting from small or moderately sized populations simply cannot be done sustainably in the long run. Like any recommendation not enforced by law, this one is unlikely to have much adoption or impact on conservation of wild ramp populations. To those who see forest farming as a means of productive conservation, generating both income and preserve vulnerable natural populations of NTFPs, deliberate cultivation of ramps may be worth considering. Ramp cultivation is not widely practised mainly because wild populations are still abundant in some locations. But the writing is on the wall - wildcrafting of natural ramp populations has already been restricted in some locations due to severe depletion (Canada, Great Smokey Mtn. National Park) and more restrictions on harvest and/or extirpation of wild populations are sure to follow.

As a result, there have now been attempts to cultivate ramps as a NTFP for forest farming. There are currently around 40 new ramp growers in North Carolina though only one, to the authors' knowledge, making income from cultivation. Given that this is a new sector, there is little information on how to cultivate ramps. Two notable exceptions are:

1. A book written by Jeanine Davis and Scott Persons (2014) titled *Growing and Marketing Ginseng, Goldenseal and Other Woodland Medicinals* at the Mountain Horticulture Crops Research and Extension Center, North Caroline State University, Mills River, NC. Davis is widely recognized for her expertise on not only ramps, but also a wide range of forest medicinals. Persons specializes on ginseng.
2. The other is a book by Glen Facemire, *Having Your Ramps, and Eating Them Too*. Facemire is a farmer in West Virginia who is one of very few commercial ramp growers. He has been growing ramps since 1998.

There are two approaches to cultivating ramps: from seeds and from transplants. Cultivation from seed has the advantage that seed can be stored for up to several years, and they are easier to sow than to transplant divisions. The drawback to seeding is that seeds can be scarce. Seed production by wild populations can vary considerably from year to year, and there is no commercial source of seed. Seeds take about 7 years to grow to harvestable size, whereas large transplants take as little as 3 years to harvest. On the other hand, transplants have to be dug from the ground, usually by digging an entire clump and separating the bulbs for transplanting. This can only be done during the narrow window (several weeks) when plants show themselves above ground.

Ramp seeds have a double dormancy which requires 18 months to overcome before seeds germinate.

14 Summary

Forest farming is a relatively low-tech agroforestry practice for cultivation of shade-tolerant NTFPs, including medicinals, mushrooms, fruits, nuts, tree syrups and/or nursery stock. It is an ecologically sustainable way for forest owners to generate income, while maintaining forest health (productive conservation). With a carefully developed forest management plan, forest farming is not inconsistent with selective logging or other forest practices like recreation, foraging or even silvopasture. The NTFPs with the greatest potential for income generation are shiitake mushrooms and American ginseng because the mushroom logs begin yielding relatively soon after its inoculation, while ginseng has greater income-generating potential, but requires many years before it begins to yield. Even crops that are not normally grown under shade may produce adequately in the forest farm under partial shade. Since many smallholder farms have both open fields and woodlots, the fields may be used for traditional cropping (corn, soy etc.), and may also include other agroforestry practices like alley cropping, while the woods may be used for forest farming.

15 References

Alexander, S. J. and R. J. McLain. 2001. An overview of non-timber forest products. In: M. Emery and R. J. McLain (Eds), *Non-timber Forest Products: Medicinal Herbs, Fungi, Edible Fruits and Nuts, and Other Natural Products from the Forest*. Haworth Press, Inc.

Anon. 2017. Wild yam. *Illinois Wildflowers*. Available at: http://www.illinoiswildflowers.info/savanna/plants/wild_yam.html (accessed 17 October 2017).

Bernhardt, E. and T. Swiecki. 1991. Guidelines for developing and evaluating tree ordinances. Prepared for: Urban Forestry Program, California Department of Forestry and Fire Protection, Sacramento, CA, 76p. Available at: http://www.isa-arbor.com/education/resources/educ_treeordinanceguidelines.pdf (accessed 4 January 2018).

Beyfuss, R. 2001. Visual site assessment & grading criteria for a woodland ginseng growing operation. Cornell Cooperative Extension, Greene County. Available at: http://www.ntfpinfo.us/docs/other/Beyfuss1998VisualAssessmentGinsengOperationSite.pdf (accessed 12 December 2018).

Beyfuss, R. 2017. The practical guide to growing ginseng. Cornell Cooperative Extension, Agroforestry Resource Center. Available at: https://ecommons.cornell.edu/handle/1813/54771 (accessed 15 November 2018).

Bruhn, J., J. Mihail and J. Pickens. 2009. Forest farming of shiitake mushrooms: An integrated evaluation of management practices. *Bioresource Technology*, 100(24): 6472–80.

Bruhn, J., N. Abright and J. Mihail. 2010. Forest farming of wine-cap *Stropharia* mushrooms. *Agroforestry Systems*, 79(2): 267–75.

Burkhart, E. and M. Jacobson. 2009. Transitioning from wild collection to forest cultivation of indigenous medicinal forest plants in eastern North America is constrained by lack of profitability. *Agroforestry Systems*, 76(2): 437–53.

Canham, C., A. Finzi, S. Pacala and D. Burbank. 1994. Causes and consequences of resource heterogeneity in forests: Interspecific variation in light transmission by canopy trees. *Canadian Journal of Research*, 24: 337–49.

Cantaluppi, C. 2016. The pawpaw: An emerging specialty crop. *Journal of the NACAA*, 9(1). Available at: https://www.nacaa.com/journal/index.php?jid=582 (accessed 19 September 2017).

Chamberlain, J. L., D. Mitchell, T. Brigham, T. Hobby, L. Zabek and J. Davis. 2009. Forest farming practices. In: H. E. Garrett (Ed.), *North American Agroforestry: An Integrated Science and Practice*. American Society of Agronomy, Madison, WI.

Chianucci, F. and A. Cutini. 2012. Digital hemispherical photography for estimating forest canopy properties: Current controversies and opportunities. *iForest* 5: 290–5.

Daily, C. and C. Wilkins. 2012. Passive water harvesting. University of Arizona Cooperative Extension. Available at: https://extension.arizona.edu/sites/extension.arizona.edu/files/pubs/az1564.pdf (accessed 28 December 2017).

Davis, J. M. and W. S. Persons. 2014. *Growing and Marketing Ginseng, Goldenseal, and other Woodland Medicinals*. New Society Publishers.

European Network for Rural Development (ENRD). 2014–20. Developing mushroom production in Finland's forests. Natural Resources Institute (LUKE), Finland.

Eurostat Statistics Explained. 2017. Forests, forestry, and logging. Available at: http://ec.europa.eu/eurostat/statistics-explained/index.php/Forests,_forestry_and_logging#Forests_and_other_wooded_land (accessed 04 November 2017).

Facemire, G. 2009. *Having Your Ramps, and Eating Them Too*. McClain Publishing.

Forestry Statistics. 2017. Chapter 1: Woodland areas and planting. Available at: https://www.forestry.gov.uk/pdf/Ch1_Woodland_FS2017.pdf/$FILE/Ch1_Woodland_FS2017.pdf (accessed 04 November 2017).

Gallagher, E., K. W. Mudge, M. Pritts and S. DeGloria. 2015. Growth and development of 'Illini Hardy' blackberry (*Rubus* subgenus *Eubatus* Focke) under shaded systems. *Agroforestry Systems*, 89: 1–17.

Grace, J. 2010. An evaluation of local isolates of *Hericium americanum* for use in mushroom production. MS thesis, Cornell University.

Grace, J. and Mudge, K. 2015. Production of *Hericium* sp. (Lion's Mane) on totem logs in a forest farming system. *Agroforestry Systems*, 89(3): 549–56.

Growing black walnut. University of Minnesota Extension. Available at: http://www.extension.umn.edu/garden/yard-garden/trees-shrubs/growing-black-walnut/(accessed 06 December 2017).

Hampson, C., A. N. Azarenko and J. Potter. 1996. Photosynthetic rate, flowering, and yield component alteration in hazelnut in response to different light environments. *Journal of the American Society for Horticultural Science*, 121: 1103–11.

Harper, L. 1998. Black walnut harvesting costs – The 50 percent factor. Available at: https://www.nrs.fs.fed.us/pubs/misc/walnut/p137_142.pdf (accessed 02 November 2017).

Kantor, S. 1994. Comparing yield with Land Equivalent Ratio (LER). Washington State University, Agriculture and Natural Resources Fact Sheet #532.

Kozak, M. E. and J. Krawczyk. 1993. *Growing Shiitake Mushrooms in a Continental Climate*. 2nd edition. ABC Printers, Marinette, WI.

Leatham, G. 1982. Cultivation of shiitake, the Japanese forest mushroom, on logs: A potential industry for the United States. *Forest Products Journal*, 32: 29–35.

Lemmon, P. E. 1957. A new instrument for measuring forest overstory density. *Journal of Forestry*, 55: 667–8.

Mathers, L. 2003. Summary of temperature stress issues in nursery containers and current methods of protection. *HortTechnology* 13: 617–24.
Mayo Clinic Staff. 2017. Herbal supplements: What to know before you buy. Available at: https://www.mayoclinic.org/healthy-lifestyle/nutrition-and-healthy-eating/in-depth/herbal-supplements/art (accessed 06 January 2018).
Mitchell, H. L. 1937. Trends in the nitrogen, phosphorous, potassium and calcium content of the leaves of some forest trees during the growing season. Black Rock Forest Paper No. 6. Cornwall Press, Cornwall, NY, 17p.
Morton Arboretum. Plants tolerant of walnut toxicity. Trees & Plants. Available at: http://www.mortonarb.org/trees-plants/tree-and-plant-advice/horticulture-care/plants-tolerant-black-walnut-toxicity
Mudge, K. 2010. Forest farming. *Arnoldia*, 67(3): 26–35.
Mudge, K. 2014. Grafting theory and practice. In: R. Trigiano and C. Beyl (Eds), *Plant Propagation Concepts and Laboratory Exercises*. 2nd edition. CRC Press, Boca Raton, FL.
Mudge, K. and S. Gabriel. 2015. *Farming the Woods*. Chelsea Green Publishers.
Mudge, K. and A. Matthews. 2011. Cultivation of shiitake mushrooms as an agroforestry crop for New England. NE SARE 2011 Annual Reports.
Mudge, K. W., E. Gallagher and R. Brinkman. 2011. Effect of tree species and log moisture content on forest cultivation of Shiitake mushrooms. *12th North American Agroforestry Conference*, 6–8 June 2011.
National Agricultural Statistical Service. 2017. US maple syrup production. Available at: https://www.uvm.edu/sites/default/files/Agriculture/maple-nass-2017.pdf (accessed 04 November 2017).
Nelson, V. 2009. Edibles you can grow in the shade. *The Oregonian/OregonLive*. Available at: http://www.oregonlive.com/hg/index.ssf/2009/07/shade_edibles_you_can_grow.html (accessed 04 December 2017).
Nowak, D. 1996. Measuring and analyzing urban tree cover. *Landscape and Urban Planning*, 36: 49–57.
Pavlis, R. 2017. Growing under walnut trees. *Garden Fundamentals*. Available at: http://www.gardenfundamentals.com/growing-under-walnut-trees/ (accessed 01 December 2017).
Pomper, K. 2009. Pawpaw research at Kentucky State University. Available at: http://www.pawpaw.kysu.edu/KSUstory.htm (accessed 04 November 2017).
Sabota, C. 1996. Strain of shiitake mushroom [*Lentinula edodes* (Berk.) Pegler] and wood species affect the yield of shiitake mushrooms. *HortTechnology*, 6(4): 388–93.
Scott, R. and W. Sullivan. 2007. A review of suitable companion crops for black walnut. *Agroforestry Systems*, 71: 185–93.
Sholto, D. J. and R. A. de J. Hart. 1985. *Forest Farming: Towards a Solution to Problems of World Hunger and Conservation*. Intermediate Technology Publishing, London, UK.
Trees that provide some shade but won't kill your lawn. Available at: http://homeguides.sfgate.com/shade-trees-wont-kill-lawn-27813.html (accessed 03 December 2018).
Westerveld, S. 2014. Thoughts on wild ginseng and replant disease. Available at: https://onspecialtycrops.wordpress.com/2013/09/24/thoughts-on-wild-ginseng-and-replant-disease/(accessed 03 October 2017).
Young, J. and C. Young. 1992. *Seeds of Woody Plants in North America*. Dioscorides Press.

Chapter 7

Modelling agroforestry systems

Paul Burgess and Anil Graves, Cranfield University, UK; Silvestre García de Jalón, Basque Centre for Climate Change (BC3), Spain; João Palma, MV Agroecology Research Centre, Portugal; Christian Dupraz, INRA-System, University of Montpellier, France; and Meine van Noordwijk, World Agroforestry Centre (ICRAF), Kenya

1 Introduction

Agroforestry can be defined as 'the practice of deliberately integrating woody vegetation (trees or shrubs) with crop and/or animal systems to benefit from the resulting ecological and economic interactions' (Burgess and Rosati, 2018). In the *Oxford English Dictionary* a 'system' is defined as 'a set of things working together as a mechanism or interconnecting network'. Hence, for the purpose of this chapter we assume that an agroforestry system is an interconnecting network of woody vegetation with crops and/or animals that work together. There are large temporal differences in the growth and development of the woody component in an agroforestry system and the crop and/or animal component. The time frame between planting a tree and it reaching maturity can be 20–100 years whereas an annual crop and some animals can reach maturity in months. Hence the growth periods for the trees and the crops and livestock in agroforestry systems are substantially different. This diversity in the range of time, and also spatial, scales and the typically non-linear ways in which the components interact mean that agroforestry is a 'complex' system (Boulton et al., 2015).

Due to the long time periods involved in tree growth, our understanding of agroforestry systems will be very restricted if it only depended on experimental data. One way to improve our understanding and management of agroforestry systems is to use models. The term 'agroforestry model' has numerous meanings.

http://dx.doi.org/10.19103/AS.2018.0041.13

Using the Scopus Search Engine (https://www.scopus.com), a literature search of peer-reviewed papers between 1990 and 2018 using the term 'agroforestry model' in the title, abstract or keywords identified 56 papers, of which 53 were accessible. Of these, 33 of the papers used the term 'agroforestry model' simply in the sense of a type of agroforestry system. This is not the focus of this chapter. Instead we use the term 'model' in the sense, defined by the *Oxford English Dictionary*, as 'a simplified description, especially a mathematical one, of a system or process, to assist calculations and predictions'. In the context of this chapter, models are representations of the real world that people have developed with the intention of making life more predictable, effective, efficient and/or enjoyable. Models can be qualitative (see Section 2.6.1), but most of the agroforestry models described in the chapter are quantitative, that is, they are based on numerical relationships.

Models of tree and crop growth are relatively new. Cheap digital computers were not widely available until the 1970s and 1980s, and Huda and Ong reported in 1987 that 'no models are yet available for simulating intercropping and agroforestry systems'. Because of the complexity, agroforestry modelling is also a time-consuming process, and hence a model user or developer is well advised to be clear about the objectives for using or developing a model. Matthews et al. (2000) identified that models were generally used by three types of end users: researchers, decision-makers, and teachers and trainers. Luedeling et al. (2016) have also highlighted five objectives for using models. Two of these relate to research: identifying knowledge gaps and testing interactions through 'virtual experiments'. A third objective relates to the provision of decision support. The final two objectives relate to teaching (and also research) in terms of using models to synthesise existing knowledge and to share that knowledge with others. The important role played by models to synthesise knowledge is also highlighted by Lusiana et al. (2011), who reported that clarity of thinking is often rated higher amongst model users than empirical precision. The use of crop simulation models as a tool in education has been reviewed by Graves et al. (2002).

The objectives of this chapter are as follows: (1) to examine the current state of modelling of agroforestry systems, (2) to give an example of how agroforestry modelling can be used to enhance sustainability, (3) to identify potential future trends and (4) to describe sources of further information.

2 Current state of agroforestry modelling

2.1 Agroforestry models as tools: what is the objective?

Models are tools that have been developed to make research, management or teaching more effective. In a conventional toolbox you may have many tools such as a hammer, a screwdriver and a saw. In a similar way, a wide range of agroforestry modelling tools is available. The most appropriate agroforestry model will depend on the problem to be addressed and, as indicated above, the first step in developing or using a model to examine an agroforestry system is to clearly define your objective. In the same way that you should not use a screwdriver to drive in a nail, or a hammer to remove a screw, using agroforestry models for situations that they were not designed for can cause problems.

A review across temperate and tropical areas suggests that there are six major groups of agroforestry model (Table 1). These are as follows: (1) allometric or regression models, (2) models where tree and crop yields are included as inputs that describe the environmental impacts of agroforestry, (3) plot-scale models where the focus is on tree and/or crop growth, (4) architecture models, (5) farm-scale management decision models and (6) landscape models. Each of these model types are considered in turn, with the greatest focus given to plot-scale models.

2.2 Allometric or regression models

Allometry is the study of how the characteristics of a living organism change with size. Allometric relationships are sometimes referred to as 'empirical' models; the *Oxford English Dictionary* defines empirical as 'based on observation or experience rather than theory or pure logic'. Whilst empirical models are useful for summarising past data and for interpolation (Boote et al., 1996), they typically 'contain no information beyond the original data' (Thornley and Johnson, 1990).

Commercial forestry makes wide use of allometric relationships such as yield tables describing, for example, how tree volume changes with tree height and tree diameter for a specified tree density (Burkhart and Tomé, 2002). For example, in the United Kingdom, there are timber yield data for poplar given at 5-year intervals for a given yield class (Christie, 1994). It is also possible to use these data to construct further polynomial regression models that allow the interpolation of timber yields for the intervening years. Although the coefficients in such polynomial curves have little physiological meaning, there are examples where there is a physiological basis for a regression. For example, the cross-sectional area of a tree trunk is a square function of its diameter. van Noordwijk and Mulia (2002) also explain how the fractal nature of some relationships means that it is also possible to develop functional branch analysis models to predict, for example, the weight of branches and leaves, from the stem diameter (van Noordwijk and Mulia, 2002; MacFarlane et al., 2014).

Allometric relationships for agroforestry systems are sometimes reported in scientific papers (e.g. Khaine and Woo, 2018; Moussa and Mahamane, 2018). They are also used in some detailed growth models (Bergez et al., 1999). Even in models that calculate the increase in tree dry matter, allometric models can be used to relate changes in tree dry weights to height, diameter and timber volume.

Table 1 Peer-reviewed papers describing the use of allometric, non-growth models, plot-scale models of tree and crop growth, architectural models, farm decision models and landscape agroforestry models

Model type	Name or description of agroforestry model	Example references
Allometric models	Models relating tree height, diameter and carbon storage	Khaine and Woo (2018), Moussa and Mahamane (2018)
Non-growth models of soil carbon	CO2FIX	Panwar et al. (2017)
	SCUAF (Soil Changes Under AgroForestry) model	Young et al. (1998), Grist et al. (1999)[a], Lojka et al. (2007)
	ICBM/N model	Salazar et al. (2011)
Plot-scale models of tree and crop growth	ALWAYS (Alternative Land-use With AgroforestrY Systems) silvopastoral model	Bergez et al. (1999), Balandier et al. (2003)
	APSim Next Generation	Keating et al. (2003), Holzworth et al. (2014), Luedeling et al. (2016), Smethurst et al. (2017), Dilla et al. (2018)
	Coffee agroforestry model	van Oijen et al. (2010)
	Hi-sAFe or STICS[b]	Talbot and Dupraz (2012), Talbot et al. (2014), Artru et al. (2017), Dupraz et al. (2018)
	HyPAR: The 'Hybrid' tree model with PARCH crop model	Mobbs et al. (1998), Lawson et al. (1995)
	WalNulCAS	van Noordwijk and Lusiana (1999), Radersma and Ong (2004), Magcale-Macandog (2014), Coulibaly et al. (2014)
	Yield-SAFE	van der Werf et al. (2007), Keesman et al. (2007), Graves et al. (2007, 2010), Keesman et al. (2011), Talbot et al. (2014), Palma et al. (2018), García de Jalón et al. (2018b)
Architectural models	'AMAP' canopy methodology	De Reffye et al. (1995), Mialet-Serra et al. (2001)
	ARCHIMED coconut model	Lamanda et al. (2007)
	FracRoot model	Ozier-Lafontaine et al. (1999)
Farm decision models	Conceptual model of farm decisions	Meylan et al. (2013)
	Knowledge-based system to identify agroforestry grants	Acuña et al. (1997)
	Model ranking agroforestry threats	Moral et al. (2014)
	BEAM and POPMOD Bio-economic models	Thomas (1991), Willis et al. (1993), Purnamasari et al. (2002)
	ARBUSTRA	Liagre (1997)
	Agroforestry Estate Model	Knowles and Middlemiss (1999)
	Agroforestry Calculator	Agriculture Western Australia and Campbell White and Associated Pty Ltd (2000)
	Farm-SAFE	Graves et al. (2007), Graves et al. (2011), García de Jalón et al. (2018b)
	Forage-SAFE	García de Jalón et al. (2018a)
	Financial model including 'real options'	Freya et al. (2013)
	Welfare maximisation model	Dhakal et al. (2007)
	Agroforestry adoption models	Lovrić et al. (2018)
Landscape	Landscape models	Jackson et al. (2013), Kay et al. (2018), Zeng et al. (2018)

[a] Combined with an economic model.
[b] Used to determine the effect of different levels of solar radiation on a barley crop.

2.3 Non-plant growth models to determine agroforestry impact

There are a number of agroforestry models that do not calculate plant growth, but can still be used to examine the effect of forestry, agricultural or agroforestry systems on environmental impacts such as changes in soil carbon, soil nutrients or water flow. One example is the Soil Changes Under AgroForestry (SCUAF) model (Young et al., 1998) which has an annual time step. The user enters the proportion of the area allocated to trees and agriculture in year one and the growth rate of the trees and crops in each year. The model can then be used to estimate the effects of the trees on changes in soil properties including soil erosion, soil organic matter and nutrients, and also on subsequent rates of plant growth. In a similar way, the CO2FIX model has been used to determine the carbon sequestration of poplar agroforestry systems in India both above and below ground (Panwar et al., 2017). With this model, the user inputs estimates of the annual growth of the trees and the anticipated crop yields.

2.4 Plot-based mechanistic models of tree and crop growth

The most widely cited agroforestry models are those that can be used to predict the effect of tree and crop interactions over time. Sinclair and Seligman (1996) defined a crop model as 'the dynamic simulation of crop growth by numerical integration of constituent processes with the aid of a computer'. Matthews et al. (2000) defined a crop model as 'a computer program describing the growth of a crop in relation to the environment operating on a time-step an order of magnitude below the length of a growing season, and with the capacity to output variables describing the state of the crop at different points in time'.

Most crop and agroforestry models include a 'mechanistic' or 'explanatory' component in that they attempt to describe behaviour at one level by using an understanding of what occurs at a sublevel (Whisler et al., 1986; Thornley and Johnson, 1990). Whisler et al. (1986) argue that it is generally unnecessary to build a crop model using an understanding on what happens at more than two levels of hierarchy lower. For example the growth of a crop can generally be determined by the interception of light by a canopy and the uptake of water by a root system, rather than by explaining what is occurring at a cellular level. The anticipated advantage of developing mechanistic models based on sub-level processes is that it should allow the model to be used in a wider range of contexts, which also makes it useful as a research tool (Table 2). However their ability to match observational data may be no better than an empirical model.

Despite the division between mechanistic and empirical models shown in Table 2, in practice there is a continuum between pure mechanistic models and pure empirical models. Boote et al. (1996) argue that most crop simulation models are a 'mixture of empiricism and mechanism', and Whisler et al. (1986) note that all models become empirical at some level. However

Table 2 A comparison of the principal differences between empirical and mechanistic models

Type	Mechanistic or explanatory models	Empirical or correlative models
Scope	Wide	Restricted
How does the model deal with time?	Dynamic	May be static
Connections to observational data	Can be tenuous	Generally good
Accuracy of predictions	Variable	Generally good
Typical use	Research and teaching	Management
Typical complexity	Complex	Simple

Source: Based on Thornley and Johnson (1990).

ideally these empirical relationships occur at a sublevel and, where possible, they are formulated in a way that has a theoretical basis. Another unobserved area in which model use depends on correlation is in the assumption that the climate variables measured in standard weather stations are pertinent to the microclimate experienced by trees or crops in a field.

An important feature of agroforestry systems is the interaction that occurs over time between the trees, the crops and the atmospheric and soil environment (Fig. 1). For this reason, most agroforestry models are 'dynamic' as they 'predict how a system unfolds with the passage of time' (Thornley and Johnson, 1990). Once the objective is clear, Connor (1983) writes that there are four steps in the development of a dynamic model.

1 There is a need to identify equations for the important flows of material and information.
2 There is a need to select values for all the parameters.
3 There is a need to set initial values for state variables.
4 The model is then run to solve the equations at appropriate time intervals.

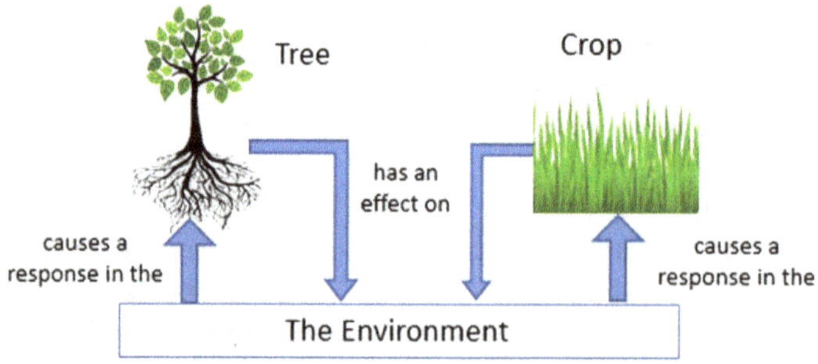

Figure 1 Schematic representation of the interaction between the trees, crops, atmosphere and soil environment that requires 'dynamic' models. Source: Adapted from Vandermeer (1989).

In the example in Fig. 1, an important flow parameter may be the flow of water. The use of soil water by the tree and the crop can be expressed by equations with parameters. The initial state variables are likely to include the initial soil water content, and the change in soil water content over future time intervals is then calculated using a series of equations.

2.4.1 How complicated should a model be?

There are arguments in favour of both 'simple' and 'complicated' models (Boote et al., 1996). However to paraphrase Albert Einstein, a common approach is that 'things should be as simple as possible, but no simpler'. Box (1979) also argued positively for the use of parsimonious models, observing that no model is true, but some can be 'illuminating and useful'. Passioura (1973) also favoured simple crop simulation models with few parameters, as they were more 'testable'. By contrast, van Noordwijk and van de Geijn (1996) argue that the minimum number of parameters may increase over time as independent measures of those parameters become available and a model is used to address more specific questions.

Leinweber (1979) defined model complexity as a simple count of the number of arithmetic operations a model performs on input data, and he specified two sources of modelling error. One source relates to uncertainties in the input data. If the potential error of these is known, then the errors can combine in a way that can be predicted mathematically. Leinweber (1979) assumed that as the number of input parameters increased the margin for error also increased. In addition, as the model becomes more complex, it can also become more difficult to describe the flows between spatial components, for example, how surplus of water in one part of the soil will be redistributed to other parts of the soil. The second form of error is a specification error where the model does not accurately describe what happens in practice. Leinweber (1979) argues that a model should only be made more complex if it is made more accurate. The assumption is that the specification error should decrease as the complexity increases. The net effect, as described by Palma et al. (2007b), is that there may be an optimum level of model complexity that minimises the total error due to simplistic assumptions and compound errors in the input data (Fig. 2).

Mechanistic models of tree and crop growth typically build on the physical and biological principles such as those described by de Wit in the Netherlands (de Wit, 1958; Bouman et al., 1996) and/or Monteith in the United Kingdom (Monteith, 1992). One of the first 'process-based' plant models called ELCROS was described by Brouwer and de Wit (1968). Jackson and Palmer (1972) described the results of a light interception model for different shapes of hedgerow orchards demonstrating the changes in diffuse and direct light

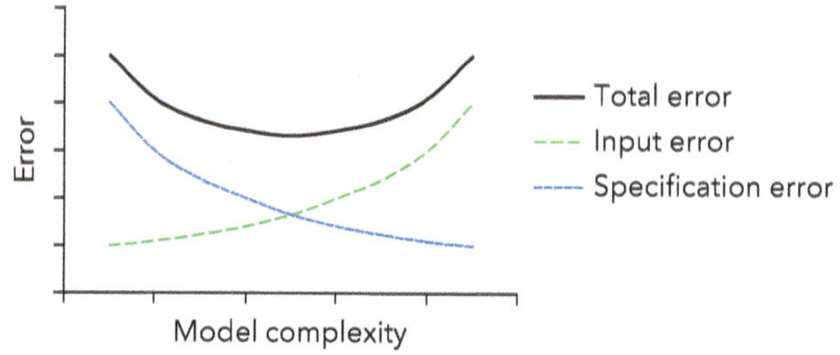

Figure 2 Relationship between model complexity and total error in the upscaling process. Source: Modified from Palma et al. (2007b) and Leinweber (1979).

interception with tree shape, alley width, season and latitude. Charles-Edwards and Thorpe (1976) also completed a similar analysis for a hedgerow apple orchard. Jackson and Palmer (1987) also described a diurnal light interception model to simulate light interception within an apple intercropping experiment.

2.4.2 Examples of a 'simple' agroforestry tree and crop growth model

One of the simplest mechanistic agroforestry models is Yield-SAFE which includes simple light capture algorithms and was originally designed to describe an alley cropping silvoarable system (van der Werf et al., 2007). As the time step for the model is a day, it is assumed that each part of the alley cropping field will receive some shade from the trees. The incoming daily solar radiation is assumed to be S (unit: MJ m^{-2} day^{-1}). The proportion of the radiation intercepted by trees (f_t) is then defined in terms of the leaf area index of the trees (LAI_t; unit: m^2 tree leaf area per m^2 plot) and a light extinction coefficient (k_t) (Eq. (1)) (van der Werf et al., 2007). Subsequent work showed that it was necessary to change the value of k_t as the trees grew (Keesman et al., 2011).

$$f_t = 1 - e^{-k_t \cdot LAI_t} \tag{1}$$

The potential radiation available below the trees (S_c) can then be defined using Eq. (2):

$$S_c = S\left(1 - f_t\right) \tag{2}$$

Within an alley cropping system it is assumed that silvoarable field is divided into a proportion that is cropped (A_c) and a proportion that is not cropped ($1-A_c$). The proportion of S_c that is intercepted by the understorey crop can therefore be derived from Eq. (3) where k_c is the radiation extinction coefficient of the crop and LAI_c is the leaf area index of the crop (m^2 crop leaf area per m^2 of cropped area).

$$f_c = A_c\left(1 - e^{-k_c \cdot LAI_c}\right) \tag{3}$$

Whereas Yield-SAFE partitions the understorey into a crop and uncropped area, an alternative approach is to partition the agroforestry system in a tree-covered and non-tree-covered area. An example of this approach is described by Bergez et al. (1999) in the ALWAYS silvopastoral model which used photosynthetically active radiation (PAR). Building on the model of Jackson and Palmer (1979), the model partitioned the system into a proportion covered by a vertical projection of the tree canopy (A_t) and a proportion that was not ($1-A_t$). The mean daily photosynthetically active radiation on the pasture (PAR_p) is then calculated as the sum of the sum of the diffuse radiation ($PAR_{diffuse}$) and the transmitted direct radiation ($PAR_{t,\ direct}$) within the tree canopy and the direct radiation (PAR_{direct}) for the remaining area (Eq. 4).

$$PAR_p = \left(PAR_{diffuse} + PAR_{t,\ direct}\right)A_t + PAR_{direct}\left(1-A_t\right) \quad (4)$$

The value of $PAR_{t,direct}$ was derived from Eq. (5) where L_d is the leaf area density in the crown and s is the height of the crown.

$$PAR_{t,\ direct} = PAR_{direct}e^{(-0.5 \cdot L_d \cdot s)} \quad (5)$$

The above five equations are examples of how simple agroforestry models can address the interception of solar radiation by a tree and a crop. Models such as Yield-SAFE and ALWAYS can also deal with water competition between the trees and understorey crops. Another example of a relatively simple agroforestry model is the coffee agroforestry model developed in MATLAB/Simulink (van Oijen et al., 2010) based on the BASic FORest simulator (BASFOR) tree model (van Oijen et al., 2005).

2.4.3 Complex agroforestry models

A limitation of the Yield-SAFE model is that it is only a one-dimensional model with some consideration of the horizontal extent of the arable crop (Fig. 3a). An example of a more complex two-dimensional model is the Water, Nutrient and Light Capture in Agroforestry Systems (WaNuLCAS) model as described by van Noordwijk and Lusiana (1999). The original version of the model included a four-layer soil profile (vertical), with four spatial zones (horizontal) (Fig. 3b), and calculation of the water and nitrogen balance and uptake by a crop and a tree. The model used a 'Stella Research' modelling shell with data inputs and outputs provided in Excel spreadsheets.

In an alley cropping context, whilst the two-dimensional WaNuLCAS model can be used to determine the effect of distance from a tree on crop yields, this particular version of the model assumes that light interception in a West-East direction will be the same as in a North-South alignment. Considerations of row alignment require a three-dimensional model. An early three-dimensional agroforestry model combined a detailed tree model called Hybrid (Friend et al.,

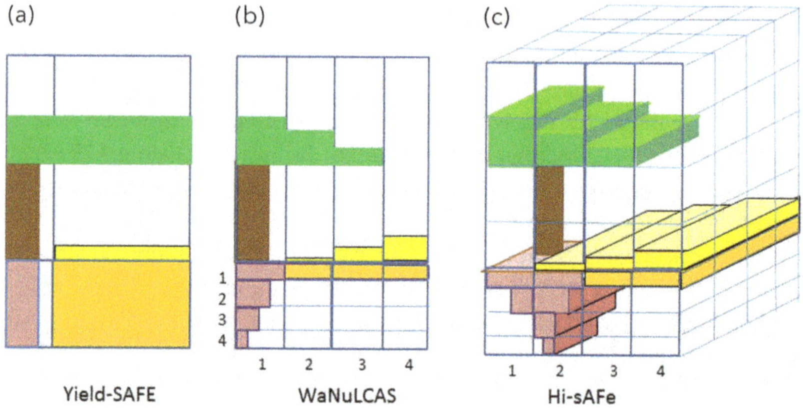

Figure 3 (a) The Yield-SAFE model comprises of two above-ground zones (crop and no crop) and one below-ground zone (soil depth). (b) The WaNuLCAS model described by van Noordwijk and Lusiana (1999) comprises four root zone depths and four distances from the tree. (c) Hi-sAFe is an example of a three-dimensional agroforestry model.

1997) with a crop model called PARCH (Lawson et al., 1995). A more recently developed example of a three-dimensional agroforestry model is Hi-sAFe which combines a tree model with the suite of STICS crop models (Dupraz et al., 2005). The Hi-sAFe model has a switchable toric symmetry module that replicates the simulated scene, allowing it to simulate isolated trees (toric symmetry fully off), tree hedges (toric symmetry off in one direction), alley cropping agroforestry (full toric symmetry) and forest edges (toric symmetry off in y and half-x axis), as well as crop and forest monocropping. In Hi-sAFe, root competition is addressed using a three-dimensional opportunistic tree root growth module based on a cellular automata algorithm. This allows the simulation of tree root systems that can adapt continuously to the soil environment as modified by the weather and the crop competition. Tree root pruning, tree branch pruning, tree stand thinning and a large panel of crop management schemes are also available. Another feature of Hi-sAFe is the possibility of considering the impact of a fluctuating seasonal water table on tree-crop competition (Lecomte et al., 2016).

The initial use of the Hi-sAFe model included an examination of the clumpiness of leaves in trees (Talbot and Dupraz, 2012) and the simulation of walnut-wheat alley cropping for the duration of the tree life cycle (Dufour et al., 2013). More recently, Dupraz et al. (2018) describe the use of the Hi-sAFe model to determine the best tree row orientations to maximise the systems outputs at various latitudes, from the tropics to boreal regions. This analysis showed that in terms of maximising understorey crop yields and uniformity, the best tree line orientation was East to West at low latitudes and North to South at high latitudes. Recent developments include the use of the model to assess the resilience of agroforestry systems to climate change, the nitrogen leaching control by the trees and the long-term C sequestration in the soils (Dupraz and Lecomte, 2017).

2.4.4 Flexibility of drawing on a wider user community

One potential advantage of the Hi-sAFe model is that the crop component is based on an established crop model called STICS, which is supported by a strong user community. Results reported by Artru et al. (2017) have demonstrated how different shading patterns affect the yields predicted using the STICS model. Luedeling et al. (2016) also describe some of the benefits in terms of flexibility in modelling agroforestry systems using the Agricultural Production Systems Simulator (APSIM) framework. The APSIM framework was initiated in 1994 and a 'next generation' of APSIM was released in October 2014 (Holzworth et al., 2018). Smethurst et al. (2017) describe the use of APSIM Next Generation framework to provide a two-dimensional (2D) representation of *Gliricidia*-maize intercropping, which is reported to describe accurately crop yield and water and soil carbon dynamics.

2.5 Architectural models of tree growth

Architectural models of tree growth are the fourth type of agroforestry model. De Reffye et al. (1995) describe the development of a canopy architecture model called AMAPmod by CIRAD in France, where tree architecture is described in terms of the form of 'growth units'. Models such as the 'FracRoot' model have also been used to describe below-ground root architecture (Ozier-Lafontaine et al., 1999). Lamanda et al. (2007) describe the use of three-dimensional architectural models to assess light availability and root bulkiness in coconut agroforestry systems in Vanuatu in the South Pacific.

2.6 Farm-scale management decision models

The fifth group of agroforestry models comprises models where the primary focus is to improve decision-making by farmers and others. Some comprise conceptual models, but many assume that profitability is an important determinant of whether a farmer will adopt a system.

2.6.1 Qualitative models of the system and farm decisions

Sinclair and Walker (1998a,b) describe a method to model indigenous ecological knowledge on agroforestry. They demonstrate that formal representation of such models can increase the coherence, consistency and completeness of our understanding of agroforestry systems. Meylan et al. (2013) also describe the development of a conceptual model to identify the constraints and trade-offs associated with four types of farmers practising agroforestry with coffee production. The model includes environmental factors (e.g. climate), management factors, agronomic and hydrological processes, and outputs such as yield, gross margin and soil erosion losses. The process results in a series of system-flow diagrams showing the relationships between the various factors, processes and outputs. They argue that this can serve as a preliminary stage before attempting to identify the key points of potential interventions in the system. In a similar way, flow diagrams were developed by Acuña et al. (1997) in their development of a knowledge-based system model to create decision trees related to agroforestry grants in Spain. Moral et al. (2014) in Spain also examine the rationale for on-farm decisions drawing on a survey of dehesa management based on a Strength, Weaknesses, Opportunities and Threats (SWOT) model. The survey identified 68 potential threats to the system, and a Rasch ranking methodology was used to develop a model that describes the most important threats.

2.6.2 Financial models

It is possible to undertake a financial cost-benefit analysis of agroforestry systems without a detailed biophysical model of tree and crop yields. For example, Dube et al. (2002) describe the financial benefits and costs of a eucalyptus agroforestry system in Brazil. However in their review of financial models of silvoarable systems, Graves et al. (2005) found five models that could incorporate yield and financial data. The models included WaNuLCAS with a financial component (van Noordwijk and Lusiana, 1999), POPMOD (Thomas, 1991), ARBUSTRA (Liagre, 1997), the Agroforestry Estate Model (Knowles and Middlemiss, 1999) and the Agroforestry Calculator (Agriculture Western Australia and Campbell White and Associates Pty Ltd, 2000). Graves et al. (2005) found that whilst ARBUSTRA and the Agroforestry Calculator relied on direct inputs of yield data, WaNuLCAS, POPMOD and the Agroforestry Estate Model contained biophysical models of differing complexities that were used to generate the yield data for the financial analyses. For WaNuLCAS and Agroforestry Estate Model, this information flowed in one direction from the biophysical component of the model to the financial component of the model. In the POPMOD model, it was possible for management feedback from the financial component of the model, to link back to the biophysical component and modify the yields obtained.

The Farm-SAFE model (Graves et al., 2011) developed during the SAFE project is an example of a financial model, developed to evaluate the profitability, practicality and feasibility of silvoarable systems from a farmer perspective. It used annual tree and crop yield data on arable, forestry and agroforestry systems, derived from the Yield-SAFE model, which was then used to compare the financial benefits and costs. As it is a farm-scale model, it can be used to model different arable, forestry and agroforestry systems across a farm reflecting, for example, contrasting levels of soil fertility. Another example of a recent model combining financial and yield data for grass-oak tree systems in Spain and Portugal is Forage-SAFE (García de Jalón et al., 2018a). This uses a relationship for grass growth at varying distances from an oak tree to calculate what the optimal stocking rate should be for livestock at different tree densities and grass productivity levels.

One of the important points in completing a useful financial analysis of an agroforestry system is the ability to complete analogous analyses of 'counterfactual' forestry and/or agricultural systems. Thus, for example, Farm-SAFE, ARBUSTRA and POPMOD provide a financial assessment of arable and forestry systems against which the financial performance of agroforestry can be compared. One effect of this is that most agroforestry financial models are effectively combined agricultural, agroforestry and forestry models, for example, they can compare a wide range of land uses. In order to account for the impact of the different systems on labour and machinery costs, it is generally accepted that the comparison should be based on net margins, rather than gross margins. Freya et al. (2013) also describes how the impact of changes in price, after the establishment of an agroforestry or forestry system, can affect management choice using a stochastic and 'real options' model. Farm-SAFE, for example, contains the option of changing future grants, prices and costs either in a continuous phased or discontinuous one-off pattern from a specified year in the future.

2.6.3 Economic and social welfare models

Whilst financial models can be used to compare the attractiveness of an agroforestry system with arable or forestry systems for a farmer, the attractiveness of agroforestry from a societal perspective requires consideration of wider societal costs and benefits. Possible benefits may be increased carbon sequestration or reduced nitrate leaching which does not have a readily available market value. Through the use of cost-benefit analysis and environmental valuation data, it is possible to impute values for many of the outputs that do not have immediate market value. Since 2014, Farm-SAFE has been developed to incorporate non-market societal costs and benefits of arable, forestry and agroforestry systems by incorporating life cycle assessment and valuation data. Life cycle assessment data can be linked with management decisions to quantify the environmental burdens associated with tree and crop production in the modelled arable, forestry and agroforestry systems. This is then linked with valuation data derived using a benefits-transfer approach in order to quantify the non-monetary value of changes in carbon sequestration, and erosion, nitrate and phosphorus emissions to water (García de Jalón et al., 2018b). Dhakal et al. (2007) also describe the use of a constrained welfare maximisation model to identify the opportunity of farmers implementing agroforestry in Nepal. The analysis was based on a household income model, a household production model and a community welfare model.

2.6.4 Adoption models

The adoption of agroforestry is not solely determined by a financial and economic analysis. Lovrić et al. (2018) describe the use of an analytic network process to quantify the driving forces between agroforestry uptake in Mediterranean Europe in environmental, economic and social areas.

2.7 Landscape models

A common question in government-sponsored research is to determine the effect of the widespread uptake of agroforestry. This can take place at a landscape, regional or national scale. Models such as ARBUSTRA and Farm-SAFE can be used as landscape models since they are able to model a range of different arable, forestry and agroforestry systems under different environmental and management conditions. The results can then be aggregated to the landscape scale by summing the outputs of the different systems. The biophysical assessments are typically based by running versions of the biophysical, financial and economic models described above within a geographical information system (GIS) framework. Such an approach was used in the SAFE project (Dupraz et al., 2005) by Palma et al. (2007a). In the AGFORWARD project (Burgess and Rosati, 2018), a similar approach was used by Kay et al. (2018). Another example of a landscape model linking GIS with other models is Polyscape (Jackson et al., 2013). The Polyscape model uses algorithms to determine the effect of land cover changes, such as agroforestry, flooding, soil erosion, carbon storage and agricultural production. In China, Zeng et al. (2018) describe the use of GIS with a land-water-environment model to determine the effect of afforestation or increased agricultural production. It is interesting that they refer to their model as an agroforestry ecosystem model - whereas in reality it is a model that can address forestry, agroforestry and agriculture.

3 Example of how agroforestry modelling can enhance sustainability

In Europe during the past two decades, there have been two large agroforestry projects that involved modelling. The Silvoarable Agroforestry For Europe (SAFE) project ran from 2001 to 2005 (Dupraz et al., 2005). The AGroFORestry that Will Advance Rural Development (AGFORWARD) project ran from 2014 to 2017 (Burgess and Rosati, 2018). Within the SAFE and AGFORWARD projects, the project aim was to determine the effect of different agroforestry systems

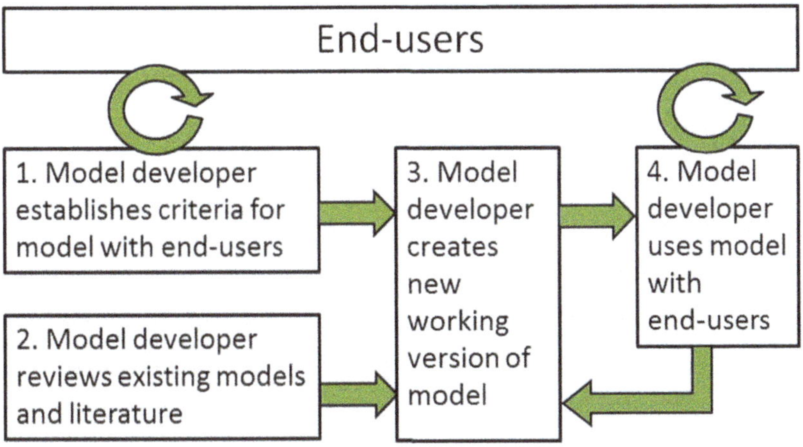

Figure 4 Schematic showing how the development and use of an agroforestry model can be based on feedback from end users. Source: Adapted from Graves et al. (2010).

at farm and landscape scales. The intention was to look at both food and fibre production and some of the most important regulating services. The stages of identifying and using an agroforestry model, in the case of the SAFE project, was described by Graves et al. (2011) and is illustrated in Fig. 4. The stages involve (1) establishing criteria for model, (2) reviewing existing models, (3) developing a new model (if required) and (4) using the model. This process is then discussed using the modelling activity described by García de Jalón et al. (2018b).

3.1 Establish criteria

Within the AGFORWARD project one of the objectives was to compare the financial attractiveness of arable, forestry and a silvoarable poplar system in the United Kingdom, and to determine the social impact of environmental externalities (Table 3). Using a framework described by Graves et al. (2005), the criteria for the model included requirements related to (1) the model background, (2) the systems to be modelled, (3) the objective of the economic analysis, (4) viewpoint of the analysis, (5) spatial scale, (6) temporal scale, (7) the generation and use of biophysical data, (8) the model platform and interface and (9) the input requirements and outputs generated.

3.2 Review existing models

As described in Section 2.6.2, Graves et al. (2005) reviewed five models of silvoarable economics at the beginning of the SAFE project. As a result of that analysis it was decided that a spreadsheet-based model building on the

Table 3 Example criteria for a societal comparison of agroforestry with forest and agricultural systems

Model characteristic	Example option
Background	
Language	English
Initial primary use	Research
Systems modelled	
Components of system	Agriculture, forestry and silvoarable
Number of systems	One agriculture, one forestry and one silvoarable system
Objectives of analysis	Comparison of different systems
Viewpoint of analysis	Micro-economic (i.e. perspective of land manager) and macro-economic (i.e. perspective of society)
Spatial scale	Comparison of 1-ha plots
Temporal scale	
Time step	Daily
Time period considered	A tree rotation of 30 years
Generation and use of biophysical data	
Biophysical-economic links	Bio-economic model with two-way information flow
Nature of biophysical model	A mechanistic model which includes growth processes
Platform and interface	
Model platform	Spreadsheet
Model interface	Spreadsheet
Inputs and outputs	
Inputs requirements	Relatively low data requirements
Outputs	Biophysical and economic

Source: Based on model characteristics described by Graves et al. (2005).

key capabilities of the POPMOD and the ARBUSTRA models needed to be developed, which was subsequently called the Farm-SAFE model (Graves et al., 2011). However, as Farm-SAFE was developed purely as a financial model, it had no means of generating yield data. Hence the biophysical Yield-SAFE model was also developed to provide annual tree and crop yield data. Based on this review, the Yield-SAFE and Farm-SAFE models were also selected for use in the AGFORWARD project. However, whereas the SAFE project focussed on the financial performance of silvoarable systems, the aim of the AGFORWARD project was to complete financial and economic analyses of a wider range of agroforestry systems.

3.3 Create new working version of the models

Since 2014 in the AGFORWARD project, researchers worked on the Yield-SAFE and Farm-SAFE models to develop new algorithms to describe important environmental externalities such as greenhouse gas emissions, carbon sequestration, soil erosion, and nitrogen and phosphorus losses (García de Jalón et al., 2018b). The method for comparing the societal effect of a silvoarable agroforestry system with forestry and arable systems comprised five stages: (1) simulation of the biophysical growth of trees and crops for an experimental plot case study using Yield-SAFE, (2) quantification of the environmental externalities (Yield-SAFE and Farm-SAFE), (3) assessment of financial performance (Farm-SAFE), (4) conversion of the environmental externalities into monetary terms (valuation data) and (5) assessment of the societal performance of the different land use systems through an aggregate financial and economic analysis (Farm-SAFE). Each stage was addressed by different models or approaches (Table 4).

Table 4 Summary of the key modelling activities to compare the societal effect of a silvoarable agroforestry system with forestry and arable systems

Modelling activity	Detailed output	Model or approach	Units	Reference
Crop and tree yields	Crop and timber yield	Yield-SAFE model	t dry matter ha^{-1} $year^{-1}$	van der Werf et al. (2007)
Financial performance	Financial performance	Farm-SAFE model	€ ha^{-1} $year^{-1}$	Graves et al. (2011)
Quantifying environmental externalities	Greenhouse gas emissions	Life cycle assessment of greenhouse gas emissions	t CO_2eq ha^{-1} $year^{-1}$	Williams et al. (2010)
	Carbon sequestration	Approach based on the simulated biomass growth	t CO_2eq ha^{-1} $year^{-1}$	Nair (2011)
	Soil loss due to erosion	Revised Universal Soil Loss Equation (RUSLE)	t soil ha^{-1} $year^{-1}$	Palma et al. (2007b)
	Nitrogen surplus	Nitrogen balance model	kg N ha^{-1} $year^{-1}$	Palma et al. (2007b) and Feldwisch et al. (1998)
	Phosphorus surplus	Phosphorus balance model	kg P_2O_5 ha^{-1} $year^{-1}$	García de Jalón et al. (2018b)
Conversion of environmental externalities to monetary terms	Societal economic benefits and costs	Benefit transfer method	€ (t CO_2eq)$^{-1}$ € (t soil)$^{-1}$ € (kg N)$^{-1}$ € (kg P_2O_5)$^{-1}$	
Economic performance	Economic performance	Farm-SAFE model	€ ha^{-1} $year^{-1}$	García de Jalón et al. (2018b)

Source: Based on García de Jalón et al. (2018b).

3.4 Use of the model

This section provides an example of the use of an agroforestry model to compare the full societal value of the agroforestry, arable and forestry systems based on an experimental case study site at Silsoe in Bedfordshire in the United Kingdom as described by Burgess et al. (2005) and García de Jalón et al. (2018b). The daily time step Yield-SAFE model was used with local weather data derived from the CliPick tool (Palma, 2017). The model was then calibrated to local conditions by adjusting the light extinction coefficient, harvest index and water use efficiency to match model predictions to observed yield and growth data obtained in local tree-only and crop-only systems (see Graves et al., 2007 for more detail). For the tree-only system, the observed data for model calibration were poplar tree heights and diameters obtained for a forest treatment at the Silsoe experimental site (see 'Observed forestry data', Fig. 5). For the crop-only system the observed data were taken from statistics on local agriculture, for example, Agro Business Consultants (2015). The model of tree height and diameter with and without intercropping is shown in Fig. 5. From this calibration to local conditions, Yield-SAFE was then used to predict the tree and crop yields for arable, forestry and silvoarable systems over a 30-year time horizon. These data were then used as inputs into the Farm-SAFE model, which was used to calculate the financial and economic outputs of the arable, silvoarable and forestry systems.

The important financial outputs of the modelling exercise are described in Table 5. The results show that, in the absence of grants, the financial profitability of the arable system expressed as an equivalent annual value (€315 ha^{-1}) was substantially higher than for the silvoarable (€120 ha^{-1}) and forestry (€62 ha^{-1}) systems (Table 5). However, after the quantification and assignment of monetary

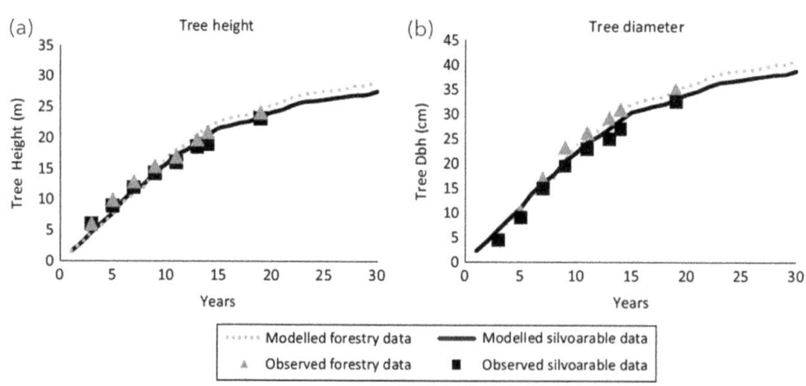

Figure 5 Observed and modelled height (a) and diameter at breast height (Dbh) (b) of the poplar trees in the forestry and silvoarable system in Bedfordshire in the United Kingdom; modelled using Yield-SAFE. Source: Adapted from García de Jalón et al. (2018b).

Table 5 Financial and economic equivalent annual value (EAV) of an arable, forestry and silvoarable system in Bedfordshire in the United Kingdom

		Silvoarable[b]			
	Arable[a]	Crop component	Tree component	Combined	Forestry[c]
Financial analysis					
EAV_F (€ ha^{-1} $year^{-1}$)	315	91	29	120	62
Environmental externalities					
CO_2eq emissions (t CO_2eq ha^{-1} $year^{-1}$)	2.4	1.1	0.0	1.1	0.0
CO_2eq sequestration (t CO_2eq ha^{-1} $year^{-1}$)	0.0	0.0	4.0	4.0	4.7
Soil erosion losses by water (t soil ha^{-1} $year^{-1}$)	0.4	-	-	0.2	0.2
Nitrogen surplus (kg N ha^{-1} $year^{-1}$)	24.6	-	-	2.2[d]	0.0
Phosphorus surplus (kg P_2O_5 ha^{-1} $year^{-1}$)	8.0	-	-	0.4	0.0
Economic analysis					
EAV CO_2eq emissions (€ ha^{-1} $year^{-1}$)	−20	−11	0	−11	0
EAV CO_2eq sequestration (€ ha^{-1} $year^{-1}$)	0	0	30	30	35
EAV soil erosion losses (€ ha^{-1} $year^{-1}$)	−3	-	-	−2	−2
EAV nitrogen surplus (€ ha^{-1} $year^{-1}$)	−186	-	-	−28	0
EAV phosphorus surplus (€ ha^{-1} $year^{-1}$)	−8	-	-	−1	0
EAV_E with externalities (€ ha^{-1} $year^{-1}$)	98	-	-	108	95

[a] Arable system was a rotation of winter wheat, winter wheat, spring barley and oilseed rape.
[b] Silvoarable system had the same rotation as the arable system with poplars planted at 156 trees ha^{-1}.
[c] Forestry system had poplars planted at a density of 156 trees ha^{-1} but no arable understorey.
[d] This is the balance of surplus from the crop minus the uptake of the tree.
Results shown for a time horizon of 30 years at a 4% discount rate without grants (García de Jalón et al., 2018b).

values to the non-market costs and benefits associated with the arable, forestry and agroforestry systems, it became evident that the arable system was also associated with high environmental costs, particularly in terms of the nitrogen surplus (−€186 ha^{-1}). The agroforestry and forestry systems on the other hand were associated with environmental benefits in the form of carbon sequestration (€30 and €35 ha^{-1} respectively), whilst also having low costs associated with the nitrogen surpluses (−€28 and €0 ha^{-1} respectively). On this basis, the net annual societal benefit of the silvoarable system (€108 ha^{-1}) was found to be slightly higher than that for the arable (€98 ha^{-1}) and the forestry systems (€95 ha^{-1}). Hence this analysis highlights that there is a rationale for providing public support for agroforestry in terms of the environmental benefits.

4 Current agroforestry modelling needs and potential trends

The above sections have examined the current state of agroforestry modelling and provided an example of the use of agroforestry models to compare the market and non-market benefits of agroforestry relative to arable and forestry systems. Looking to the future, this final section considers current agroforestry modelling needs and potential trends. A first point to make is that agroforestry models have a role beyond just agroforestry. The other points considered include the accessibility of agroforestry models, the benefits of long-term support, the need to know when to stop model development, the availability of data and the societal assessments.

4.1 Agroforestry models are particularly useful as they are forest to agriculture continuum models

A particular strength of agroforestry models is that they can model the whole continuum from pure forestry via a range of agroforestry systems to rotational and monoculture agriculture. They have a use beyond just agroforestry. In fact agroforestry models can be used to compare land use decisions across a range of land uses that include both agriculture and forestry as described by Burgess et al. (2012) for a case study area in the United Kingdom.

4.2 Publicising and making models available

Matthews et al. (2002) argue that crop models can be considered to pass through infancy, juvenile, adolescence and maturity stages. Hence they question at what stage is a model ready to find gainful employment beyond the environment where it was initially raised? As described earlier, able agroforestry modellers should be open to use models that they have not developed. When someone needs to extract a screw, and they only have a hammer in their toolbox, this does not mean that the hammer is the best and only way for extracting the screw. Instead, the individual would be well advised to obtain a screwdriver to make the job successful. It is therefore important, as argued by Boote et al. (1996), that agroforestry modellers should be clear about the capabilities and limitations of both their own models and other alternative models on the market. It also means that models should be publicised and be made as widely available as possible. Details on how to access some of the models described in this chapter are provided in the final section.

4.3 The use of models depends on the research communities

The development, use and longevity of agroforestry models as a research tool seem to be dependent on long-term research projects and/or funding. For example, the development and use of the Hi-sAFe, Yield-SAFE and Farm-SAFE models in Europe has been dependent on the SAFE (2001-05) and the AGFORWARD (2014-17) projects. Another example is the AgMIP models developed from the International Consortium for Agricultural Systems Applications (ICASA) project between 1993 and 2011 (Jones et al., 2017).

4.4 The need to know when to stop developing a model

The second Utz law of computer programming states that 'if a programme is useful it will have to be changed' (The Economist, 1986). It can be difficult to know when it is time to stop developing and calibrating a computer model and when to validate and use it (Boote et al., 1996). Calibration refers to the fine-tuning of model parameters to make it appropriate for a particular location or design. Validation refers to the testing of the model outputs, after calibration, using independent datasets. Passioura (1973) has argued that comprehensive crop models are works of art, quoting a conference comment by de Wit that developing crop models are 'the most cumbersome and subjective technique of curve fitting ... that can be imagined'. Passioura also argues that it can be easy in model development to generate new parameters rather than new hypotheses. This chapter has highlighted those agroforestry models that have been developed, used and written-up in the form of a scientific report or paper. There are probably agroforestry models that have been developed but have not been reported.

4.5 Agroforestry models and data

The use of agroforestry models is nearly always constrained by a lack of accurate data either in the forms of inputs, for parameters or for calibration and validation of the model. One potential development which may help address this is the increasing availability of detailed remote sensing data. Time series of such data may be useful to help calibrate agroforestry models. In addition to a lack of biophysical data, the wide range of agroforestry systems mean that it can also be difficult to find financial data related to tree management. A difficulty in finding financial information of the timber value of trees is a particular problem (Burgess et al., 2017). Hence, one benefit of agroforestry modelling is that it provides a means by which biophysical and economic knowledge on agroforestry systems can be generated and shared. A major achievement would be to collect such data in a format that is transparent and easily used by farmers.

4.6 Societal assessments

An increasing need is to broaden the scope of agroforestry model assessments so that the full costs and benefits of land use decisions can be identified. Such modelling is challenging because of its breadth and it is still in its early stages, for example, García de Jalón et al. (2018b). Such evaluations are likely to be of increasing relevance to society given current and future societal pressures on the environment.

5 Acknowledgements

Some of the modelling work was funded through the AGFORWARD (www.agforward.eu) project funded the European Commission, Directorate General for Research & Innovation under grant agreement 613520.

6 Where to look for further information

The four main agroforestry models described in this chapter are the Yield-SAFE suite of models, Hi-sAFe, WaNuLCAS and APSIM. Details on the source of these models are outlined below:

The Yield-SAFE and Farm-SAFE models used in the SAFE and AGFORWARD project is available from staff at Cranfield University (P.Burgess@cranfield.ac.uk) and from the AGFORWARD plot-scale modelling web page: http://www.agforward.eu/index.php/en/wp6.html.

The three-dimensional Hi-sAFe model is developed by INRA, France and is freely available at: https://www1.montpellier.inra.fr/wp-inra/hi-safe/en/. Registration is required to download the model. A new version including the APEX crop models is currently being prepared. The Hi-sAFer toolbox is available to process easily the model outputs (Wolz, 2018).

WaNuLCAS (https://www.worldagroforestry.org/output/wanulcas-model-water-nutrient-and-light-capture-agroforestry-systems) is an open source, free access model that combines a spreadsheet data pre-processing part with a STELLA-based dynamic model.

The APSIM (https://www.apsim.info/AboutUs.aspx) and DSSAT Cropping System Model http://dssat.net/downloads/dssatv46) are both open source and allow free access to model source code to enable community-based development. The framework is freely available for non-commercial use and it can be operated using Microsoft Visual C++.

7 References

Acuña, S. T., Juristo, N. and Recio, B. 1997. Knowledge-based system for generating administrative grant alternatives applying the IDEAL methodology. *Computers and Electronics in Agriculture* 18(1), 1–28. doi:10.1016/S0168-1699(97)00015-X.

Agriculture Western Australia and Campbell White and Associates Pty Ltd. 2000. Agroforestry calculator user manual. A report for the Rural Industries Research and Development Corporation (RIRDC), Land and Water Resource Research and Development Corporation (LWRRDC), and Forest and Wood Products Research and Development Corporation (FWPRDC). RIRDC Publication No. 99/154. RIRDC Project No. DAW-84A. Western Australia, Australia: Department of Agriculture.

Agro Business Consultants. 2015. *The Agricultural Budgeting and Costing Book* (82nd edn.). Melton Mowbray, UK: Agro Business Consultants.

Artru, S., Garré, S., Dupraz, C., Hiel, M.-P., Blitz-Frayret, C. and Lassois, L. 2017. Impact of spatio-temporal shade dynamics on wheat growth and yield, perspectives for temperate agroforestry. *European Journal of Agronomy* 82, 60–70. doi:10.1016/j.eja.2016.10.004.

Balandier, P., Bergez, J.-E. and Etienne, M. 2003. Use of the management-oriented silvopastoral model ALWAYS: Calibration and evaluation. *Agroforestry Systems* 57(3), 159–71. doi:10.1023/A:1024863408559.

Bergez, J.-E., Etienne, M. and Balandier, P. 1999. ALWAYS: A plot-based silvopastoral system model. *Ecological Modelling* 115(1), 1–17. doi:10.1016/S0304-3800(98)00153-7.

Boote, K. J., Jones, J. W. and Pickering, N. B. 1996. Potential uses and limitations of crop models. *Agronomy Journal* 88(5), 704–16. doi:10.2134/agronj1996.00021962008800050005x.

Boulton, J. G., Allen, P. M. and Bowman, C. 2015. *Embracing Complexity: Strategic Perspectives for an Age of Turbulence*. Oxford, UK: Oxford University Press, 219pp.

Bouman, B. A. M., van Keulen, H., van Laar, H. H. and Rabbinge, R. 1996. The 'School of de Wit' crop growth simulation models: A pedigree and historical overview. *Agricultural Systems* 52(2–3), 171–98. doi:10.1016/0308-521X(96)00011-X.

Box, G. E. P. (1979). Robustness in the strategy of scientific model building. Mathematics Research Centre, Technical Summary Report 1954. Madison, WI: University of Wisconsin. Available at: https://apps.dtic.mil/dtic/tr/fulltext/u2/a070213.pdf.

Brouwer, R. and de Wit, C. T. 1968. A simulation model of plant growth with special attention to root growth and its consequences. *Proceedings of the 15th Easter School of Agricultural Science*. University of Nottingham, pp. 224–42. Available at: https://core.ac.uk/download/pdf/29391856.pdf.

Burgess, P. J., Incoll, L. D., Corry, D. T., Beaton, A. and Hart, B. J. 2005. Poplar growth and crop yields within a silvoarable agroforestry system at three lowland sites in England. *Agroforestry Systems* 63(2), 157–69. doi:10.1007/s10457-004-7169-9.

Burgess, P. J., Rivas-Casado, M., Gavu, J., Mead, A., Cockerill, T., Lord, R., van der Horst, D. and Howard, D. C. 2012. A framework for reviewing the trade-offs between renewable energy, food, feed and wood production at a local level. *Renewable and Sustainable Energy Reviews* 16(1), 129–42. doi:10.1016/j.rser.2011.07.142.

Burgess, P. J. and Rosati, A. 2018. Advances in European agroforestry: Results from the AGFORWARD project. *Agroforestry Systems* 92(4), 801–10. doi:10.1007/s10457-018-0261-3.

Burgess, P. J., Schmutz, U., Balaguer, F., Boosten, M., Csikvari, J., Hannanchi, Y., Pecenka, R., Poza Llorente, J., Sepp, M. and Vityi, A. 2017. Financial impact of agroforestry.

EIP-AGRI Focus Group Agroforestry. Available at: https://ec.europa.eu/eip/agriculture/sites/agri-eip/files/fg22_mp6_financials_2017_en.pdf.

Burkhart, H. E. and Tomé, M. 2002. *Modelling Forest Trees and Stands*. Dordrecht, The Netherlands: Springer, 458pp.

Charles-Edwards, D. A. and Thorpe, M. R. 1976. Interception of diffuse and direct-beam radiation by a hedgerow apple orchard. *Annals of Botany* 40(3), 603–13. doi:10.1093/oxfordjournals.aob.a085170.

Christie, J. M. 1994. Provisional yield tables for poplar in Britain. Forestry Commission Technical Paper 6. Edinburgh: Forestry Commission, 35pp.

Connor, D. J. 1983. Crop models: Components of and contributors to models of agroforestry plant associations. In: Huxley, P. A. (Ed.), *Plant Research and Agroforestry*. Nairobi: ICRAF, pp. 249–56.

Coulibaly, Y. N., Mulia, R., Sanou, J., Zombre, G., Bayala, J., Kalinganire, A. and van Noordwijk, M. 2014. Crop production under different rainfall and management conditions in agroforestry parkland systems in Burkina Faso: Observations and simulation with WaNuLCAS model. *Agroforestry Systems* 88(1), 13–28. doi:10.1007/s10457-013-9651-8.

De Reffye, P., Houllier, F., Blaise, F., Barthelemy, D., Dauzat, J. and Auclair, D. 1995. A model simulating above- and below-ground tree architecture with agroforestry applications. *Agroforestry Systems* 30(1–2), 175–97. doi:10.1007/BF00708920.

de Wit, C. T. 1958. Transpiration and crop yields. Mededeling 56. Wageningen: Wageningen University.

Dhakal, B., Bigsby, H. R. and Cullen, R. 2007. The link between community forestry policies and poverty and unemployment in rural Nepal. *Mountain Research and Development* 27(1), 32–9. doi:10.1659/0276-4741(2007)27[32:TLBCFP]2.0.CO;2.

Dilla, A., Smethurst, P. J., Barry, K., Parsons, D. and Denboba, M. 2018. Potential of the APSIM model to simulate impacts of shading on maize productivity. *Agroforestry Systems* 92(6), 1699–709. doi:10.1007/s10457-017-0119-0.

Dube, F., Couto, L., Silva, M. L., Leite, H. G., Garcia, R. and Araujo, G. A. A. 2002. A simulation model for evaluating technical and economic aspects of an industrial eucalyptus-based agroforestry system in Minas Gerais, Brazil. *Agroforestry Systems* 55(1), 73–80. doi:10.1023/A:1020240107370.

Dufour, L., Metay, A., Talbot, G. and Dupraz, C. 2013. Assessing light competition for cereal production in temperate agroforestry systems using experimentation and crop modelling. *Journal of Agronomy and Crop Science* 199(3), 217–27. doi:10.1111/jac.12008.

Dupraz, C., Blitz-Frayret, C., Lecomte, I., Molto, Q., Reyes, F. and Gosme, M. 2018. Influence of latitude on the light availability for intercrops in an agroforestry alley-cropping system. *Agroforestry Systems* 92(4), 1019–33. doi:10.1007/s10457-018-0214-x.

Dupraz, C., Burgess, P. J., Gavaland, A., Graves, A. R., Herzog, F., Incoll, L. D., Jackson, N., Keesman, K., Lawson, G., Lecomte, I., et al. 2005. SAFE (Silvoarable Agroforestry for Europe) synthesis report. SAFE Project (August 2001 – January 2005). Available at: http://www1.montpellier.inra.fr/safe/english/index-report.htm.

Dupraz, C. and Lecomte, I. 2017. Update and use directions of the Hi-sAFe agroforestry model. INRA, UMR System, University of Montpellier, 236pp. Available at: https://www1.montpellier.inra.fr/wp-inra/hi-safe/en/publications/.

Feldwisch, N., Frede, H. and Hecker, F. 1998. Verfahren zum Abschätzen der Erosions und Auswaschungsgefahr. In: Frede, H. and Dabbert, S. (Eds), *Handbuch zum*

Gewässerschutz in der Landwirtschaft. Landsberg: Ecomed Publishers, pp. 22–57.

Freya, G. E., Mercer, D. E., Cubbage, F. W. and Abt, R. C. 2013. A real options model to assess the role of flexibility in forestry and agroforestry adoption and disadoption in the Lower Mississippi Alluvial Valley. *Agricultural Economics* 44(1), 73-91. doi:10.1111/j.1574-0862.2012.00633.x.

Friend, A. D., Stevens, A. K., Knox, R. G. and Cannell, M. G. R. 1997. A process-based, terrestrial biosphere model of ecosystem dynamics (Hybrid v 3.0). *Ecological Modelling* 95(2–3), 249–87. doi:10.1016/S0304-3800(96)00034-8.

García de Jalón, S., Graves, A., Moreno, G., Palma, J. H. N., Crous-Durán, J., Kay, S. and Burgess, P. J. 2018a. Forage-SAFE: A model for assessing the impact of tree cover on wood pasture profitability. *Ecological Modelling* 372, 24–32. doi:10.1016/j.ecolmodel.2018.01.017.

García de Jalón, S., Graves, A., Palma, J. H. N., Williams, A., Upson, M. A. and Burgess, P. J. 2018b. Modelling and valuing the environmental impacts of arable, forestry and agroforestry systems: A case study. *Agroforestry Systems* 92(4), 1059–73. doi:10.1007/s10457-017-0128-z.

Graves, A. R., Hess, T., Matthews, R. B., Stephens, W. and Mason, T. 2002. Crop simulation models as tools in education. *Journal of Natural Resources and Life Sciences Education* 31, 48–54.

Graves, A. R., Burgess, P. J., Liagre, F., Terreaux, J. -P. and Dupraz, C. 2005. Development and use of a framework for characterising computer models of silvoarable economics. *Agroforestry Systems* 65(1), 53–65. doi:10.1007/s10457-004-5545-0.

Graves, A. R., Burgess, P. J., Palma, J. H. N., Herzog, F., Moreno, G., Bertomeu, M., Dupraz, C., Liagre, F., Keesman, K., van der Werf, W., et al. 2007. Development and application of bio-economic modelling to compare silvoarable, arable and forestry systems in three European countries. *Ecological Engineering* 29(4), 434–49. doi:10.1016/j.ecoleng.2006.09.018.

Graves, A. R., Burgess, P. J., Liagre, F., Terreaux, J. -P., Borrel, T., Dupraz, C., Palma, J. and Herzog, F. 2011. Farm-SAFE: The process of developing a plot- and farm-scale model of arable, forestry and silvoarable economics. *Agroforestry Systems* 81(2), 93–108. doi:10.1007/s10457-010-9363-2.

Graves, A. R., Burgess, P. J., Palma, J., Keesman, K. J., van der Werf, W., Dupraz, C., van Keulen, H., Herzog, F. and Mayus, M. 2010. Implementation and calibration of the parameter-sparse Yield-SAFE model to predict production and land equivalent ratio in mixed tree and crop systems under two contrasting production situations in Europe. *Ecological Modelling* 221(13–14), 1744–56. doi:10.1016/j.ecolmodel.2010.03.008.

Grist, P., Menz, K. and Nelson, R. 1999. Multipurpose trees as improved fallow: An economic assessment. *International Tree Crops Journal* 10(1), 19–36. doi:10.1080/01435698.1999.9752989.

Holzworth, D. P., Huth, N. I., deVoil, P. G., Zurcher, E. J., Herrmann, N. I., McLean, G., Chenu, K., van Oosterom, E. J., Snow, V., Murphy, C., et al. 2014. APSIM–evolution towards a new generation of agricultural systems simulation. *Environmental Modelling and Software* 62, 327–50. doi:10.1016/j.envsoft.2014.07.009.

Holzworth, D., Huth, N. I., Fainges, J., Brown, H., Zurcher, E., Cichota, R., Verrall, S., Herrmann, N. I., Zheng, B. and Snow, V. 2018. APSIM Next Generation: Overcoming challenges in modernising a farming systems model. *Environmental Modelling and Software* 103, 43–51. doi:10.1016/j.envsoft.2018.02.002.

Huda, A. K. S. and Ong, C. K. 1987. Crop simulation models and some implications for agroforestry systems. In: Reifsnyder, W. E. and Darnhofer, T. O. (Eds), *Meteorology and Agroforestry. Proceedings of an International Workshop on the Application of Meteorology to Agroforestry Systems Planning and Management*, Nairobi, 9–13 February 1987. Nairobi, Kenya: ICRAF, pp. 115–24.

Jackson, J. E. and Palmer, J. W. 1972. Interception of light by model hedgerow orchards in relation to latitude, time of year and hedgerow configuration and orientation. *Journal of Applied Ecology* 9(2), 341–58. doi:10.2307/2402436.

Jackson, J. E. and Palmer, J. W. 1987. Light availability at the tree/crop interface. In: Reifsnyder, W. S. and Darnhofer, T. O. (Eds), *Meteorology and Agroforestry. Proceedings of an International Workshop on the Application of Meteorology to Agroforestry Systems Planning and Management*, Nairobi, 9–13 February 1987. Nairobi, Kenya: ICRAF, pp. 391–400.

Jackson, J. E. and Palmer, J. W. 1979. A simple model of light transmission and interception by discontinuous canopies. *Annals of Botany* 44(3), 381–3. doi:10.1093/oxfordjournals.aob.a085744.

Jackson, B., Pagella, T. F., Sinclair, F., Orellana, B., Henshaw, A., Reynolds, B., McIntyre, N., Wheater, H. and Eycott, A. 2013. Polyscape: A GIS mapping framework providing efficient and spatially explicit landscape-scale valuation of multiple ecosystem services. *Landscape and Urban Planning* 112, 74–88. doi:10.1016/j.landurbplan.2012.12.014.

Jones, J. W., Antle, J. M., Basso, B., Boote, K. J., Conant, R. T., Foster, I. H., Godfray, H. C. J., Herrero, M., Howitt, R. E., Janssen, S., et al. 2017. Brief history of agricultural systems modeling. *Agricultural Systems* 155, 240–54. doi:10.1016/j.agsy.2016.05.014.

Kay, S., Crous-Duran, J., Ferreiro-Domínguez, N., García de Jalón, S., Graves, A., Moreno, G., Mosquera-Losada, M. R., Palma, J. H. N., Roces-Díaz, J. V., Santiago-Freijanes, J. J., et al. 2018. Spatial similarities between European agroforestry systems and ecosystem services at the landscape scale. *Agroforestry Systems* 92(4), 1075–89. doi:10.1007/s10457-017-0132-3.

Keating, B. A., Carberry, P. S., Hammer, G. L., Probert, M. E., Robertson, M. J., Holzworth, D., Huth, N. I., Hargreaves, J. N. G., Meinke, H., Hochman, Z., et al. 2003. An overview of APSIM, a model designed for farming systems simulation. *European Journal of Agronomy* 18(3–4), 267–88. doi:10.1016/S1161-0301(02)00108-9.

Keesman, K. J., Graves, A., van der Werf, W., Burgess, P. J., Palma, J., Dupraz, C. and van Keulen, H. 2011. A system identification approach for developing and parameterising an agroforestry system model under constrained availability of data. *Environmental Modelling and Software* 26(12), 1540–53. doi:10.1016/j.envsoft.2011.07.020.

Keesman, K. J., van der Werf, W. and van Keulen, H. 2007. Production ecology of agroforestry systems: A minimal mechanistic model and analytical derivation of the land equivalent ratio. *Mathematical Biosciences* 209(2), 608–23. doi:10.1016/j.mbs.2007.04.001.

Khaine, I. and Woo, S. Y. 2018. Exploration of the aboveground carbon sequestration and the growth estimation models of four species in agroforestry system of semi-arid region, Myanmar. *Agroforestry Systems* 92, 183–94.

Knowles, L. and Middlemiss, P. 1999. *Evaluating Agroforestry Options: A Continuing Professional Development Course*. Rotura, New Zealand: Forest Research.

Lamanda, N., Dauzat, J., Jourdan, C., Martin, P. and Malézieux, E. 2007. Using 3D architectural models to assess light availability and root bulkiness in coconut agroforestry systems. *Agroforestry Systems* 72(1), 63–74. doi:10.1007/s10457-007-9068-3.

Lawson, G. J., Crout, N. M. J., Levy, P. E., Mobbs, D. C., Wallace, J. S., Cannell, M. G. R. and Bradley, R. G. 1995. The tree-crop interface: Representation by coupling forest and crop process-models. *Agroforestry Systems* 30(1-2), 199-221. doi:10.1007/BF00708921.

Lecomte, I., Dupraz, C., Gosme, M. and Blitz, C. 2016. Improvement of the Hi-sAFe model. Milestone 30 (6.5) for EU FP7 Research Project: AGFORWARD 613520 (21 July 2016), 7pp. Available at: https://www.agforward.eu/index.php/en/improvement-of-the-hi-safe-model.html.

Leinweber, D. 1979. Models, complexity and error. A Rand Note prepared for the US Department of Energy. Available at: https://www.rand.org/content/dam/rand/pubs/notes/2007/N1204.pdf.

Liagre, F. 1997. ARBUSTRA Manuel de l'utilisateur. Montpellier, France: User manual for ARBUSTRA, Centre Régional de la Propriété Forestière (CRPF) and l'Institut National de la recherche Agronomique (INRA), 71pp.

Lojka, B., Lojkova, J., Banout, J., Polesny, Z. and Preininger, D. 2007. Performance of an improved fallow system in the Peruvian Amazon–Modelling approach. *Agroforestry Systems* 72(1), 27–39. doi:10.1007/s10457-007-9079-0.

Lovrić, M., Rois-Díaz, M., den Herder, M., Pisanelli, A., Lovrić, N. and Burgess, P. J. 2018. Driving forces for agroforestry uptake in Mediterranean Europe: Application of the analytic network process. *Agroforestry Systems* 92(4), 863–76. doi:10.1007/s10457-018-0202-1.

Luedeling, E., Smethurst, P. J., Baudron, F., Bayala, J., Huth, N. I., van Noordwijk, M., Ong, C. K., Mulia, R., Lusiana, B., Muthuri, C., et al. 2016. Field-scale modeling of tree-crop interactions: Challenges and development needs. *Agricultural Systems* 142, 51–69. doi:10.1016/j.agsy.2015.11.005.

Lusiana, B., van Noordwijk, M., Suyamto, D., Mulia, R., Joshi, L. and Cadisch, G. 2011. Users' perspectives on validity of a simulation model for natural resource management. *International Journal of Agricultural Sustainability* 9(2), 364–78. doi:10.1080/14735903.2011.582362.

MacFarlane, D. W., Kuyah, S., Mulia, R., Dietz, J., Muthuri, C. and van Noordwijk, M. 2014. Evaluating a non-destructive method for calibrating tree biomass equations derived from tree branching architecture. *Trees* 28, 807–17. doi:10.1007/s00468-014-0993-2.

Magcale-Macandog, D. B. 2014. Agroforestry models for promoting effective risk management and building sustainable communities. In: Kaneko, N., Yoshiura, S. and Kobayashi, M. (Eds), *Sustainable Living with Environmental Risks*. Tokyo, Japan: Springer.

Matthews, R., Stephens, W., Hess, T., Mason, T. and Graves, A. 2000. Applications of crop/soil simulation models in developing countries. Final Report. DFID Contract PD82. Cranfield University, 174pp.

Matthews, R. B., Stephens, W., Hess, T., Mason, T. and Graves, A. R. 2002. Application of crop soil simulation models in tropical agricultural systems. *Advances in Agronomy* 17, 31–124.

Meylan, L., Merot, A., Gary, C. and Rapidel, B. 2013. Combining a typology and a conceptual model of cropping system to explore the diversity of relationships between ecosystem services: The case of erosion control in coffee-based agroforestry systems in Costa Rica. *Agricultural Systems* 118, 52–64. doi:10.1016/j.agsy.2013.02.002.

Mialet-Serra, I., Dauzat, J. and Auclair, D. 2001. Using plant architectural models for estimation of radiation transfer in a coconut-based agroforestry system. *Agroforestry Systems* 53(2), 141–9. doi:10.1023/A:1013320419289.

Mobbs, D. C., Cannell, M. G. R., Crout, N. M. J., Lawson, G. J., Friend, A. D. and Arah, J. 1998. Complementarity of light and water use in tropical agroforests. I. Theoretical model outline, performance and sensitivity. *Forest Ecology and Management* 102(2–3), 259–74. doi:10.1016/S0378-1127(97)00167-9.

Monteith, J. L. 1992. Principles of resource capture by crop stands. In: Monteith, J. L., Scott, R. K. and Unsworth, M. H. (Eds), *Resource Capture by Crops*. Nottingham, UK: University of Nottingham.

Moral, F. J., Rebollo, F. J., Paniagua, M. and Murillo, M. 2014. Using an objective and probabilistic model to evaluate the impact of different factors in the dehesa agroforestry ecosystem. *Ecological Indicators* 46, 253–9. doi:10.1016/j.ecolind.2014.06.040.

Moussa, M. and Mahamane, L. 2018. Allometric models for estimating aboveground biomass and carbon in *Faidherbia albida* and *Prosopis africana* under agroforestry parklands in drylands of Niger. *Journal of Forestry Research* 29(6), 1703–17. doi:10.1007/s11676-018-0603-z.

Nair, P. K. R. 2011. Methodological challenges in estimating carbon sequestration potential of agroforestry systems. In: Kumar, B. M. and Nair, P. K. R. (Eds), *Carbon Sequestration Potential of Agroforestry Systems: Opportunities and Challenges*. Heidelberg: Springer, pp. 3–16.

Ozier-Lafontaine, H., Lecompte, F. and Sillon, J. F. 1999. Fractal analysis of the root architecture of *Gliricidia sepium* for the spatial prediction of root branching, size and mass: Model development and evaluation in agroforestry. *Plant and Soil* 209(2), 167–79. doi:10.1023/A:1004461130561.

Palma, J. H. N., Graves, A. R., Bunce, R. G. H., Burgess, P. J., de Filippi, R., Keesman, K. J., van Keulen, H., Liagre, F., Mayus, M., Moreno, G., et al. 2007a. Modeling environmental benefits of silvoarable agroforestry in Europe. *Agriculture, Ecosystems and Environment* 119(3–4), 320–34. doi:10.1016/j.agee.2006.07.021.

Palma, J. H. N., Graves, A. R., Burgess, P. J., Keesman, K. J., van Keulen, H., Mayus, M., Reisner, Y. and Herzog, F. 2007b. Methodological approach for the assessment of environmental effects of agroforestry at the landscape scale. *Ecological Engineering* 29(4), 450–62. doi:10.1016/j.ecoleng.2006.09.016.

Palma, J. H. N., Crous-Duran, J., Graves, A. R., Garcia de Jalon, S. G., Upson, M., Oliveira, T. S., Paulo, J. A., Ferreiro-Domínguez, N., Moreno, G. and Burgess, P. J. 2018. Integrating belowground carbon dynamics into Yield-SAFE, a parameter sparse agroforestry model. *Agroforestry Systems* 92(4), 1047–57. doi:10.1007/s10457-017-0123-4.

Palma, J. H. N. 2017. CliPick – Climate change web picker. A tool bridging daily climate needs in process based modelling in forestry and agriculture. *Forest Systems* 26(1), 4p. doi:10.5424/fs/2017261-10251.

Panwar, P., Chauhan, S., Kaushal, R., Das, D. K., Ajite Arora, G., Chaturvedi, O. P., Jain, A. K., Chaturvedi, S. and Tewari, S. 2017. Carbon sequestration potential of poplar-based agroforestry using the CO2FIX model in the Indo-Gangetic Region of India. *Tropical Ecology* 58, 439–44.

Passioura, J. B. 1973. Sense and nonsense in crop simulation. *Journal of Australian Institute of Agricultural Science* 39, 181-3.

Purnamasari, R., Cacho, O. and Simmons, P. 2002. Management strategies for Indonesian rubber production under yield and price uncertainty: A bioeconomic analysis. *Agroforestry Systems* 54(2), 121-35. doi:10.1023/A:1015080728895.

Radersma, S. and Ong, C. K. 2004. Spatial distribution of root length density and soil water of linear agroforestry systems in sub-humid Kenya: Implications for agroforestry models. *Forest Ecology and Management* 188(1-3), 77-89. doi:10.1016/j.foreco.2003.07.021.

Salazar, O., Casanova, M. and Kätterer, T. 2011. The impact of agroforestry combined with water harvesting on soil carbon and nitrogen stocks in central Chile evaluated using the ICBM/N model. *Agriculture, Ecosystems and Environment* 140(1-2), 123-36. doi:10.1016/j.agee.2010.11.019.

Sinclair, F. L. and Walker, D. H. 1998a. Acquiring qualitative knowledge about complex agroecosystems. Part 1: Representation as natural language. *Agricultural Systems* 56(3), 341-63. doi:10.1016/S0308-521X(97)00048-6.

Sinclair, F. L. and Walker, D. H. 1998b. Acquiring qualitative knowledge about complex agroecosystems. Part 2: Formal representation. *Agricultural Systems* 56(3), 365-86. doi:10.1016/S0308-521X(97)00049-8.

Sinclair, T. R. and Seligman, N. G. 1996. Crop modelling: From infancy to maturity. *Agronomy Journal* 88, 698-703.

Smethurst, P. J., Huth, N. I., Masikati, P., Sileshi, G. W., Akinnifesi, F. K., Wilson, J. and Sinclair, F. 2017. Accurate crop yield predictions from modelling tree-crop interactions in Gliricidia-maize agroforestry. *Agricultural Systems* 155, 70-7. doi:10.1016/j.agsy.2017.04.008.

Talbot, G., Roux, S., Graves, A., Dupraz, C., Marrou, H. and Wery, J. 2014. Relative yield decomposition: A method for understanding the behaviour of complex crop models. *Environmental Modelling and Software* 51, 136-48. doi:10.1016/j.envsoft.2013.09.017.

Talbot, G. and Dupraz, C. 2012. Simple models for light competition within agroforestry discontinuous tree stands: Are leaf clumpiness and light interception by woody parts relevant factors? *Agroforestry Systems* 84(1), 101-16. doi:10.1007/s10457-011-9418-z].

The Economist. 1986. *The Economist Style Guide*. London, UK: Economist Publications Limited, 144pp.

Thomas, T. H. 1991. A spreadsheet approach to the economic modelling of agroforestry systems. *Forest Ecology and Management* 45, 207-35.

Thornley, J. H. N. and Johnson, I. R. 1990. *Plant and Crop Modelling: A Mathematical Approach to Plant and Crop Physiology*. Oxford, UK: Clarendon Press, 669pp.

Vandermeer, J. 1989. *The Ecology of Intercropping*. Cambridge, UK: Cambridge University Press, 237pp.

van der Werf, W., Keesman, K., Burgess, P. J., Graves, A. R., Pilbeam, D., Incoll, L. D., Metselaar, K., Mayus, M., Stappers, R., van Keulen, H., et al. 2007. Yield-SAFE: A parameter-sparse process-based dynamic model for predicting resource capture, growth and production in agroforestry systems. *Ecological Engineering* 29(4), 419-33. doi:10.1016/j.ecoleng.2006.09.017.

van Noordwijk, M. and Lusiana, B. 1999. WaNuLCAS, a model of water, nutrient and light capture in agroforestry systems. *Agroforestry Systems* 43, 217-42. doi:10.1007/978-94-017-0679-7_14.

van Noordwijk, M. and Mulia, R. 2002. Functional branch analysis as tool for fractal scaling above and belowground trees for their additive and non-additive properties. *Ecological Modelling* 149, 41–51.

van Noordwijk, M. and van de Geijn, S. C. 1996. Root, shoot and soil parameters required for process-oriented models of crop growth limited by water or nutrients. *Plant and Soil* 183(1), 1–25. doi:10.1007/BF02185562.

van Oijen, M., Dauzat, J., Harmand, J. M., Lawson, G. and Vaast, P. 2010. Coffee agroforestry systems in Central America: II. Development of a simple process-based model and preliminary results. *Agroforestry Systems* 80(3), 361–78. doi:10.1007/s10457-010-9291-1.

van Oijen, M., Rougier, J. and Smith, R. 2005. Bayesian calibration of process-based forest models: Bridging the gap between models and data. *Tree Physiology* 25(7), 915–27. doi:10.1093/treephys/25.7.915.

Whisler, F. D., Acock, B., Baker, D. N., Fye, R. E., Hodges, H. F., Lambert, J. R., Lemmon, H. E., McKinion, J. M. and Reddy, V. R. 1986. Crop simulation models in agronomic systems. *Advances in Agronomy* 40, 141–208. doi:10.1016/S0065-2113(08)60282-5.

Williams, A. G., Audsley, E. and Sandars, D. L. 2010. Environmental burdens of producing bread wheat, oilseed rape and potatoes in England and Wales using simulation and system modelling. *International Journal of Life Cycle Assessment* 15(8), 855–68. doi:10.1007/s11367-010-0212-3.

Willis, R. W., Thomas, T. H. and van Slycken, J. 1993. Poplar agroforestry: A re-evaluation of its economic potential on arable land in the United Kingdom. *Forest Ecology and Management* 57(1–4), 85–97. doi:10.1016/0378-1127(93)90164-I.

Wolz, K. J. 2018. HisAFer: An R toolbox for the Hi-sAFe biophysical agroforestry model. R package version 1.5.0. Available at: https://github.com/kevinwolz/hisafer.

Young, A., Menz, K., Muraya, P. and Smith, C. 1998. SCUAF Version 4: A model to estimate soil changes under agriculture, agroforestry and forestry. ACIAR Technical Reports Series No. 41, 49pp. Available at: https://www.aciar.gov.au/node/7226.

Zeng, X., Li, T., Chen, C., Si, Z., Huang, G., Guo, P. and Zhuang, X. 2018. A hybrid land-water-environment model for identification of ecological effect and risk under uncertain meteorological precipitation in an agroforestry ecosystem. *Science of the Total Environment* 633, 1613–28. doi:10.1016/j.scitotenv.2018.03.224.

Chapter 8

Tree planting and management in agroforestry

Lydie Dufour, INRA, France

1 Introduction

Agroforestry is currently being promoted as a land use strategy which is capable of providing additional products and income in addition to environmental and conservation benefits. Such systems create a wide range of biophysical interactions between the associated plants. These interactions may consist of competition for water, light and mineral elements resources (Cannell et al., 1998; Bellow and Nair, 2003; Dufour et al., 2013). However, the different species can also create improvements in their shared environment: for example, a tree may improve the microclimate in its understory. This is known as 'facilitation' (Vandermeer, 1989). As plant species differ in their temporal and spatial patterns of resource acquisition, an intercrop may capture the available resources of light, water and nutrients more completely than a single species (Yu et al., 2015). The deep roots of trees serve as a 'safety net' below the root zone of herbaceous crops (Allen et al., 2004) and trees in agroforestry systems are able to recycle soil nutrients which leach down through the herbaceous crop rooting zone, thereby reducing groundwater contamination and increasing nutrient use efficiency in the system (Rowe et al., 1999; Wolz et al., 2018). Deep rooted trees may also take up water resources which are not accessible to the crop, thus increasing resource use by the association (Fernandez et al., 2008). If the development of the two

http://dx.doi.org/10.19103/AS.2018.0041.14

associated species is not simultaneous, one may use the available light while the other is not present or is in a leafless stage. This increased efficiency of resources use is termed 'complementarity' (Gathumbi et al., 2002). The key to success in agroforestry systems lies in choosing the optimum combination of trees and crops for the exploitation of spatial and temporal complementarity or facilitation in resource use (Cannell et al., 1996; Descheemaeker et al., 2013).

Thus the plantation design (i.e. tree density, tree rows orientation, alley cropping width), the chosen trees and crop species, together with tree and soil management, are the levers which can be applied in interaction with pedoclimatic conditions.

Modern agroforestry in temperate areas generally utilises alley cropping systems, defined as the practice of planting annual crops between widely spaced rows of woody plants or trees (Van Lerberghe, 2017d). Before planting, a detailed analysis is necessary to evaluate the benefits expected from the introduction of trees. These benefits are ecosystem services (Pantera et al., 2018) and can be divided into four groups:

1 Environmental services (Quinkenstein et al., 2009): the trees protect the soil against erosion, introduce organic matter into the soil through leaf fall and the turnover of fine roots, store carbon in deep layers of the soil, improve water quality by intercepting lixiviated nitrogen (Allen et al., 2004; Andrianarisoa et al., 2016) or other inputs, improve vegetal and animal biodiversity (Torralba et al., 2016) and moderate the microclimate (Jose, 2009).
2 Economic outputs (Pantera et al., 2018): the exploitation of tree products including timber, fuelwood, fodder, honey and fruits increases productivity and diversifies the farm resources.
3 Patrimonial investment: trees can be capital for the next generation or for the farmer's retirement.
4 Cultural and societal services, including improved landscape aesthetics, enhanced countryside value, increased opportunities for employment and education and the maintenance of cultural heritage (Pantera et al., 2018) (Fig. 1a and b).

The farmer must also be aware of the constraints of introducing trees. The plantation and maintenance of trees requires time and training if good results are to be achieved. The climatic conditions and soil properties must also be carefully considered before choosing the tree species and undertaking planting.

In order to produce high-quality timber, it is necessary to plant suitable seedlings, to protect them appropriately and to carry out pruning and other maintenance operations.

Figure 1 Agroforestry landscape (a) compared to monoculture (b).

2 Choice of tree species

2.1 One or several species?

The plantation of several different species in one field reduces the economic and sanitary risks. As the wood market changes, one timber species with an initially high economic value may depreciate within a few years. Monocropping also tends to facilitate the spread of diseases with trees, as with other species. The association of species or varieties reduces the likelihood of spreading diseases or pests. It also improves biodiversity through increasing variety within auxiliaries habitat.

Complementarity between species in the use of resources is enabled, for instance, when a phenological lag exists between them. Thus wherever it is compatible with the plantation goals, a mixture of species, or at least of varieties, is to be preferred over a monospecific plantation.

2.2 Species choice

The choice of tree species must take into account (1) the commercial priorities of the owner, (2) the site characteristics and (3) the intercrop requirements:

- The first-choice criterion is the objective of the project. If timber production is the priority, the different rates of growth among species can guide the choice. For instance, if the farmer wants fast-maturing trees, poplars (*Populus* sp.) will be a good choice, with harvest around 20 years after plantation whereas sorb trees (*Sorbus domestica* L.) will require 100 years of growth before harvesting.
- Not every tree will be suitable for a given area (Fig. 2). A biophysical assessment is essential before choosing the species. Each tree species has specific requirements with regard to soil, climate and topography. Table 1 shows the requirements of the main tree species likely to be used in temperate alley cropping systems. Many soil characteristics have an

Figure 2 Sunburn on the trunk of a young wild cherry tree.

influence on the development of tree roots: depth, texture, amount of stones, structure, pH, waterlogging, nutrients and organic matter content. A soil analysis can give valuable information on these factors. Temperature and rainfall are the two critical climatic variables which mostly impact tree survival and growth. The resistance of trees to drought is also linked with the water retention capacity and depth of the soil. The topography of the field will also significantly modify the effect of the local climate on the soil surface properties. For example, a steep slope will increase water run-off, thus decreasing water penetration in the soil; a south-facing field is often warmer and dryer than a north-facing one. Complementarity between trees and crops is the key to success. Many combinations are possible but if the associated plants (i.e. crop(s) and tree species) have different phenological cycles, depth of root systems and specific pests, the negative interactions will be reduced and complementarity or facilitation can exist. For example, the entire vegetative growth of winter cereals occurs during the period when deciduous trees have shed their leaves, causing a very low shade. Winter cereals also cause a deeper development of the tree roots as they dry the upper soil layer before tree budbreak (Fig. 3).

Guides are available to help match tree species to the edaphic and climatic conditions (Ellis et al., 2005; Orwa et al., 2015).

Table 1 Main tree species usable for temperate agroforestry plantations and their characteristics

Species	Soil properties	Climate	Qualities	Harvest year
Sorbus torminalis	Not shallow nor dry soils; pH = 5.5–8; clay soils, even calcareous, bears temporary waterlogging	Oceanic or continental; poor resistance to cold; not sensitive to late frost; good resistance to summer drought	Hardwood, brown to red; high value; used for cabinet making, veneer	50–70
Alnus glutinosa	Deep; pH = 5–7; wet clay soils	Temperate, not in Mediterranean area; not sensitive to late frost	Softwood, pink; used to make objects and guitars, paper and energy; the tree fix nitrogen; melliferous	15–25
Alnus cordata	Not shallow nor dry soils; pH = 5–6; light soils	Mediterranean; thrives in dryer soils than other alders	Softwood, reddish-orange; used to make objects; the tree fix nitrogen	15–25
Castanea sativa	Deep light and humid soils; pH = 5–7; intolerant of lime	Oceanic, Western continental, Mediterranean; the year-growth is sensitive to late spring and early autumn frosts	Mid-hardwood, light brown; used to make furniture, barrels and roof beams, posts; nuts, fodder, coppicing; melliferous	20 years for nuts, 50 years for wood exploitation
Quercus suber	Deep light dry soils; pH = 5–7; intolerant of lime	Mediterranean, Western continental; thick, insulating bark that makes it well adapted to forest fires	Bark used to make cork plugs, fodder, heating; hardwood, yellow to brown, with very good mechanical properties, but difficult to work with. Used to make furniture, barrels, floor and roof beams	90–120
Quercus robur	Deep and humid soils, prefers fertile soil; tolerates flooding; pH = 5.5–6.5	Northern oceanic, continental; poor resistance to cold and late frosts	Hardwood, yellow to brown, with very good mechanical properties, but difficult to work with. Used to make furniture, barrels, floor and roof beams	60–100
Quercus pubescens	Adapted to dry shallow calcareous soils; pH = 6.5–8	Mediterranean; sensitive to late frosts	Hardwood, yellow to brown, with very good mechanical properties, but difficult to work with. Used to make furniture, barrels, floor and roof beams	120–200
Quercus rubra	Deep, light and humid soils; pH = 4.5–6.5	Oceanic and mountain, eventually continental	Hardwood, brown to reddish, with very good mechanical properties, easy to work with; used to make furniture, floor and roof beams; acorns for animals	45–70

(Continued)

Table 1 *(Continued)*

Species	Soil properties	Climate	Qualities	Harvest year
Quercus petraea	All type of soil texture, but not too dry nor too shallow; pH = 5-7	Usual in continental area, can be in oceanic climate; poor resistance to cold and late frosts	Hardwood, yellow to brown, very good mechanical properties, but difficult to work with; used to make furniture, barrels, floor and roof beams, fodder (leaves and acorns)	60–100
Sorbus domestica	Light soils, without clay; bears dry and shallow soils; pH = 6.5-8; adapted to very calcareous soils	Oceanic and continental, found in Mediterranean area; barely tolerant to gold, but good resistance to late frosts	Very hard coloured wood with very good mechanical properties; used to make furniture, one of the most expensive wood in Europe; edible fruits; melliferous	100–120
Cupressus sempervirens	Bears dry, calcareous and shallow soils; pH = 6–8	Mediterranean, found in oceanic area; sensitive to heavy frosts	Very hard pinkish, rot-resistant wood; used to make furniture and roof beams; production of essential oil	90–100
Acer campestre	Compact soils with good clay-humus balance; not on dry and shallow soils; pH = 5.5-8	Oceanic and continental	White to yellowish mid-hardwood, with good mechanical properties, but not stable; used to make tools; melliferous	50–60
Acer platanoides	Soils with good clay-humus balance; not on compact nor too light soils; bears shallow and dry soils; pH = 5.5-8	Eastern continental and mountain; very good resistance to late frosts	White hardwood, with good mechanical properties; Used for wood turning, lutherie, veneer, posts; melliferous	60-80
Acer pseudoplatanus	Soils with good clay-humus balance; bears compact, but not light and shallow soils; pH = 6–8	Oceanic, continental and mountain; good tolerance to cold and late frosts	White to yellowish mid-hardwood, with good mechanical properties, used for wood turning, lutherie, veneer, posts; melliferous	40–60
Gleditsia triacanthos	Light soils; not compact, but bears shallow soils; pH = 5–8	Mediterranean and Western oceanic; good tolerance to cold	Brown to reddish hardwood; Used to make furniture, crossbars posts and roof beams; fodder (pods), melliferous, dyeing; thorny tree, thornless forms exist	60-90

Fraxinus excelsior	Adapted to light and compact, but not dry nor shallow soils; pH = 5–8	Oceanic, continental and mountain; poor resistance to late frosts	White to yellowish mid-hardwood, with good mechanical properties but difficult to work with; used to make furniture, tools, objects and barrels; fodder (branches)	40–65
Larix decidua	Light soils but bears compact and dry, but not shallow soils; pH = 6–7	Mountain and continental, found in oceanic and Mediterranean areas; very good resistance to cold, not to late frosts	Brown to red softwood, with good mechanical properties; used to make roof beams, posts, objects	40–60
Prunus avium	Light deep soils, intolerant to compact, dry and shallow soils and to waterlogging; pH = 5–8	Oceanic and continental, found in well-exposed mountain and in Mediterranean area. Bad tolerance to cold, but good to late frosts; susceptible to sunburn on the trunk	Reddish mid-hardwood, with good mechanical resistance easy to work with; used to make furniture, carving; edible fruits; melliferous	40–65
Juglans regia	Light deep soils, intolerant to compact, dry and shallow soils; pH = 6–8	Oceanic, Western continental and Mediterranean, susceptible to heavy frosts	Light colour to grey half-hardwood, easy to work with, not very resistant; used for veener, wood turning, furniture, luxury objects; edible fruits (nuts)	60–80
Juglans nigra	Light deep soils, intolerant to compact, dry and shallow soils; pH = 6–7.5	Continental and oceanic, found in Mediterranean area, susceptible to late frosts	Half-hardwood, purplish, darker than the *regia* wood, easy to work with, used for furniture, veener	60–80
Hybrid: *Juglans nigra X regia or Juglans major X regia*	Light deep soils, intolerant to compact, dry and shallow soils; pH = 6–7.5	Oceanic, continental and Mediterranean, susceptible to cold and late frosts	Half-hardwood, darker than the *regia* wood, easy to work, rather resistant; used for veener, wood turning, furniture. High phenotypical variability between trees	40–50
Olea europaea	Light soils, adapted to dry and shallow soils, but not compact nor too wet; pH = 6–8	Mediterranean, susceptible to cold and late frosts	Brown, marbled, very hardwood; with good mechanical properties, very stable and easy to work with; used for wood turning, marquetry, tools and objects; edible fruits, after treatment	90–110

(*Continued*)

Table 1 *(Continued)*

Species	Soil properties	Climate	Qualities	Harvest year
Ulmus campestris	Deep and fresh soils; pH = 7–8	Oceanic, continental and Mediterranean, susceptible to cold but not to late frosts	Brown reddish hardwood, very easy to use; used for veener, furniture, roof beams; very susceptible to Dutch elm disease	40–60
Populus sp.	Light deep, humid soils (alluvial); intolerant to compact, dry and shallow soils; pH = 5–8	Continental and oceanic, found in Mediterranean area, resistant to cold and late frosts	White to very light colour, very softwood; used for packaging, paper, matches, coppice, energy; can be used to dry humid areas	15–25
Pirus sp.	Light deep, humid soils; intolerant to compact, dry and shallow soils; pH = 5–8	Continental and oceanic, found in Mediterranean area, poor tolerance to cold and late frosts	Pinkish to brown hardwood, rather easy to work with; highly sought for marquetry, furniture, lutherie, toys, objects; melliferous, difficult to prune because of its thorns	90–110
Malus sp.	Light deep, humid soils; intolerant to compact, dry and shallow soils; pH = 5–8	Oceanic and continental, found at high altitude, resistant to cold and late frosts	Light reddish coloured hardwood, easy to work with; used for marquetry, wood turning, carving, fodder, melliferous, fruits	80–100
Robinia pseudoacacia	All kinds of soils; pH = 6.5–7.5	Oceanic, continental and Mediterranean, resistant to cold and late frosts	Yellow to ochre hardwood, very good mechanical properties, rot-resistant, but difficult to work with; used to make tools, posts, garden furniture; fodder, melliferous; N-fixing tree; difficult to prune because of its thorns	50–60
Tilia cordata	Deep and humid soils, intolerant to dry soils; pH = 7–8	Continental and oceanic, found in Mediterranean area, poor tolerance to cold but good resistance to late frosts	White to pale yellow softwood, very easy to work with; used for wood turning and carving; flowers for infusion, melliferous	60–80
Liriodendron tulipifera	Light deep, humid soils; tolerant to compact but not dry nor shallow soils; pH = 7–8	Oceanic, mountain and continental, very resistant to frost, but intolerant to frosts when it is young	Pale yellow to greenish softwood, very easy to work with; used for veneer, furniture, lutherie	70–80

Figure 3 Association of pea and hybrid walnut in April. The pea begins flowering while the tree budburst is not beginning.

3 Tree planting

3.1 Planting pattern

From a landscape point of view, the line layout of the plantation may appear dull. This may be improved by irregularly spaced planting on the tree rows, by staggered repartition between the rows or by mixing tree species.

3.1.1 Tree density

The choice of tree density depends on the desired balance between trees and crops. The higher the tree density, the stronger the competition with crops will be. Research suggests that agroforestry systems are profitable with 50–100 trees ha^{-1}. However, 100–150 trees ha^{-1} is recommended in order to allow for thinning, only keeping the good trees (Van Lerberghe, 2017d).

A clear area should be maintained at the end of each line to allow machines to turn without damaging trees.

3.1.2 Tree rows orientation

Four criteria must be taken into account when deciding on orientation:

- The shape of the plot: a rectangular shape allows more trees to be planted. Where a plot is almost square, it is possible to choose the orientation (Fig. 4).
- Availability of light for the crop growing under the tree shade: a north-south orientation, if the plot shape makes it possible, allows homogenous

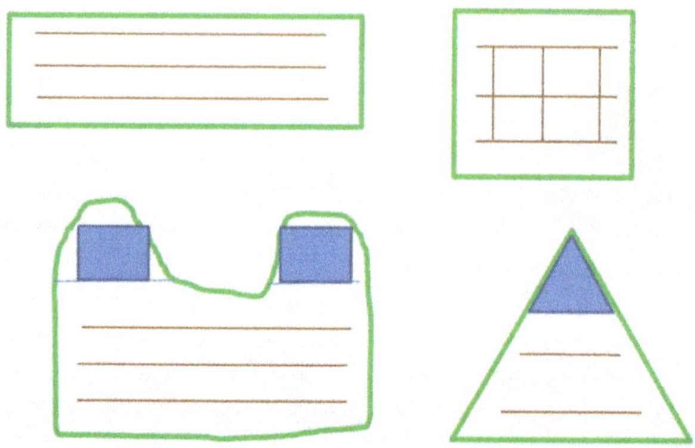

Figure 4 Possible tree rows (brown lines) orientations according to the plot shape. The blue areas are lost surfaces for agroforestry, but can be wooded.

incident radiation on the crop. With this plantation scheme, each location of the alley cropping will receive light during the daily sun path. However, in an east-west-oriented field, the crop situated at the north of the tree lines will receive no light in the summer.

- Protection against the prevailing winds: in a windy area, the tree rows should be placed at right angles to the wind axis. A windbreak hedge can improve the protection effect.
- Anti-erosion efficiency: the vegetation of tree rows improves water infiltration, thus decreasing run-off. In sloping fields, planting tree rows along contour lines is recommended to facilitate the movement of agricultural machinery. In flooded areas, tree rows can be arranged across the current in order to slow it, to reduce soil erosion and to improve the fertile alluvium deposit. If the field is drained, the tree rows should not interfere with the drainage ditches.

3.1.3 Distance between the rows

The distance between the rows determines the balance between crop and trees. If the intention is to maintain arable cropping through the whole tree rotation, the distance between the tree lines must be at least twice the height of the adult trees, that is, 30–40 m (Lawson et al., 2018). Narrower spacing, which gives increased timber production, is possible if the arable crop is replaced by grass lays in the last third of the rotation. The line spacing should also be compatible with the agricultural machinery used on the farm, allowing the use of the largest tools (spray booms for example) and being a multiple of the narrowest one (the seed drill for example). However, a problem may arise during the tree cycle where equipment changes over time. Figure 5 shows the example of a 13 m wide cropping alley which was designed for the equipment of 20 years ago but is no longer adapted to the modern combine width, so causing overlapping.

Figure 5 Harvest in an agroforestry stand.

The non-cropped strip at the foot of the trees must be integrated into the calculation of the distance between tree lines. This strip is 1-4 m wide. A narrow strip requires great precision in the use of machinery, but gives a greater crop area and reduces the development of weeds which are often more competitive for young trees than the crop.

3.1.4 Distance between the trees on rows

The distance between the trees on rows generally varies between 4 m and 10 m. If it is <4 m, the competition between trees will be too strong and an early thinning will be necessary. If the distance is >10 m, the low density will not allow for choosing the best trees through thinning, and dead trees must be replaced to maintain a sufficient density (around 50 trees ha^{-1}).

3.2 Preparing the soil

The previous crop may affect the development of young trees though cultivated plots are generally easier to plant than unmanaged land. Where the soil of a cropped field was tilled in the previous year and the weeds were kept under control, no specific weeding is needed. In this case, ploughing the stubble is carried out after the crop is harvested. Weeding must be done before planting on grassland or fallow land to avoid competition which would cause a reduction of tree growth and possible mortality.

The weeding can be carried out by mechanical or chemical means; in the latter case, the remanence time must be taken into account. The false seedbed

Figure 6 Tool to relieve soil compaction to a depth of 80 cm.

technique may be used, which consists of working the soil at the end of summer as for normal sowing. This causes the seeds from previous crops and weeds to germinate. A second tillage three or four weeks later destroys the emerging plants, thus reducing the seed stock of annual weeds. However, this method is not efficient on perennials.

The soil must be prepared by subsoiling to a depth of >30 cm, especially if a plough pan is present (Fig. 6).

Finally, an additional shallow tillage or disking may be done on the tree plantation lines. This operation is necessary for mulching which requires a regular surface. A biodegradable or natural mulching (e.g. ramial chipped wood, straw, mowing residues) is preferable to a plastic film. Mulching improves seedling growth by reducing the development of weeds and limiting water loss through evaporation.

3.3 Picketing the plot

Picketing is an important operation for marking the location of planted seedlings. The distance between tree lines must be regular and precisely

Figure 7 Picketing the field with decameters and posts.

measured to avoid problems or damage during mechanical management of the crop. A GPS can be used for staking or it may be done with decameters and strings (Fig. 7).

3.4 Seedling quality

The key rule for a successful tree planting is to choose good quality planting material which will ensure good root development. As only a small number of trees are planted when compared to a forestry plantation, these must be of good quality. Seedlings from a specified region will be better adapted to local soil and climatic conditions.

3.4.1 Genetic quality

Ideally, seedlings should be sourced from a certified supplier who will ensure the material is healthy and compliant with the expected genetic quality (Van Lerberghe, 2017a). The OECD Forest Seed and Plant Scheme defines four broad categories of forest reproductive material which are recognised for certification and identified by coloured tags:

Source-identified material: the information is limited to geographical origin,

Selected material: harvested from selected stands in well-delimited regions of provenance,

Qualified material: the seeds are sourced from orchards where trees are individually selected according to vigour, shape, wood quality and disease resistance,

Tested material: harvested from stands or seed orchards where the superiority of particular characteristics has been proved through testing.

3.4.2 Types of plants

Four kinds of packaging are available for seedlings:

1. Bare-rooted, conditioned by size, in bunches 2–4 years after sowing. These seedlings are more fragile than the others, the advantage is that the sanitary condition and shape of the roots is visible, giving a good idea of a plant's state.
2. In single containers with an appropriate substrate. However, the root development may be limited by the shape of the container, so some nurseries offer seedlings in 'anti-bun' containers.
3. In clods: each seedling grown in a clump of earth without a container. This causes less disturbance to the root system but these seedlings are more expensive than those grown in a container. The transplantation stress is reduced for these two kinds of seedlings and when compared to bare-rooted plants, they are especially recommended for spring plantations.
4. Cuttings: for certain tree species (e.g. poplars and willows), unbranched shoots can be planted directly into the soil for rooting. In this case, there is no stress and the sanitary state is easy to check. Poplar cuttings should be strong and >1 m in length (Fig. 8a–d).

3.4.3 Age and dimensions

Nursery catalogues are precise as to the age (3 years maximum), the height and the collar diameter of the seedlings they sell. When possible, visiting the nursery prior to purchase gives a good idea of the seedlings quality and offers

Figure 8 Different types of seedlings packaging: bare-rooted (a), container (b), clod (c) and cuttings (d).

an opportunity to seek expert advice. In selecting seedlings, it is advisable to choose plants as young as possible as these will have a high regeneration capacity, produce new roots quickly and are easier to handle than bigger plants.

3.4.4 Morphological quality

The vigour, architecture of the stem and roots if visible (for bare-rooted seedlings) are a guide to successful planting as well as to the sanitary state of the seedlings. Figure 9 shows a good stem shape (Van Lerberghe, 2017a).

The balance between root and stem volumes must be considered. The roots should be abundant, intact and have a good shape (see Fig. 10a–d).

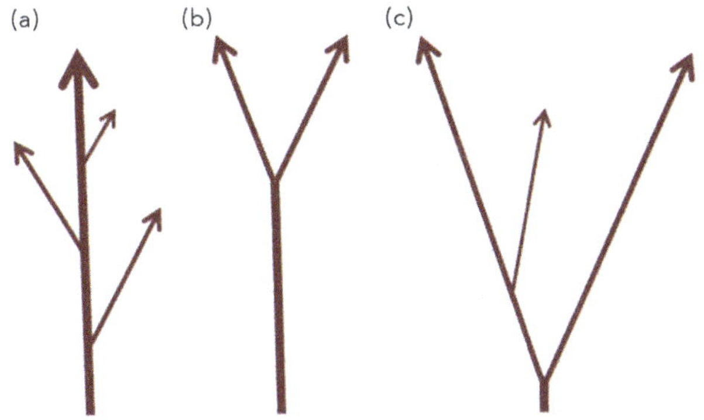

Figure 9 Good shape of seedling stem and branches (a) vs split, with weak apical buds and frail stem (b and c).

Figure 10 Good root system shape (a), vs damaged collar (b), upgoing roots (c), hunting horn shape (d) (Van Lerberghe, 2017a).

3.5 Seedlings preparation

If the seedlings are to be planted more than two days after receipt, they must be stocked in a cool, shady place, sheltered from the wind. Bare-rooted plants must be kept in a trench, deep enough to take the roots and refilled with damp sand or soil. Cuttings should be placed vertically in a bucket of water.

Damaged or overlong roots must be cut off before planting. This helps the seedling to recover and should be done with sharp pruning shears. Any container must be removed before planting and the root hairs spread if necessary.

3.6 Planting

Bare-rooted seedlings and cuttings may be planted from November to March. Spring planting is preferable for heavy and humid soils, and autumn planting for light and sandy soils. Seedlings grown in containers or clods may be planted up to the end of April, but always before budburst. Cuttings are planted in the autumn. Snow, frost, wind or heavy rain are not favourable conditions for planting (Van Lerberghe, 2017c). The soil must be well prepared, neither too dry nor too wet, in order to avoid root rot.

If surface vegetation is still present, it should be removed within a 1 m^2 area around the planting areas. Square holes of 30 cm × 30 cm × 30 cm are dug with a spade. Bare-root seedlings must be settled in soft soil with enough space for the roots which should be well spread out in the planting hole. The roots can then be covered with fine earth.

To improve the contact of roots with the soil, place the collar around 2 cm below ground level and pull gently on the seedling stem after refilling the hole, so that the collar rises just above the soil surface. This improves the spread of the roots and avoids any risk of deformation which could hamper the growth and future stability of the tree.

When planting seedlings grown in containers or clods, the size of the hole should be approximately 10 cm larger than the root ball. The clods need to be immersed in water a few minutes before planting and re-covered with 2 cm of soil to prevent the substrate from drying out after planting (Fig. 11).

For cuttings, the hole may be made with a simple garden planter or an auger, according to their size (Fig. 12a and b).

Points to keep in mind:

- The stem must be vertical;
- The collar must be placed at ground level;
- The roots must be well spread in the planting hole;
- The earth covering the roots must be fine and slightly tamped down around the stem without damaging the collar;
- Watering is only necessary in very dry conditions.

Figure 11 How to plant a bare-root seedling (Dupraz and Liagre, 2008).

Figure 12 Planting small (a) or big (b) poplar cuttings.

3.7 Young tree protection

Individual protection is essential in agroforestry systems because where there are only a few trees, they must all be preserved. There are a variety of guards with different properties. Mesh guard are less expensive, but also less safe than rigid tree shelters. All the devices need to be strengthened by a post that also supports the seedling. The height must be adapted to the risk: 0.5 m to protect against rabbits, 1.2 m for hares or only to segregate the tree from the crop, 1.7 m for sheep and roe deer and up to 2.2 m for cattle and deer.

Both ventilated and unventilated shelters have a greenhouse effect. They may be made of green, white or clear material and the ventilation system may consist of holes or slots. Ventilation at the bottom of the pipe (around 20 cm above the ground) ensures a regular renewal of the air inside it by the 'chimney effect' and provides the CO_2 necessary for young tree growth. These shelters have several functions:

- Protection of the tree stem against wild or domestic animals which can damage them by eating buds, leaves, branches or bark; by rubbing on the trunk. Deer may rub against tree stems to remove new antler velvet and rodents may gnaw bark at the collar or at the base of the trunk to wear down their incisors.
- Protection against contact with herbicides that may be sprayed on the crops or on the non-cropped strip.
- Identification of the tree location for the guiding of agricultural workers.
- Improvement of seedling growth, as a result of the greenhouse effect and the limitation of weed growth near the young tree.

Regular checking of the protection devices is important in order to straighten, repair or replace them where necessary. Checks must also be made to ensure that the protection does not hinder the tree growth or cause injury (Fig. 13a–d).

4 Plantation maintenance

4.1 Weeding

During the first years of tree growth, the presence of weeds near the seedlings must be avoided in order to reduce competition for water and nutrients.

4.1.1 Mulching

The practice of mulching has been widely used as a management tool in many parts of the world. It reduces the influence of environmental factors on soil by increasing its temperature, controlling its diurnal/seasonal fluctuations (Ghosh

Figure 13 Different kinds of seedlings protection: ventilated shelters (a and b), non-ventilated shelter (c) and mesh guard (d).

et al., 2006) and restricting weed development. If a bio-mulch is used, the thickness of the mulch layer should be maintained at a minimum of 10 cm and renewal is recommended at least every 2 years. The advantages of bio-mulch are: (1) it is wind-proof, (2) it reduces the overall cost, (3) it is environmentally friendly because it reduces chemical or mechanical weeding or replaces plastic mulch which is non- or only slowly biodegradable, (4) it reduces erosion, (5) it provides organic matter when decomposing, (6) it increases soil temperature and water content and (7) consequently, it improves tree growth.

The major disadvantage of this system is that mulching may increase or decrease the available N in the soil, depending on the C/N ratio of the mulching material. If the mulch has a high C/N ratio (>25 e.g. wheat straw), the immobilisation of N by microflora is likely to exceed mineralisation and the mineral N levels in the soil will decrease. The use of mature (at least 1-year-old) material can ameliorate this problem (Fig. 14).

4.1.2 Chemical weeding

If there is no mulch, chemical weeding may be done in early spring in order to stop weed development as soon as possible. A second intervention may be necessary at the beginning of summer. The treatment must be applied on a day free from wind and rain, preferably early in the morning or in the evening, when the temperatures are not too high, and within a radius of 50 cm around

Figure 14 Ramial chipped wood used as mulching.

each tree. The presence of non-perforated shelters will prevent seedlings from accidental contact with the product. Alternatively, a cover on the sprayer head can be used to direct the product stream away from the seedling.

4.1.3 Mechanical weeding

Mechanical weeding should be done manually to avoid possible damage to the seedlings by the use of tools. This is achievable where the tree density is low and the total plot area is not too large.

4.2 Non-crop strips management

Non-crop strips occupy a significant area of the field (3–10%), so there will be a decrease in production due to area loss. However, sub-tree vegetation can also host weed species or pests which may spread to the crop. This has to be balanced with the fact that the vegetation contributes significantly to biodiversity conservation in providing habitats and resources for pollinators and pest predators.

If the sub-tree vegetation is not systematically removed, some intrusive species, such as thistle, bramble or bindweed, will need to be controlled. Cutting the vegetation during spring with a brush cutter will prevent weed seeds spreading in the crop.

One method of limiting losses caused by the understory strips is to work the soil and plant the crop as close to the trees as possible. The tree roots are then confined under this surface and under the plough pan and will not be disturbed by successive soil management actions.

Another option is to cultivate the understory strips by sowing melliferous species which are attractive for some insects. Legumes may also be sown to enhance tree growth by fixing nitrogen and providing it to the soil. The understory strips can also be used to cultivate secondary crops such as berries and aromatic or medicinal plants.

4.3 Irrigation

Irrigating trees can be harmful if done frequently. This is especially the case with drop-by-drop irrigation which causes a limited and shallow root system to form, which will be undersized in relation to the aerial part of the tree. Regularly irrigated trees are very vulnerable.

However, supplemental watering may be necessary to ensure survival if there is drought during the 2 years after planting. It should be done mainly during the spring, using large amounts of water that will reach the deep layer of soil and make the roots grow downwards. Summer watering is only recommended where there may be a risk of the tree dying.

If the crop is irrigated, the trees will benefit from this water supply. However, it is unsafe to associate trees with irrigated crops over several consecutive years as this will modify the spatial distribution of the tree root system, and as a consequence, the trees will then have difficulty in surviving without watering.

Finally, the quality of the wood is affected by irrigation which causes fast growth without heartwood formation.

4.4 Fertilisation

In agroforestry systems, the tree does not require fertilisation as it exploits the fertiliser which is applied to the understory strips (if there is some) and their deep roots will capture the fertilisers that have leached below the crop root zone. Fertilisers may be applied to young trees during the first year after planting, provided an accurately calibrated boom sprayer that will not over-spray onto the non-cropped strips used.

5 Tree pruning and thinning

5.1 Tree pruning

5.1.1 General guidelines

The purpose of pruning is to improve the tree shape at its mature stage and to increase the production of high-quality wood without knots. Agroforestry trees are widely spaced in planting so they naturally develop a short trunk with many vigorous branches, while in forests, competition decreases the strength of the lower branches and branches in shade may die. Moreover, in an agroforestry plot, long branches may hamper the use of agricultural machinery.

The key reasons for pruning are:

- to improve the mechanical resistance of the tree and its ability to withstand wind,
- to preserve tree health by removing dead, broken or diseased branches,
- to lift the tree crown by cutting low branches to ease the farming operation,
- to provide a clean bole (the part of the trunk located between the soil and the first branches) of around 30–50% of the tree height and to allow the timber to be exploited as knot-free veneer logs or high-quality carpentry logs. This increases the value of the timber. The bole represents 90% of the tree value and must be straight and knotless. The knots result from the healing of pruning wounds. These cause deformations during drying and are problem areas for future use of the wood,
- to increase light penetration to crops.

However, it must be kept in mind that the branches carry leaves which feed the whole tree. Therefore, pruning must be gradual, frequent and precise (Chesney, 2012).

To prune a branch effectively, the cut must be clean, leaving no torn bark around it. It should be at a slight angle to the trunk and not too close in order to allow the underlying tissues to develop a healing callus. But if the cut is done too far from the trunk, it will leave a dead stump which will take several years to disappear, generating a high risk of infection.

Although dead branches can be removed at any time, it is recommended that trees should not be pruned during frost or periods of heavy sap flow. Living branches must never be cut during budburst, nor while the sap flow is going downwards (from the end of August to the leaf fall) in order to avoid depleting tree's resources. From late June, a moderate summer pruning of young trees is advisable. This summer pruning allows a good healing of wounds with a high resistance to pathogens and decreases epicormic growth. Pruning can also be undertaken in winter, when it is more easy to observe the underlying tree architecture. However, the epicormic growth is then likely to be more vigorous during the following spring.

Cut branches may be ground and used as mulch for the uncropped strip. They may also be exported as ramial chipped wood or as fuel.

5.1.2 Shape pruning

The purpose of shape pruning is to form the trunk axis. It consists of the removal of forks which divide the trunk in two axes, multiple, broken or defective top branches and those which are too big and compete with the main stem. The objective is to get the highest single, straight, cylindrical trunk possible, while giving the crown a shape that will allow the tree to withstand wind and ensure growth (Van Lerberghe, 2017b).

Pruning usually starts 2 years after planting or when the young trees appear out of the shelters. If a fork has formed in the shelter, it has to be cut by using hand pruning shears through the shelter or by lifting it. If there are shoots inside the shelter, these are removed at the same time. Then the shelter is re-positioned. It is not necessary to repeat this operation because the apical dominance of the upper part of the tree will prevent branches from growing inside the shelter. Weak seedlings which have difficulties in establishing or growing should not be pruned as this deprives them of a part of the foliage needed for development of the root system.

Shape pruning is carried out each year until the trunk is straight, without a fork and has reached the desired height. This will take around 10–25 years, depending on the tree growth. A moderate pruning every year is more efficient than a heavy pruning every 2 or 3 years.

Shape pruning is carried out from the top of the tree. Top branch defects are corrected as they represent a threat to the straightness of the trunk. Forks must be cut as soon as they appear. Lower vigorous branches are cut as soon as their diameter exceeds 3 cm. However, no more than 30% of living branches carrying photosynthetically active leaves must be removed annually. It is possible to cut larger branches (with a diameter between 3 cm and 6 cm) that were missed in previous years, but in order to obtain a clean wound and avoid ripping the bark this must be done in two successive steps: a first cut 30 cm away from the trunk followed by a second cut near the trunk. Branches with a diameter of 7 cm or above should not be cut as this could threaten the health of the tree and greatly reduce its growth (Fig. 15).

If a tree has a very bad shape or if the main stem is broken, it is wise to cut it back to around 5 cm from the ground before the beginning of spring. During the following summer, the most vigorous shoot growing from the stump should be selected to form a new trunk and the others removed.

Figure 15 Three operations for a good shape pruning: removing (1) the top fork (red), (2) the biggest branches (blue) and (3) the branches with an acute angle (green) (Van Lerberghe, 2017b).

5.1.3 Silvicultural pruning

When a tree has been pruned into the desired shape, further pruning consists of the removal of lower branches to produce a quality bole, without defects, as high and cylindrical as possible and with the greatest proportion of knot-free wood. Knots should only be present at the core of the trunk, in a cylinder of 10–15 cm diameter, at the most.

Larger branches should not be cut if the grower wishes to avoid the formation of knots for this reason; silvicultural pruning should begin a few years after the shape pruning when trees have reached 3 m in height for walnuts and oaks and 4 m for other species. Trees must also be checked frequently, ideally each year, to deal with small branches of 3–4 cm in diameter at the most. These branches should not be cut before reaching 2 cm in diameter.

The smallest branches are left as these contribute to the tree growth, reducing the development of new shoots by acting as sap drawers. Pruning must be gradual to avoid unbalancing the tree, or the growth of epicormics. After the first pruning, the height of the branchless trunk should not exceed one-third of the total tree height.

Unlike shape pruning, silvicultural pruning is done from bottom to top, although hazardous large branches located at the top of the tree may be cut before lower ones with a smaller diameter. Branches are considered hazardous when they grow with an angle of <30° from the main stem (these may overtake the apical stem and create a fork) or when they reach a diameter equal to or greater than half that of the trunk.

Pruning facilitates the work of machinery in reducing the space occupied by branches, but branches of mature trees may spread out over the alley cropping. In this case, cutting can be undertaken with a mower bar. This is easy and quickly made and reduces the canopy without prejudice to the timber wood as the trunk is not affected. It also encourages the trunk to swell.

5.1.4 Specific case of pollards

Pollarding is a traditional practice which consists of periodically topping the trunk of a tree and maintaining steady pruning of all the branches (Chesney, 2012) without reducing the tree bole. This pruning system, which involves the removal of the upper branches of a tree, promotes a dense head of foliage and branches, producing a distinctive thick and bushy appearance.

5.1.4.1 Productions generated by pollarding

The tree size and leaf area are significantly reduced during the first year after pollarding which historically was frequently practised for fodder or firewood production (Sjolund and Jump, 2013). For instance, pollarding ash trees produces 'pollard hay', which was traditionally used as livestock feed. Trees were pruned at intervals of 2–6 years in order to maximise leafy material.

Wood-producing pollards were pruned at longer intervals of 8–15 years, a pruning cycle that allowed the production of upright poles favoured for fence rails, posts and boat construction. Supple young willow or hazel branches could be harvested as material for weaving baskets, fences and garden constructions such as bowers.

Pollarding can be undertaken to combine biomass and timber wood production. The branches and leaves harvested at each pollarding may be used as fodder (ash tree), fire wood, ramial chipped wood or chips.

5.1.4.2 Pollarding practice

Many tree species can be pollarded to provide a range of products (Colin et al., 2017). Large pollards of fast-growing trees, such as willow or ash, will produce up to 30 kg of biomass per tree and per year.

Pollarding can be carried out every 5–15 years depending on the tree growth and biomass use. All the branches are removed when there is no sap rise, often in winter. However, pollarding may also be carried out in summer in order to provide fresh fodder during the period of grass shortage (Fig. 16).

5.1.4.3 Advantages and disadvantages of agroforestry pollarding

In an alley cropping system, pollarding may reduce the light and water competition between trees, and consequently may enhance crop yield (Dufour et al., 2016). Provided growth in the trunk diameter of timber trees is not too severely impacted, pollarding offers an effective way of obtaining three products simultaneously in an agroforestry field: annual crops, branch and leaf biomass for energy, fodder or wooden chips (Valipour et al., 2014) and timber wood. Although pollarding is an age-old practice that arose from the experience of farmers and foresters (Thomas, 2000), the growth and physiological responses of pollards have rarely been studied, particularly in the context of agroforestry.

The advantage of pollarding in an agroforestry system is the reduction of competition with the crop during the first years after pruning. However, especially in vigorous species, the trees recover their canopy quickly and a few years after pollarding, may produce a heavier shade than non-pollarded trees (Dufour et al., 2018).

Figure 16 Pollarded ash trees.

5.1.4.4 An original experiment with pollards

An experiment was conducted to assess the impact of pollarding on the growth, both in trunk diameter and branch biomass, and development (phenology) of adult hybrid walnut trees, which are not traditionally pollarded, in an alley cropping agroforestry system (Fig. 17a-g).

The distance between tree rows is 13 m. In 2013, the walnut trees were 15 years old and their mean height was 10 m. In the same year, the crop yield was on average 25% lower in the agroforestry alleys than in the full crop control. Fifty trees were pollarded in December 2013 and their growth and development were compared to those of control trees, regularly pruned. The main results of this experiment were as follows:

- Pollarding changed the tree phenology: budburst was delayed during the first, third and fourth year after pollarding and the leaf fall was noticeably delayed for pollards when compared to control trees in the second year

Figure 17 Pollarded walnut trees, at the first pollarding, in December 2013 (a), in December 2014 (b), in December 2015 (c), in December 2016 (d), in October 2017, just before (e) and after (f) the second pollarding, in May 2018, at the beginning of budburst (g).

after pollarding. As a result, the leaf-bearing time was extended by 26 days during 2015. This increase of leaf presence extends the photosynthetic assimilation of the tree and allows it to balance a part of the loss due to leaf area reduction.

- The pollards showed a significantly greater growth in height than control trees and except during the first year, their growth in diameter was not much slower than that of the control trees: cumulative growth from 2013 to 2017: 4.3 mm for pollards, 4.6 mm for control.
- The number of branches cut at the second pollarding, in October 2017, was 29.6 ± 0.4 (mean, standard error) per tree, with a mean branch diameter of 4.5 ± 0.07 cm. An allometric equation between the circumference of the branches and their dry biomass allowed an estimation of the biomass of branches produced between the two pollardings to be 80.9 ± 7.6 kg $tree^{-1}$, that is around 20 kg $trees^{-1}$ $year^{-1}$.
- The crop yield was only enhanced in the first 2 years because, as the photographs show, the trees quickly produce new vigorous branches and from the third year, the shade near the pollards is higher than that of the near regularly pruned trees.

Uncertainty remains as to how the trees will respond to repeated pollardings. The soil nutrients may be depleted due to the export of nutrient pools in the removed branches. This may increase the competition for nutrients with the crop. In addition, the rapid regrowth of branches creating very dense shade increases light competition with the crop as soon as the third year after pollarding. More studies are needed to optimise the trade-off between trunk (timber wood), branches (fuelwood or fodder) and crop (agricultural production) productivities, depending on the frequency of pollarding and the density of pollards on the field.

5.2 Tree thinning

Early elimination of poorly formed or weak trees which cannot be improved is important for successful agroforestry and the final density must allow the remaining trees to grow freely and without competition.

No real thinning is necessary when initial density is <50 trees ha^{-1}. For densities of 50–100 trees ha^{-1}, one tree out of two can be removed, and for higher densities, two out of three.

The thinning aims are: (1) to promote beautiful trees and to reduce the competition between trees, (2) to spend the working time only on crop trees and (3) to reduce the competition with the intercrop.

Thinning is undertaken as soon as possible to select the most beautiful trees as soon as the bole is formed. The bole is considered formed when forks have been removed and when the height of the tree is about twice that of the bole. There is no advantage in waiting for the bole to be totally pruned as the benefit of thinning to save pruning time will be lost.

The first thinning is carried out between 5 and 12 years after planting and if necessary, a second thinning may be done between 15 and 20 years after planting. After thinning, the final density is between 25 and 100 trees ha^{-1}. Thinning is selective, not systematic: several beautiful trees may be consecutive within a row, and they must be kept. Two or three trees can grow together even if their canopies touch. To achieve the most effective spacing, thinning may be undertaken in groups of three to six consecutive trees in the row. Half of the trees, the least beautiful in a group, will be removed.

Care must be taken when several tree species are planted, as growth rates differ greatly between species. It is important to take this characteristic into account and to keep trees of each species during the thinning.

The wood produced by early thinning has a low marketable value but may be ground or used as energy wood.

Many species will produce clump shoots after they are cut (e.g. walnut, poplar, maple, chestnut). The shoots can be used for biomass (ramial chipped wood, energy), otherwise the stump should be removed.

6 Conclusion and future trends

A decision to set an agroforestry plot requires an evaluation of the feasibility of the project in terms of land availability and fit, of skilled labour availability and of the correct tools. Success with an agroforestry plantation depends on the skill, experience and enthusiasm of the farmer who must also be knowledgeable in managing trees. These requirements could be met by two individuals, for instance the owner of the field takes care of the trees and a farmer manages the crops. This will require good coordination between the two roles.

Financial feasibility and return are key issues which farmers and land owners must consider. Regional differences in yields, prices and government grants might lead to substantial variations in revenue and costs within Europe and even within a given country. It is difficult to estimate the proceeds of tree sales due to the long period of time before harvesting, during which the selling price will fluctuate.

Graves et al. (2011) developed a new economic model: Farm-SAFE. This model allows comparison of arable farming, forestry and agroforestry systems at both plot- and farm-scale. The comparisons include net margins, net present values, infinite net present values, equivalent annual values and labour requirements. The new and free version of Farm-SAFE can be uploaded from: https://www.agforward.eu/index.php/fr/1683.html.

7 Where to look for further information

In many countries, associations of agroforestry can help the farmers to prepare and to realise their plantation project.

In the United States, the Association for Temperate Agroforestry (AFTA, https://www.aftaweb.org/), formed in 1991, promotes the wider adoption of agroforestry by landowners in temperate regions of North America. It is a private, non-profit, organization based at the University of Missouri Center for Agroforestry at Columbia. Also, the USDA developed a National Agroforestry Center (NAC). This agency accelerates the application of agroforestry through a national network of partners. They conduct research, develop technologies and tools, coordinate demonstrations and training, and provide useful information to natural resource professionals. For example, the centre published '*Profitable Farms and Woodlands: A Practical Guide to Agroforestry for Landowners, Farmers and Ranchers*' and the 'Agroforestry practices' section (https://www.fs.usda.gov/nac/practices/index.shtml).

In Europe, the EURopean Agroforestry Federation (EURAF, http://www.eurafagroforestry.eu/welcome) gives examples of agroforestry best practices. This association also reports on the place of agroforestry in the Common Agricultural Policy in the EU (http://www.eurafagroforestry.eu/action/policy/Agroforestry_in_the_new_CAP). EURAF sustained AGFORWARD project (AGroFORestry that Will Advance Rural Development), which was a 4-year research project funded by the European Union's Seventh Framework Programme for Research and Technological Development. One of the project outputs was the creation of two sets of leaflets on Best Practices and Innovations in agroforestry (https://www.agforward.eu/index.php/en/best-practices-leaflets.html and http://www.agforward.eu/index.php/en/Innovation-leaflets.html). On this topic, Rigueiro-Rodríguez, A., McAdam, J., Mosquera-Losada, M. R. published *Agroforestry in Europe: Current Status and Future Prospects* (2008; books.google.com).

Some books can also provide some guidance for the plantation and management of an agroforestry field: Dupraz, C. and Liagre, F. *Agroforesterie – Des arbres et des cultures* (book + DVD; Editions France Agricole, 2008) or for the tree species choice: Orwa, C., Mutua, A., Kindt, R., Jamnadass, R. and Simons, A. *Agroforestree Database: A Tree Reference and Selection Guide* (2015; http://www.worldagroforestry.org/af/treedb/). Finally, *Temperate Agroforestry Systems*, edited in 2018 by Gordon et al. brings together many examples of temperate agroforestry and will make valuable reading for all those working in this area as researchers, practitioners and policy makers.

8 References

Allen, S. C., Jose, S., Nair, P. K. R., Nkedi-Kizza, P. and Ramsey, C. L. (2004). Safety-net role of tree roots: Evidence from a pecan (*Carya illinoensis* K. Koch)-cotton (*Gossypium hirsutum* L.) alley cropping system in the southern United States. *For. Ecol. Manage.* 192, 395–407. doi:10.1016/j.foreco.2004.02.009.

Andrianarisoa, K. S., Dufour, L., Bienaime, S., Zeller, B. and Dupraz, C. (2016). The introduction of hybrid walnut trees (*Juglans nigraXregia* cv. NG23) into cropland reduces soil mineral N content in autumn in southern France. *Agrofor. Syst.* 90(2), 193–205. doi:10.1007/s10457-015-9845-3.

Bellow, J. G. and Nair, P. K. R. (2003). Comparing common methods for assessing understory light availability in shaded-perennial agroforestry systems. *Agric. For. Meteorol.* 114, 197–211. doi:10.1016/S0168-1923(02)00173-9.

Cannell, M. G. R., Van Noordwijk, M. and Ong, C. K. (1996). The central agroforestry hypothesis: The trees must acquire resources that the crop would not otherwise acquire. *Agrofor. Syst.* 34, 27–31. doi:10.1007/BF001229630.

Cannell, M. G. R., Mobbs, D. C. and Lawson, G. J. (1998). Complementarity of light and water use in tropical agroforests II. Modelled theoretical tree production and potential crop yield in arid to humid climates. *For. Ecol. Manage.* 102(2/3), 275–82. doi:10.1016/S0378-1127(97)00168-0.

Chesney, P. E. K. (2012). Shoot pruning and impact on functional equilibrium between shoots and roots in simultaneous agroforestry systems. In: M. L. Kaonga (Ed.), *Agroforestry for Biodiversity and Ecosystem Services – Science and Practice*. Croatia, www.intechopen.com. ISBN: 978-953-51-0493-3.

Colin, J., Van Lerberghe, P. and Balaguer, F. (2017). Farming with pollards – A productive way of pruning. *Agroforestry Innovation Leaflets*. Agforward Research Project. https://www.agforward.eu/index.php/en/best-practices-leaflets.html.

Descheemaeker, K., Bunting, S. W., Bindraban, P., Muthuri, C., Molden, D., Beveridge, M., van Brakel, M., Herrero, M., Clement, F., Boelee, E. and Jarvis, D. I. (2013). Increasing water productivity in agriculture. In: E. Boelee (Ed.), *Managing Water and Agroecosystems for Food Security*. Book Series: Comprehensive Assessment of Water Management in Agriculture Series. Vol. 10, pp. 104–23. doi:10.1079/9781780640884.0000.

Dufour, L., Metay, A., Talbot, G. and Dupraz, C. (2013). Assessing light competition for cereal production in temperate agroforestry systems using experimentation and crop modelling. *J. Agron. Crop Sci.* 199, 217–27. doi:10.1111/jac.12008.

Dufour, L., André, J., Le Bec, J. and Dupraz, C. (2016). Influence of tree pollarding on crop yield in a mediterranean agroforestry system. *3rd European Agroforestry Conference*. Montpellier, 23–25 May 2016.

Dufour, L., Gosme, M., Le Bec, J. and Dupraz, C. (2018). Impact of pollarding on growth and development of adult agroforestry walnut trees. *4th European Agroforestry Conference*, Nijmegen, 28–30 May 2018.

Dupraz, C. and Liagre, F. (2008). *Agroforesterie – Des arbres et des cultures*. Editions France Agricole, 413p. ISBN: 13-198-2-85557-150-8.

Ellis, E. A., Nair, P. K. R. and Jeswani, S. D. (2005). Development of a web-based application for agroforestry planting and tree selection. *Comput. Electron. Agric.* 49, 129–41. doi:10.1016/compag.2005.02.008.

Fernandez, M. E., Gyenge, J., Licata, J, Schlichter, T. and Bond, B. (2008). Belowground interactions for water between trees and grasses in a temperate semiarid agroforestry system. *Agrofor. Syst.* 74, 185–97. doi:10.1007/s10457-008-9119-4.

Gathumbi, S. M., Ndufa, J. K., Giller, K. E. and Cadish, G. (2002). Do species mixtures increase above- and belowground resource capture in woody and herbaceous tropical legumes? *Agron. J.* 94(3), 518–26. doi:10.2134/agronj2002.5180.

Ghosh, P. K., Dayal, D., Bandyopadhyay, K. K. and Mohanty, M. (2006). Evaluation of straw and polythene mulch for enhancing productivity of summer ground nut. *Field Crops Res.* 99, 76–86. doi:10.1016/j.fcr.206.03.004.

Graves, A. R., Burgess, P. J., Liagre, F., Terreaux, J. P., Borrel, T., Dupraz, C., Palma, J. and Herzog, F. (2011). Farm-SAFE: The process of developing a plot-and farm-scale model of arable, forestry, and silvoarable economics. *Agrofor. Syst.* 81(2), 93–108. doi:10.1007/s10457-010-9363-2.

Jose, S. (2009). Agroforestry for ecosystem services and environmental benefits: An overview. *Agrofor. Syst.* 76(1), 1–10. doi:10.1007/s10457-009-9229-7.

Lawson, G., Dupraz, C. and Watté, J. (2018). Can silvoarable systems maintain yield, resilience, and diversity in the face of changing environments? In: G. Lemaire, P. Carvalho, S. Kronberg and S. Recous (Eds), *Agroecosystem Diversity: Reconciling Contemporary Agriculture and Environmental Quality* (1st edn.). Academic Press, pp. 145–68. eBook ISBN: 9780128110515; Paperback ISBN: 9780128110508.

Orwa, C., Mutua, A., Kindt, R., Jamnadass, R., Simons, A. (2015). Agroforestree database: A tree reference and selection Guide Version 4.0. 2009. http://www.worldagroforestry.org/af/treedb/.

Pantera, A., Burgess, P. J., Mosquera Losada, R., Moreno, G., Lopez-Diaz, M. L., Corroyer, N., McAdam, J., Rosati, A., Papadopoulos, A. M., Graves, A., Rigueiro Rodriguez, A., Ferreiro-Dominguez, N., Fernández Lorenzo, J. L., Gonzáles-Hernández, M. P., Papanastasis, V. P., Mantzanas, K., Van Lerberghe, P. and Malignier, N. (2018). Agroforestry for high value tree systems in Europe. *Agrofor. Syst.* 92, 945–59. doi:10.1007/s10457-017-0181-7.

Quinkenstein, A., Wöllecke, J., Böhm, C., Grünewald, H., Freese, D., Schneider, B. U. and Hüttl, R. F. (2009). Ecological benefits of the alley cropping agroforestry system in sensitive regions of Europe. *Environ. Sci. Policy* 12, 1112–21. doi:10.1016/j.envsci.2009.08.008.

Rowe, E. C., Hairiah, K., Giller, K. E., van Noordwijk, M. and Cadish, G. (1999). Testing the safety-net role of hedge row tree roots by 15N placement at different soil depths. *Agrofor. Syst.* 43, 81–93. doi:10.1023/A:1022123020738.

Sjolund, M. J. and Jump, A. S. (2013). The benefits and hazards of exploiting vegetative regeneration for forest conservation management in a warming world. *Forestry* 86(5), 503–13. doi:10.1093/forestry/cpt030.

Thomas, P. (2000). *Trees: Their Natural History*. Cambridge University Press, Cambridge, UK, 286pp. doi:10.1017/CBO9780511790522.

Torralba, M., Fagerholm, N., Burgess, P. J., Moreno, G. and Plieninger, T. (2016). Do European agroforestry systems enhance biodiversity and ecosystem services? A meta-analysis. *Agric. Ecosyst. Environ.* 230, 150–61. doi:10.1016/j.agee.2016.06.002.

Valipour, A., Plieninger, T., Shakeri, Z., Ghazanfari, H., Namiranian, M. and Lexer, M. J. (2014). Traditional silvopastoral management and its effects on forest stand structure in northern Zagros, Iran. *For. Ecol. Manage.* 327, 221–30. doi:10.1016/j.foreco.2014.05.004.

Van Lerberghe, P. (2017a). Choosing quality planting material. *Agroforestry Best Practice Leaflets*. Agforward Research Project. https://www.agforward.eu/index.php/en/best-practices-leaflets.html.

Van Lerberghe, P. (2017b). Shaping the trees. *Agroforestry Best Practice Leaflets*. Agforward Research Project. https://www.agforward.eu/index.php/en/best-practices-leaflets.html.

Van Lerberghe, P. (2017c). Planting the trees. *Agroforestry Best Practice Leaflets.* Agforward Research Project. https://www.agforward.eu/index.php/en/best-practices-leaflets.html.

Van Lerberghe, P. (2017d). Planning an agroforestry project. *Agroforestry Best Practice Leaflets.* Agforward Research Project. https://www.agforward.eu/index.php/en/best-practices-leaflets.html.

Vandermeer, J. (1989). *The Ecology of Intercropping.* Cambridge University Press, Cambridge, UK, 237pp. ISBN: 0-521-34592-8.

Wolz, K. J., Branham, B. E. and DeLucia, E. H. (2018). Reduced nitrogen losses after conversion of row crop agriculture to alley cropping with mixed fruit and nut trees. *Agric. Ecosyst. Environ.* 258, 172–81. doi:10.1016/j.agee.2018.02.024.

Yu, Y., Stomph, T.-J. and Makowski, D. (2015). Temporal niche differentiation increases the land equivalent ratio of annual intercrops: A meta-analysis. *Field Crops Res.* 184, 133–44. doi:10.1016/j.fcr.2015.09.010.

Part 2

Agroforestry ecosystem services

Chapter 9

A holistic approach to sustainable agriculture: trees, science and global society

Roger R. B. Leakey, International Tree Foundation, UK

1 Introduction

How many people understand the connection between trees, science and global society? Sadly, the answer is probably 'very few', yet arguably the future of our planet probably depends on getting the connections and the relationships between them correct. Trees, in fact, are a greatly underutilized resource that play crucial environmental roles of great importance to the sustainability and the productivity of global ecosystems. Trees, especially from the tropics/subtropics, are also the source of numerous useful products, which again have been overlooked by science and society and could be part of a hugely diversified set of new industries of great importance to the global economy. Both these roles are poorly understood by international and national policy makers, with the role of trees in the productivity of agriculture being grossly misunderstood and overlooked, as will be explained later. Indeed, the cultural and traditional use of a wide range of tree products by local people has been poorly recognized right back to colonial times (Maathai, 2009), despite correspondence by early missionaries in Africa to colonial administrators (e.g. Harry Leakey Archive in the Bodleian Library, Oxford).

An obvious conclusion from these statements is that there is 'work to be done' to harness these overlooked benefits for both humankind and the planet. That of course leads us to the need for scientific investigations about how to

http://dx.doi.org/10.19103/AS.2018.0041.25

best capture these benefits so that global society can take advantage of the overlooked opportunities and use them to support the increasingly demanding needs of a growing human population and improve the equity in the wealth derived from them. Greater equity might even lead to much greater harmony between the peoples of our divided and conflict-ridden world of extremely rich and extremely poor. This chapter reviews recent progress in this direction through agroforestry and looks at the opportunities for further progress.

2 Trees: an underutilized resource for agriculture

The definition of a tree is far from easy, but there are said to be 60 065 tree species in the world (Beech et al., 2017). The difficulties in definition are due to the grey line between what is a 'tree' and a 'shrub'; the importance of 'woodiness', and the importance of the distinction between monocotyledons and dicotyledons. Typically, agroforestry involves trees for timber, wood, fruits and nuts, medicines and a very wide range of other useful products and environmental/ecological services. In this chapter the word 'tree' is used to refer to trees and shrubs, all woody perennials, palms and bamboo, in accordance with the Corporate Strategy 2017–26 of the World Agroforestry Centre (https://www.worldagroforestry.org/sites/default/files/users/admin/Strategy%20Report_2017.pdf).

Of the approximately 250 000 plant species in the world about 20 000 (4%) have edible products, many of which are trees, yet only about 0.3% of these edible plants are cultivated in agriculture (Leakey and Tomich, 1999), with just 12 plant species contributing about 75% of our food (FAO, 1999). Despite this apparently low use of edible species, there is good evidence from ethnobotany of the much wider use of plant species by indigenous peoples around the world for food and many other day-to-day uses (Cunningham, 2001). It was the discrepancy between the agricultural and the ethnobotanical use of plants that gave rise to the term 'Cinderella species' to describe those species which could be formally domesticated to 'make the land more productive and . . . contribute to domestic, regional and international trade' (Leakey and Newton, 1994). To get 'Cinderella' to this 'Ball' it was realized that there were social, political, economic and biological issues to be resolved.

Over the last 25 years much progress has been made to bring underutilized indigenous species, especially fruits and nuts, into formal domestication (Leakey et al., 2012). This process builds on ethnobotany and the traditional knowledge of local people (Franzel et al., 1996; Cunningham, 2001). Key factors driving this process have been (1) the upsurge of biological and social research on the importance of these little-known Cinderella species to local people (Schreckenberg et al., 2002; Leakey, 2012) and (2) the development and implementation of a participatory process of decentralized domestication which puts the communities in the driving seat, as well as making them the recipients of the economic and social benefits (Leakey et al., 2003; Leakey, 2014a). The success of this participatory process has led to the recognition of these farmer-developed horticultural cultivars as 'socially modified organisms' (Leakey, 2017a) with the capacity to address the very serious 'cycle of land degradation and social deprivation' that underlies the environmental, social and economic constraints to food production and rural development in the tropics and subtropics (Leakey, 2013).

3 Trees: prerequisites for the use of underutilized indigenous species

The agroforestry research underpinning the domestication of these little-known tree species has been multidisciplinary.

3.1 Vegetative propagation

Development of vegetative propagation techniques is essential so that the phenotypic genetic characteristics of elite trees can be captured as horticultural cultivars (Leakey, 2014b). Until the 1970s the vegetative propagation of many trees was severely constrained by a lack of fundamental knowledge of the processes determining success; despite centuries of successful propagation by 'green-fingered' horticulturalists. Since then, more rigorous studies have evaluated the importance of pre-severance and post-severance factors culminating in some widely applicable general principles (Leakey, 2014b). It has emerged that, through its impact on the morphology and physiology of subsequently severed cuttings, rigorous pre-severance management of stockplants is fundamental to the achievement of a robust and productive programme of propagation. In parallel with this, the development of low-technology propagation facilities appropriate for use in remote rural areas without piped water or electricity have allowed the establishment of simple village-level tree nurseries at very low cost (Leakey, 1996, 2014b).

3.2 Characterization of tree-to-tree variation

This clonal approach to domestication has used simple techniques to determine the extent of tree-to-tree variation in characteristics that are important for the quality and marketability of the tree products. Such studies have now been done for many tropical tree species entering domestication programmes (e.g. *Canarium indicum*, Leakey et al., 2008) and included traits such as fruit/nut morphology, nutritional/medicinal properties, chemical composition of oils/fats, essential oils and so on. The results of these studies have then led to the formulation of 'ideotypes' to meet the needs of consumers and new industries (Leakey and Page, 2006). Importantly, around 85% of this tree-to-tree variation has been found to occur at different sites (de Smedt et al., 2011; Pauku et al., 2010). This indicates that the decentralized approach to tree domestication has the potential to achieve very significant improvements in the selected traits, as well as lowering the risk of seriously narrowing the genetic diversity of these species (Leakey, 2012).

3.3 Technical support

Practical success depends on the provision of technical support for participatory domestication programmes. This has been achieved by the creation of community-based Rural Resource Centres (Degrande et al., 2015) in partnership with community-based organizations (CBOs) or non-governmental organizations (NGOs). These provide the hands-on training, information and knowledge that allows farmers to develop small-scale tree nurseries, set up simple vegetative propagation facilities, initiate elite tree selection, manage multiplication gardens and establish agroforestry systems on their own farms. Among these skills is the well-known agroforestry practice of using leguminous trees and shrubs in agroforestry to fix atmospheric nitrogen in their root nodules and so to improve soil fertility (Sileshi et al., 2014). Putting these newfound skills together, farmers are empowered to address the big constraints that have been limiting the productivity of their farming systems for decades, namely soil infertility and poverty, in association with hunger and malnutrition (Leakey, 2013).

In addition to these practical field skills, the training provided by Rural Resource Centres also provides knowledge in community management, financial management, marketing and trade, use of microfinance, local infrastructure developments and so on. In parallel with this capacity-building programme, research on the development process was initiated in order to enhance and upscale the concepts and initiatives, with a focus on farmer livelihood strategies (Schreckenberg et al., 2006; Degrande et al., 2006); community capacity-building processes (Degrande et al., 2012; Takoutsing et al., 2014; Franzel et al., 2015); farmers' rights (Gyau et al., 2014); product uses and markets (Cosyns et al., 2011, 2013; Facheux et al., 2006, 2007); adoption of innovation (Mbosso et al., 2015); incentives to plant and national law (Foundjem-Tita et al., 2013); rural development issues (Foundjem-Tita et al., 2011, 2012; Gyau et al., 2012), processing and value chains (Degrande et al., 2014); and impact (Facheux et al., 2012) have been investigated. These advances in the social sciences of community engagement have led to Rural Resource Centres now being adopted in other African countries.

Once equipped with the technical knowledge and engaged in a self-help process with obvious livelihood benefits, these farmers put it into practice in their own farms in whatever way best meets their own needs. The knowledge acquired in Rural Resource Centres provide these farmers with the skills to implement generic steps towards more productive and sustainable agriculture and better lives, namely (Leakey, 2013):

- the use of leguminous trees and shrubs to restore soil nitrogen,
- the domestication of locally important indigenous trees and
- commercialization of tree nurseries and tree products.

Together with local processing and value-adding, the last of these steps creates urban as well as rural employment (Lombard and Leakey, 2010; Leakey and van Damme, 2014).

These steps can be applied in sequence or simultaneously (Leakey, 2012, 2013) to address the key problem facing tropical/subtropical agriculture – the yield gap between the potential yield of modern staple crop varieties and the actual yield achieved by farmers. Yield gaps are the outcome of the 'cycle of land degradation and social deprivation' caused by the inaccessibility to modern technologies faced by ultra-poor subsistence farmers with only 2–3 ha of degrading land. So, with the training they have received, farmers can apply the three agroforestry steps mentioned above, improve their crop yields, produce more nutritious food and become wealthier (Asaah et al., 2011a; Leakey, 2012, 2013, 2017a). The new culturally important agroforestry crops domesticated by these farming communities underpin this self-help approach to self-sufficiency and a better lifestyle and have been described as the 'socially modified organisms/crops' (Leakey, 2017b).

The above approach to agroforestry also has very important agroecological and environmental benefits resulting from the diversification of the farming system with both nitrogen-fixing trees and the socially modified crops. The former ameliorate soil nitrogen deficiencies, while the latter produce marketable products. Both these sets of long-lived woody perennials can therefore be considered as the ecological 'heart' of agricultural and natural ecosystems (Leakey, 2014c). This role is fundamental to agroforestry which has been defined as a means to creating an agroecological succession, akin to that in natural ecosystems (Leakey, 1996, 2014c). A healthy, functioning ecosystem is important for the maintenance of the nutrient, water and carbon cycles: the stability of the ecosystem through the completion of food chains and life cycles. These roles together regulate the natural balance between different organisms and reduce the risk of weed, pest and disease outbreaks (Leakey, 2017c). They are also important for pollination and seed dispersal which maintain the reproductive health and longevity of the component organisms. When these ecological functions are destroyed by deforestation and the cultivation of crop monocultures, conventional modern agriculture depends on artificial fertilizers and pesticides to control pest, disease and weed outbreaks: inputs which are inaccessible to poor subsistence farmers. Additional outcomes from mainstream agriculture are the breakdown of the hydrological and carbon cycles, which lead to increased run-off to waterways, reduced recharge of groundwater, reduced advection of moisture to the atmosphere and importantly to the emission of greenhouse gas which promote climate change.

From all the above, and indeed the contents of this and many other books (e.g. Nair and Garrity, 2012; Atangana et al., 2014; Leakey, 2017a), as well as thousands of research publications, it is clear that trees in agroforestry systems are fundamental to sustainable agriculture, especially in the tropics and subtropics. This then leads our discussion in two important directions:

- what research is needed to further improve these approaches to agroforestry?
- how can the social, political and economic constraints that hinder the implementation of more sustainable agroforestry practices be resolved?

These issues are discussed in more detail below.

4 Science: priorities in agroforestry research

Agroforestry is a relatively new science, with early advocates recognizing the need for trees in sloping field systems to minimize soil erosion (Leakey, 1949). Institutionally, agroforestry was established in the 1970s with the creation of the International Council for Research in Agroforestry in Nairobi, Kenya. From the outset, as the approach to problem solving, it used participatory techniques to determine the needs and priorities of local farmers – 'Diagnosis & Design' (Raintree, 1987). In more recent years, a participatory approach has been used to ensure that tree domestication is effectively adopted by local farmers (Leakey, 2014a). This contrasts with much top-down modern agricultural research in which scientists select an interesting research project that seems to fit a topical problem.

A review of the science agenda for agroforestry in humid West Africa at the end of the twentieth century identified tree domestication and its linkages with commerce as priority research issues; together with the integration of domesticates within multi-strata agroforests for both their agroecological benefits and income generation (Leakey, 1998). More recently, advances in agroecology have indicated how agroforestry provides ecological niches above- and below-ground for colonization by a wide array of organisms that promote wildlife conservation in addition to ecological functions such as nutrient cycling, soil health and groundwater management (Garbach et al., 2014; Lavelle et al., 2014; Leakey, 1996, 2014c, 2017c; Hobley et al., 2017).

However, scientific knowledge and understanding of the numerous interacting factors is poor and is a challenging frontier for much more detailed research (Leakey, 2014c).

As large, long-lived, perennials, trees have adapted to the many environments found in both pioneer and mature ecosystems. There is therefore also a need for much better understanding of tree biology, especially the physiology of phase change during maturation and the genetics of adaptation to a very wide range of climates and soils. Tree biology and the enhancement of their role in agroecology thus represent the ultimate challenges for scientists wishing to develop new tree crops (Leakey et al., 2012; Leakey, 2014c).

4.1 Advances in vegetative propagation

The successful rooting of stem cuttings has been built on the development of robust and appropriate vegetative propagation techniques using simple, low-technology propagators (Leakey et al., 1990). These were developed for use by rural communities living in remote villages without electricity and piped water using practical protocols based on sound principles. Special importance has been focused on an understanding of the numerous pre- and post-severance factors determining the successful vegetative propagation by leafy stem cuttings (Leakey, 2004, 2014b), such as the differences in leaf size and internode length, and the effects of nutrition, irradiance and light quality regimes on different stockplant shoots (Leakey, 2014b). Community-level training in these techniques has allowed many species to be propagated vegetatively as cultivars in village nurseries (Tchoundjeu et al., 2002; Ngo Mpeck et al., 2003) from selected individual elite trees identified by the farmers themselves (Mbile et al., 2004).

To capture the mature phase of the tree crown and so achieve early fruiting (Fig. 1), it is easiest to use grafting or air layering techniques as the rooting ability of cuttings is often said to be negatively affected by the development of sexual maturity. The rooting ability of mature cuttings is a topic which in my view suffers from flawed understanding of tree development. Consequently, it is a topic where our knowledge of how to manipulate the rooting ability of shoots from within the crown of mature trees needs to be greatly improved (Dick and Leakey, 2006). This is perhaps the greatest challenge for plant physiologists engaged in tree domestication.

Despite our improved knowledge of the rooting process, some trees remain a domestication challenge - such as *Allanblackia* species. In this case, success is hindered by slow rooting and the formation of very few roots (Tsobeng et al., 2016). Furthermore, the shoots of rooted cuttings of these species tend to grow plagiotropically - that is not vertically - as these species are characterized by weeping branches. This is an epigenetic trait associated with

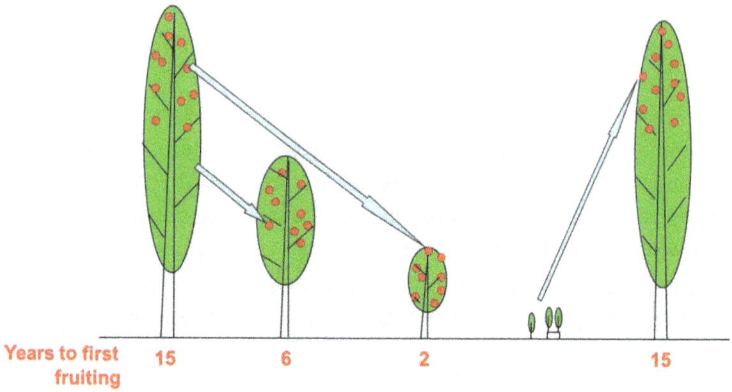

Figure 1 Effects of vegetatively propagating shoots from the adolescent, mature or juvenile parts of a tree on the time to fruiting (After: Leakey, 2017a).

topophysis, nevertheless ways can sometimes be found to release accessory buds capable of orthotropic, or vertical, growth (Leakey, 1990). Plagiotropism is often associated with cuttings from the sexually mature crown of trees - that is those with the capacity to immediately flower and produce fruits. This is a trait of great importance when domesticating fruit and nut trees typically captured by the techniques of air layering or marcotting, and grafting (Mialoundama et al., 2002; Tchoundjeu et al., 2010; Ofori et al., 2015; Asaah et al., 2011b). However, there is a need for much greater understanding of the numerous factors determining propagation success by these techniques. Associated with these research topics is the need for rootstock selection to control tree growth (dwarfing) and fruit yield.

4.2 Improving the understanding of tree biology

As the discussion of research priorities in agroforestry and vegetative propagation suggests, there is an urgent need for a better understanding of reproductive and other aspects of tree biology. These issues include:

- Understanding and managing processes of sexual maturation
- Understanding and managing physiological variation
- Controlling seasonality in production

Addressing these challenges will be critical in ensuring commercial production of a consistent product suited to market needs.

Earlier we saw that the physiological processes of sexual maturation in trees is a research topic important for vegetative propagation, but of course many aspects of reproductive biology are also important in tree breeding and seed production. Among these are questions about the tree-to-tree variation in sexuality. Trees are typically reported to be either monoecious or dioecious but actually there may in fact be the full range of dioecious sexuality from 100% male to 100% female, and everything in between, as monoecious/hermaphrodite species can be seen to vary in their male:female ratios, affecting pollination success, which obviously requires both sexes. When unproductive male fruit trees are cut down, as has been reported in *Sclerocarya birrea*, poor pollination seems to result in fewer kernels per nut (Leakey et al., 2005b), but it is not known how many male trees are needed to effectively fertilize a predominantly female stand. This may also be manipulated by the appropriate location of beehives to boost pollination success.

Time to first flowering is one of the biggest constraints to tree breeding, affecting generation time. Thus, controlled breeding would be eased if ways can be found to induce the capacity to flower precociously and produce viable seeds in sexually juvenile shoots, like seedlings. Some success has been achieved in the manipulation of the physiology of sexual maturation in trees (Longman, 1978; Leakey et al., 1981), but it is still very poorly understood, making it difficult to induce and manipulate flowering for the purposes of rapid tree breeding.

Another related issue is the physiological variation associated with developing a mature tree. In timber tree species correlative inhibition is thought to regulate the proleptic and sylleptic branching processes that control tree form and branching characteristics (Leakey, 1986), and so determine log volume and stem form (Ladipo et al., 1991). Research has, however, indicated that even within a single tree species the tree-to-tree variation in these processes is approximately 10-fold. Thus, in *Triplochiton scleroxylon* for example, the development of cultivars with very strong correlative inhibition would be appropriate for lumber, while those at the other extreme would be best for edible leaf production from managed hedges.

Physiological processes like branching and stem elongation, leaf thickness, photosynthesis and so on are strongly affected by the environment (especially nutrients, light quality, irradiance, temperature, day length etc.) affecting growth and the partitioning of dry matter affecting harvest index (Ladipo et al., 1992). However, there has been very little research to optimize the capture of physiological and intraspecific genetic variation, and their interactions.

Finally, there is much to learn about the ripening of fruits. This is important for the marketing of fruits which is often severely constrained by the 'shelf life' of the products. While techniques can be developed to preserve the products, there is also the opportunity to select cultivars which are inherently 'in season' at times outside the typical natural range. For example, the cultivar 'Noël' of *Dacryodes edulis* fruits in December when the main season is May–September (Leakey et al., 2003). Controlled pollination may be needed to breed for trees fruiting throughout the year, and hence develop them as cultivars.

4.3 Capturing genetic variation

A horticultural approach to tree domestication based on the vegetative propagation of elite trees raises the question of how to identify the candidate trees? This in turn raises questions about how much intraspecific variation is available, and what is the structure of this variation in wild populations of different species from different environments?

Three approaches have been engaged to seek answers to these questions. The first and easiest is to allow the farmers to use their local knowledge about the size of the fruits, their flavour or some other attributes, like oiliness - and how this affects the price paid in the local marketplace. Secondly, simple measurements of key traits can quantify tree-to-tree intraspecific variability (Leakey et al., 2000). This data can be used to calibrate local knowledge, and to assess the magnitude of the variation in the local population - for example in 100 trees of Marula (Leakey et al., 2005a,b), Canarium (Leakey et al., 2008), Allanblackia (Atangana et al., 2011) and Sandalwood (Page et al., 2010). This approach can also be used to assess more complex traits, such as nutrient content, fatty acid profiles, food thickening properties, wood quality and so on in a laboratory (Leakey et al., 2012). Measuring the extent of this intraspecific diversity allows the identification of multi-trait 'ideotypes' meeting different combinations of market-oriented traits (Leakey and Page, 2006). Thirdly, DNA analysis can examine genetic diversity with regard to the relatedness of different trees and different populations and the identification of genetic markers for selection, or to determine priorities for genetic conservation (Lowe et al., 2000; Jamnadass et al., 2009; Muchugi et al., 2008; Pauku et al., 2010).

Some of the above qualitative traits will become increasingly important as the use of ideotypes expands to meet more sophisticated opportunities, such as the independent traits of viscosity and the drawability of polysaccharides determining the food thickening properties of *Irvingia gabonensis* (Leakey et al., 2005c). In the future there will probably be a greater need to capture desirable traits, such as biochemical profiles affecting nutrition and anti-nutrients. It is important in this connection to recognize that due to the considerable tree-to-tree variation in these traits the species average is a 'very blunt instrument' when trying to select trees to address the problem of malnutrition. The same is true when quantifying sensory traits (taste and smell) such as astringency, sweetness, acidity and so on which can be linked to molecular genetic profiles to aid tree breeding.

In the future, molecular approaches to studies of variation will probably become applicable to the assessment of epigenetic variation within an individual organism - such things as plagiotropism, sexuality and fecundity, probably with extensive tree-to-tree variation.

The importance of tree-to-tree variation raises the broader issue of conserving genetic diversity. Conservationists are often concerned about the potential (and frequently real) loss of genetic diversity associated with domestication. No quantitative study has been made of the impact of decentralized participatory domestication of the indigenous trees being cultivated in agroforestry systems. However, a domestication strategy has been published (Leakey and Akinnifesi, 2008) which emphasizes the need for long-term availability of genetically superior plants - seed sources and elite horticultural cultivars. Importantly, studies have identified very high proportion of the intraspecific tree-to-tree variation found in different populations (de Smeldt et al., 2011; Leakey et al., 2005a,b) and that the trees with superior traits are typically unrelated (Pauku et al., 2010). Thus, it seems that decentralized domestication should minimize the loss of genetic diversity. Nevertheless, it is important to ensure that communities engaged in domestication are aware of the risks and the need to maximize genetic diversity within their area.

4.4 Testing cultivars to meet market needs

A further set of issues relate to how best to exploit the commercial potential of currently underutilized tree species. As has been discussed, despite the progress in tree domestication described above, much remains to be done to underpin the science, and to test that the cultivars are delivering the expected benefits to remote communities. Many farmers in Cameroon have already created elite clonal cultivars of *Dacryodes edulis, Irvingia gabonensis, Ricinodendron heudelotii, Garcinia kola, Cola nitida, Allanblackia* spp. and so on, but none of these clones have been evaluated in formal field trials to determine if they meet either the selection criteria, or the ideotypes required by urban markets or processing industries. These tests are also required to provide the information needed to register the clones for the protection of intellectual property rights (Lombard and Leakey, 2010). A prerequisite of these evaluations, and subsequent marketing, is the rigorous identification of the clonal material by a name or number and the maintenance of information on its geographic origin and ownership.

The field testing of cultivars is critical to building on the earlier success of ICRAF's work in Cameroon and the intellectual, physical and social capital already invested in tree domestication. It is also essential for the expansion of the technology into a new generation of genetic traits of great importance for the marketability of AFTPs.

4.5 Post-harvest processing

Almost no work has been done to improve the shelf-life of perishable African fruits by developing post-harvest product processing and value addition by drying, bottling or canning. This is important as markets demand greater uniformity, quality and reliability of supply to make further progress in the value chain and as the legislation on food safety becomes stricter. It is for these reasons that the processes of domestication and commercialization should run in parallel (Leakey and van Damme, 2014). This processing also expands the nutritional benefits of these traditional foods to consumers throughout the year and allows their transportation to more remote markets. Year-round production would also benefit producers because of the higher prices they could charge for making produce available out of season. Opportunities also exist for more complex post-harvest processing to improve market appeal of fruit pulp, kernels or leaves. Fermentation, for example, has been a critical factor in the commercial success of coffee, tea and cocoa. Research on these possibilities should take advantage of multi-trait ideotypes capturing rare trait combinations present in individual trees within both wild or bred populations.

To ensure many of the above initiatives are implemented effectively, there is the hope that 'big business' will see the benefits to develop partnerships with local communities engaged in domestication and cottage industries in tropical and subtropical countries so that communities can benefit from the increased economic returns derived from value-addition (Lombard and Leakey, 2010). These companies will in turn derive benefits from a wider customer base and from new food and non-food products derived from appropriately protected traditional and cultural knowledge. There is growing recognition of the importance of wild foods in the diet, nutrition and health of the peoples of tropical and subtropical countries (Schnorr et al., 2014). Thus, unlike conventional staple foods, it seems likely that the decentralized domestication of a wide range of indigenous food species could have important implications for the dietary health of global consumers, and perhaps have positive impacts in current campaigns against obesity and food allergies.

4.6 Capacity building in applied science

Local communities need the resources and infrastructure to develop new crops and build on these market opportunities. As knowledge progresses from the ethnobotany of wild species to that of socially modified crops (Leakey, 2017b), there will be a need to greatly enhance the capacity of local communities to engage in tree domestication, agroforestry planting and management, marketing and use of microfinance, infrastructure development, community finance management and so on. Much of this has already been initiated by the development of Rural Resource Centres (Degrande et al., 2015). However, as these Centres spread, new training programmes and skills will be needed to ensure that the production of AFTPs effectively promotes income generation, rural and urban livelihoods, as well as the formation ecological niches in support of agroecosystem functions (Ong and Leakey, 1999). This should include compliance with government legislation on trade and product processing, intellectual property rights and genetic conservation (Leakey, 2017a; Leakey and Prabhu, 2017).

5 Global society: meeting the need for sustainable intensification

The UN Millennium and Sustainable Development Goals have set out targets for international development, while numerous international reports (MEA, GEO, IAASTD etc.) have called for more sustainable farming systems in the tropics and subtropics. However, a practical, appropriate and effective solution to food and nutritional insecurity has been elusive for about 100 years. Conventional thinking and policy has been based on the concept that what works in temperate latitudes must work in the tropics and subtropics. Unfortunately, this fails to recognize that the biophysical, social and economic conditions in industrialized countries dominant in temperate latitudes differ dramatically from the developing countries of the tropics and subtropics, especially in Africa. The result of this has been a two-tier and dysfunctional world exceeding its 'planetary boundaries' threatened by climate change (Campbell et al., 2017) and divided between the rich and the extremely poor (Fig. 2).

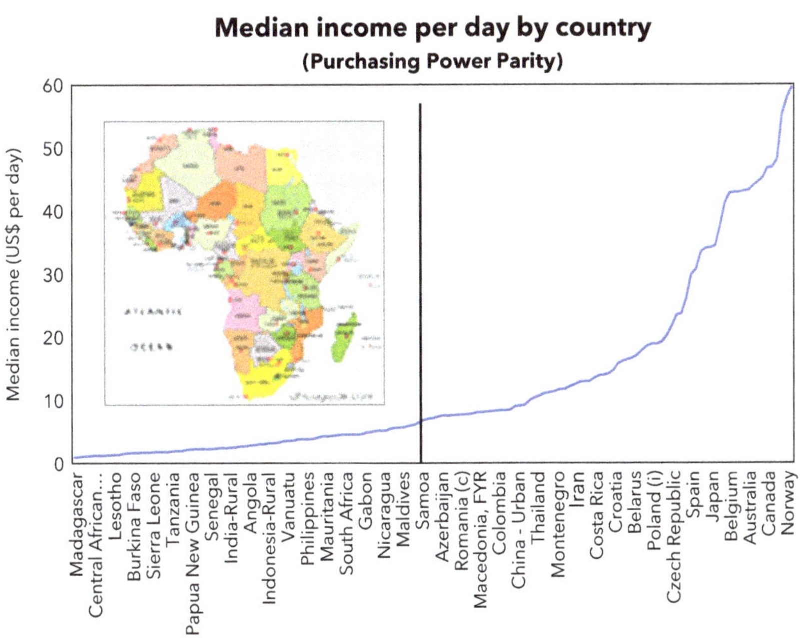

Figure 2 The median income per day (adjusted for purchasing power parity) of countries across the world in US$ per day (Diofasi and Birdsall, 2016). Note: All African countries fall below the median of these medians.

There are two critical issues here: firstly, hundreds of millions of households living on the brink of the cash economy depend on small farms (1–5 ha) to meet all their daily needs without any social support systems. Secondly, these people are unable to purchase the essential technical inputs necessary to cultivate monocultures of staple food crops. This results in deforestation and a downward spiral of declining soil fertility with the loss of agroecological functions and so to declining crop yields, hunger and malnutrition. This land degradation also traps these deprived rural households in poverty and social inequity (Leakey, 2012).

Part of this scenario is the view, held by many from industrialized countries, that trees are a hindrance to cultivation and should be removed. This view fails to recognize that trees are crucial environmentally as the keystone of mature ecosystems, and socially because many are the source of culturally important food and non-food products (www.mongabay.com). This leads to serious negative trade-offs impacting on local culture, human welfare and global society, and occurs despite the vastly greater collective knowledge in the population, perhaps because the knowledge of modern individuals tends to be limited to one or two facets of life (Harari, 2011). Following this train of thought, maybe part of the problem of modern agriculture arises from the exclusion of the broader wisdom of earlier generations living a close symbiotic relationship with their natural environment.

The integrated approach to rural development and agricultural intensification proposed in this chapter aims to harmonize the use of modern agrichemicals and genetics with traditional knowledge about tree products. However, in doing so, we must recognize that the public, and specifically the policy makers, donors and agribusiness companies, will not be on our wavelength. This means that more effective communication and better public relations are needed when seeking support for agroforestry. It is interesting, in this connection, that while modern conventional wisdom sees wild fruits and nuts as 'famine foods' of no relevance to life in the twenty-first century, recent evidence from Cameroon indicates that people in the peri-urban population have a strong affinity and desire for indigenous fruits, without any social stigma or shame attached to consumption of wild fruits (Tata epse Ngome, 2015). Furthermore, this study found that the intensification of agriculture on cleared forest lands was associated with hunger and poor diets and with negative impacts on nutritional security and health in the rural population (Tata epse Ngome, 2015).

The socially modified crops developed through multidisciplinary and participatory agroforestry research described in this chapter demonstrate an unconventional approach to agricultural intensification that addresses the environmental, social and economic problems that constrain the productivity of smallholder farming systems in the tropics with potential relevance to the rest of Africa (Leakey, 2017b). It is a low-cost approach which is both easy to apply and has rapid impact (Leakey and Prabhu, 2017). Furthermore, it appears to convert many of the negative trade-offs typically associated with agricultural intensification into positive 'trade-ons' (Leakey, 2018). The generic model has been tested with over 10 000 farmers in 500 communities (Leakey and Asaah, 2013) and could probably be equally well implemented more widely in areas with similar problems, especially as most locations have numerous local species which could be candidates for domestication. Capturing this opportunity would build on the lessons learned from the IAASTD and be in accordance with the 10 principles for sustainable agriculture and the 11 action-oriented targets presented by Leakey and Prabhu (2017). In addition, this upscaling should include capacity building in urban and industrialized societies so that policy makers and agribusiness understand how the needs of farmers and local entrepreneurs can also include better linkages between commerce, industry and international trade. This integrated and multifunctional approach to rural development may also make more effective use of development dollars than current monocultural approaches.

Upscaling agroforestry is about much more than just increasing crop yields. This approach to the sustainable intensification of agriculture simultaneously addresses the many causes of hunger, malnutrition and poverty by combining elements of agroecology, organic farming with conventional agrichemical inputs (Leakey, 2012, 2013), in numerous tree/crop configurations and densities at the plot level, and in landscape mosaics (Fig. 3). It involves the domestication of numerous tree species across sites with very different climates, soils, ethnic groups and indigenous tree species.

Upscaling agroforestry has immediate relevance to current political initiatives in Africa, such as the AFR100 initiative of the African Union to restore 100 million ha of deforested and degraded land (www.wri.org/our-work/project/AFR100/about-afr100). An initiative of this scale will require numerous stakeholder partnerships and alliances engaging the many NGOs and other agencies already active in agroforestry and related approaches to rural development across the continent of Africa (Carbon Farming/Climate Smart Agriculture/Conservation Agriculture/Ecoagriculture/Evergreen Agriculture/Fair Trade/Organic Agriculture/Permaculture etc.). To improve the coordination and common understanding of appropriate strategies and techniques, a new initiative aims to create an overarching alliance – the Agroforestry Alliance for Africa – AAfA) for upscaling sustainable approaches to land rehabilitation and creating more productive and wildlife-friendly farming systems that reduce the pressure on forest/nature reserves.

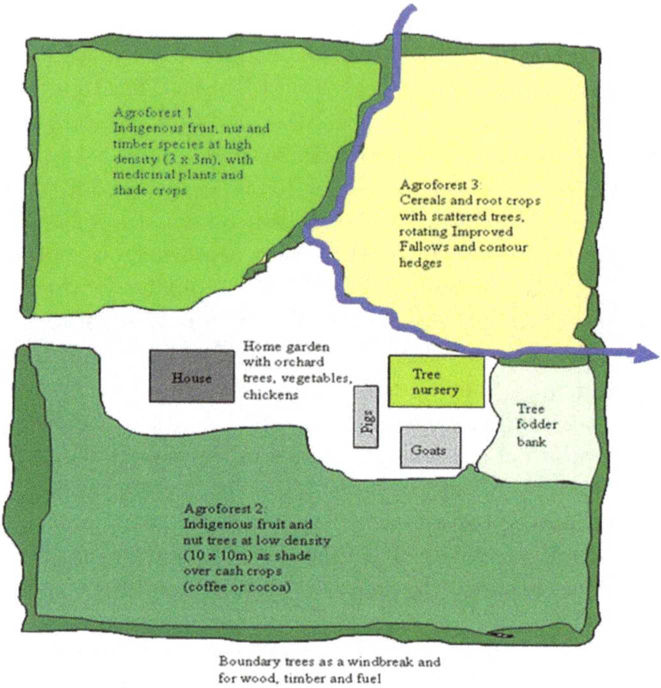

Figure 3 A diagrammatic representation of 'whole farm' agroforestry in which different practices have come together as an integrated approach to maximize economic, social and agroecological benefits (After: Leakey, 2017a).

5.1 Truly sustainable farming systems must avoid economic, social and environmental trade-offs

Conventional thinking accepts that even the sustainable intensification of agriculture often comes hand in hand with many inevitable trade-offs, so the idea that they can be avoided is novel. Quantifying the ability of agroforestry to increase crop production and have economic, social and environmental benefits depends on accurately assessing the impact of its upscaling. This should be based on data collected using randomized and replicated blocks and implemented in partnership with a wide range of NGOs and CBOs through Rural Resource Centres (Degrande et al., 2015).

Typically, tropical smallholders farm primarily for self-sufficiency. Consequently, assessments of economic and social impact should be made on wealth; education; nutritional and medical health; gender equity; land tenure; communal and cultural stability; community cohesion; employment; access to finance, communications, transport, markets and trading associations; and so on as inputs for a risk-averse 'home production' economic model for maximum production efficiency (Ellis and Allison, 2004).

Based on the experience with Rural Resource Centres in Cameroon (Degrande et al., 2015), it seems that smallholder farmers are willing to invest time and effort in appropriate and relevant practical innovations, especially when young people see an opportunity to avoid the need to migrate to big towns and cities for employment.

There are a number of ways of assessing environmental impact. To quantify the restoration of land degradation and ecosystem function, assessments should be made on soil fertility; biodiversity above and below ground; soil structure; carbon sequestration above and below ground; groundwater and freshwater resources; microclimate; and so on. It will also be important to assess differences in the loss of genetic diversity between 'decentralized' village-level and conventional 'centralized' domestication.

In addition to functional biodiversity, there is also great interest in the capacity of agroforestry to conserve threatened species (Perfecto and Vandermeer, 2008) – as an alternative to the creation of protected areas and nature reserves. Likewise, agroforestry can be an effective approach to soil protection and conservation (Saiz et al., 2016) and watershed management. A cost:benefit analysis could then determine the effectiveness of this approach to resolving Africa's environmental, social and economic development. By closing the 'yield gap' arising from conventional agriculture this study might determine how agroforestry can even enhance the returns on past investments in the Green Revolution.

Conventional wisdom suggests that 'trade-offs' between the intensification of agriculture for enhanced food production and its impacts on the environment are inevitable. However, recent evidence has recognized that human well-being and biodiversity can be synergistic (Bennett et al., 2015). However, rather than just accepting this synergy as a desirable indirect benefit, the approach advocated in this chapter suggests that 'trade-ons' achieved by direct action based on innovative and unconventional approaches to agricultural intensification should be deliberately sought as new policy interventions (Leakey, 2018). Research is however needed to further quantify the interactions between all the driving factors.

Unconventional approaches to agricultural intensification are particularly important in Africa. Africa continues to suffer from social and economic deprivation with its associated implications on illegal migration, civil disturbance and perhaps even terrorism. This is a stagnating factor in the global economy and also has serious impacts on the global environment, especially climate change. Much of this can be attributed to agriculture in the tropics and subtropics. A new political mind-set based on relieving the environmental, social and economic constraints to agricultural productivity in Africa could be based on deliberately seeking 'trade-ons', as mentioned above. This would stimulate the local economic development 'in-country', rather than supporting the export-oriented 'globalized trade' model, which in contrast enriches the economy of industrialized countries.

6 Conclusions

Unlike in the industrialized countries, agriculture has failed to be the engine of economic growth in Africa and some other parts of the tropics and subtropics due to failure to recognize the interacting social, economic and environmental that constrain productivity in smallholder farming systems on the brink of the cash economy. To address these constraints there is a need to harness and combine multidimensional environment and production benefits within a global intensification strategy that combines the keystone agroecological role of trees with the ability of local communities to generate income from new business and local industries based on the post-harvest processing, value-addition and trade from traditionally important but overlooked tree products.

It seems that the conventional philosophy of agricultural intensification has blinded policy and decision makers to the opportunities for sustainable intensification based on alternative cropping systems. Change from the current construct which has tried to squeeze tropical smallholders into economic theories designed for large, capital-intensive, businesses open to 'economies of scale' will only become a reality if policy and decision makers 'buy-in' to new concepts and approaches. This is despite numerous reports (e.g. MEA, 2005; IAASTD, 2009; Royal Society, 2009) stating that 'business as usual' cannot be the future of global agriculture.

A new policy mind-set should recognize that global sustainability has to embrace organisms at all trophic levels and scales in agroecological function. Thus, *Homo sapiens* cannot remain outside of the agroecosystem as an external force when intensifying agriculture. This has been recognized by Pimbert (2018) who links agroecology with food sovereignty and biocultural diversity. Meeting the needs of agriculture in this way sees that addressing the overexploitation of planetary boundaries must involve a new approach to intensification. One that combines the needs of people and the planet as suggested in this chapter, should boost food and nutritional security and bring all people around the world into the cash economy.

For this to happen, I suggest that a new international mind-set for tropical agriculture should be based on new socially modified tree crops integrated into smallholder farming systems for both agroecological and income generation. There are numerous, underutilized, 'Cinderella' tree species (Leakey and Newton, 1994) appropriate for domestication and commercialization in this way. This could stimulate local businesses and so raise the median per capita income in developing economies (Fig. 2), and also trigger vibrant new economies in Africa. Interestingly, there are already new food, nutraceutical and pharmaceutical products emerging from the increased awareness of the potential of tropical tree products, which could be expanded if more tropical trees were domesticated using ideotypes targeting new and emerging markets and industries (Fig. 4). This could also be expected to have 'spin-off'

Figure 4 Some of the 50 tree species being domesticated as new crops and some of the products being produced and marketed from new tree crops. Fruits: 1. *Dacryodes edulis*, 2. *Terminalia kaernbachii*, 3. *Allanblackia stuhlmannii*, 4. *Barringtonia procera*, 5. *Vitellaria paradoxa*, 6. *Sclerocarya birrea*, 7. *Adansonia digitata*. Products: 8. Mixed herbs, 9. *Paullinia cupana*, 10. *Bactris gasipaes* fruits, 11. *Bactris gasipaes* palm heart, 12. *Uncaria tomentosa*, 13. *Prunus africana*, 14–15. *Vitellaria paradoxa*, 16. *Santalum austrocaledonicum*, 17. *Prunus africana*, 18. *Pometia pinnata*, 19–20. *Sclerocarya birrea*, 21. *Paullinia cupana*, 22. *Canarium indicum*, 23. *Warburgia salutaris*.

benefits on food sovereignty and the protection of traditional knowledge and related culture.

The outcome from this vision could be a 'greener,' carbon-neutral and more sustainable world feeding and supporting a population well in excess of limits currently anticipated. In addition to addressing the major global issues of hunger, malnutrition and poverty, this could be expected to mitigate against the current levels of social disillusionment, resentment and jealousy that underpin the international security issues of illegal migration, human trafficking and terrorism. The realization of this dream is, however, very unlikely if we continue to plunder natural capital, destroy the local and global environment, and marginalize poor people living in areas where agriculture is failing to meet their social and economic needs.

In this direction, great progress has been made by agroforestry research over the last 25 years with the development and testing of a simple, practical, highly adaptable and generic, three-step model for multifunctional agriculture. This is best implemented in participatory mode by local communities and can reverse the downward spiral of the land degradation and social deprivation cycle, so closing the commonly found yield gap in modern staple food crops.

The approach proposed in this chapter brings together critical elements of modern tree science and the long-known traditional knowledge and culture of local people in a new and innovative approach to resolving many of the big global issues of modern society. Fundamentally these are agricultural reform issues with big environmental, social and economic implications for the planet and humanity which are not addressed by conventional wisdom (e.g. Long, 2015). Thus, it is suggested that this integrated and holistic approach is highly relevant to the simultaneous attainment of many of the 2030 Sustainable Development Goals.

7 References

Asaah, E. K., Tchoundjeu, Z., Leakey, R. R. B., et al. 2011a. Trees, agroforestry and multifunctional agriculture in Cameroon. *International Journal of Agricultural Sustainability* 9, 110–19.

Asaah, E. K., Tchoundjeu, Z., Ngahane, W., et al. 2011b. *Allanblackia floribunda* a new oil tree crop for Africa: Amenability to grafting. *New Forests* 41, 389-98.

Atangana, A. R., van der Vlis, E., Khasa, D. P., et al. 2011. Tree-to-tree variation in stearic and oleic acid content in seed fat from *Allanblackia floribunda* from wild stands: Potential for tree breeding. *Food Chemistry* 126: 1579-85.

Atangana, A. R., Khasa, D., Chang, S., et al. 2014. *Tropical Agroforestry*. Springer Science, Dordrecht, the Netherlands, 380pp.

Beech, E., Rivers, M., Oldfield, S., et al. 2017. GlobalTreeSearch: The first complete global database of tree species and country distributions. *Journal of Sustainable Forestry* 36: 454-89.

Bennett, E. M., Cramer, W., Begossi, A., et al. 2015. Linking biodiversity, ecosystem services and human wellbeing: Three challenges for designing research for sustainability. *Current Opinion in Environmental Sustainability* 14, 76-85.

Campbell, B. M., Beare, D. J., Bennett, E. M., et al. 2017. Agriculture production as a major driver of the earth system exceeding planetary boundaries. *Ecology and Society* 22(4): 8.

Cosyns, H., Degrande, A., de Wulf, R., et al. 2011. Can commercialization of NTFPs alleviate poverty? A case study of *Ricinodendron heudelottii* (Baill.) Pierre ex Pax. Kernel marketing in Cameroon. *Journal of Agriculture and Rural Development in the Tropics and Subtropics* 112: 45-56.

Cosyns, H., van Damme, P., de Wulf, R., et al. 2013. Can rural development projects generate social capital? A case study of *Ricinodendron heudelotii* kernel marketing in Cameroon. *Small-Scale Forestry* 13: 163-82.

Cunningham, A. B. 2001. *Applied Ethnobotany: People, Wild Plant Use and Conservation*. Earthscan Publications Ltd., London, UK, 300pp.

Degrande, A., Schreckenberg, K., Mbosso C., et al. 2006. Farmers' fruit tree growing strategies in the humid forest zone of Cameroon and Nigeria. *Agroforestry Systems* 67: 159-75.

Degrande, A., Franzel, S., Yeptiep, Y. S., et al. 2012. Effectiveness of grassroots organisations in the dissemination of agroforestry innovations. In: Kaonga, E. (Ed.), *Agroforestry for Biodiversity and Ecosystem Services - Science and Practice* (pp. 141-64). Elsevier.

Degrande, A., Gyau, A., Foundjem-Tita, D., et al. 2014. Improving smallholders' participation in tree product value chains: Experiences from the Congo Basin. *Forests, Trees and Livelihoods* 23: 102-15.

Degrande, A., Tchoundjeu, Z., Kwidja, A., et al. 2015. Rural resource centres: A community approach to extension. Note 10. In: *GFRAS Good Practice Notes for Extension and Advisory Services*. GFRAS, Lindau, Switzerland.

de Smedt, S., Alaerts, K., Kouyaté, A. M., et al. 2011. Phenotypic variation of baobab (*Adansonia digitata* L.) fruit traits in Mali. *Agroforestry Systems* 82: 87-97.

Dick, J. McP. and Leakey, R. R. B. 2006. Differentiation of the dynamic variables affecting rooting ability in juvenile and mature cuttings of cherry (*Prunus avium*). *Journal of Horticultural Science and Biotechnology* 81: 296-302.

Diofasi, A. and Birdsall, N. 2016. *The World Bank's Poverty Statistics Lack Median Income Data*. Center for Global Development, Washington DC. https://www.cgdev.org/blog

/world-bank-poverty-statistics-lack-median-income-data-so-we-filled-gap-ourselves-download-available

Ellis, F. and Allison, E. 2004. Livelihood diversification and natural resources access. *Livelihood Support Program Working Paper* 9. FAO, Rome, Italy.

Facheux, C., Tchoundjeu, Z., Foundjem, D., et al. 2006. From research to farmer enterprise development in Cameroon: Case study of kola nuts. *Acta Horticulturae (ISHS)* 699: 181-7.

Facheux, C., Tchoundjeu, Z., Foundjem-Tita, D., et al. 2007. Optimising the production and marketing of NTFPs. *African Crop Science Society Conference Proceedings* 8: 1248-54.

Facheux, C., Gyau, A., Foundjem-Tita, D., et al. 2012. Comparison of three modes of improving benefits to farmers within agroforestry product market chains in Cameroon. *African Journal of Agricultural Research* 7: 2336-43.

FAO. 1999. Women: Users, preservers and managers of agrobiodiversity. www.fao.org/FOCUS/E/Women/Biodiv-e.htm.

Foundjem-Tita, D., D'Haese, M., Degrande, A., et al. 2011. Farmers' satisfaction with group market arrangements as a measure of group market performance: A transaction cost analysis of non-timber forest products' producer groups in Cameroon. *Forest Policy and Economics* 13: 545-53.

Foundjem-Tita, D., Degrande, A., D'Haese, M., et al. 2012. Building long-term relationships between producers and trader groups in the non-timber forest product sector in Cameroon. *African Journal of Agricultural Research* 7: 230-9.

Foundjem-Tita, D., Tchoundjeu, Z., Speelman, S., et al. 2013. Policy and legal frameworks governing trees: Incentives or disincentives for smallholder tree planting decisions in Cameroon? *Small-scale Forestry* 12: 489-505.

Franzel, S., Jaenicke, H. and Janssen, W. 1996. Choosing the right trees: Setting priorities for multipurpose tree improvement. *ISNAR Research Report* 8. ISNAR, The Hague, the Netherlands, 87p.

Franzel, S., Degrande, A., Kiptot, E., et al. 2015. Note 7: Farmer-to-Farmer Extension. *GFRAS Good Practice Note for Extension and Advisory Services*. Global Forum for Rural Advisory Services, Lindau, Switzerland, 4p.

Garbach, K., Milder, J. C., Montenegro, M., et al. 2014. Biodiversity and ecosystem services in agroecosystems. In: van Alfen, N. (Editor-in-chief), *Encyclopedia of Agriculture and Food Systems*, Vol. 2 (pp. 21-40). Elsevier Publishers, San Diego, CA.

Gyau, A., Takoutsing, B., Degrande, A., et al. 2012. Producers' motivation for collective action for kola production and marketing in Cameroon. *Journal of Agriculture and Rural Development in the Tropics and Subtropics* 113: 43-50.

Gyau, A., Faith, A. N., Foundjem-Tita, D., et al. 2014. Small-holder farmers' access and rights to land of Njombé in the Littoral region of Cameroon. *Afrika Focus* 27: 23-39.

Harari, Y. N. 2011. *Sapiens - A Brief History of Humankind*. Penguin Random House, London, UK, 498pp.

Hobley, E., Garcia-Franco, N., Hübner, R., et al. 2017. Reviewing our options: Managing water-limited soils for conservation and restoration. *Land Degradation and Development* 29(4): 1041-53.

IAASTD. 2009. *Agriculture at a Crossroads: International Assessment of Agricultural Science and Technology for Development Global Report*. McIntyre, B. D., Herren, H. R., Wakhungu, J. and Watson, R. T. (Eds). Island Press, Washington DC, 590pp.

Jamnadass, R., Lowe, A. and Dawson, I. K. 2009. Molecular markers and the management of tropical trees: The case of indigenous fruits. *Tropical Plant Biology* 2: 1-12.

Ladipo, D. O., Leakey, R. R. B. and Grace, J. 1991. Clonal variation in a four-year-old plantation of *Triplochiton scleroxylon* K. Schum. and its relation to the predictive test for branching habit. *Silvae Genetica* 40: 130-5.

Ladipo, D. O., Leakey, R. R. B. and Grace, J. 1992. Variations in bud activity from decapitated, nursery-grown plants of *Triplochiton scleroxylon* in Nigeria: Effects of light, temperature and humidity. *Forest Ecology and Management* 50: 287-98.

Lavelle, P., Moreira, F. and Spain, A. 2014. Biodiversity: Conserving biodiversity in agroecosystems. In: van Alfen, N. (Editor-in-chief), *Encyclopedia of Agriculture and Food Systems*, Vol. 2 (pp. 41-60). Elsevier Publishers, San Diego, CA.

Leakey, D. G. B. 1949. Changes in systems of cultivation aimed at limiting soil degradation by development of the cultivation of perennial tree crops in the Central Province of Kenya Colony. *Bulletin Agricole du Congo Belge* 40: 2164-72.

Leakey, R. R. B. 1986. Prediction of branching habit of clonal *Triplochiton scleroxylon*. In: Tigerstedt, P. A., Puttonen, P. and Koski, V. (Eds), *Crop Physiology of Forest Trees* (pp. 71-80). University of Helsinki, Finland.

Leakey, R. R. B. 1990. *Nauclea diderrichii*: Rooting of stem cuttings, clonal variation in shoot dominance and branch plagiotropism. *Trees* 4: 164-9.

Leakey, R. R. B. 1996. Definition of agroforestry revisited. *Agroforestry Today* 8(1): 5-7.

Leakey, R. R. B. 1998. Agroforestry in the humid lowlands of West Africa: Some reflections on future directions for research. *Agroforestry Systems* 40: 253-62.

Leakey, R. R. B. 2004. Physiology of vegetative propagation in trees. In: Burley, J., Evans, J. and Youngquist, J. A. (Eds), *Encyclopedia of Forest Sciences* (pp. 1655-68). Academic Press, London, UK.

Leakey, R. R. B. 2012. *Living with the Trees of Life - Towards the Transformation of Tropical Agriculture*. CABI, Wallingford, UK, 200pp.

Leakey, R. R. B. 2013. Addressing the causes of land degradation, food/nutritional insecurity and poverty: A new approach to agricultural intensification in the tropics and sub-tropics. In: Hoffman, U. (Ed.), *Wake up before it is Too Late: Make Agriculture Truly Sustainable Now for Food Security in a Changing Climate* (pp. 192-8). UNCTAD Trade and Environment Review 2013. UN Publications, Geneva, Switzerland.

Leakey, R. R. B. 2014a. Agroforestry - Participatory domestication of trees. In: van Alfen, N. (Editor-in-chief), *Encyclopedia of Agriculture and Food Systems*, Vol. 1 (pp. 253-69). Elsevier Publishers, San Diego, CA.

Leakey, R. R. B. 2014b. Plant cloning: Macro-propagation. In: van Alfen, N. (Editor-in-chief), *Encyclopedia of Agriculture and Food Systems*, Vol. 4 (pp. 349-59). Elsevier Publishers, San Diego, CA.

Leakey, R. R. B. 2014c. The role of trees in agroecology and sustainable agriculture in the tropics. *Annual Review of Phytopathology* 52: 113-33.

Leakey, R. R. B. 2017a. *Multifunctional Agriculture: Achieving Sustainable Development in Africa*. Academic Press, San Diego, CA, 502pp.

Leakey, R. R. B. 2017b. Socially modified organisms in multifunctional agriculture - addressing the needs of smallholder farmers in Africa. *Archives of Crop Science* 1: 20-9

Leakey, R. R. B. 2017c. The role of trees in agroecology. In: Hunter, D., Guarino, L., Spillane, C., et al. (Eds), *Handbook of Agricultural Biodiversity* (pp. 238-52). Routledge Publishers, London, UK.

Leakey, R. R. B. 2018. Greatly enhanced food security in Africa without high-level trade-offs: Multiple environmental, economic and social benefits from 'Socially Modified Crops'. *Food Security* 10: 505-24.

Leakey, R. R. B. and Akinnifesi, F. K. 2008. Towards a domestication strategy for indigenous fruit trees in the tropics. In: Akinnifesi, F. K., Leakey, R. R. B., Ajayi, O. C., et al. (Eds), *Indigenous Fruit Trees in the Tropics: Domestication, Utilization and Commercialization* (pp. 28-49). CAB International, Wallingford, UK.

Leakey, R. R. B. and Asaah, E. K. 2013. Underutilised species as the backbone of multifunctional agriculture – The next wave of crop domestication. *Acta Horticulturae* 979, 293-310.

Leakey, R. R. B. and Newton, A. C. 1994. Domestication of 'Cinderella' species as the start of a woody-plant revolution. In: Leakey, R. R. B. and Newton, A. C. (Eds), *Tropical Trees: The Potential for Domestication and the Rebuilding of Forest Resources* (pp. 3-4). HMSO, London, UK.

Leakey, R. R. B. and Page, T. 2006. The 'ideotype concept' and its application to the selection of 'AFTP' cultivars. *Forests, Trees and Livelihoods* 16: 5-16.

Leakey, R. R. B. and Prabhu, R. 2017. Towards multifunctional agriculture - an African initiative. In: *Multifunctional Agriculture: Achieving Sustainable Development in Africa* (pp. 395-416). Academic Press, San Diego, CA.

Leakey, R. R. B. and Tomich, T. P. 1999. Domestication of tropical trees: From biology to economics and policy. In: *Agroforestry in Sustainable Ecosystems* (pp. 319-38). CRC Press and Lewis Publishers, New York, NY.

Leakey, R. R. B. and van Damme, P. 2014. The role of tree domestication in value chain development. *Forests, Trees and Livelihoods* 23: 116-26.

Leakey, R. R. B., Ferguson, N. R. and Longman, K. A. 1981. Precocious flowering and reproductive biology of *Triplochiton scleroxylon* K. Schum. *Commonwealth Forestry Review* 60: 117-26.

Leakey, R. R. B., Mesén, J. F., Tchoundjeu, Z., et al. 1990. Low-technology techniques for the vegetative propagation of tropical trees. *Commonwealth Forestry Review* 69: 247-57.

Leakey, R. R. B., Fondoun, J.-M., Atangana, A., et al. 2000. Quantitative descriptors of variation in the fruits and seeds of *Irvingia gabonensis*. *Agroforestry Systems* 50: 47-58.

Leakey, R. R. B., Schreckenberg, K. and Tchoundjeu, Z. 2003. The participatory domestication of West African indigenous fruits. *International Forestry Review* 5: 338-47.

Leakey, R. R. B., Shackleton, S. and du Plessis, P. 2005a. Domestication potential of Marula (*Sclerocarya birrea* subsp. *caffra*) in South Africa and Namibia: 1. Phenotypic variation in fruit traits. *Agroforestry Systems* 64: 25-35.

Leakey, R. R. B., Pate, K. and Lombard, C. 2005b. Domestication potential of Marula (*Sclerocarya birrea* subsp *caffra*) in South Africa and Namibia: 2. Phenotypic variation in nut and kernel traits. *Agroforestry Systems* 64: 37-49.

Leakey, R. R. B., Greenwell, P., Hall, M. N., et al. 2005c. Domestication of *Irvingia gabonensis*: 4. Tree-to-tree variation in food thickening properties and in fat and protein contents of dika nut. *Food Chemistry* 90: 365-78.

Leakey, R. R. B., Fuller, S., Treloar, T., et al. 2008. Characterization of tree-to-tree variation in morphological, nutritional and chemical properties of *Canarium indicum* nuts. *Agroforestry Systems* 73: 77-87.

Leakey, R. R. B., Weber, J. C., Page, T., et al. 2012. Tree domestication in agroforestry: Progress in the second decade. In: Nair, P. K. R. and Garrity, D. (Eds), *Agroforestry - The Future of Global Land Use* (pp. 145-73). Springer, Dordrecht, the Netherlands.

Lombard, C. and Leakey, R. R. B. 2010. Protecting the rights of farmers and communities while securing long term market access for producers of non-timber forest products: Experience in southern Africa. *Forests, Trees and Livelihoods* 19, 235–49.

Long, N. 2015. Activities, actants and actors: Theoretical perspectives on development practice and practitioners. In: Milone, P., Ventura, F. and Ye, J. (Eds), *Constructing a New Framework for Rural Development: Research in Rural Sociology*. Emerald Publishing, London, UK.

Longman, K. A. 1978. Control of flowering for forest tree improvement and seed production. *Scientific Horticulture* 30, 1–10.

Lowe, A. J., Gillies, A. C. M., Wilson, J., et al. 2000. Conservation genetics of bush mango from central/west Africa: Implications from random amplified polymorphic DNA analysis. *Molecular Ecology* 9, 831–41.

Maathai, W. 2009. *The Challenge for Africa*. Random House Inc., New York, NY.

MEA. 2005. *Ecosystems and Human Well-Being*. Island Press, Washington DC.

Mbile, P., Tchoundjeu, Z., Degrande, A., et al. 2004. Non-mist vegetative propagation by resource-poor, rural farmers of the forest zone of Cameroon: Some technology adaptations to enhance practice. *Forests, Trees and Livelihoods* 14: 43–52.

Mbosso, C., Degrande, A., Villamor, G. B., et al. 2015. Factors affecting the adoption of agricultural innovation: The case of *Ricinodendron heudelotii* kernel extraction machine in southern Cameroon. *Agroforestry Systems* 89: 799–811.

Mialoundama, F., Avana, M.-L., Youmbi, E., et al. 2002. Vegetative propagation of *Dacryodes edulis* (G. Don) H. J. Lam by marcots, cuttings and micropropagation. *Forests, Trees and Livelihoods* 12: 85–96.

Muchugi, A., Dawson, I. A. K. and Jamnadass, R. H. 2008. Molecular markers for tropical trees: A practical guide to principles and procedures. *ICRAF Technical Manual No. 9*. World Agroforestry Centre, Nairobi, Kenya.

Nair, P. K. R. and Garrity, D. (Eds). 2012. *Agroforestry - The Future of Global Land Use*. Springer, Dordrecht, the Netherlands.

Ngo Mpeck, M. L., Tchoundjeu, Z. and Asaah, E. 2003. Vegetative propagation of *Pausinystalia johimbe* (K. Schum) by leafy stem cuttings. *Propagation of Ornamental Plants* 3: 11–18.

Ofori, D. A., Asomaning, J. M., Peprah, T., et al. 2015. Addressing constraints in propagation of *Allanblackia* species through seed sectioning and air layering. *Journal of Experimental Biology and Agricultural Sciences* 3: 89–96.

Ong, C. K. and Leakey, R. R. B. 1999. Why tree-crop interactions in agroforestry appear at odds with tree-grass interactions in tropical savannahs, *Agroforestry Systems* 45: 109–29.

Page, T., Southwell, I., Russell, M., et al. 2010. Geographic and phenotypic variation in heartwood and essential oil characters in natural populations of *Santalum austrocaledonicum* in Vanuatu. *Chemistry and Biodiversity* 7: 1990–2006.

Pauku, R. L., Lowe, A. and Leakey, R. R. B. 2010. Domestication of indigenous fruit and nut trees for agroforestry in the Solomon Islands. *Forests, Trees and Livelihoods* 19: 269–87.

Perfecto, I. and Vandermeer, J. 2008. Biodiversity conservation and tropical agroecosystems: A new conservation paradigm. *Annals of New York Academy of Sciences* 1134: 173–200.

Pimbert, M. P. 2018. *Food Sovereignty, Agroecology and Biocultural Diversity - Constructing and Contesting Knowledge*. Routledge Publishers, London and New York, 338p.

Raintree, J. B. 1987. The state of the art of agroforestry diagnosis and design. *Agroforestry Systems* 5: 219–50.

Royal Society. 2009. *Reaping the Benefits: Science and the Sustainable Intensification of Global Agriculture* (Rep. 11/09 RS1608). Royal Society, London, UK.

Saiz, G., Wandera, F. W., Pelster, D. E., et al.2016. Long-term assessment of soil and water conservation measures (Fanya-juu terraces) on soil organic matter in South Eastern Kenya. *Geoderma* 274: 1–9.

Schnorr, S. L., Candela, M., Rampelli, S., et al. 2014. Gut microbiome of the Hadza hunter-gatherers. *Nature Communications* 5: 3654. doi:10.1038/ncomms4654.

Schreckenberg, K., Degrande, A., Mbosso, C., et al. 2002. The social and economic importance of Dacryodes edulis (G.Don) H. J. Lam in southern Cameroon. *Forests, Trees and Livelihoods* 12: 15–40.

Schreckenberg, K., Awono, A., Degrande, A., et al. 2006. Domesticating indigenous fruit trees as a contribution to poverty reduction. *Forests, Trees and Livelihoods* 16: 35–51.

Sileshi, G. W., Mafongoya, P., Akinnifesi, F. K., et al. 2014. Agroforestry: Fertilizer trees. In: Van Alfen, N. K. (Editor-in-chief), *Encyclopedia of Agriculture and Food Systems*, Vol. 1 (pp. 222–34). Elsevier, San Diego, CA.

Takoutsing, B., Tchoundjeu, Z., Degrande, A., et al. 2014. Scaling-up sustainable land management practices through the concept of the rural resource centre: Reconciling farmers' interests with research agendas. *Journal of Agricultural Education and Extension* 5: 463–83.

Tata epse Ngome, P. I. 2015. The contribution of fruit from trees to improve household food security in the context of deforestation in Cameroon. *PhD thesis*, Rhodes University, Grahamstown, 195pp.

Tchoundjeu, Z., Avana, M. L., Leakey, R. R. B., et al. 2002. Vegetative propagation of *Prunus africana*: Effects of rooting medium, auxin concentrations and leaf area. *Agroforestry Systems* 54: 183–92.

Tchoundjeu, Z., Tsobeng, A. C., Asaah, E., et al. 2010. Domestication of *Irvingia gabonensis* (Aubry Lecomte) by air layering. *Journal of Horticulture and Forestry* 2: 171–9.

Tsobeng, A., Ofori, D., Tchoundjeu, Z., et al. 2016. Improving growth of stockplants and rooting ability of leafy stem cuttings of *Allanblackia floribunda* Oliver (Clusiaceae) using different NPK fertilizers and periods of application. *New Forests* 47(2): 179–94.

Chapter 10

The role of biodiversity in agroforestry and other types of smallholder farming

Mary Ng'endo, World Agroforestry Centre (ICRAF), Kenya; and Shonil A. Bhagwat, The Open University, UK

1 Introduction

For the past 50 years, there has been a consistent increase in global supply (Khoury et al. 2014) and consumption of food from plant and animal sources (Kearney 2010). Despite the global agricultural surpluses, more than one billion people still remain hungry (Ponisio et al. 2015), with rising numbers facing micronutrient deficiencies (von Grebmer et al. 2014). This paradox is partially explained by agricultural growth that is increasingly accompanied by a narrowing of the existing diversity – factors that jeopardise food and nutrition security efforts (Khoury et al. 2014; Puma et al. 2015) leaving a billion people without adequate access to diverse healthy foods (Kremen et al. 2012). Globally, around 94 plant species account for 90% of the world's food crops (Khoury et al. 2014) and just three crops (maize, rice and wheat) account for close to 60% of the calories obtained from plants (Tilman et al. 2011). With the total livestock diversity of about 40 species also being eroded (Pattison et al. 2007), the over-reliance on only a handful of crops and even fewer animal species and breeds for a large proportion of people's calorific intake makes the global food system highly unsustainable in numerous ways. At the global level, the focus on reducing food insecurity in the developing world has been on increased crop

http://dx.doi.org/10.19103/AS.2018.0041.09

productivity, improvement of staples and other market-oriented crops (Thrupp 2000). However, locally, in many developing countries, there is an abundance of less well-known agricultural species currently underutilised (Gruere et al. 2006) but with significant potential to contribute to both food security and environmental sustainability.

The need to feed the world's growing population by 2030 in line with the 2030 UN Sustainable Development Goals brings into focus a threat of large-scale destruction of natural resources (Johns and Sthapit 2004) including biodiversity, in and around agricultural land (Thrupp 2000). The narrowing of diversity in crop species contributing to the world's food supplies has been considered a potential threat to food security (Khoury et al. 2014). The high vulnerability of the global food system makes it imperative to explore agricultural production alternatives to ensure food and nutrition security without causing further environmental degradation (Tscharntke et al. 2012). Agricultural biodiversity, commonly referred to as agrobiodiversity, is biological diversity in agricultural systems that is of relevance to food and agriculture (Frison et al. 2006; Baidu-Forson et al. 2012) from both cultivated and wild plant and livestock species (Powell et al. 2015). Hitherto neglected and underutilised, agrobiodiversity can provide alternatives to conventional agricultural crops. Conservation of agrobiodiversity can harness the available biological resources managed in agricultural landscapes (Kawa et al. 2015) to make food production more environmentally sustainable. It has been recognised specifically for its importance to smallholder farming communities in rural areas, not only to conserve important ecosystem services but also to increase productivity (Baidu-Forson et al. 2012). Agrobiodiversity includes not only a wide variety of species and genetic resources but also the many ways in which farmers can exploit biodiversity to produce and manage crops, with such practices serving as a basis for survival and livelihood for millions of people (Thrupp 2000). For many rural families, the sustainable use of local agrobiodiversity is their key to survival, since it allows them to exploit marginal lands and ensure a minimum level of food production even when faced with extremely harsh living conditions (Ekesa et al. 2009). There is a growing realisation worldwide that biodiversity is fundamental to agricultural production and food security, as well as a valuable ingredient of environmental conservation (Thrupp 2000). Through an extensive review of the existing literature, this chapter discusses the extent and usefulness of agrobiodiversity on smallholder farming systems. It relates this information to household food security outcomes.

2 The multiple dimensions of food security

The widely accepted definition of food security is 'physical and economic access by all people at all times to sufficient, safe and nutritious food for an active, healthy life' (FAO 1996). This definition emphasises the greater importance of increased food access over increased food production (Ingram 2011). It is acknowledged that sufficient global food production is not synonymous with guaranteed food security for all (Beddington 2011). While lack of or poor access to food is the main cause of food insecurity in general, some parts of the world, especially Africa, face chronic hunger mainly due to low food production (Ingram 2011). Although famines in Africa mainly persist because of under-production of food coupled with frequent and unpredictable production shocks, low food production played a minor role in the famines in Ethiopia, Niger and Malawi between 1998 and 2006. In Ethiopia, for example, famine in the lowland pastoral areas occurred in the face of sufficient food availability in other crop-producing highland areas (Devereux 2009), which is often the case in many other African countries. This suggests that relying solely on increased food production is insufficient to ensure food security for all (Misselhorn et al. 2012).

Hunger results from a failure in any of the three main interconnected food entitlement sources: own production, exchange (in-kind exchange or purchase) and household transfers (Sen 1981), with lack of economic and social access to food as the main cause (Ericksen et al. 2010; Garnett and Godfray 2012). There is also the need to support more than access to sufficient calories (Misselhorn et al. 2012). While the FAO food security definition also captures the nutritional dimension of food (Ericksen et al. 2010), food security as commonly used in the development discourse emphasises food quantity more than food quality, hence the term 'nutrition security' is now commonly used to capture the quality dimension (Smith et al. 2013). Stability is an equally important dimension of food security (Ericksen et al. 2010, Ingram 2011; Garnett and Godfray 2012), as all food security components have to be satisfied and stable over time for food security to be met (Ingram 2011). The multiple dimensions of food security necessitate research on a more integrated food security concept which focuses not only on food production (Ingram 2011), but also on other aspects such as food access, allocation and nutritional adequacy. As an outcome of multiple factors operating at household up to international levels, whether or not food security exists for any unit of analysis requires the consideration of multiple factors (Ericksen et al. 2010).

3 Food security in the context of global environment change: the food system concept

Though food systems and agricultural production are closely linked (Hawkesworth et al. 2010), enhancing food security needs to occur without further environmental degradation (Ericksen et al. 2010; Foley et al. 2011; Ingram 2011). This is especially so for low-income households who depend on food systems that are sensitive to changing climatic conditions (Vermeulen et al. 2012). Given the multifaceted nature of food security that increases its complexity, especially in the context of global environmental change (Ingram 2011), solving food insecurity needs to centre on a comprehension of complex interactions. System-based approaches can facilitate this understanding (Ericksen et al. 2010). Such approaches which address the multiple dimensions of food security on a single framework while at the same time considering its impacts on global environmental change should be considered holistic.

In the late 1990s, as research interest on interactions between global environmental change and food security increased (Ingram 2011), the Global Environmental Change and Food Systems (GECAFS) project was set out to research on strategies of enhancing food security without further environmental degradation (Ericksen et al. 2010). One of GECAFS's main research outputs, the food system concept, served the purpose of enhancing interdisciplinary research on the two-way interactions between food security efforts and global environmental change (Ericksen et al. 2010; Ingram 2011). Food systems entail a number of activities which give rise to a number of food security outcomes (Ingram 2011). Essentially, a food system includes the set of activities involved in producing, processing, packaging, distributing, retailing and consuming food (Ericksen et al. 2010). Undertaking these activities leads to a number of outcomes, which not only contribute to food security but also relate to environmental outcomes (Ingram 2011).

3.1 Application of the food system concept at the smallholder farming household level

The food system concept can be adapted to serve as a conceptual framework, as a rationale for considering household food security outcomes and as a determinant of household food security activities mediated by sustainable utilisation of an environmental resource, local agrobiodiversity (Fig. 1).

The food security outcome at the household level can be explained in terms of three components: food availability, accessibility and utilisation (Fig. 1). Food availability is the amount, type and quality of food that a household has to consume (Ericksen et al. 2010). Food availability, in terms of calories, proteins and micronutrients (Foley et al. 2011), can be achieved through local production as well as distribution channels and mechanisms to exchange money, labour or other items of value for food (Ericksen et al. 2010; Ingram 2011). In sub-Saharan Africa, the consistently large numbers of subsistence producers are evidencing the importance that they attach to such production, as it contributes directly to household food security as a supply of food (Aliber and Hart 2009). Secondly, access to food is the ability to obtain the type, quality and quantity of food required. This ability can be assessed by the affordability of food that is available and how well allocation mechanisms such as markets work (Ericksen et al. 2010; Ingram 2011). Rural farming households predominantly access their foods from subsistence production, markets and household transfers (Baiphethi and Jacobs 2009) such as gifts and food aid (Negatu 2004). Lastly, food utilisation refers to the ability to consume and benefit from food and one of the factors determining this is the nutritional value of food (Ericksen et al. 2010; Ingram 2011). Nutrition security is a function of both food and non-food factors (Klennert 2009), as it goes beyond food access to food consumption and dietary choices (Committee on World Food Security 2012; Acharya et al. 2014), coupled with health status determinants, such as access to adequate health services, sanitary environment as well as maternal/ childcare practices (Committee on World Food Security 2012). Food utilisation is best captured using household-level data (Remans et al. 2014), so as to allow a more accurate prediction of overall nutritional diversity, since both food and non-food consumption factors determine utilisation of food to ensure dietary quality is achieved.

The main household food system activities comprise of food production, mainly for food consumption but also for commercial purposes, where the surplus of products is distributed and retailed, for example in local markets (Fig. 1). Local agrobiodiversity, as a set of food system activities, can contribute to dietary improvement (through on-farm food diversity, dietary

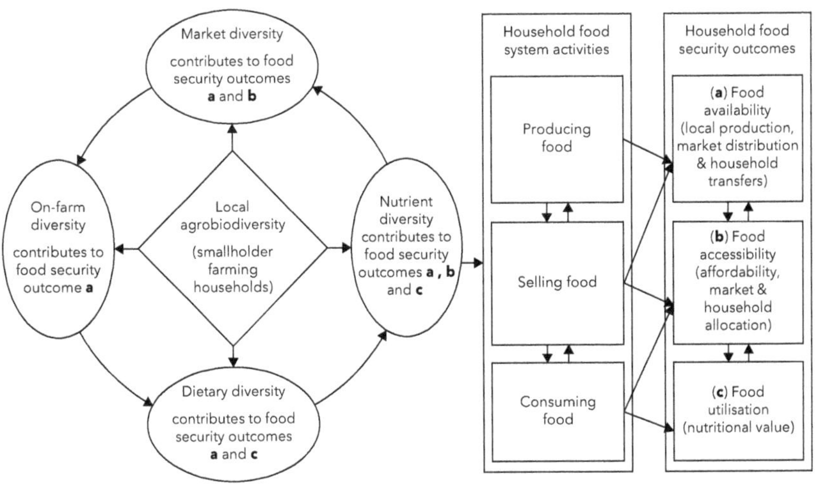

Figure 1 Conceptual framework investigating the economic and nutrition contribution of local agrobiodiversity among smallholder farmers (through market, on-farm, dietary and nutritional diversity) leading to household food system activities and ultimately household food security outcomes (food system activities and food security outcomes adapted from Ericksen et al. (2010)).

and nutrient diversity) and economic gain (through income generation) (Fig. 1). Dietary and economic improvement may be mutually reinforcing of local agrobiodiversity conservation (Johns and Sthapit 2004). This is because when people perceive that agrobiodiversity has greater value through positive impacts on both health and income, they are more likely to maintain and protect it (Frison et al. 2004). Healthy diets for populations depend on the availability and accessibility of a variety of plant and animal foods, and both of these resources are characteristic of traditional food systems where the diversity of indigenous crops and wild plant and animal species available in tropical countries provides essential nutrients (Johns and Sthapit 2004). Though the availability of processed calorie-rich foods at affordable prices contributes to food security and has generally reduced the number of undernourished people (Johns and Maundu 2006), agrobiodiversity plays a pivotal role since even if dietary energy needs can be met without dietary diversity, micronutrient needs cannot be met without diversity (Fanzo et al. 2011). Agrobiodiversity is also the basis of dietary diversity, the preferred choice for nutrition and health (Frison et al. 2004; Johns and Maundu 2006; Fanzo et al. 2011). Agrobiodiversity within food systems not only provides a wide and varied range of nutrient-rich foods and dietary components with important health properties, it is also a resource that is locally available (Fanzo et al. 2011), hence, quality and quantity of food need not be mutually exclusive (Johns and Sthapit 2004). In addition to its role on dietary improvement, linking agrobiodiversity and health can be a way to direct economic growth in a positive manner. Smallholder farmers and local communities can manage and use traditional agrobiodiversity to comparative economic advantage to improve their livelihoods by developing products that can be marketed to meet demands of local food cultures and offer nutritional and cultural benefits to both rural and urban consumers (Johns and Sthapit 2004; Johns and Maundu 2006). An additional route to economic growth is business and trade in agricultural produce, which provides income-generating opportunities that even those with limited land can take up (De Weerdt 2009).

Figure 1 thus provides a visual aid on the multiple contributions of local agrobiodiversity and how they map onto the conceptual framework in a cyclical fashion, eventually closing the 'local agrobiodiversity circle'. The complex relationship between the 'food system activities' and 'food security outcomes' is depicted by two-way arrows that show interdependence of the food security components (Fig. 1). Thus, food security outcomes-agrobiodiversity connections explained through multiple food system activities better illuminate the extent of food and nutrition security at household level. This visual representation also allows a clearer representation of the multidimensional nature of the local food system among smallholder farming households in a developing country context. As opposed to examining a few of the food system elements in isolation, a detailed understanding of the interconnected nature of food system components is necessary if agricultural research is to lead to holistic food and nutrition security (Noack and Pouw 2014; McDermott et al. 2015).

While agrobiodiversity is the central plank of the food system conceptual framework, it is important to note that agrobiodiversity is just one of the many food security strategies, and it has its limitations and a political dimension. Thus, this chapter as a whole offers a critical perspective on limited agrobiodiversity thinking, while also advocating greater attention for agrobiodiversity. Insights resulting from this multidimensional view offer both a strong agricultural perspective, as well as one that moves beyond the agricultural production dimension and takes on a food system view. Thus, this chapter contributes to the literature on agrobiodiversity and food security by showing that the multidimensional perspectives on agrobiodiversity need to not only build the food system concept but also highlight the need for taking the food system conceptual framework down to the local level and grounding it with empirical household data. The benefit of focusing on the household level is that it allows for one to effectively address one of the key gaps identified in literature about the present global food system, namely that there is a disconnect between the global understanding and the local potential (Cash et al. 2006) of attaining food and nutrition security.

4 Multifactor food security promotion

A multifactor approach is essential to foster agrobiodiversity and food security. Six recommendations for organisations and policy makers that emerge from these agrobiodiversity findings are as follows.

4.1 Support for smallholders

Organisations working on conservation of agrobiodiversity and plant genetic resources at the local level need to support smallholder farmers to concurrently diversify production into both high-value nutrient-dense crops as well as livestock species. One of the main forms of support needed by farmers as they transition into this high-risk venture is a regular contact point for advice and information tailored to smallholder farmer needs and constraints. The mainstream service providers, agricultural extension officers, belong in a rural development sector facing perennial budgetary constraints in many developing countries (Fanzo et al. 2015). For example, in Kenya, underfunded extension operations result in their limited reach, and smallholder farmers who are poor, with small farm sizes and living in remote areas, are often excluded (Muyanga and Jayne 2006). Despite this, alternative agents are emerging, such as private agricultural extension officers (Muyanga and Jayne 2006), agriculture-trained community members and nutrition-trained community health workers from the public sector, community-based organisations or non-governmental organisations (Fanzo et al. 2015). These service providers could be the determining change agents in moving the food and nutrition security agenda forward.

4.2 Discussions of food and nutrition security at local level

A possible strategy for organisations to integrate smallholder farmers better into the food system is by ensuring discussions on food and nutrition security happen more at the local level. This is because sociocultural and socio-economic factors shaping food consumption patterns may be more important in dictating a household's dietary quality than prescribing predetermined portfolios of nutrient diverse food crop and animal species for farmers to incorporate in their farming systems. Among the service providers are community-based personnel who possess various attributes that have enabled them to contribute positively to the food and nutrition security agenda. In addition to (sometimes) volunteering their services, they often have gained extensive cultural awareness, community trust and local reach (Fanzo et al. 2015), providing a vital skill set that organisations can tap into. So far, successful nutrition education interventions spearheaded by community health workers have resulted in increased nutritional knowledge among persons in charge of cooking food (usually mothers) and consumption of diversified foods by their children in various rural areas, including western Kenya (Waswa et al. 2015), western Uganda (Kabahenda et al. 2011) and southern Malawi (Hotz and Gibson 2005). Farmer-led agricultural initiatives in various countries in Africa, Asia and Latin America have also been found to be taken up by community members (Waters-Bayer et al. 2015). To increase the efficiency of these personnel, there is the need for organisations working with community service providers to target not only persons in charge of cooking but also extended family members (Fanzo et al. 2015; Waswa et al. 2015). These social networks are likely to provide supportive environments that catalyse sustained behaviour change, a crucial ingredient needed if interventions are to translate from knowledge transfer to long-term consumption of diversified foods (Waswa et al. 2015). Besides the need for balanced household targeting, there is an urgent need for these organisations to offer training that goes beyond technical and soft skills to an integrated training that enables the agriculture and health personnel to better understand each other's disciplinary approaches and 'speak a common language' (Fanzo et al. 2015; Lachat et al. 2015). This interdisciplinary understanding is likely to foster their advisory capacity and reduce duplication in their operations so that they can further extend their reach.

4.3 Encouraging longer-term commitments from institutions

Policy makers need to create incentives that influence decision-making to encourage longer-term commitments by institutions organising smallholder farmers to fill in the local market gaps and to add value to their surplus perishable products. The primary organisations championing this cause are the private sector and other development agencies, and efforts to stimulate their business activities as they support smallholder farmers could be a strategic win–win mechanism. For instance, while the institutional and policy environment for natural resource management in Kenya is generally encouraging (Yatich et al. 2007), in most cases the policies are either poorly implemented or not targeted to smallholder farmers (Ng'endo et al. 2013). When policies are explicitly designed to benefit smallholder farmers, they often remain poorly enforced, such as the initiative by the Kenyan government and the private sector to standardise Irish potato packaging that did not result in tangible benefits for smallholder farmers (Kaguongo et al. 2014). Besides poor targeting, implementation and enforcement, another major drawback in the existing policies identified by FARM-Africa, a development agency linking smallholder farmers to profitable markets in Kenya, Uganda and Tanzania, is multiple taxations at different government administrative levels (Nyang et al. 2010). As in Kenya, the local taxation regime in rural Uganda and Tanzania is such that trading of crops and livestock attracts taxes and fees by public agencies at different levels (Ellis and Bahiigwa 2003; Ellis and Mdoe 2003). The likelihood of multiple and incremental costs in a single transaction further discourages smallholder farmers from selling food crops and livestock in formal markets (Ellis and Bahiigwa 2003). However, organised and trained smallholder farmer groups may have better capacity to influence government policies. For example, a coffee farmer association in Tanzania, initiated by FARM-Africa, is actively lobbying with central and local governments to harmonise coffee tax and cess (Nyang et al. 2010).

4.4 Role of government

With the proliferation of private and non-governmental organisations' initiatives offering support to rural smallholder farmers, sometimes with contrasting interests and power dynamics, the strength of government in designing and implementing protective policies that improve transparency and capacity in contractual agreements and entrepreneurial support could enable risk-averse smallholder farmers to negotiate profitable commercialised production systems (Kilelu et al. 2013). Integrating smallholder farmers into value chains for non-traditional cash crops such as fruits and vegetables will not only increase household incomes, but also contribute to an improved supply of these nutrient-dense foods at local markets, which are an important food procurement source for rural households.

4.5 Stimulating high demand for diverse local foods

There remains a need to stimulate high demand for diverse local foods of high nutritional value so that even if they are not necessarily desirable by rural households in a locality, they could still be sold in other rural and urban markets. Nutrition education may lead to a better understanding of nutritional value of locally available foods, especially niche on-farm products, which could in turn lead to better use of agrobiodiversity, and potentially to better markets as well. Nutrition education could be initiated by formal and semi-formal organised smallholder farmers interested in promoting specific niche products, and it is likely that once they start the nutrition education and marketing process from the local level, other interested stakeholders are more likely to support in building market integration for smallholders based on their niche products. The media is also a potential support system for the widespread dissemination of nutrition education and they may be more willing to assist proactive smallholder farmers. In Kenya, for example, the media is increasingly supportive of both group and individual smallholder farmers in a range of ways, from featuring stories of smallholder farmers who have succeeded as entrepreneurs based on their niche products (Chao 2013) to airing weekly episodes of smallholder farmer visits that have improved the farmers' production, value-addition and marketing potential (Shamba Shape Up 2015).

4.6 Risks of building new markets for diverse niche products

There remains a considerable risk of building new markets for diverse niche products. As an illustration, this has been the case for Kenyan farmers rearing quails, small domestic birds considered to have more nutrient-dense eggs as compared to chicken eggs. As the awareness of their nutritional benefits rapidly spread across the country, the small-sized quail eggs retailed at several times higher than the larger-sized chicken eggs. Subsequently, quail farming sharply picked up in early 2014 as more farmers engaged in the lucrative business (Xinhua 2014), only for the prices to plummet soon afterwards as the quail market saturated (Murumba 2014). While this was a part of the normal business cycle governed by the supply and demand principles, the sharp price fall left many start-up farmers with immense financial losses (Outdoors Africa 2014). Needless to say, the repercussions must have been worse for smallholder farmers with no back-up financial strategies. Thus, while better markets for nutritious products may be created, compensation mechanisms need to be put in place for farmers to buffer the imminent market shocks. An example of a market price guarantee agricultural insurance scheme for smallholder farmers that has been successfully implemented by the Brazilian government is the Family Farming Price Guarantee Programme (FAO 2015). However, in sub-Saharan Africa, agricultural insurance is rare (FAO 2015), thus posing an immense challenge to the use of diverse local agrobiodiversity for better markets in the long-term. There are also other high risks that apply to local products that have already penetrated different markets. For example, despite the high market demand for indigenous vegetables in Kenya (Gotor and Irungu 2010; Muhanji et al. 2011; Chege et al. 2015), these vegetable smallholder farmers still face exploitation by middlemen and lack access to centralised market information (Muhanji et al. 2011), issues that could be best tackled not only by the creation of the financially challenging compensation schemes but also by implementation and enforcement of existing government policies. Hence, by creating an enabling environment for smallholder farmers to take up opportunities to sell their products in markets, the incomes raised could contribute to meeting high input costs as well as having extra money to buy other produce that is available in markets year-round but at different accessibility levels, contributing to enhanced food and nutrition security, income generation and to more sustainable and resilient food systems.

5 Conclusion

In conclusion, in the coming decades improving food security, reducing poverty and improving environmental sustainability will be closely linked to the development of strategies that are relevant and appropriate for smallholder farmers (FAO and PAR 2011) in light of UN Sustainable Development Goals for 2030. To this end, enhanced sustainable utilisation of the most readily available resource, local agrobiodiversity, can have positive benefits by both enhancing the environmental sustainability of agricultural land and improving the livelihoods of smallholder farmers.

6 Where to look for further information

Two of the books providing a standard introduction to the subject include:

- Carolan, M. (2012). *The Sociology of Food and Agriculture*. Available from: https://www.crcpress.com/The-Sociology-of-Food-and-Agriculture/Carolan/p/book/9781138946255
- Ingram, J., P. Ericksen and D. Liverman (Eds) (2010). *Food Security and Global Environmental Change*. London: Earthscan. https://cgspace.cgiar.org/handle/10568/2671

Two of the centres of expertise on the subject are:

- Food and Agriculture Organization of the United Nations (FAO), including their annual state of food and agriculture publication series (http://www.fao.org/publications/sofa/the-state-of-food-and-agriculture/en/)
- The Consultative Group of International Agricultural Research (CGIAR) centres: https://www.cgiar.org/

7 References

Acharya, T., J. Fanzo, D. Gustafson, J. Ingram and B. Schneeman (2014). Assessing sustainable nutrition security: The role of food systems. Washington DC: The International Life Sciences Institute, Research Foundation, Center for Integrated Modeling of Sustainable Agriculture and Nutrition.

Aliber, M. and T. G. Hart (2009). Should subsistence agriculture be supported as a strategy to address rural food insecurity? *Agrekon* 48(4): 434–58.

Baidu-Forson, J. J., T. Hodgkin and M. Jones (2012). Introduction to special issue on agricultural biodiversity, ecosystems and environment linkages in Africa. *Agriculture, Ecosystems and Environment* 157(1): 1–4.

Baiphethi, M. N. and P. T. Jacobs (2009). The contribution of subsistence farming to food security in South Africa. *Agrekon* 48(4): 459–82.

Beddington, J. R. (2011). The future of food and farming: Challenges and choices for global sustainability. Final project report of the UK Government foresight global food and farming futures, Government Office for Science.

Cash, D. W., W. N. Adger, F. Berkes, P. Garden, L. Lebel, P. Olsson, L. Pritchard and O. Young (2006). Scale and cross-scale dynamics: Governance and information in a multilevel world. *Ecology and Society* 11(2): 8.

Chao, S. (2013). Rabbit keeper makes fast cash from her dream job. *Business Daily*, 26 August. Available from: https://www.businessdailyafrica.com/Rabbit-keeper-makes-fast-cash-from-her-dream-job/-/1248928/1968766/-/item/1/-/mqda81z/-/index.html (Accessed 24 November 2015).

Chege, C. G. K., C. I. M. Andersson and M. Qaim (2015). Impacts of supermarkets on farm household nutrition in Kenya. *World Development* 72(1): 394–407.

Committee on World Food Security (2012). Coming to terms with terminology: Food security, nutrition security, food security and nutrition, food and nutrition security. Rome: FAO. 39th session.

Devereux, S. (2009). Why does famine persist in Africa? *Food Security* 1(1): 25–35.

De Weerdt, J. (2009). Moving out of poverty in Tanzania: Evidence from Kagera. *The Journal of Development Studies* 46(2): 331–49.

Ekesa, B., M. Walingo and M. Onyango (2009). Accessibility to and consumption of indigenous vegetables and fruits by rural households in Matungu division, western Kenya. *African Journal of Food, Agriculture, Nutrition and Development* 9(8): 1725–38.

Ellis, F. and G. Bahiigwa (2003). Livelihoods and rural poverty reduction in Uganda. *World Development* 31(6): 997–1013.

Ellis, F. and N. Mdoe (2003). Livelihoods and rural poverty reduction in Tanzania. *World Development* 31(8): 1367–84.

Ericksen, P., B. Stewart, J. Dixon, D. Barling, P. Loring, M. Anderson and J. Ingram (2010). The value of a food system approach. *In* J. Ingram, P. Ericksen and D. Liverman (Eds), *Food Security and Global Environmental Change*. London: Earthscan, pp. 25–45.

Fanzo, J., M. Holmes, P. Junega, E. Musinguzi, I. F. Smith, B. Ekesa and N. Bergamini (2011). Improving nutrition with agricultural biodiversity: A manual on implementing food systems field projects to assess and improve dietary diversity, nutrition and health outcomes. Rome: Bioversity International.

Fanzo, J., Q. Marshall, D. Dobermann, J. Wong, R. I. Merchan, M. I. Jaber, A. Souza, N. Verjee and K. Davis (2015). Integration of nutrition into extension and advisory services: A synthesis of experiences, lessons, and recommendations. *Food and Nutrition Bulletin* 36(2): 120–37.

FAO (Food and Agriculture Organization of the United Nations) (1996). Rome declaration and world food summit plan of action. Rome: FAO.

FAO (Food and Agriculture Organization of the United Nations) (2015). Social protection and agriculture: Breaking the cycle of rural poverty. Rome: FAO. The State of Food and Agriculture 2015.

FAO (Food and Agriculture Organization of the United Nations) and PAR (Platform on Agrobiodiversity Research) (2011). Biodiversity for food and agriculture: Contributing to food security and sustainability in a changing world. G. Galluzzi, C. van Duijvendijk, L. Collette, N. Azzu and T. Hodgkin. Rome: FAO and PAR.

Foley, J. A., N. Ramankutty, K. A. Brauman, E. S. Cassidy, J. S. Gerber, M. Johnston, N. D. Mueller, C. O'Connell, D. K. Ray and P. C. West (2011). Solutions for a cultivated planet. *Nature* 478(7369): 337–42.

Frison, E. A., J. Cherfas, P. B. Eyzaguirre and T. Johns (2004). Biodiversity, nutrition and health: Making a difference to hunger and conservation in the developing world. *Seventh Meeting of the COP 7 (Conference of the Parties to the Convention on Biological Diversity)*. Kuala Lumpur, 9–20 February 2004.

Frison, E. A., I. F. Smith, T. Johns, J. Cherfas and P. B. Eyzaguirre (2006). Agricultural biodiversity, nutrition, and health: Making a difference to hunger and nutrition in the developing world. *Food and Nutrition Bulletin* 27(2): 167–79.

Garnett, T. and C. Godfray (2012). Sustainable intensification in agriculture. Navigating a course through competing food system priorities. Oxford: Food Climate Research Network and the Oxford Martin Programme on the Future of Food.

Gotor, E. and C. Irungu (2010). The impact of Bioversity International's African leafy vegetables programme in Kenya. *Impact Assessment and Project Appraisal* 28(1): 41–55.

Gruere, G., A. Giuliani and M. Smale (2006). Marketing underutilized plant species for the benefit of the poor: A conceptual framework. Washington DC: IFPRI (International Food Policy Research Institute). EPT (Environment and Production Technology Division) Discussion Paper 154.

Hawkesworth, S., A. D. Dangour, D. Johnston, K. Lock, N. Poole, J. Rushton, R. Uauy and J. Waage (2010). Feeding the world healthily: The challenge of measuring the effects of agriculture on health. *Philosophical Transactions of the Royal Society B: Biological Sciences* 365(1554): 3083–97.

Hotz, C. and R. S. Gibson (2005). Participatory nutrition education and adoption of new feeding practices are associated with improved adequacy of complementary diets among rural Malawian children: A pilot study. *European Journal of Clinical Nutrition* 59(2): 226–37.

Ingram, J. (2011). A food systems approach to researching food security and its interactions with global environmental change. *Food Security* 3(4): 417–31.

Johns, T. and P. Maundu (2006). Forest biodiversity, nutrition and population health in market-oriented food systems. *Unasylva* 57(2): 34–40.

Johns, T. and B. R. Sthapit (2004). Biocultural diversity in the sustainability of developing-country food systems. *Food and Nutrition Bulletin* 25(2): 143–55.

Kabahenda, M., R. M. Mullis, J. G. Erhardt, C. Northrop-Clewes and S. Y. Nickols (2011). Nutrition education to improve dietary intake and micronutrient nutriture among children in less-resourced areas: A randomised controlled intervention in Kabarole district, western Uganda. *South African Journal of Clinical Nutrition* 24(2): 83–8.

Kaguongo, W., G. Maingi and S. Giencke (2014). Post-harvest losses in potato value chains in Kenya: Analysis and recommendations for reduction strategies. Bonn and Eschborn: Deutsche Gesellschaft für Internationale Zusammenarbeit (GIZ).

Kawa, N. C., J. A. C. Michelangeli and C. R. Clement (2015). Household agrobiodiversity management on amazonian dark earths, oxisols, and floodplain soils on the lower Madeira river, Brazil. *Human Ecology* 43(2): 339–53.

Kearney, J. (2010). Food consumption trends and drivers. *Philosophical Transactions of the Royal Society B: Biological Sciences* 365(1554): 2793–807.

Khoury, C. K., A. D. Bjorkman, H. Dempewolf, J. Ramirez-Villegas, L. Guarino, A. Jarvis, L. H. Rieseberg and P. C. Struik (2014). Increasing homogeneity in global food supplies and the implications for food security. *Proceedings of the National Academy of Sciences* 111(11): 4001–6.

Kilelu, C. W., L. Klerkx and C. Leeuwis (2013). How dynamics of learning are linked to innovation support services: Insights from a smallholder commercialization project in Kenya. *The Journal of Agricultural Education and Extension* 20(2): 213–32.

Klennert, K. (2009). Achieving food and nutrition security: Actions to meet the global challenge. Feldafing: InWEnt (Internationale Weiterbildung und Entwicklung GmbH).

Kremen, C., A. Iles and C. Bacon (2012). Diversified farming systems: An agroecological, systems-based alternative to modern industrial agriculture. *Ecology and Society* 17(4): 44.

Lachat, C., E. Nago, A. Ka, H. Vermeylen, J. Fanzo, L. Mahy, M. Wustefeld and P. Kolsteren (2015). Landscape analysis of nutrition-sensitive agriculture policy development in Senegal. *Food and Nutrition Bulletin* 36(2): 154–66.

McDermott, J., N. Johnson, S. Kadiyala, G. Kennedy and A. J. Wyatt (2015). Agricultural research for nutrition outcomes – rethinking the agenda. *Food Security* 7(3): 593–607.

Misselhorn, A., P. Aggarwal, P. Ericksen, P. Gregory, L. Horn-Phathanothai, J. Ingram and K. Wiebe (2012). A vision for attaining food security. *Current Opinion in Environmental Sustainability* 4(1): 7–17.

Muhanji, G., R. L. Roothaert, C. Webo and M. Stanley (2011). African indigenous vegetable enterprises and market access for small-scale farmers in East Africa. *International Journal of Agricultural Sustainability* 9(1): 194–202.

Murumba, S. (2014). Quail farmers facing challenges as market saturates. *Daily Nation*, 24 January. Retrieved 24 November 2015, from: https://www.nation.co.ke/counties/nyeri/Quail-farmers-challenges/1954190-2158522-kxuv0k/index.html

Muyanga, M. and T. S. Jayne (2006). Agricultural extension in Kenya: Practice and policy lessons. Nairobi: Tegemeo Institute of Agricultural Policy and Development. Working Paper 26.

Negatu, W. (2004). Reasons for food insecurity of farm households in south Wollo, Ethiopia: Explanations at grassroots. Wisconsin, MI: BASIS CRSP (Broadening Access and Strengthening Input Market Systems Collaborative Research Support Group).

Ng'endo, M., D. Catacutan, J. Kung'u, J. Muriuki, J. Kariuki and J. Mowo (2013). The policy environment of Conservation Agriculture With Trees (CAWT) in eastern Kenya: Do small scale farmers benefit from existing policy incentives? *African Journal of Agricultural Research* 8(23): 2924–39.

Noack, A. L. and N. R. M. Pouw (2014). A blind spot in food and nutrition security: Where culture and social change shape the local food plate. *Agriculture and Human Values* 32(2): 169–82.

Nyang, M. N., C. Webo and R. L. Roothaert (2010). The power of farmers organisations in smallholder agriculture in East Africa: A review of 5 project initiatives of the maendeleo agricultural technology fund. London: FARM-Africa. Working Paper 13.

Outdoors Africa (2014). What quail farmers have taught start-ups about business cycles. Retrieved 21 November 2015, from http://www.outdoorsafrica.co.ke/what-quail-farmers-have-taught-start-ups-about-business-cycles/

Pattison, J., A. Drucker and S. Anderson (2007). The cost of conserving livestock diversity? Incentive measures and conservation options for maintaining indigenous pelón pigs in Yucatan, Mexico. *Tropical Animal Health and Production* 39(5): 339–53.

Ponisio, L. C., L. K. M'Gonigle, K. C. Mace, J. Palomino, P. de Valpine and C. Kremen (2015). Diversification practices reduce organic to conventional yield gap. *Proceedings of the Royal Society B: Biological Sciences* 282(1799): 20141396.

Powell, B., S. H. Thilsted, A. Ickowitz, C. Termote, T. Sunderland and A. Herforth (2015). Improving diets with wild and cultivated biodiversity from across the landscape. *Food Security* 7(3): 535–54.

Puma, M. J., S. Bose, S. Y. Chon and B. I. Cook (2015). Assessing the evolving fragility of the global food system. *Environmental Research Letters* 10(2): 024007.

Remans, R., S. A. Wood, N. Saha, T. L. Anderman and R. S. DeFries (2014). Measuring nutritional diversity of national food supplies. *Global Food Security* 3(3): 174–82.

Sen, A. (1981). *Poverty and Famines: An Essay on Entitlement and Deprivation*. Oxford, Oxford University Press.

Shamba Shape Up (2015). Shamba Shape Up: Practical advice on improving your shamba. Retrieved 18 November 2015, from http://www.shambashapeup.com/.

Smith, J., K. Sones, D. Grace, S. MacMillan, S. Tarawali and M. Herrero (2013). Beyond milk, meat, and eggs: Role of livestock in food and nutrition security. *Animal Frontiers* 3(1): 6–13.

Thrupp, L. A. (2000). Linking agricultural biodiversity and food security: The valuable role of agrobiodiversity for sustainable agriculture. *International Affairs* 76(2): 265–81.

Tilman, D., C. Balzer, J. Hill and B. L. Befort (2011). Global food demand and the sustainable intensification of agriculture. *Proceedings of the National Academy of Sciences* 108(50): 20260–4.

Tscharntke, T., Y. Clough, T. C. Wanger, L. Jackson, I. Motzke, I. Perfecto, J. Vandermeer and A. Whitbread (2012). Global food security, biodiversity conservation and the future of agricultural intensification. *Biological Conservation* 151(1): 53-9.

Vermeulen, S. J., B. M. Campbell and J. S. Ingram (2012). Climate change and food systems. *Annual Review of Environment and Resources* 37(1): 195–222.

von Grebmer, K., A. Saltzman, E. Birol, D. Wiesmann, N. Prasai, S. Yin, Y. Yohannes, P. Menon, J. Thompson and A. Sonntag (2014). 2014 Global hunger Index: The challenge of hidden hunger. Bonn, Washington DC and Dublin: Welthungerhilfe, International Food Policy Research Institute and Concern Worldwide.

Waswa, L. M., I. Jordan, J. Herrmann, M. B. Krawinkel and G. B. Keding (2015). Community-based educational intervention improved the diversity of complementary diets in western Kenya: Results from a randomized controlled trial. *Public Health Nutrition* 18(18): 3406–19.

Waters-Bayer, A., P. Kristjanson, C. Wettasinha, L. van Veldhuizen, G. Quiroga, K. Swaans and B. Douthwaite (2015). Exploring the impact of farmer-led research supported by civil society organisations. *Agriculture & Food Security* 4(4): 1–7.

Xinhua (2014). Craze for quail farming hits Kenya. *Standard Digital*, 15 January. Retrieved 24 November 2015, from: https://www.standardmedia.co.ke/business/article/2000102356/craze-for-quail-farming-hits-kenya

Yatich, T., A. Awiti, E. Nyukuri, J. Mutua, A. Kyalo, J. Tanui and D. Catacutan (2007). Policy and institutional context for NRM in Kenya: Challenges and opportunities for landcare. Nairobi: ICRAF. ICRAF (World Agroforestry Centre) Working Paper 43.

Chapter 11

Agroforestry: a system for improving soil health

S. H. Anderson and R. P. Udawatta, University of Missouri, USA

1 Introduction

Soil health or soil quality has received renewed interest in recent years as land managers have become aware of the importance of the resilience of the soil for future sustainability and for long-term human survival. Soil health is critical for plant and animal production, ecosystem services delivery and human health/well-being. Long-term sustainable soil use is vital for human health as these systems can suppress disease-causing soil organisms and influence the quantity and quality of the food we consume, the air we breathe and the water we use (Wall et al., 2015).

Definitions of soil quality or soil health are varied, but most include the soil's ability to function in sustaining plant and animal productivity as well as maintain environmental quality. The formal definition of soil quality from the Soil Science Society of America is 'the capacity of a soil to function within ecosystem boundaries to sustain biological productivity, maintain environmental quality, and promote plant and animal health' (Soil Science Society of America, 2008). The USDA-NRCS defines soil quality or soil health as 'the capacity of a specific kind of soil to function, within natural or managed ecosystem boundaries, to sustain plant and animal productivity, maintain or enhance water and air quality, and support human health and habitation. In short, the capacity of the soil to function' (USDA-NRCS, 2017). Most scientists' efforts to define soil

http://dx.doi.org/10.19103/AS.2018.0041.10

health include the use of a suite of properties including biological, chemical and physical to assess the current state of the soil and methods to improve its performance.

Agroforestry can play a significant role in improving soil health for future sustainable production and ecosystem services (Gold and Hanover, 1987). Agroforestry's role in improving soil properties on the landscape for agroecosystems can influence soil health. Sustaining future plant and animal production is critical for the future of humans due to anticipated population growth (Godfray et al., 2010). Agroforestry will play a vital role in sustaining this future by improving soil health.

Soil and nutrient loss is considered a major challenge for maintaining agricultural productivity (Lal, 2001; Yost et al., 2016; Conway et al., 2017). Some land managers view that restoring native perennial plants to the landscape will restore the soil to a healthy state; however, significant soil erosion and, thus, the loss of valuable topsoil prevent the soil from returning to its native condition (Chandrasoma et al., 2016). This issue must be considered in evaluating soil health. Increased demand for food, fibre and fuel will necessitate maintaining and improving productivity and is a significant challenge for agriculture (Godfray et al., 2010; McLaughlin and Kinzelbach, 2015).

Future sustainable production requires conservation of natural resources in order to prevent the loss of productive soil. Agroforestry buffers have been shown to be a significant conservation tool (Akdemir et al., 2016). Conservation buffers, including agroforestry buffers, are considered a good strategy to reduce water runoff and associated soil erosion (Evans and Sadler, 2008).

One significant practice for conserving soil and water resources which deserves wide application is conservation buffer systems. These buffer approaches are utilized to improve agricultural production and protect soil and water quality (Lowrance et al., 2002). Soils susceptible to erosion can be benefitted by these conservation buffers which include any species of woody perennials (trees or shrubs) or grass grown to decrease soil erosion and improve water quality; these buffers also enhance water infiltration and diversify the landscape (Kumar et al., 2012).

A land management system which utilizes trees and/or shrubs as well as traditional row crop and pasture crops simultaneously in the landscape for economic and environmental benefits is agroforestry (Nair, 1993; Gold and Hanover, 1987). Conservation buffers, which include agroforestry buffers, have been shown to improve soil health through the enhancement of levels of organic matter as well as improved soil microbial activity (Paudel et al., 2012).

Buffer conservation strips are narrow strips of permanent vegetation widely prescribed to increase water infiltration and reduce surface runoff (Schmitt et al., 1999). These conservation buffer systems include agroforestry buffers as well as grass buffers, which have perennial plants to decrease soil erosion from land under row crop production (Seobi et al., 2005). The influence of perennial root systems from shallower grass roots and deeper tree roots provide soil quality benefits from these buffer systems.

The purpose of this chapter is to review how agroforestry practices improve soil health. Soil health can be assessed by evaluating management effects on soil physical, chemical and biological properties. Several studies are reviewed in this presentation to illustrate how soil health can be improved using agroforestry.

2 Biological properties critical for soil health

2.1 Soil carbon

The most critical soil property which strongly influences soil health and is directly affected by agroforestry systems is soil carbon content. Several studies have shown improvement in soil carbon concentration due to enhanced root development from perennial vegetated systems such as agroforestry (Kumar et al., 2010a). Kumar et al. (2010a) evaluated differences in root length density, as well as root and soil carbon content within agroforestry buffer, grass buffer, rotationally grazed pasture and continuously grazed pasture treatments. Eastern cottonwood trees (*Populus deltoids* Bartr. ex Marsh.) were used in this buffer study. The treatments also included pasture and grass buffer areas with red clover (*Trifolium pratense* L.) and lespedeza (*Kummerowia stipulacea* Maxim.) planted into fescue (*Festuca arundinacea* Schreb.).

Root length density values were 4.5 times higher for buffer treatments compared to those for pasture treatments within a 1-m profile; the agroforestry buffer had the highest value (174 cm per 100 cm^3), while the continuously grazed pasture the lowest value (11 cm per 100 cm^3). Agroforestry buffers had about 3% higher root carbon compared to rotationally grazed pastures. These buffers also had about 115% higher soil carbon relative to grazed pasture areas. Landscapes were improved by including agroforestry buffers in watersheds with grazed pasture, and they showed significant increases in soil carbon content (Kumar et al., 2010a).

A review carried out by Udawatta and Jose (2012) showed that agroforestry has the potential to sequester 640 Tg C $year^{-1}$ in riparian buffers, silvopasture, windbreaks and alley cropping in North America. This review emphasizes that long-term maintenance of agroforestry, proper species selection and strategic placement of suitable practices are key aspects to increasing carbon sequestration. There are three examples from the United States, Canada and Russia on soil carbon sequestration and shelterbelts (Fig. 1). President Franklin Roosevelt's Prairie States Forestry Project helped stabilize the blowing soil and improved land productivity and soil carbon sequestration in the United States thanks to the inclusion of the shelterbelts. From North Dakota to Texas,

Figure 1 Major areas of (a) shelterbelt planting in the United States, (b) shelterbelts in Canada and (c) shelterbelts at the Kamennaya Steppe Research Station, southern Russia. Adapted from Udawatta and Jose (2012).

223 million trees were planted in 300 000 km^2 of shelterbelts throughout the plains states. Agriculture Canada has estimated an increase in soil C by more than 218 mega tons by planting more than 610 million shelterbelt trees in Canada. Other goods and services from these shelterbelts were valued over \$600 million. The 'Great Plan for the Transformation of Nature' of Joseph Stalin in the USSR planted about 5.7 million ha of windbreaks to defeat drought and improve crop yields and soils.

2.2 Biological properties including soil enzymes

In addition to soil carbon storage, soil enzymes are important indicators of soil health. Paudel et al. (2011, 2012) found that water quality from grazed pasture systems, as well as soil health, can be improved by the establishment of buffers and incorporation of trees and shrubs. This study examined the effects of vegetative agroforestry buffers and grass buffers on several biological enzymes, water-stable aggregates, soil carbon and soil nitrogen for grazed pastures and row crop systems. Enzymes included in the study were fluorescein diacetate (FDA) hydrolase, dehydrogenase, β-glucosidase and β-glucosaminidase. Their study consisted of four treatments: grazed pastures, row crops, agroforestry buffers and grass buffers. Each treatment was evaluated at two soil depths: 0–10 cm and 10–20 cm. Enzymes β-glucosidase and β-glucosaminidase showed significantly lower activity for the row crop treatment compared to the other treatments. This effect was also present for water-stable aggregates. Dehydrogenase activity was significantly higher in the grazed pasture system compared to the row crop treatment. Measurements of soil organic carbon and total nitrogen strongly supported the results of the water-stable aggregates and enzyme activities. This study illustrates that perennial vegetation enhances accumulation of organic matter, which benefits the ecosystem.

Environmental soil health, including soil enzyme activity, can be enhanced with agroforestry and grass buffers. Udawatta et al. (2008b, 2009) conducted studies to examine management and landscape effects on water-stable soil aggregates, enzyme activity and soil carbon content. Their studies included agroforestry buffers, grass waterways, grass buffers and row crop treatments. Agroforestry buffers had pin oak (*Quercus palustris* Muenchh.) trees distributed down the centre of grass buffers consisting of a mixture of redtop (*Agrostis gigantea* Roth), brome grass (*Bromus* spp.) and birdsfoot trefoil (*Lotus corniculatus* L.). The following enzymes were evaluated: FDA hydrolase, β-glucosidase, glucosaminidase and dehydrogenase. Landscape positions and treatments significantly affected water-stable aggregates. The stability of aggregates increased among treatments from row crop to grass buffer to agroforestry buffer to grass waterways. Perennial grass buffer management improved soil carbon levels compared to the row crop treatment. Relative to row crop management, agroforestry and grass buffers significantly increased FDA, β-glucosidase and glucosaminidase enzyme activity. Dehydrogenase activity was found to be better in grass buffer compared to row crop treatments. These studies show that soil health improved with increased water-stable aggregates, enzyme activities and soil carbon after agroforestry and grass buffer establishment.

3 Physical properties critical for soil health

Agroforestry systems have been found to improve soil physical properties. These properties include porosity, water retention, pore size distributions, macroporosity, hydraulic conductivity, infiltration, seasonal water use and thermal conductivity.

3.1 Density and porosity

Agroforestry production systems have been introduced in temperate regions to diversify farm income and improve soil and water quality. Agroforestry and grass-legume buffer effects on soil bulk density and porosity were evaluated by Seobi et al. (2005). Their experiment was conducted on a Vertic Albaqualf (claypan soil) in northeastern Missouri, USA. Treatments included no-till row crop management [corn (*Zea mays* L.) - soybean (*Glycine max* L.) rotation], agroforestry buffers and grass-legume buffers. Grass buffers, 4.5 m wide and 36.5 m apart, consisted of redtop, brome and birdsfoot trefoil. Agroforestry buffers were of the same size and consisted of pin oak, swamp white oak (*Q. bicolor* Willd.) and bur oak (*Q. macrocarpa* Michx.) trees planted in redtop, brome and birdsfoot trefoil grass buffer areas. Assessments of bulk density and porosity were conducted using 7.6-cm-diameter and 7.6-cm-long soil cores taken from four depth increments (0–10, 10–20, 20–30 and 30–40 cm) 6 years after buffer establishment. Results from their work showed that agroforestry and grass buffers decreased bulk density by 2.3% and increased porosity by 3%

compared to the row crop treatment. Buffers may provide benefits by reducing runoff from row crop management.

These same agroforestry and grass-legume buffers were also evaluated 17 years after establishment by Akdemir et al. (2016). Undisturbed soil cores were extracted from the soil from four 10-cm depth treatments. Significant differences ($p < 0.05$) were found among the treatments for bulk density, with the row crop treatment having higher values compared with the buffer treatments. Buffers had 8.5% lower density in the 0–10 cm depth position. Across all four sampling depths, the porosity was 4.7% higher for the agroforestry and grass buffer treatments compared to the row crop treatment. The agroforestry buffers showed slightly improved bulk density and porosity values 17 years after establishment, but most of the improvement occurred during the first 6 years (Akdemir et al., 2016).

Studying deep loess soils in central Missouri, USA, Kumar et al. (2008) assessed the effects of agroforestry and grass buffers compared to rotationally grazed pasture and continuously grazed pasture treatments. Agroforestry buffers included Eastern cottonwood trees planted into fescue while pasture and grass buffer areas included red clover and lespedeza planted into fescue. Similar to the Seobi et al. (2005) and Akdemir et al. (2016) studies, soil core samples were removed from four selected 10-cm depths in order to assess density and porosity. Soil bulk density was 13% higher for the rotationally grazed and continuously grazed treatments (1.41 and 1.45 g cm^{-3}) compared to the grass buffer and agroforestry buffer treatments (1.25 and 1.29 g cm^{-3}). Soil porosity was greater in the buffer treatments relative to the other treatments for the 0–10-cm soil depth. Results from this study by Kumar et al. (2008) illustrate that agroforestry buffers maintain better porosity and reduce bulk density compared to grazed pasture systems.

3.2 Water retention

The effects of agroforestry buffers on soil water retention were also evaluated by the Seobi et al. (2005) project cited above. Assessment of soil water retention was conducted on 7.6 cm diameter × 7.6 cm long soil cores sampled in four 10-cm depth increments 6 years after buffer establishment. Water was extracted from these soil cores at 0.0, −0.4, −1.0, −2.5, −5.0, −10.0, −20.0 and −30.0 kPa soil water pressures. Increased water storage of 9.0 and 11.0 mm was found for the grass and agroforestry buffer treatments throughout the 30-cm depth relative to the row crop treatment. This project showed that agroforestry buffers appropriately located on the landscape may provide ecosystem benefits by reducing runoff from row crop management areas.

These same agroforestry and grass-legume buffers were also evaluated 17 years after establishment by Akdemir et al. (2016). Additional undisturbed soil cores were extracted from the soil from four 10-cm depth treatments. Similar results were found compared to the earlier study (Seobi et al., 2005) with only slight changes. Thus, the significant initial changes in soil water retention caused by the tree buffers during the first 6 years were only slightly changed over the subsequent 10 years.

Evaluating deep loess soils in central Missouri, USA, Kumar et al. (2008) assessed the effects of agroforestry and grass buffers on soil water retention compared to rotationally grazed pasture and continuously grazed pasture treatments. Agroforestry buffers included Eastern cottonwood trees planted into fescue while pasture and grass buffer areas included red clover and lespedeza planted into fescue. For this study within the 10-cm soil depth, soil water retained at high soil water pressures (0.0 and −0.4 kPa) was higher for the agroforestry and grass buffer treatments compared to the rotationally grazed and continuously grazed pastures.

3.3 Pore size distributions

The effects of agroforestry buffers on soil pore size distributions estimated from soil water retention data were evaluated by Seobi et al. (2005). These researchers investigated agroforestry and grass-legume buffers relative to their influence on water retention in soils for a Vertic Albaqualf (claypan soil) compared to row crop management. Assessment of pore size distributions was conducted in four 10 cm depth increments 6 years after buffer establishment. Coarse mesoporosity (60 to 1000 μm diameter) and total porosity were 33% and 3% higher, respectively, for the grass buffer and agroforestry buffer treatments compared to the row crop treatment. No significant differences were found among treatments for soil macroporosity (>1000 μm diameter pores).

Kumar et al. (2008) evaluated the effects of agroforestry and grass buffers on soil pore size distributions compared to rotationally grazed pasture and continuously grazed pasture treatments in deep loess soils. These researchers found agroforestry buffer, grass buffer and rotationally grazed pastures had 5.7, 4.5 and 3.9 times higher soil macroporosity compared to the continuously grazed pasture treatment. The grazed pasture treatments had higher microporosity (<10 μm diameter) compared to the agroforestry and grass buffer treatments, but lower macroporosity (>1000 μm diameter), coarse mesoporosity (60–1000 μm diameter) and fine mesoporosity (10–60 μm diameter).

3.4 Computed tomography-measured macroporosity

Although soil health is highly dependent on the degree of porosity, amounts for selected soil pore sizes are also critical (Udawatta et al., 2006). Soil porosity can be divided into several pore classes with two major groups – macropores

(<1000 μm diameter) and micropores (<10 μm diameter). Macropores and mesopores (10–1000 μm diameter) are critical for water drainage while micropores are critical for storing plant available water. Mesopores, pores of sizes between macropores and micropores, are also important for water and gas transport.

A relatively recent tool for assessing biopores in the macropore and mesopore size range is x-ray computed tomography (CT). Kumar et al. (2010b) assessed differences in macroporosity and coarse mesoporosity (200–1000 μm diameter) for agroforestry and grass buffers relative to grazed pastures utilizing x-ray CT image analysis. Their study included four treatments: continuously grazed pastures, rotationally grazed pastures, grass buffers and agroforestry buffers. Pasture and grass buffer areas included red clover and Korean lespedeza planted into fescue, while agroforestry buffers included Eastern cottonwood trees planted into fescue. At a deep loess site in central Missouri, USA (Kumar et al., 2010b), soil cores were removed from five 10-cm depth increments, and five equally spaced images were acquired from each core and analysed for macropores and coarse mesopores. CT-measured soil macroporosity was 13 times lower (0.004 m^3 m^{-3}) for the pasture treatments than the agroforestry buffer treatments (0.053 m^3 m^{-3}) for the surface 0- to 10-cm soil depth. Results showed that saturated hydraulic conductivity was positively correlated with all CT-measured pore parameters except macropore circularity. This research by Kumar et al. (2010b) illustrates that agroforestry buffers improve soil macroporosity critical for infiltration of water.

Other studies (Udawatta et al., 2006; Udawatta and Anderson, 2008) also investigated the effects of agroforestry buffers on soil pore sizes using CT. These investigators evaluated the effects of vegetative tree-grass and grass contour buffers on CT-measured macropore (>1000 μm) and coarse mesopore (200–1000 μm) parameters and examined the relationships between CT-measured pore parameters and saturated hydraulic conductivity of the soil. The treatments investigated in this study included a no-till corn–soybean rotation, a cool season grass–legume buffer, and a pin oak buffer. Relatively undisturbed soil cores were evaluated using procedures similar to the Kumar et al. (2010b) study. The tree and grass buffer treatments had better pore parameters compared to the row crop treatment (pore parameters included macroporosity, mesoporosity, area of largest pore and number of pores). Row crop, grass buffer and tree buffer treatments had on the average 44, 87 and 207 CT-measured pores within a 3600 mm^2 area, respectively. CT-measured macropores explained about 64% of the variation in saturated hydraulic conductivity.

Udawatta et al. (2008a) evaluated soil pore sizes from these same buffer treatments using high-resolution x-ray CT. The study was conducted on a Putnam silt loam (fine, smectitic, mesic Vertic Albaqualf) claypan soil with treatments of agroforestry buffer (pin oak trees); grass buffer (redtop, brome, birdsfoot trefoil) and a no-till corn–soybean rotation. Core samples (7.6 cm diameter × 7.6 cm long) were removed from the upper 10-cm depth and scanned with a high-resolution scanner at a voxel size of 73 × 73 × 84 μm. Three-Dimensional Medial Axis software was used to analyse images. Agroforestry and grass buffer treatments had pore paths five and three times greater than the row crop treatment, respectively. The row crop treatment had higher pore path tortuosity values (112% higher) than the buffer treatments and lower characteristic pore path length constant values (0.46 value of buffers). The study by Udawatta et al. (2008a) illustrated that three-dimensional pore structure differences were probably caused by variations in organic matter accumulation and root growth for these buffer systems.

3.5 *Saturated hydraulic conductivity* (Ksat)

The influence of agroforestry buffers on saturated hydraulic conductivity (*Ksat*) was evaluated by Seobi et al. (2005). Saturated hydraulic conductivity is an assessment measure of how fast water flows through the soil when all the soil pores are full of water and under laboratory controlled conditions. These researchers investigated agroforestry and grass–legume buffers relative to their influence on *Ksat* for a Vertic Albaqualf (claypan soil) compared to row crop management. The grass and agroforestry buffers had 3 and 14 times higher *Ksat* values compared to row crop management, respectively (Fig. 2). For these types of soils, the claypan horizon dominates the surface hydrology, but these tree buffers may provide some benefit by reducing runoff from row crop management.

Akdemir et al. (2016) studied the same agroforestry and grass–legume buffers 17 years after establishment. They found higher *Ksat* in the upper 10 cm and 30–40 cm depths for the agroforestry buffer treatment compared with the row crop treatment. After 17 years, these buffer treatments slightly improved the hydraulic properties of this soil.

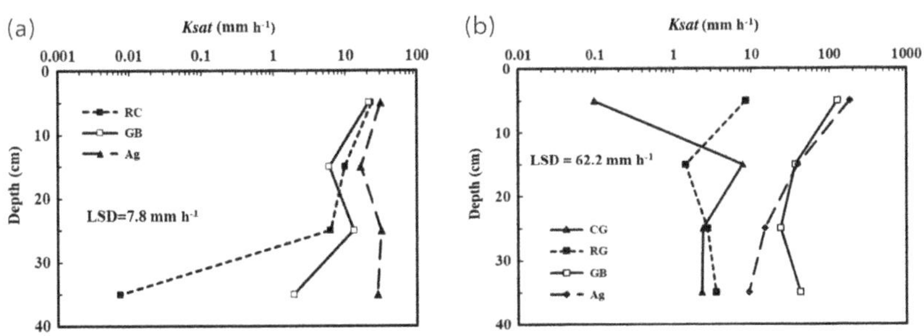

Figure 2 Saturated hydraulic conductivity (*Ksat*) differences in agroforestry (Ag or AgB) for row crop (a) and silvopasture (b) sites in Missouri as compared to row crop (RC), grass buffer (GB), continuously grazed pasture (CG) and rotationally grazed pasture (RG). Source: Seobi et al. (2005) and Kumar et al. (2008).

Evaluating deep loess soils, Kumar et al. (2008) assessed the effects of agroforestry buffers on *Ksat* compared to rotationally grazed pasture and continuously grazed pasture treatments. Values for the *Ksat* parameter (Fig. 2) were 17 times higher in the agroforestry and grass buffer treatments (57 versus 61 mm h^{-1}) compared with rotationally grazed and continuously grazed pasture systems (4.0 versus 3.1 mm h^{-1}).

3.6 Infiltration

Agroforestry buffers may enhance soil health by increasing water infiltration. Soil water infiltration was measured by Anderson et al. (2009) for a Putnam soil (fine, smectitic, mesic Vertic Albaqualf) to assess whether agroforestry buffers influenced water infiltration. Ponded water infiltration was measured in agroforestry and grass buffers and row crop areas in 2003 6 years after buffer establishment. Quasi-steady infiltration rates were not found to be different among the treatments 6 years after buffer establishment. However, differences were found for the same treatments 17 years after establishment (Sahin et al., 2016). Therefore, adoption of agroforestry buffer practices may reduce runoff and soil loss from watersheds in row crop management. Fitting infiltration data to the Green-Ampt infiltration parameters (Green and Ampt, 1911) helped to assess differences among treatments. During 2014, the water sorptivity parameter for the fitted infiltration equations was highest for the agroforestry buffers compared to row crop management. Sorptivity is an infiltration parameter indicating how fast water moves into the soil under initial field conditions, when the soil is less than saturated. During 2015, the saturated hydraulic conductivity parameter for the fitted equations was highest for the buffer treatments relative to row crop management. Thus, the data for claypan soils are somewhat variable throughout tree growth as a function of years, but some differences in infiltration are observed for these tree buffer systems (Sahin et al., 2016).

Kumar et al. (2012) evaluated the influence of agroforestry and grass buffer systems compared to rotationally stocked and continuously stocked pasture systems on water infiltration measured using ponded infiltration. Soils at the site were Menfro silt loam (fine-silty, mixed, superactive, mesic Typic Hapludalf) which develop on deep loess. Infiltration rates were measured using ponded ring infiltration units for the four treatments. They used the Green-Ampt and Parlange (Parlange et al., 1982) infiltration equations to fit the infiltration data. The field-*Ksat* values estimated from these data indicated that buffers had 46 times higher values compared to the grazed pasture systems. The steady infiltration flux during the first measurement year for the grass buffer treatment was the highest (221 mm h^{-1}) and the value for the continuously grazed pasture treatment was the lowest (3.7 mm h^{-1}). The sorptivity and *Ksat* parameters were higher for the buffer areas relative to grazed pasture areas. Thus, grazed pasture systems were found to reduce infiltration compared to agroforestry buffers.

These changes in infiltration properties as influenced by perennial vegetation can allow the soil system to be more resilient to extreme changes in rainfall. Zaibon et al. (2017) in working with switchgrass on eroded soils showed these perennial grasses significantly improved infiltration compared to row crop management and reduced simulated runoff from extreme storms.

3.7 Seasonal plant water use

Assessment of profile soil water use is important for assessment of the hydrologic cycle (Topp and Ferré, 2002). Agroforestry practices allow trees to consume water from deeper in the profile which shallow rooted plants cannot access (Jose et al., 2000). Scientists have evaluated the early growth of oak species and their associated root system in an agroforestry practice designed for watershed protection (Udawatta et al., 2005). In comparing swamp white oak, bur oak and pin oak trees, the total proximal root cross-sectional area was lowest for the swamp white oak, and similar for the other two species. While bur oak species may compete with adjoining crops for soil water, pin oak roots are limited to buffer areas (Udawatta et al., 2005).

Agroforestry buffers may enhance soil health by enhancing seasonal water use to increase water infiltration and reduce runoff. Soil water content, water use and recharge as affected by agroforestry buffers were monitored over several years by Anderson et al. (2009), Udawatta et al. (2011a) and Sahin et al. (2016) for a Putnam soil (fine, smectitic, mesic Vertic Albaqualf). Soil water content in agroforestry and row crop areas at 5-, 10-, 20- and 40-cm depths were monitored throughout several years. For agroforestry buffers, soil water content decreased more rapidly during the growing season compared to row crop-managed treatments. Soil water content was lower for agroforestry buffers compared to row crop management during most of the growing season. Better recharge during post-infiltration events occurred within the agroforestry buffers (Anderson et al., 2009; Sahin et al., 2016). Due to the lower growing season water content and better infiltration within agroforestry buffers, higher water content after post-infiltration events and better water recharge occurred for the agroforestry buffers compared to row crop areas (Sahin et al., 2016). Thus, goals of reduced runoff and soil loss suggest greater utilization of agroforestry buffer practices for row crop-managed watersheds.

3.8 Thermal properties

Soil health includes thermal properties in addition to other physical and hydraulic properties. Less information is available about the effects of agroforestry systems on soil thermal properties. Adhikari et al. (2014) evaluated agroforestry buffer effects on soil thermal properties compared to row crop management. This study assessed soil heat capacity and thermal conductivity for five treatments: agroforestry buffers, native prairie, restored prairie, grass buffers and corn-soybean rotation. Soil cores were collected from 10-cm depth increments and thermal properties were determined at 0, −33, −100 and −300 kPa soil water pressures. Results showed that organic carbon content was negatively correlated with thermal conductivity and positively correlated with heat capacity. Thermal conductivity was significantly higher and heat capacity was significantly lower under row crop management compared to the buffer and prairie systems. A higher amount of soil organic carbon decreases thermal conductivity due to the insulating characteristics of organic matter; thus higher organic carbon levels in the perennial vegetation treatments act as a barrier to heat transport. The agroforestry buffer, grass buffer and prairie treatments had lower thermal conductance to deeper soil depths which helps to conserve more soil water as well as increase the longevity of organic carbon in the soil compared to row crop management. Results from the Adhikari et al.'s (2014) study suggested that buffers and perennial vegetation can help reduce heat flow by increasing the thermal heat capacity and thereby help crops adapt to the effects of climate change. Improved thermal properties can help make soils and associated ecosystems be more resilient to extreme temperature events.

4 Chemical properties critical for soil health

4.1 Chemical adsorption

Agroforestry practices can benefit chemical properties of soils to improve soil health. Improved soil adsorption of veterinary antibiotics may occur within vegetated buffers including agroforestry systems. Chu et al. (2010, 2013a) evaluated the benefits of vegetated agroforestry and grass buffers for retention of veterinary antibiotics. They investigated two veterinary antibiotics (oxytetracycline and sulfadimethoxine) for sorption to cropland, agroforestry buffer and grass buffer soils. These researchers found significantly higher solid/solution partition coefficients for these antibiotics for agroforestry buffers compared to cropland soils. Due to the enhanced sorption properties, these studies indicated that agroforestry and buffer strips can reduce antibiotic losses from agricultural management practices to water systems. Integration of agroforestry may also help enhance degradation of these chemicals within the soil and production of less harmful degradation products (Lin et al., 2010, 2011). Their study has shown that combinations of trees, shrubs and grasses were more effective in degrading chemicals than an individual species. Fast-growing trees of agroforestry practices that produce high amounts of biomass can accumulate large amounts of chemicals as compared to small and slow-growing species (Gomes, 2012).

Transport of sulfamethazine (veterinary antibiotic) in soil from agroforestry systems was investigated by Chu et al. (2013b). These scientists discovered that transport of antibiotics was governed by chemical non-equilibrium processes. A three-site model including two chemical adsorption reversible sites (instantaneous and kinetic) and one chemical adsorption irreversible site were needed to adequately describe the antibiotic transport in dynamic flow studies. These dynamic flow studies showed that soil from an agroforestry buffer retained more antibiotic compared to soil from a cropland site due to enhanced sorption attributed to higher levels of organic carbon.

4.2 Reduced chemical transport

Agroforestry practices have the potential to reduce non-point source pollution and various other chemicals from land to benefit water quality. Within conventional agricultural systems, less than half of the applied nutrient fertilizers of nitrogen and phosphorus are taken up by plants (Jose, 2009). The use of agroforestry practices in watersheds helps to reduce runoff losses. These practices reduce water runoff and runoff of plant nutrients such PO_4^{-2}, NO_3^- and NH_4^+ which affect soil and water quality (Walker and Branham, 1992). Several research studies have shown the benefits of agroforestry for reducing nutrient and pesticide runoff from cultivated agroecosystems (Lovell and Sullivan, 2006; Lowrance et al., 2002; Stone et al., 2004; Udawatta et al., 2002, 2011b).

Agroforestry practices can be potentially used to remove chemical contaminants from selected sites. Selected fast-growing tree species produce significant biomass and deep roots which accumulate significant amounts of contaminants and assist in phytoremediation (Gomes, 2012). Phytoremediation of heavy metals and organic compounds by various tree species in the United States has been summarized by Rockwood et al. (2004).

Amounts of nutrient, herbicide and antibiotic chemicals transported to water bodies are also affected by degradation pathways. Degradation patterns vary with species and multi species buffers as established in agroforestry; multi species buffers have better rhizodegradation processes for antibiotics and herbicides than individual species (Lin et al., 2010, 2011). Thus, agroforestry buffers are a method to reduce leaching and runoff losses of antibiotics from agroecosystem watersheds.

5 Conclusion

Soil health can be enhanced with the use of agroforestry management in temperate ecosystems and is mainly related to increased soil carbon and microbiological activity. Improved soil physical and hydraulic properties have occurred through use of agroforestry buffers. These properties include decreased soil bulk density, increased porosity and macroporosity, increased hydraulic conductivity, increased infiltration, increased heat capacity and reduced thermal conductivity. Improved infiltration properties as influenced by perennial agroforestry vegetation can allow the soil system to be more resilient to extreme changes in temperature and rainfall. Integration of agroforestry in row crop and livestock management further enhances ecosystem services by increased sorption and rhizodegradation of various chemicals including antibiotics and herbicides and eventually producing less harmful degradation products. Thus, as soil health improves, the agroecosystem will be more resilient.

6 Future trends

Meeting the demands of food, fibre and energy of the expected 9 billion people while maintaining the quality of water, soil and air will be a significant challenge in the coming years. Study after study has shown the importance of soil health and its role in providing for the needs of people and protecting environmental quality. Soil health is strongly linked with land management practices and quality of water and air. Although agroforestry is identified as a land management practice that can improve ecosystem services including provisional, regulating, cultural and economic services, the role of agroforestry in improving soil health is not well understood. Therefore, many well-designed long-term studies in all geographic regions are required to collect data on health of soil, water and air as influenced by integration of agroforestry. For example, newer techniques such molecular, metagenomics and metabolomics as well as satellite and remote sensing techniques can be included in addition to conventional research methodologies to acquire information at the molecular to watershed scales. This information will help with the development of guidelines for adoption of agroforestry for improved food, fibre and energy supply and protect soil, water and air.

7 Where to look for further information

There are a number of new books and book chapters on agroforestry which are available in bookstores and libraries:

Garrett, H. E., ed. 2009. *North American Agroforestry: An Integrated Science and Practice*, 2nd Edition. American Society of Agronomy, Madison, WI, 379pp.

Gordon, A. M., S. M. Newman and R. R. W. Coleman, eds. 2018. *Temperate Agroforestry Systems*, 2nd Edition. CABI, Wallingford, UK, 328pp.

Schoeneberger, M., G. Bentrup and T. Patel-Weynand, eds. 2018. Agroforestry: Enhancing resiliency in U.S. agricultural landscapes under changing conditions. Gen. Tech. Rep. WO-96. U.S. Department of Agriculture, Forest Service, Washington Office, Washington DC, 214pp.

Udawatta, R. P., C. J. Gantzer and S. Jose. 2017. Agroforestry practices and soil ecosystem services, pp. 305–34. *In* M. M. Al-Kaisi and B. Lowery (Eds), *Soil Health and Intensification of Agroecosytems*. Elsevier Science, Amsterdam, the Netherlands.

Agroforestry institutes at universities and the United States Department of Agriculture provide updated information on soil health benefits of agroforestry. Information on a number of current research projects supported by government and private agencies can be found on their websites:

The Center for Agroforestry – University of Missouri. http://www.centerforagroforestry.org/
National Agroforestry Center. https://www.fs.usda.gov/nac/index.shtml
World Agroforestry Centre. http://www.worldagroforestry.org/

8 References

Adhikari, P., R. P. Udawatta and S. H. Anderson. 2014. Soil thermal properties under prairies, conservation buffers and corn/soybean land use systems. *Soil Science Society of America Journal* 78:1977–86.

Akdemir, E., S. H. Anderson and R. P. Udawatta. 2016. Influence of agroforestry buffers on soil hydraulic properties relative to row crop management. *Soil Science* 181:368–76.

Anderson, S. H., R. P. Udawatta, T. Seobi and H. E. Garrett. 2009. Soil water content and infiltration in agroforestry buffer strips. *Agroforestry Systems* 75:5–16.

Chandrasoma, J. M., R. P. Udawatta, S. H. Anderson, A. L. Thompson and M. A. Abney. 2016. Soil hydraulic properties as influenced by prairie restoration. *Geoderma* 283:48–56.

Chu, B., K. W. Goyne, S. H. Anderson, C. H. Lin and R. P. Udawatta. 2010. Veterinary antibiotic sorption to agroforestry buffer, grass buffer and cropland soils. *Agroforestry Systems* 79:67–80.

Chu, B., K. W. Goyne, S. H. Anderson, C. H. Lin and R. N. Lerch. 2013a. Sulfamethazine sorption to soil: Vegetative management, pH, and dissolved organic matter effects. *Journal of Environmental Quality* 42:794–805.

Chu, B., S. H. Anderson, K. W. Goyne, C. H. Lin and R. N. Lerch. 2013b. Sulfamethazine transport in agroforestry and cropland soils. *Vadose Zone Journal* 12(2):1–14.

Conway, L. S., M. A. Yost, N. R. Kitchen and K. A. Sudduth. 2017. Using topsoil thickness to improve site-specific phosphorus and potassium management on claypan soil. *Agronomy Journal* 109:2291–301.

Evans, R. G. and E. J. Sadler. 2008. Methods and technologies to improve efficiency of water use. *Water Resources Research* 44:1–15.

Godfray, H. C. J., J. R. Beddington, I. R. Crute, L. Haddad, D. Lawrence, J. F. Muir, J. Pretty, S. Robinson, S. M. Thomas and C. Toulmin. 2010. Food security: The challenge of feeding 9 billion people. *Science* 327: 812–18.

Gold, M. A. and J. W. Hanover. 1987. Agroforestry systems for temperate zone. *Agroforestry Systems* 5:109–21.

Gomes, H. I. 2012. Phytoremediation for bioenergy: Challenges and opportunities. *Environmental Technology Reviews* 1:59–66.

Green, W. H. and G. A. Ampt. 1911. Studies on soil physics: 1. Flow of air and water through soils. *Journal of Agricultural Science* 4(1):1–24.

Jose, S. 2009. Agroforestry for ecosystem services and environmental benefits: An overview. *Agroforestry Systems* 76(1):1–10.

Jose, S., A. R. Gillespie, J. R. Seifert and D. J. Biehle. 2000. Defining competition vectors in temperate alley cropping system in the Midwestern USA: 2. Competition for water. *Agroforestry Systems* 48(1):41–59.

Kumar, S., S. H. Anderson, L. G. Bricknell, R. P. Udawatta and C. J. Gantzer. 2008. Soil hydraulic properties influenced by agroforestry and grass buffers for grazed pasture systems. *Journal of Soil and Water Conservation* 63:224–32.

Kumar, S., R. P. Udawatta and S. H. Anderson. 2010a. Root length density and carbon content of agroforestry and grass buffers under grazed pasture systems in a Hapludalf. *Agroforestry Systems* 80:85–96.

Kumar, S., S. H. Anderson and R. P. Udawatta. 2010b. Agroforestry and grass buffer influences on macropores measured by computed tomography under grazed pasture systems. *Soil Science Society of America Journal* 74:203–12.

Kumar, S., S. H. Anderson, R. P. Udawatta and R. L. Kallenbach. 2012. Water infiltration influenced by agroforestry and grass buffers for a grazed pasture system. *Agroforestry Systems* 84:325–35.

Lal, R. 2001. Soil degradation by erosion. *Land Degradation & Development* 12: 519–39.

Lin, C. H., K. W., Goyne, R. J. Kremer, R. N. Lerch and H. E. Garrett. 2010. Dissipation of sulfamethazine and tetracycline in the root zone of grass and tree species. *Journal of Environmental Quality* 39:1269–78

Lin, C.-H., R. N. Lerch, K. W. Goyne and H. E. Garrett. 2011. Reducing herbicides and veterinary antibiotic losses from agroecosystems using vegetative buffers. *Journal of Environmental Quality* 40:791–9.

Lovell, S. T. and W. C. Sullivan. 2006. Environmental benefits of conservation buffers in the United States: Evidence, promise, and open questions. *Agriculture, Ecosystems and Environment* 112(4):249–60.

Lowrance, R., S. Dabney and R. Schultz. 2002. Improving water and soil quality with conservation buffers. *Journal of Soil and Water Conservation* 57:37A–43A.

McLaughlin, D. and W. Kinzelbach. 2015. Food security and sustainable resource management. *Water Resources Research* 51: 4966–85.

Nair, P. K. R. 1993. *An Introduction to Agroforestry*. Kluwer Academic Publishers, Norwell, MA.

Parlange, J. Y., I. Lisle, R. D. Braddock and R. E. Smith. 1982. The three-parameter infiltration equation. *Soil Science* 133(6):337–41.

Paudel, B. R., R. P. Udawatta and S. H. Anderson. 2011. Agroforestry and grass buffer effects on soil quality parameters for grazed pasture and row-crop systems. *Applied Soil Ecology* 48:125–32.

Paudel, B. R., R. P. Udawatta, R. J. Kremer and S. H. Anderson. 2012. Soil quality indicator responses to row crop, grazed pasture, and agroforestry buffer management. *Agroforestry Systems* 84:311–23.

Rockwood, D. L., C. V. Naidu, D. R. Carter, M. Rahmani, T. A. Spriggs, C. Lin, G. R. Alker, J. G. Isebrands and S. A. Segrest. 2004. Short-rotation woody crops and phytoremediation: Opportunities for agroforestry? *Agroforestry Systems* 61:51–63.

Sahin, H., S. H. Anderson and R. P. Udawatta. 2016. Water infiltration and soil water content in claypan soils influenced by agroforestry and grass buffers compared to row crop management. *Agroforestry Systems* 90:839–60.

Schmitt, T., M. Dosskey and K. Hoagland. 1999. Filter strip performance and processes for different vegetation, widths, and contaminants. *Journal of Environmental Quality* 28:1479–89.

Seobi, T., S. H. Anderson, R. P. Udawatta and C. J. Gantzer. 2005. Influence of grass and agroforestry buffer strips on soil hydraulic properties for an Albaqualf. *Soil Science Society of America Journal* 69:893-901.

Soil Science Society of America. 2008. *Glossary of Soil Science Terms*. Soil Science Society of America, Madison, WI.

Stone, K. C., P. G. Hunt, J. M. Novak, M. H. Johnson, D. W. Watts and F. J. Humenik. 2004. Stream nitrogen changes in an eastern coastal plain watershed. *Journal of Soil and Water Conservation* 59:66-72.

Topp, G. C. and P. A. Ferre. 2002. The soil solution phase, pp. 417-573. *In* W. A. Dick (Ed.), *Methods of Soil Analysis. Part 4*. Soil Science Society of America, Madison, WI.

Udawatta, R. P. and S. H. Anderson. 2008. CT-measured pore characteristics of surface and subsurface soils influenced by agroforestry and grass buffers. *Geoderma* 145:381-9.

Udawatta, R. P. and S. Jose, 2012. Agroforestry strategies to sequester carbon in temperate North America. *Agroforestry Systems* 86:225-42.

Udawatta, R. P., J. J. Krstansky, G. S. Henderson and H. E. Garrett. 2002. Agroforestry practices, runoff, and nutrient loss: A paired watershed comparison. *Journal of Environmental Quality* 31: 1214-25.

Udawatta, R. P., P. Nygren and H. E. Garrett. 2005. Growth of three oak species during establishment of an agroforestry practice for watershed protection. *Canadian Journal of Forest Research* 35:602-9.

Udawatta, R. P., S. H. Anderson, C. J. Gantzer and H. E. Garrett. 2006. Agroforestry and grass buffer influence on macropore characteristics. *Soil Science Society of America Journal* 70:1763-73.

Udawatta, R. P., C. J. Gantzer, S. H. Anderson and H. E. Garrett. 2008a. Agroforestry and grass buffer effects on high resolution X-ray CT-measured pore characteristics. *Soil Science Society of America Journal* 72:295-304.

Udawatta, R. P., R. J. Kremer, B. W. Adamson and S. H. Anderson. 2008b. Variations in soil aggregate stability and enzyme activities in a temperate agroforestry practice. *Applied Soil Ecology* 39:153-60.

Udawatta, R. P., R. J. Kremer, H. E. Garrett and S. H. Anderson. 2009. Soil enzyme activities and physical properties in a watershed managed under agroforestry and row-crop systems. *Agriculture, Ecosystems & Environment* 131:98-104.

Udawatta, R. P., S. H. Anderson, P. P. Motavalli and H. E. Garrett. 2011a. Calibration of a water content reflectometer and soil water dynamics for an agroforestry practice. *Agroforestry Systems* 82:61-75.

Udawatta, R. P., H. E. Garrett and R. L. Kallenbach. 2011b. Agroforestry buffers for non-point source pollution reductions from agricultural watersheds. *Journal of Environmental Quality* 40:800-6.

USDA-Natural Resources Conservation Service. 2017. Soil health glossary. Website: https://www.nrcs.usda.gov/wps/portal/nrcs/detailfull/soils/health/?cid=nrcs142p2_053848 (accessed on 26 December 2017).

Walker, W. J. and B. Branham. 1992. Environmental impacts of turfgrass fertilization, pp. 105-219. *In* J. C. Balogh and W. J. Walker (Eds), *Golf Course Management and Construction: Environmental Issues*. CRC Press, Boca Raton, FL.

Wall, D. H., U. N. Nielsen and J. Six. 2015. Soil biodiversity and human health. *Nature* 528: 69-76.

Yost, M. A., N. R. Kitchen, K. A. Sudduth, E. J. Sadler, C. Baffaut, M. R. Volkmann and S. T. Drummond. 2016. Long-term impacts of cropping systems and landscape positions on claypan soil grain crop production. *Agronomy Journal* 108:713–25.

Zaibon, S., S. H. Anderson, A. L. Thompson, N. R. Kitchen, C. J. Gantzer and S. I. Haruna. 2017. Soil water infiltration affected by topsoil thickness in row crop and switchgrass production systems. *Geoderma* 286:46–53.

Part 3

Agroforestry products

Chapter 12

Agroforestry for hardwood timber production

J. W. 'Jerry' Van Sambeek, formerly of USDA Forest Service Northern Research Station and University of Missouri Center for Agroforestry, USA

1 Introduction

Managing hardwood trees for high-quality sawlogs within the agroforestry practices of alley cropping, silvopasture, forested riparian buffers and upland (windbreaks) buffers means intensively managing relatively wide-spaced trees and a ground cover as a companion crop on the same unit of land. These agroforestry practices provide the opportunity to produce logs and other tree products as a long-term investment while obtaining regular income through the production of companion crops or animals to offset some of the carrying costs associated with establishing the practices. The introduction of hardwoods into agricultural cropping systems also provides important ecosystem services and environmental benefits that conservation farming practices alone do not provide such as moderating air and soil temperatures and increasing biological diversity both above and belowground (Stamps et al., 2002; Jose, 2009; Kremer and Kussman, 2011). By moderating the microclimate, animals stay cooler and may graze longer in a silvopastural practice (Jose et al., 2017). Although forage yields may decrease in the shaded environment, the duration of forage production may be longer (Lehmkuhler et al., 1999; Kallenbach,

http://dx.doi.org/10.19103/AS.2018.0041.16

2009). In addition, forage quality will be improved, especially for crude protein content (Lin et al., 2001; Kallenbach et al., 2006; Pang et al., 2019b).

Alley cropping in the temperate regions will typically involve managing an annual grass or forage crop between rows of trees possibly with a perennial ground cover within the trees' rows. The relatively open tree spacing should allow adequate sunlight to reach the companion crops. Open-grown trees (ideally between 40% and 60% canopy coverage) are managed so that there is adequate light for the forage and vertical distribution of roots to minimize root competition for water and soil nutrients. Recent research indicates 50% of ambient sunlight for many crops and forages results in statistically similar yields as growing in full sun (Lin et al., 1999; Semchenko et al., 2012; Pang et al., 2019a). Differences in shade tolerance when measured under tree canopies are apparently less likely to be in response to reduced light and more likely a function of root competition for water and nutrients, especially available soil nitrogen (Jose et al., 1997). The addition of nitrogen-fixing shrubs within-tree rows may help by adding nitrogen to the soil and providing side shade to improve stem quality of the crop trees (Schlesinger and Williams, 1984).

Silvopasture deliberately combines forage and cattle with trees as part of a larger, managed intensive grazing system that includes open pasture (Kallenbach, 2009). Silvopasture designs should include open pastures as a place to move livestock while grazed forages are allowed to recover or when soils are high in available water to minimize soil compaction and damage to tree roots. Agroforestry practices that include rotational grazing or haying can reduce the rooting zone of the forages and expand the rooting zone for the deeper tree roots (Jose et al., 2017). Overgrazing by removing 50–80% of a plant's forage rapidly reduces root growth by 5–100%, respectively, and substantially increases the time (normally 3–4 weeks) needed for rest and regrowth (USDA-NRCS, 2005).

Buffers can also be managed for timber production. Forested riparian buffers managed with open-grown trees and a grass understory are frequently more effective in reducing surface run-off and capturing nutrients and sediments than buffers with dense tree plantings next to the waterway and an adjacent grass strip. Effective windbreaks with three or more rows of tall trees and an ideal canopy density of 40–60% will allow sufficient light penetration to maintain an herbaceous ground cover (Cunningham, 1988). Sheltered zones between two and twenty times tree heights can result in overall higher crop yields by reducing wind speeds and creating favourable microclimates especially under droughty conditions (Kort, 1988; Rivest and Vezina, 2015). Open-grown trees will require more human intervention to produce high-quality logs in buffers unlike in densely spaced tree plantings where side shade forces vertical growth on a straight central stem and early death and shedding of relatively small diameter branches.

2 Impact of ground cover on tree growth

The ground cover seeded as a crop or forage can substantially affect the growth of hardwood trees in an agroforestry practice with differences due to the type of cover, the competitiveness of the trees and soil or climatic conditions (Van Sambeek and Garrett, 2004). Usually the best tree growth is achieved with removal of all understory vegetation either by using chemical or mechanical methods or by mulching. Most agroforestry practices are installed on unlevelled ground, so vegetation-free approaches are not ecologically viable approaches. Establishment of some legumes as living mulches can be almost as effective as chemical or mechanical methods to reduce ground cover competition, especially with hardwoods that respond well to high-nitrogen fertilizers (Van Sambeek et al., 1986; Alley et al., 1999). In contrast, some perennial grasses established as living mulches can result in low foliage nitrogen concentrations in the trees and very slow growth (Schlesinger and Van Sambeek, 1986; Miller et al., 1987).

In a study to determine how competitive herbaceous legumes and grasses were, Van Sambeek and Garrett (2004) synthesized the results from over 110 research reports or papers that reported on hardwood tree growth with different ground covers. To be included in the database, these reports or papers had to include the growth rate for one or more tree species for stem diameter and/or height for 2 or more years after establishing a single species ground cover (Van Sambeek, 2010). Reports also had to include information from control plots with trees growing in vegetation-free areas or in areas with a living mulch of mixed vegetation including short-lived herbaceous legumes, non-legume broadleaved forbs and grasses. A living mulch of weeds reduces tree growth to approximately 60% of that reported for trees growing in adjacent vegetation-free plots with minor differences among the hardwoods grown for high-quality logs and wood products (Table 1). Insufficient data were found to compare effects of controlled grazing although it is likely similar to mowing of mixed vegetation types that further reduces tree growth an additional 16% for black walnut and an additional 10% for other hardwoods (Van Sambeek and Garrett, 2004).

Depending on the legume species and agroforestry practice, hardwood tree growth can either be impacted quite severely or slightly improved with a legume companion crop (Table 1). A living mulch of forage legumes has less of an impact on hardwood tree growth than does a living mulch of weeds (White et al., 1981; Van Sambeek and Garrett, 2004). Annual legumes such as soybeans (*Glycine max*) usually have less of an impact than perennial legumes with some exceptions like crownvetch (*Coronilla varia*), arrowleaf (*Trifolium vesiculosum*) and subterranean (*T. subterraneum*) clovers. Although available soil nitrogen is the nutrient most likely to be deficient, nitrogen-fixation rates of legumes

Table 1 Average tree growth rate and standard deviation as percent of growth rates when grown with and without different ground covers and average-reported nitrogen-fixation rates for the different ground covers

Legume or grass ground cover		Growth reduction (%)	N-fixation rates (kg/ha)
Trifolium vesiculosum	Arrowleaf clover	122 + 54 (10)[a]	???
Glycine max	Soybean	103 + 18 (8)	160
Trifolium subterraneum	Subterranean clover	94 + 54 (18)	145
Coronilla varia	Crownvetch	93 + 43 (28)	230
Mixed genera and species	Annual grasses	93 + 20 (12)	NA
Brassica napus	Kale	89 + 20 (12)	NA
Muhlenbergia schriberi	Nibblewill	88 + 25 (15)	NA
Trifolium incarnatum	Crimson clover	81 + 44 (34)	95
Vicia sativa	Hairy vetch	79 + 33 (172)	135
Agrostis giganta	Redtop	78 + 18 (10)	NA
Festuca rubra	Red fescue	73 + 26 (82)	NA
Poa pratensis	Bluegrass	72 + 27 (27)	NA
Bromus inermis	Bromegrass	72 + 14 (12)	NA
Trifolium pratense	Red clover	71 + 28 (122)	125
Trifolium ambiguum	Kura clover	71 + 13 (13)	???
Triticum and *Hordeum* spp.	Winter cereal grains	68 + 39 (15)	NA
Lespedeza cuneata	Sericea lespedeza	67 + 16 (33)	175
Trifolium repens	White clover	67 + 19 (56)	170
Mixed genera and species	Mix of grasses and legumes	67 + 24 (80)	???
Phleum pratense	Timothy	66 + 13 (11)	NA
Dactylis glomerata	Orchard grass	65 + 28 (103)	NA
Lotus corniculatus	Bird's foot trefoil	65 + 23 (44)	146
Trifolium fragiferum	Strawberry clover	65 + 33 (57)	???
Mixed genera and species	Mix of forbs without legumes	64 + 19 (17)	NA
Kummerowia striata	Striate lespedeza	63 + 14 (5)	???
Mixed genera and species	Mix grasses, forbs, legumes	63 + 29 (195)	???
Mixed genera and species	Perennial grasses	63 + 20 (20)	NA
Mixed genera and species	Mowed mixed vegetation	61 + 25 (38)	???
Mixed genera and species	Mix of legumes only	59 + 22 (28)	???
Lolium perenne	Ryegrass (annual/perennial)	59 + 17 (10)	NA
Kummerowia stipulacea	Korean lespedeza	59 + 12 (36)	???
Mixed genera and species	Native legumes	55 + 24 (24)	50
Mixed genera and species	Ferns	54 + 19 (11)	NA
Mixed genera and species	Mix of grasses only	52 + 22 (14)	NA
Medicago sativa	Alfalfa	51 + 24 (15)	190
Kummerowia striata	Common lespedeza	45 + 20 (10)	???
Schedonorus arundinaceus	Tall fescue	43 + 28 (141)	NA

[a] (#) = number of comparisons found in the literature, field reports and unpublished datasets.
NA - not applicable; ??? - insuffcent information to provide an estimated rate of atmospheric nitrogen fixation for these legumes.

are poorly correlated with average reduction in tree growth (Table 1). Few studies have measured response of walnut or other hardwoods to harvesting of legumes for forage, although it can be assumed much of the fixed nitrogen is removed with the forage resulting in no net gain in available soil nitrogen to the trees (Van Sambeek, 2017).

In general, grass sods are more competitive than forage legumes as living mulches, although there is considerable variation among the different species of grasses and some variation among hardwood species (Van Sambeek and Garrett, 2004). When compared to tree growth in vegetation-free plots, reductions in tree growth with grasses average 47% for black walnut compared to only 32% for other hardwoods. In general, the least competitive grasses including several turf grasses and the annual cereal grains reducing hardwood growth as little as 30% compared to growth for trees in vegetation-free plantings. It is not uncommon to find reports of hardwood tree growth being reduced by more than 70% of the potential growth with a companion crop of tall fescue (*Schedonorus arundinaceus*) (Todhunter and Beineke, 1979; Dey et al., 1987; Van Sambeek et al., 1989; Van Sambeek and McBride, 1991; Alley et al., 1999). Reducing hardwood tree growth to 30% of their potential without competing vegetation has the effect of increasing rotation length by threefold, that is instead of 50–60 years (assuming 1 cm dbh/year) for harvestable veneer logs, rotation lengths will likely exceed to 170–200 years.

It is unclear why tall fescue, a relatively deep-rooted, sod-forming grass used in many silvopastoral practices, is so competitive, although there is evidence the dense fibrous root system is quite competitive for available soil nitrogen essentially starving the trees for nitrogen (Van Sambeek et al., 1989). There is good evidence that phytotoxic compounds may also be released from tall fescue roots or decomposing litter that slow growth of black walnut and white ash seedlings (Rink and Van Sambeek, 1985, 1987; Preece et al., 1991). Reduced growth of hardwoods by phytotoxic compounds for other grasses and forbs is also relatively well documented (Larsen and Schwarz, 1980; Ponder, 1986; Adler and Chase, 2007; Van Sambeek, 2017). Likewise, reduced growth of companion crops by trees producing phytotoxins is also well documented (Rietveld, 1982, 1983; De Scisciolo et al., 1990; Jose and Gillespie, 1998; Rizvi et al., 1999; Willis, 2000; von Kiparski et al., 2007).

3 Growing-space requirements

Unless thinned, hardwood stands will tend towards a condition called fully stocked meaning the site is being fully utilized. Stocking (or density) is a measure of the number of trees per ha compared with an optimal density for balanced health and growth (Stewart and Dawson, 2013). Stocking guides for upland temperate hardwoods when displayed graphically indicate if the trees are fully utilizing a site as overstocked, fully stocked or understocked (Gingrich, 1967). Unfortunately stocking charts are not very useful for evaluating tree-to-tree competition in most agroforestry practices because they would be considered understocked if an herbaceous understory is to be maintained. In addition, stocking guides were developed for stands where height growth is forced by side shade and trees have relatively small crowns compared to open-grown trees due to early death of shaded small diameter limbs in the lower crown.

Tree spacing within an agroforestry practice will likely be a compromise between having open-grown trees with minimal shading of the companion crop and denser spacing providing side shade to force tree height growth on a single stem and early death of small diameter branches. Krajicek et al. (1961), for planted trees, proposed that the ratio of the sum of the theoretical crown areas of all trees in a defined area when dividing by that area would better quantify tree competition than would stocking charts. Krajicek (1966) also found there was a strong linear relationship between stem diameter at breast height (DBH) and the crown width (CW) on open-grown temperate hardwoods (Table 2). For example, the CW in metres for open-grown black walnut trees is:

$$\text{CW (in m)} = (0.239)\,(\text{DBH in cm}) + 1.49$$

as derived from the English version of this equation:

$$\text{CW (in feet)} = (1.99)\,(\text{DBH in inches}) + 4.87$$

The crown competition factor (CCF) was coined by Krajicek et al. (1961) to describe the ratio between the sums of the open-grown crown area of all trees divided by the area the trees occupy. Because original calculations were made with crown area as the per cent of the area, the ratio is now multiplied by 100 to give per cent of area covered by tree crowns. The theoretical crown area rather than the actual crown area is used because actual crown areas when canopies

Table 2 Equations for determination of crown widths in metres as a function of diameter at breast height (DBH, 1.4 m tree height) in centimetres for several open-grown temperate agroforestry timber species (converted from English units in listed sources)

Species	Crown width (m), DBH (cm)	R^2	Sources
Black walnut	0.24 DBH + 1.49	0.99	Krajicek (1967), Krajicek (1966)
White oak	0.23 DBH + 0.55	0.99	Krajicek et al. (1961)
Red/black oaks	0.20 DBH + 1.38	0.98	Krajicek et al. (1961)
Pin oak	0.18 DBH + 2.76	0.91	Krajicek (1967)
Mixed oak/hickory	0.22 DBH + 0.95	0.98	Krajicek et al. (1961)
Sweetgum	0.24 DBH + 0.81	0.93	Krajicek (1967)
Bald cypress	0.24 DBH + 2.53	0.89	Krajicek (1967)

close adjust to the growing space and are no longer linearly related to stem diameter. For example, the CCF for a 2-ha silvopasture paddock with 50 black walnut trees with an average DBH of 40 cm is calculated as:

$$\text{CCF} = \frac{(100)(3.14)(\#\ \text{trees})(0.12\ \text{DBH} + 0.75)^2}{(10000)(\#\text{ha})}$$

where # is the number of trees and ha is the planting area. The resulting CCF is 25 indicating 25% of the paddock is covered by the sum of the theoretical crown areas for the 50 trees or the ratio of 0.25 m^2 of tree crowns per m^2 of pasture.

Schlesinger (1997) compared the CCF and ratio of stem diameter growth of walnut in close-spaced plantings to diameter growth in adjacent wide-spaced planting to estimate when tree competition began. Regression analyses indicated diameter growth started to be affected at a CCF of approximately 80, well under the original assumptions made by Krajicek et al. (1961) that between-tree competition begins when canopies close. Because root spread can be two to three times greater than canopy spread, most likely tree root competition for soil water and nutrients is more important than canopy competition for light in determining effects of tree competition (Schlesinger, 1988b).

Management recommendations as to when to thin can be made with reference to an appropriate range of CCF values. For example, maintaining black walnut stocking between CCFs of 100 and 120 is recommended for veneer production to keep trees growing near maximum rates while having some competition to promote good stem form and the slow growth and early death of lower limbs (USDA-FS, 1981). For nut production, CCFs of between 80 and 100 are recommended to maintain open canopies with good light penetration to the side branches (Jones et al., 1998; Reid et al., 2009).

Schlesinger (unpublished) found a CCF = 285 is equivalent to 100% stocking (net annual grow is zero) in pure walnut stands and a CCF of 100 is equivalent to 35% stocking. At 35% stocking, Blizzard et al. (2013) found transmitted light averaged approximately 40% of full sunlight. This is above light levels where shading alone decreases forage production (Pang et al., 2019a). Gardner et al. (1985) reported that photosynthesis equals or exceeds daily respiration for most forages at about 10% of full sun with light saturation for cool- and warm-season forages at 50% and 85% of full sun. From the above relationships, charts have been produced to graphically determine CCF (Fig. 1) or crown cover using the stocking chart format developed by Gingrich (1967) to relate tree density, tree DBH and basal area to CCF or tree crown cover as a per cent of the total area (USDA-FS, 1981; Law et al., 1994).

In some agroforestry practices, tree spacing can have a greater distance between rows than within rows. In this situation, tree competition affecting

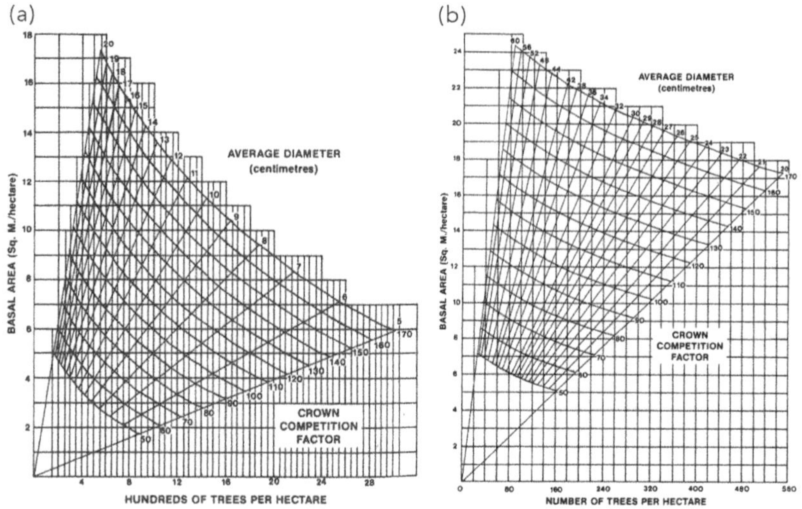

Figure 1 Charts for graphical determination of CCF or crown competition factor (USDA-FS, 1981). (a) Average tree diameters 5–20 cm. (b) Average tree diameters 20–60 cm.

stem growth is assumed to begin when the crowns begin to touch. When tree crowns touch, the distance between trees in metres is approximately 2.4 m plus 0.11 times the average of their DBHs in cm. For example, two trees 10 and 20 cm DBH if more than 4.1 m apart should still be putting on maximum diameter growth. The relationship can also be used to determine the distance between trees when thinning to allow trees to sustain maximum growth until the next planned thinning. For example, if trees are annually growing 1 cm in diameter and the next thinning is anticipated in 10 years, the above trees would need to be thinned to 5.2 m apart to sustain current growth rates to the next thinning.

In most agroforestry practices, trees will be thinned before canopy closure and substantial between-tree competition occurs to maintain adequate light to the understory crops or forages. Before thinning, the CCF should be calculated to determine how intense between-tree competition is. When managing hardwoods for high value lumber and timber products, CCF should be maintained between 90 and 120. When CCFs are near 120 or higher, removing one-third of the trees by selective thinning of the slower growing, poorly formed trees is usually sufficient to allow remaining trees to grow near maximum rates for another 5–10 years. The higher the CCF above 120 when thinned, the longer it may take for trees to respond with increased growth rates. Because of wide genetic variability in growth rates among individual trees of the same species, row thinning is not recommended for hardwoods unless trees have been grafted with superior timber or timber/nut scions. A double-row alley-cropping system has been suggested for managing grafted pecan and walnut during establishment that has narrow perennial grass alleyways allowing season-long access to the trees altering with wide alleyways for crops or forages (Van Sambeek and Reid, 2017). The double-row design allows row thinning on the diagonals to maintain open-crowns for nut production and light penetration into the cropped alleyways.

An alternative method to thinning hardwoods in a silvopastural or multirow buffer practice is crop tree management. This system focusses on managing the growing space around the canopy by releasing individual trees using a crown touching technique (Perkey et al., 1994; Miller et al., 2007). Theoretically up to 70–100 of the faster-growing trees with the potential for a high-quality log that are approximately 10 m or more apart are selected per ha and marked for release. Release is done simply by looking up into the crown of each crop tree and removing any adjacent trees that are touching the crop tree canopy, or that are likely to touch the crown of the crop tree before the next release. Crowns of temperate hardwoods typically expand 0.3 m/year. If the next release will be in 10 years, then the crown of the crop tree should be given 3 m or more distance from the crowns of the nearest competitive trees to permit crown expansion on three or four sides. If two crop trees are very close together, they can be treated as 'one' tree so both trees are released on three sides. Trees outside the release area around crop trees are left to compete with each other in part to keep crop trees from leaning to capture sunlight and to provide side shade to minimize production of epicormics sprouts on the crop trees.

4 Pruning recommendations and practices

In natural stands of close-spaced trees, side shade limits growth and longevity of lateral branches and slows stem diameter growth that allows time for dead branches to be shed and wounds to callus limiting defects to the central core. In any agroforestry practices with open-canopy trees, tree growth rates will be higher, branches larger in diameter and wound callus is likely to include dead stubs if branches are not artificially pruned. Pruning will be essential for production of high-quality logs of most species in any agroforestry practice. The objective should be clear, knot-free logs that peel easily for veneer that produce veneer sheets easily bookend-matched and have an attractive appearance due to straight grain and uniform cathedral patterns (Kesner, 1986). In production of high-grade construction lumber, knot-free timber is stronger with more uniform strength properties along with being easier to machine and finish.

To determine if artificial pruning is necessary or can be economically justified, Schlesinger (1986) analysed data on hardwoods in terms of their ability to shed branches naturally, how likely they were to produce epicormics sprouts in response to pruning and whether the average price difference between log grades (Prime $/mbf - Number 1 $/mbf and Number 1 $/mbf - Number 2 $/mbf) justified pruning. Schlesinger's tabulated results are given in Table 3 for Appalachian hardwoods with price differentials adjusted to present values. Additional hardwood species important to the Midwestern United States were added based on the research by Stubbs (1986), Meadows (1995), Miller et al. (2007) and the author's own experiences.

Tree species with persistent live or dead branches must be artificially pruned to produce knot-free timber (Table 3, Shedding, Category D and E). Artificial pruning must be considered an investment to improve log quality so that it will command a price high enough to more than cover the compounded costs for pruning that occur early in the stand rotation. For most hardwoods, the price differential between log grades is not sufficient to justify artificial pruning for log quality only (Table 3). For hardwoods where logs graded as prime can be sold as veneer logs, the price differential may justify pruning, especially in an agroforestry practice where trees should be regularly pruned to increase light penetration to the understory vegetation. Kesner (1986) reported hardwood logs sold for veneer sell for three to five times more than if sold as high-quality sawlogs. Because of the price differential between veneer and sawlogs and the increased costs of pruning above 3 m, pruning operations should concentrate on producing 2.5–3 m long logs for veneer rather than 5.5–6 m for sawlogs. Other advantages of pruning for one high-quality veneer log on open-grown trees include being less susceptible to wind damage and more easily managed for nut production.

Wide variation exists among the hardwood species in the production of epicormics branches following pruning (Table 3). The probability that a tree will produce epicormics branches after pruning is a function of its position in the stand canopy, tree density before thinning, the clear bole to crown ratios and the amount of canopy removed. The probability of epicormic sprouting increases substantially when clear bole to crown ratios exceed a 40:60 ratio or if more than a quarter of the potential canopy is removed in a single pruning operation. Hardwood species listed in Table 3 that have a high probability of producing epicormics branches may require less intense, more frequent pruning when grown in an agroforestry practice to minimize bole sprouting.

How and when to prune hardwood trees is not without controversy. Current recommendations are to first concentrate on any necessary pruning to remove forks to maintain a central, straight stem to the height of a harvestable log(s), and then begin removing the largest diameter lateral branches (Schlesinger and Shigo, 1989). In dense plantings few trees may need to be pruned as poorly formed trees will be removed during an early thinning. In most agroforestry practices, however, fewer trees are planted and most hardwood trees may need to be repeatedly pruned to produce a knot-free central stem and maximize light penetration to the understory plants.

Table 3 Hardwood species' characteristics relative to their ability to shed branches naturally, produce epicormics sprouts following pruning, log-grade price differential and whether cost-effective to prune

Genus species	Common name	Shedding[a]	Sprouting[b]	Price[c]	Prune
Acer rubrum	Red maple	B	B/C	C	No
Acer saccharinum	Silver maple	-	B	B	No
Acer saccharum	Sugar maple	D	B/C	C	-
Betula alleghaniensis	Yellow birch	B	C	-	-
Betula lenta	Sweet birch	B	C	B	No
Carya glabra	Pignut hickory	B	C	B	No
Carya illinoensis	Pecan	B	B/C	B	No
Carya laciniosa	Shellbark hickory	B	C	B	No
Carya ovata	Shagbark hickory	B	C	B	No
Carya tomentosa	Mockernut hickory	B	C	B	No
Celtis occidentalis	Hackberry	-	-	-	No
Fagus grandifolia	Beech	B	A	-	No
Fraxinus americana	White ash	B	D	C	-
Fraxinus pennsylvanica	Green ash	B	D	C	No
Juglans cinerea	Butternut	C	-	E	Yes
Juglans nigra	Black walnut	E	B	F	Yes
Liriodendron tulipifera	Yellow poplar	B	C	B	No
Liquidambar styraciflua	Sweetgum	-	B	-	-
Magnolia acuminata	Cucumbertree	B	-	-	No
Nyssa sylvatica	Black gum	C	-	-	No
Platanus occidentalis	Sycamore	B	B/C	-	No
Populus spp.	Poplar/aspen	A	C	-	No
Prunus serotine	Black cherry	C	B	D	Yes
Quercus alba	White oak	C	A	E	Yes
Quercus bicolor	Swamp white oak	D	B	D	-
Quercus coccinea	Scarlet oak	D	-	D	Yes
Quercus falcata	Cherrybark oak	D	A/B	E	Yes
Quercus macrocarpa	Bur oak	C	-	D	-
Quercus michauxii	Chestnut oak	C	B	D	No
Quercus rubra	Northern red oak	B	A/B	D	-
Quercus shumardii	Shumard oak	C	B	D	No
Quercus velutina	Black oak	C	-	D	Yes
Robinia pseudoacacia	Black locust	B	-	-	No
Sassafras albidum	Sassafras	B	-	-	No
Tilia americana	Basswood	B	B	A	No
Ulmus Americana	American elm	B	A	-	No

[a] Natural shedding categories: A = excellent, B = good under side shade, C = okay, D = poor and E = very poor.
[b] Categories for producing epicormics sprouts: A = very high, B = high, C = moderate and D = low.
[c] Price differential lumber/MBF were A = US$0-50, B = US$50-100, C = US$100-150, D = US$150-200, E = US$200-250 and F = >US$250.
Source: modified from Schlesinger (1986).

A fork usually consists of two or more closely spaced, vertically oriented stems with diameters that are within half the diameter of the largest stem. If forking occurs in the current year's growth (sometimes called a crow's nest), artificial pruning may not be required (Schlesinger and Bey, 1978). Schlesinger (1982) evaluated effectiveness of pruning a crow's nest to a single terminal shoot in black walnut saplings and found no differences in percentage of trees that eventually developed straight, single stems whether pruned or left to naturally straighten. Most of the competing terminal shoots in a crow's nest became lateral branches in the next growing season. Saplings with vertically oriented stems having two or more years of older tissues are less likely to naturally straighten and will benefit from corrective pruning to a single terminal shoot (Beineke, 1977). Unwanted vertical stems should be cut diagonally between the unwanted stem and the branch bark ridge (described below) and exiting at a depth equal to the base of the branch bark ridge (Shigo, 1989).

In some cases, especially when open grown, saplings develop multiple forks and crooks in response to disease, injury or poor genetics. For these saplings, coppicing is frequently recommended; however, there is no consensus whether to coppice the stem close to the root collar or to high coppice just below the deformed stem. Because coppiced saplings have well-developed root systems, coppice growth can be very rapid (and may require staking) with little net loss in height growth. Initially new shoots of coppiced saplings will have excellent form but most saplings will again develop form issues after the first or second growing season. Results are also mixed whether the coppice cut should be a horizontal cut or a sloping cut made just above a dormant bud. A sloping high-coppice cut behind a bud tends to produce the fewest sprouts especially if made in the region of stem between where height growth ended each year. On pole-sized trees that have been cut horizontally for bark grafts, the stem naturally abscises a triangular piece of stem immediately behind a successful graft suggesting a sloping cut should be used for a high coppice.

Pruning should concentrate secondly on removing the fastest-growing or largest diameter lower lateral branches before removing the lowest lateral branches to raise the base of the crown (Shigo, 1989). For veneer-quality logs, it is best if lateral branches can be removed before they are 3–5 cm in diameter. Branches tend to cluster in the areas immediately below where a terminal bud was set at the end of each growing season. Multiple large diameter lateral branches in the same area should be slowly removed over multiple years to reduce the possibility of developing ring shake in the future log. If possible also avoid pruning large lateral branches in vertical alignment within 1 m of each other. Also removing more than a quarter of the next growing seasons' anticipated foliage is likely to substantially slow tree growth and induce epicormic sprouts to replace the lost crown (Schlesinger, 1988a). If more large diameter branches exist than can be removed in one pruning operation without harming the tree, diameter growth can be slowed by cutting or heading back branches to approximately a third of their original length. Numerous dormant and lateral buds will elongate the following spring feeding the entire tree with minimal increase in branch diameter. Also, if branch and stem are similar in diameter, head back the branch and wait a few years before removing stubbed branches to minimize main stem breakage.

When pruning lateral branches, there is the temptation to 'flush cut' or cut vertically close to the stem to make the stem smooth and begin laying down straight-grained wood sooner. Shigo et al. (1979) found flush-cut pruning tends to lead to substantial internal decay and other defects. The alternative is 'target' or 'collar' pruning. In his article on how branches are attached to stems, Shigo (1985) reported that in the spring the branch produces new xylem first which extends back along parts of the stem before the stem produces new xylem that extends into the branch (Fig. 2). Where the two tissues overlap is a slightly swollen area or collar around the branch. Within the branch collar are cells pre-programmed to rapidly form a compartmentalization zone that naturally blocks entrance of most microorganisms and insects into the stem when branches either die or are pruned (Shigo, 1984).

Proper target pruning just outside the branch bark ridge and branch collar (line AB in Fig. 3) results in a full circle of visible callus along the edges of the wound that will eventually cover the wound. O'Hara (2007) argues pruning

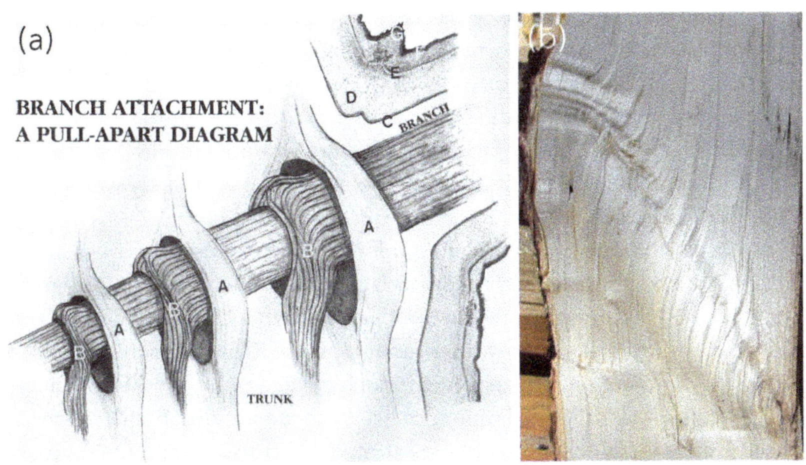

Figure 2 (a) Exploded diagram showing how branches are attached to stems when new branch wood or collar (B) is overlaid by new stem wood or collar (A) resulting in a swollen branch collar laid down by the cambium (C) and the branch bark ridge (G) (Shigo, 1996) that forms as branch and stem tissues meet above the branch. D, E and F are the phloem, bark cambium (phellogen) and dead bark tissues, respectively. (b) Maple stem split through the region near a branch showing the overlapping growth rings.

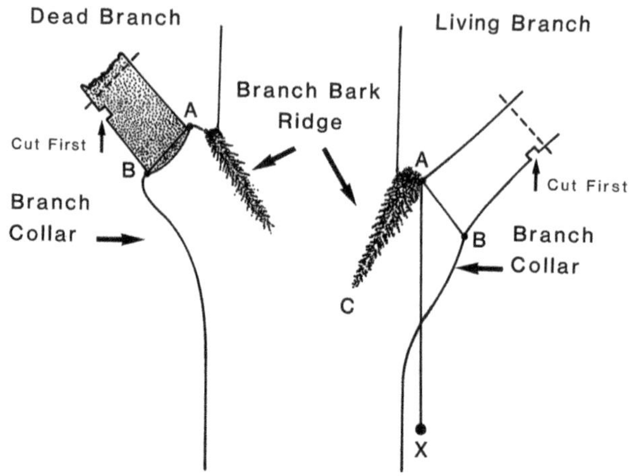

Figure 3 Diagram of natural target pruning where pruning cuts are made at A–B just outside the branch collar on dead branches and along A–B on live branches where A–X is a vertical line just outside the branch bark ridge and the angle XAC is equal to XAB.

just outside branch bark ridge with vertical cuts through the branch collar (line from A emerging between X and B in Fig. 3) may produce a larger wound, but it will callus over faster and begin laying down clear wood sooner than target pruning especially on small diameter branches. Flush cutting (line AX in Fig. 3 if placed before the branch bark ridge) creates an even larger wound and removes the natural defence tissues found in the branch collar and delays callus formation leaving wood tissues exposed to pathogens for the longest period of time. Stub cutting, or pruning well away from the branch collar, will delay wound closure and increase the number of growth rings that must be laid down before straight wood is produced over the wound. Wound dressings are no longer recommended because they are expensive and rarely work as well as the tree's natural defences (Shigo and Shortle, 1984).

Although there is little published information on when the best time to prune is, the consensus is that hardwoods can be pruned anytime of the year except possibly the period between budburst and when the first leaves are fully mature (April–May when energy reserves are lowest). Dead branch pruning can be done anytime if living tissues in the branch collar are not cut creating a fresh wound (Schlesinger, 1988a). Pruning of live branches during the late dormant season usually leads to the least amount of wood discolouration and internal decay and the most rapid wound closure (Smith, 1980; Armstrong et al., 1981). Trees of some species should not be pruned when insects are present that are attracted to open wounds (Bedker et al., 2012). Although it should be avoided, there is little evidence that pruning trees that bleed after sap starts rising in the spring will result in additional discolouration or slow tree growth.

5 Log and wood quality

Straight, cylindrical butt logs consisting of perfect wood with straight grain, evenly spaced growth rings ought to be the objective for hardwood trees managed in an agroforestry planting. Few studies have apparently been done to see if we can achieve this objective. Based on what little information was published before 2003, Garrett (2003) suggested that the quality of hardwoods grown under alley cropping if properly managed would not vary substantially from forest-grown trees but the time required to grow trees to market size in an agroforestry practice would be greatly reduced. Theoretically, research comparing wood quality of planted hardwoods maintained at relatively low CCFs to trees harvested from natural stands should provide a clue as to what differences, if any, are likely to occur when open-grown hardwoods are managed for quality in an agroforestry practice. As is the case with many of these studies, Phelps and Chen (1989) found the open-grown planted hardwoods exhibited more defects mainly due to lack of intensive management.

Because of the price differential between veneer and sawlogs for most hardwoods, Kesner (1986) suggested managing trees for veneer should be more profitable even if the quality must be higher than managing for sawlogs. He proposed that logs marketed for veneer must meet six general requirements that include having a nearly cylindrical shape with a well-centred heart, a minimum of internal defects which are concentrated in either the central core or slab zone, uniform wood colour and texture and properly sawed to length without embedded soil and other debris. The last requirement is controlled by the logger at the time of harvest and will not be addressed in this section.

In an agroforestry practice with open-grown trees, the shape of the butt log is largely a function of regular pruning to maintain a central leader by removing forks. Cutter (1991) reported only minor differences in stem eccentricity of black walnut logs harvested from an alley-cropping practice where trees with elliptical crowns were grown in close spacings within rows and wide spacings between rows. Because eccentricity was measured as the radius between pith and outer growth ring, it can be assumed these trees also had a well-centred heart. With frequent and regular pruning, branch wounds should callus over quickly laying down straight-grained wood, thus limiting defects associated with branches that extend to the central core. Crook and sweep, log defects associated with forks, tend to disappear over time due to differential growth along the stem, and may potentially yield a round cylindrical stem with irregular grain limited to the central core along with the wavy pith centre. In his review on stem form, Larson (1963) concluded open-grown trees are unfortunately likely to have greater clear bole taper which will affect the grain appearance in lumber and veneer. He also concluded that open-grown trees in response to wind may exhibit more butt flare, so veneer-quality butt logs may need to be bucked a little longer if only straight-grained veneer is desired.

Carpenter et al. (1989) described a wide range of defects that can occur in hardwood logs including limbs, bumps and overgrowths; suppressed buds and epicormic sprouts; shakes and splits; holes including birdpeck; and decay and discolourations. Most of the important blemishes in wood are associated with abnormal features visible in the bark. As an aid to growers and timber buyers, Rast co-authored several photographic guides that included stereo pairs of photographs of external indicators of the above internal defects for northern red oak (Rast, 1982), black cherry (Rast and Beaton, 1985), black walnut (Rast et al., 1988), white oak (Rast et al., 1989), yellow poplar (Rast et al., 1990a), sugar maple (Rast et al., 1990b) and yellow birch (Rast et al., 1990c).

The most frequent bark distortions are bumps and overgrowths (catfaces) covering knots from lateral branches. Bumps are protuberances usually projecting over sound or rotting limb stubs while overgrowths appear as roughly concentric rings in the bark formed by the callus that has grown over knots in the wood. As the knot is buried deeper into the stem, overgrowths gradually disappear, especially knots from small diameter limbs. More bumps and overgrowths are reported on logs from open-grown trees than trees from natural stands (Phelps and Chen, 1989). On open-grown trees, lateral branches will persist for a longer time and have greater diameter growth without artificial pruning resulting in a larger area to callus over producing more prominent overgrowths that can remain visible in the bark when harvested even though straight-grained wood has been laid down over the wound callus.

Suppressed buds and epicormic branches can also result in bumps and overgrowths. Most suppressed buds are formed in the cambium at the base of each leaf at the time of shoot development (Kormanik and Brown, 1967). Although most axillary buds die, a few buds remain dormant within the meristem that can be observed as raised pimples on the wood surface when the bark is peeled off. Most of these dormant buds have a vascular trace that extends back to the primary xylem surrounding the pith. In cross section, the vascular trace is considered a defect in high-end veneers. Dormant buds or pinknots can also develop adventitiously following injury to the cambium. Overpruning, damage to the crown or other stresses can cause these suppressed buds to divide forming pinknot clusters, or the buds to elongate and develop into epicormics sprouts. Chen et al. (1995) found walnut lumber cut from an alley-cropping practice that was frequently pruned averaged four pinknots and four clusters of two to four pinknots per metre with minor differences between the basal and apical ends of the future veneer logs. Pinknots within clusters were usually within 3 cm of each other and usually associated with knots from pruned branches. Minor variation existed from lumber cut from trees of seed selected for nut production or timber production suggesting death of suppressed buds is controlled by environment and not genetics. Epicormic sprouts from dormant buds will leave small tapered knots in the stem until the sprout is shed or pruned off and the wound calluses over.

Ring and radial shakes are major defects because they cause wood to separate during processing for veneer or lumber. Both are associated with compartmentalization of wounds (Shigo, 1984). Wood in ring shakes separate along the circumferential plane when wood in a new growth ring becomes infected and separates from the underlying wood as the tree forms barrier zones to compartmentalize the infection. Shigo et al. (1979) reported that ring shake was associated with 11 of 17 pruning wounds following 'flush-cut' pruning of both live and dead branches. Ring shake is frequently found following removal of the branch collars that form around the bases of dying and dead branches on older trees.

Radial shakes are also often associated with wounds resulting from infected knots, insect attacks or mechanical injury. Some decay commonly extends upwards and downwards from the wound into the wood-producing longitudinal radial separations in the wood. When the wood is stressed such as during a drought or sudden changes in temperature, these separations are weak areas and will open resulting in vertical cracks in the bark followed by wound callus to close the crack (Butin and Shigo, 1981). Wound callus from repeated cracking when frozen wood is rapidly thawed by direct sunlight results in vertical bulges referred to as frost cracks. In an agroforestry practice with open-grown trees, stems of some species may need to be painted white to reflect sunlight during the winter when stems are not shaded by leaves. Trees with seams will not produce veneer-quality logs and should be removed when thinning. When young, trees with vertical seams should be coppiced below the seam hoping the new wounds will callus over without leaving weak areas that will become new radial shakes.

Holes are wounds that expose the sapwood and remain open or become occluded with wound callus leaving a depression in the bark (Carpenter et al., 1989). When logs are rotary-sliced, wounds do not show concentric growth rings in the wood as do overgrowths from knots. Common causes of wounds include mechanical damage from logging activities, increment borers, tap holes, boring insects and birdpeck. Trees in an agroforestry practice have an increased risk for developing holes associated with use of equipment or livestock to manage the understory crops and forages. New holes are usually accompanied by stain in the wood of the exposed growth ring which may separate leading to ring shake (Shigo, 1963, 1996). Sapsuckers are a common cause of holes as they drill into the sapwood to feed on bleeding sap and the sapwood frequently creating bands or rows of holes around the stem. Sapsuckers sample the cambium and sapwood of many trees and usually select particular trees to repeatedly tap for multiple years. Garrett et al. (1994) recommended that trees heavily damaged in an agroforestry practice by sapsuckers be retained when thinning to minimize risk to sapsuckers feeding on neighbouring trees after thinning.

Decay and discolouration in the wood are also defects that can be minimized through stand management. Smith (1980) found spring pruning wounds on walnut resulted in much smaller columns of discoloured wood and much more diffuse dark margins than did fall wounds. The dark discolouration is maintained in the heartwood and is a serious source of degrade in finished veneer. Eisner et al. (2002) suggested the amount of discoloured wood that develops following branch removal decreases as branch orientation changes from vertical to horizontal. The major cause of decay in stems and branches is infection of wounds by microorganisms that grow through the protective chemical barriers formed in the wood around wounds (Shigo, 1984; Hart, 1982).

Hoffard et al. (1977) and Hart et al. (1986) examined the economic impact and causes of decay in black walnut in the central United States and found presence of decay decreased log value by 66%. When stands were examined for probable cause of decay, over 40% had some damage caused by livestock, especially swine because swine scrape and peel the bark at the base of trees allowing decay to enter and develop. Kesner (1986) also indicated prolonged presence of livestock around trees usually results in dark stains in the wood and many small scars around the lower portion of the stem. Proper silvopastoral management using rotational grazing should minimize discolouration and decay in the wood associated with livestock, but not swine. Lehmkuhler et al. (2003) found no damage to four hardwood species by livestock when trees were planted into open pastures and protected by electric fencing used to create multiple paddocks for rotational grazing.

Few studies have been done comparing the appearance (texture) and properties of wood from forest-grown trees to open-grown trees in plantations or in an agroforestry practice (Cutter et al., 1997). Kesner (1986) and Cassens (2004) suggested the most attractive hardwood veneer is sliced from logs with six to ten rings per inch. Within agroforestry practices where tree CCF is maintained below 120 to minimize fluctuations in growth ring widths, texture is likely to be reduced. Presumably the price reduction for too few rings per inch will be offset by the increased volume of clear wood in trees from managed agroforestry practices. One option to improve texture is to plant seed from or graft saplings to selections from trees displaying unique figure such as 'bird's eye', 'fiddle-back' or 'curly grain' (Beals and Davis, 1977; Bragg and Stokke, 1999). Unfortunately, McKenna et al. (2015) found curly grain figure does not propagate true-to-type in grafts on open-grown walnut or within seedlings from parent trees with figure.

Agroforestry practices are unlikely to produce substantial changes in the colour of hardwood lumber or veneer. Nelson et al. (1969) found soil properties affected colour more than growth rate for fast- and slow-grown black walnut trees on good and poor sites in Indiana and Missouri. Later, Phelps (1989) reported that walnut heartwood from faster-grown trees is usually somewhat lighter in colour (luminance) than that from slow-grown trees. There is some indication that in heartwood an increase in specific gravity is likely to cause a darker colour.

The best indicators of easily measure parameters for accessing changes in important wood properties are probably wood-specific gravity or density (Phelps and Chen, 1999). In his review on stem form, Larson (1963) concluded that wood density is lower in the butt log of open-grown trees than in forest-grown trees. Cutter and Garrett (1993), however, found that height, diameter and wood-specific gravity was equal or higher for alley-cropped trees than forest-grown trees. Cutter (1991) found that specific gravity in the heartwood of 15-year-old black walnuts grown in an alley-cropping practice gradually increased from the pith towards the sapwood followed by a substantial decrease in specific gravity in the sapwood. Similarly, Phelps and Chen (1989) reported that specific gravity gradually increased in the heartwood with distance from the pith with a marked decrease in the sapwood although volumetric shrinkage of lumber cut from open-grown and forest-grown walnut was similar.

Although few studies have been done quantifying log and wood quality from open-grown hardwoods grown in an agroforestry practice, results comparing open-grown trees to forest-grown (slow-growth) trees would indicate that logs from an agroforestry practice are likely to have similar wood properties and colour, a less desirable texture (figure) and show more defects. Most defects could be limited to the central core if trees were properly pruned on a regular basis. At this point in time there is no reason not to accept Garrett's original conclusion in 2003 that the quality of hardwoods grown under an agroforestry practice (if properly managed) would not vary substantially from forest-grown trees but the time required to grow them to market size would be greatly reduced.

6 Summary

This chapter synthesizes available information on managing hardwood trees for production of veneer and high-quality sawlogs. Managing timber within the agroforestry practices of alley cropping, forested riparian and upland (windbreaks) buffers, and silvopasture typically involves planting hardwood trees at lower densities, and their development in more open conditions than which occurs in natural stands. Choice of crops or forages can substantially impact the growth rate of trees in the different agroforestry practices. Trees will need intensive management with frequent pruning and thinning to produce high-quality hardwood logs. Frequent thinning will be required to maintain open tree canopies to provide sufficient sunlight for crop and forage production. Frequent thinning will also be required to minimize within-tree species competition to promote production of wood that has evenly spaced growth rings and uniform texture. Target pruning should be done in the late dormant season to minimize wood discolouration and to create small wounds that callus over quickly, thus reducing the risk of infection and subsequent internal decay. Overall, high-quality open-grown hardwood trees can be produced in an agroforestry practice with intensive management, and in a relatively shorter time frame than trees grown in natural forests.

7 Future trends in research

Research needs include a more intensive search of existing literature reporting on tree growth as influenced by different ground covers to add to the existing database for more reliable estimates on what grasses and legumes produce quality forage and have the least impact on hardwood tree growth and seed production. Few studies have evaluated the impacts of harvesting crops and forages on growth of hardwood trees. Even fewer studies have been done to determine impacts of pruning outside the dormant season. More information is also needed on the impacts of rotational grazing of cattle, sheep and goats in silvopastoral practices on soil properties and tree quality. More evaluation of wood properties and defects are needed as trees mature in established agroforestry practices.

8 Where to look for further information

An excellent source of general information on hardwoods that could be managed in an agroforestry practice is the *Silvics of North America - Hardwoods* (Burns and Honkala, 1990). *Agroforestry Systems* (http://www.springer.com/life+sciences/forestry/journal/10457) is the journal most likely to publish new information in response to temperate hardwood trees within the five agroforestry practices.

The Training Manual for Applied Agroforestry Practices (Gold et al., 2015b) and *Handbook for Agroforestry Planning and Design* (Gold et al., 2015a) prepared by the University of Missouri Center for Agroforestry staff as training guides for their Agroforestry Academy are excellent introductions to the agroforestry practices and their applications. Because most agroforestry practices involve planting open-grown trees and shrubs, *Fruit and Nut Production* by Olcott-Reid and Reid (2007) is an excellent textbook with guidelines for managing the woody component as horticultural crops within these practices. A complementary textbook to this agroforestry book is *North American Agroforestry: an Integrated Science and Practice* edited by Garrett (2009).

There are a number of organizations with annual or biannual meetings that include technical presentations on managing hardwoods for timber production including:

- The Central Hardwood Forest Conference (website changes with each meeting)
- The Association for Temperate Agroforestry (http://www.aftaweb.org)
- The Walnut Council (http://walnutcouncil.org/)
- The Northern Nut Growers Association (https://nutgrowing.org/)
- International Walnut Symposium sponsored by International Society for Horticultural Science (http://www.ishs.org/symposium/)

Centres of expertise include:

- The University of Missouri Center for Agroforestry (http://www.centerforagroforestry.org/)
- Hardwood Tree Improvement and Regeneration Center (https://htirc.org/)
- Dale Bumpers Small Farms Research Center (https://www.ars.usda.gov/southeast-area/booneville-ar/dale-bumpers-small-farms-research-center/)
- National Agroforestry Center (NAC) (https://www.fs.usda.gov/nac/)

9 References

Adler, M. J. and Chase, C. A. (2007). Comparison of the allelopathic potential of leguminous summer cover crops: cowpea, sunn hemp, and velvetbean. *HortSci.* 42(2), 289–93.

Alley, J. L., Garrett, H. E., McGraw, R. L., Dwyer, J. P. and Blanche, C. A. (1999). Forage legumes as living mulches for trees in agroforestry practices - preliminary results. *Agrofor. Syst.* 44, 289–91.

Armstrong, J. E., Shigo, A. L., Funk, D. T., McGinnes Jr., E. A. and Smith, D. E. (1981). A macroscopic and microscopic study of compartmentalization and wound closure after mechanical wounding of black walnut trees. *Wood Fiber* 13, 275–91.

Beals, H. O. and Davis, T. C. (1977). *Figure in Wood - an Illustrated Review*. Alabama Agric. Exp. Stn. Bull. #486. Auburn, AL, 5pp. Available at: https://aurora.auburn.edu/handle/11200/2414.

Bedker, P. J., O'Brien, J. G. and Mielke, M. E. (2012). *How to Prune Trees*. NA-FR-01-95. USDA For. Serv. Northeast. Area St. & Private For., 30pp. Available at: https://www.fs.usda.gov/naspf/resources/tree-care-how-prune-trees-english.

Beineke, W. F. (1977). *Corrective Pruning of Black Walnut for Timber Form*. Forestry and Natural Resources Woodland Management FNR-76. Purdue University, West Lafayette, IN. Available at: http://www.agcon.purdue.edu/AgCom/Pubs/FNR/FRN-76.html.

Blizzard, E. M., Kabrick, J. M., Dey, D. C., Larsen, D. R., Pallardy, S. G. and Gwaze, D. P. (2013). Light, canopy closure, and overstory retention in upland Ozark forests. In: *Proceedings of the 15th Biennial Southern Silvicultural Research Conference* (Ed.

Guildin, J. M.), pp. 73-9. e-Gen. Tech. Rep. SRS-GTR-175. USDA For. Serv. South., Asheville, NC. Available at: http://www.fs.usda.gov/treesearch/pubs/43478.

Bragg, D. C. and Stokke, D. D. (1999). Annotated bibliography on 'birdseye' figure grain. Gen. Tech. Rep. NE-263. USDA For. Serv., Radnor, PA, 15pp. Available at: https://www.fs.usda.gov/treesearch/pubs/3795.

Burns, R. M. and Honkala, B. H. (Eds) (1990). *Silvics of North America - Hardwoods* (vol. 2). Agric. Handb. 654. USDA For. Serv., Washington DC. Available at: https://www.na.fs.fed.us/spfo/pubs/silvics_manual/volume_2/vol2_Table_of_contents.htm.

Butin, H. and Shigo, A. L. (1981). *Radial Shakes and 'frost cracks' in Living Oak Trees*. Res. Paper NE-478. USDA For. Serv., Broomall, PA, 21pp. Available at: http://www.fs.usda.gov/treesearch/pubs/15026.

Carpenter, R. D., Sonderman, D. L., Rast, E. D. and Jones, M. J. (1989). *Defects in Hardwood Timber*. Agric. Handb. 678. USDA For. Serv., Washington DC, 88pp. Available at: http://www.fs.usda.gov/treesearch/pubs/13903.

Cassens, D. L. (2004). Factors affecting the quality of walnut lumber and veneer. In: *Black Walnut in a New Century* (Eds. Michler, C. H., Pijut, P. M., Van Sambeek, J. W., Coggeshall, M. V., Siefert, J., Woeste, K., Overton, R. and Ponder Jr., F.), pp. 161-7. Gen. Tech. Rep. NC-243. USDA For. Serv., St. Paul, MN. doi:10.2737/NC-GTR-243. Available at: https://www.fs.usda.gov/treesearch/pubs/14727.

Chen, P. Y. S., Bodkin, R. E. and Van Sambeek, J. W. (1995). Variation in pin knot frequency in black walnut lumber cut from a small provenance/progeny test. In: *Proceedings of the 10th Central Hardwood Forest Conference* (Eds. Gottschalk, K. W. and Fosbroke, S. L. C.), p. 488. Gen. Tech. Rep. NE-197. USDA For. Serv., Broomall, PA. Available at: http://www.fs.usda.gov/treesearch/pubs/12819.

Cunningham, R. A. (1988). Genetic improvement of trees and shrubs used in windbreaks. *Agric. Ecosyst. Environ.* 22/23, 483-98.

Cutter, B. E. (1991). Tree and wood quality in 15-year-old eastern black walnut (*Juglans nigra* L.) grown in agroforestry regimes. In: *Proceedings of the 2nd Conference on Agroforestry in North America* (Ed. Garrett, H. E.), pp. 116-23. Association for Temperate Agroforestry and the University of Missouri, Columbia, MO.

Cutter, B. E. and Garrett, H. E. (1993). Wood quality in alley-cropped eastern black walnut. *Agrofor. Syst.* 22, 25-32.

Cutter, B. E., Phelps, J. E. and Stokke, D. D. (1997). The silviculture-wood quality connection in eastern black walnut. In: *Knowledge for the Future of Black Walnut* (Ed. Van Sambeek, J. W.), pp. 224-30. Gen. Tech. Rep. NC-191. USDA For. Serv. North Central For. Exp. Sta., St. Paul, MN. Available at: https://www.fs.usda.gov/treesearch/pubs/10255.

De Scisciolo, B., Leopold, D. J. and Walton, D. C. (1990). Seasonal patterns of juglone in soil beneath *Juglans nigra* (black walnut) and influence of *J. nigra* on understory vegetation. *J. Chem. Ecol.* 16, 1111-30.

Dey, D. C., Conway, M. R., Garrett, H. E., Hinckley, T. S. and Cox, G. S. 1987. Plant-water relationships and growth of black walnut in a walnut-forage multicropping regime. *For. Sci.* 33(1), 70-80. Available at: http://www.fs.usda.gov/treesearch/pubs/12337 or 12338.

Eisner, N. J., Gilman, E. F. and Grabosky, J. C. (2002). Branch morphology impacts compartmentalization of pruning wounds. *J. Arbor.* 28(2), 99-105.

Gardner, F. P., Pearce, B. B. and Mitchell, R. L. (1985). *Physiology of Crop Plants*. Iowa State University Press, Ames, IA.

Garrett, H. E. (2003). Roles for agroforestry in hardwood management and natural-stand management. In: *Proceedings of the 13th Central Hardwood Forest Conference* (Eds. Van Sambeek, J. W., Dawson, J. O., Ponder Jr., F., Loewenstein, E. F. and Fralish, J. S.). Gen. Tech. Rep. NC-234, pp. 33–43. USDA For. Serv., St. Paul, MN. Available at: http://www.fs.usda.gov/treesearch/pubs/15722.

Garrett, H. E. (Ed.) (2009). *North American Agroforestry: an Integrated Science and Practice*. American Society of Agronomy, Inc., Madison, WI, 379pp.

Garrett, H. E., Neelan, C. and Jones, J. E. (1994). Incidence of yellow-bellied sapsucker damage under walnut agroforestry. In: *Proceedings of the 3rd North American Agroforestry Conference* (Eds. Schultz, R. C. and Anc Colletti, C. P.), pp. 25–8. Association for Temperate Agroforestry and Iowa State University, Ames, IA.

Gingrich, S. F. (1967). Measuring and evaluating stocking and stand density in upland hardwood forests in the central states. *For. Sci.* 13, 38–53.

Gold, M., Cernusca, M. and Hall, M. (Eds) (2015a). *Handbook for Agroforestry Planning and Design*. The Center for Agroforestry, University of Missouri, Columbia, MO, 80pp. Available at: http://www.centerforagroforestry.org/pubs/training/HandbookP&D13.pdf.

Gold, M., Hemmelgarn, H. and Mori, G. (Eds) (2015b). *Training Manual for Applied Agroforestry Practices*. The Center for Agroforestry, University of Missouri, Columbia, MO, 385pp. Available at: http://www.centerforagroforestry.org/pubs/training/.

Hart, J. H. (1982). Variation in inherent decay resistance of black walnut. In: *Black Walnut for the Future*, pp. 12–17. Gen. Tech. Rept. NC-74. USDA For. Serv. N. Cent. For. Exp. Sta., St. Paul, MN. Available at: https://www.fs.usda.gov/treesearch/pubs/10138.

Hart, J. H., Baughan, R. and Jennings, N. E. (1986). Economic impact of decay on black walnut. *North. J. Applied For.* 3, 116–18.

Hoffard, W. H., Heflin, E. L. and Anderson, R. L. (1977). *Economic Impact of Defects in Missouri-grown black walnut*. Impact Survey D-19-77. USDA For. Serv. Northeastern For. Exp. Sta., Broomall, PA.

Jones, J. E., Mueller, R. and Van Sambeek, J. W. (Eds) (1998). *Nut Production Handbook for Eastern Black Walnut*. Southwest Missouri Res. Conserv. & Dev., Inc., Springfield, MO, 147pp. Available at: https://www.fs.usda.gov/treesearch/pubs/10852.

Jose, S. (2009). Agroforestry for ecosystem services and environmental benefits: an overview. *Agrofor. Syst.* 76, 1–10.

Jose, S, and Gillespie, A. R. (1998). Allelopathy in black walnut (*Juglans nigra* L.) alleycropping: II. Effects of juglone on hydroponically grown corn (*Zea mays* L.) and soybean (*Glycine max* L. Merr.) growth and physiology. *Plant Soil* 203, 199–205. doi:10.1023/A:1004353326835.

Jose, S., Benjamin, T., Stall, T., Gillespie, A. R., Hoover, W. L., Mengel, D. B., Seifert, J. R. and Biehle, D. J. (1997). Biology and economics of a black walnut-corn alley cropping system. In: *Knowledge for the Future of Black Walnut* (Ed. Van Sambeek, J. W.), pp. 203–8. Gen. Tech. Rep. NC-191. USDA For. Serv., St. Paul, MN. Available at: https://www.fs.usda.gov/treesearch/pubs/10255.

Jose, S., Walter, D. and Kumar, B. M. (2017). Ecological considerations in sustainable silvopasture design and management. *Agrofor. Syst.* 93(1), 317–31. doi:10.1007/s10457-016-0065-2.

Kallenbach, R. (2009). Integrating silvopastures into current forage-livestock systems. In: *Agroforestry Comes of Age: Putting Science into Practice* (Eds. Gold, M. A. and Hall, M. M.), pp. 455–61. Proceedings of the 11th North American Agroforestry

Conference. Association for Temperate Agroforestry and University of Missouri, Columbia, MO.

Kallenbach, R. L., Kerley, M. S. and Bishop-Hurley, G. J. (2006). Forage accumulation, quality and livestock production from an annual ryegrass and cereal rye mixture in a pine-walnut silvopasture. *Agrofor. Syst.* 66, 43-53.

Kesner, A. L. (1986). Characteristics of and management for veneer quality hardwood logs. In: *Guidelines for Managing Immature Appalachian Hardwood Stands Proceedings* (Eds. Smith, H. C. and Eye, M. C.), pp. 22-32. SAF Publ. 86-02. West Virginia Univ. Books, Morgantown, WV. Available at: https://www.fs.usda.gov/treesearch/pubs/50049.

Kormanik, P. P. and Brown, C. L. 1967. Epicormic branching: a real problem in plus tree selection. In: *Proceedings of the Ninth Southern Forest Tree Improvement Conference* (Ed. Tennessee Valley Authority), pp. 69-74. Eastern Tree Seed Laboratory, Macon, GA. Available at: https://rngr.net/publications/tree-improvement-proceedings/sftic/1967.

Kort, J. (1988). Benefits of windbreaks to field and forage crops. *Agric. Ecosyst. Environ.* 122(23), 165-91.

Krajicek, J. E. (1966). Growing space requirements. In: *Black Walnut Culture*, pp. 47-9. Gen. Tech. Rep. NC-4. USDA For. Serv., St. Paul, MN.

Krajicek, J. E. (1967). Maximum use of minimum acres. In: *Proceedings of the 9th Southern Conference for Forest Tree Improvement* (Ed. Tennessee Valley Authority), pp. 35-7. Eastern Tree Seed Laboratory, Macon, GA. Available at: https://rngr.net/publications/tree-improvement-proceedings/sftic/1967.

Krajicek, J. E., Brinkman, K. A. and Gingrich, S. F. (1961). Crown competition -a measure of density. *For. Sci.* 7, 35-42.

Kremer, R. J. and Kussman, R. D. (2011). Soil quality in a pecan-kura clover alley cropping system in the Midwestern USA. *Agrofor. Syst.* 83, 213-23.

Larsen, M. M. and Schwarz, E. L. (1980). Allelopathic inhibition of black locust, red clover, and black alder by six common herbaceous species. *For. Sci.* 26(3), 511-20.

Larson, P. R. (1963). *Stem Form Development of Forest Trees*. For. Sci. Monograph 5, 42pp. Society of American Foresters, Washington DC.

Law, J. R., Johnson, P. S. and Houf, G. (1994). *A Crown Cover Chart for Oak Savannas*. Technical Brief TB-NC-2. USDA For. Serv. North Cent. For. Exp. Sta., St. Paul, MN. Available at: https://www.fs.usda.gov/treesearch/pubs/10999.

Lehmkuhler, J. W., Kerley, M. S., Garrett, H. E., Cutter, B. E. and McGraw, R. L. (1999). Comparison of continuous and rotational silvopastoral systems for establishing walnut plantations in southwest Missouri, USA. *Agrofor. Syst.* 2-3, 267-79.

Lehmkuhler, J. W., Felton, E. E. D., Schmidt, D. A., Bader, K. J., Moore, A., Huck, M. B., Garrett, H. E. and Kerley, M. S. (2003). Tree protection methods during the silvopastoral systems establishment in midwestern USA: cattle performance and tree damage. *Agrofor. Syst.* 59, 35-42.

Lin, C.-H., McGraw, R. L., George, M. F. and Garrett, H. E. (1999). Shade effects on forage crops with potential in temperate agroforestry practices. *Agrofor. Syst.* 44, 109-19.

Lin, C.-H., McGraw, R. L., George, M. F. and Garrett, H. E. (2001). Nutritive quality and morphological development under partial shade of some forages with agroforestry potential. *Agrofor. Syst.* 53, 1-13.

McKenna, J. R., Geyer, W. A., Woeste, K. E. and Cassens, D. L. (2015). Propagating figured wood in black walnut. *Open Journal of Forestry* 5, 518–25. doi:10.4236/ojf.2015.55045. Available at: https://www.fs.usda.gov/treesearch/pubs/48657.

Meadows, J. S. (1995). Epicormic branches and lumber grade of bottomland oak. In: *Proceedings of the 23rd Annual Hardwood Symposium* (Eds. Lowery, G. and Meyer, D.), pp. 19–25. National Hardwood Lumber Association, Memphis, TN. Available at: https://www.fs.usda.gov/treesearch/pubs/3267.

Miller, L. A., Wray, P. H. and Mize, C. W. (1987). Response of stagnated black walnut to chemical release. *North. J. Applied For.* 4, 93–5.

Miller, G. W., Stringer, J. W. and Mercker, D. C. (2007). *Technical Guide to Crop Tree Release in Hardwood Forests*. SREF-FM-100. Southern Regional Extension Pub SREF-FM-100, 24pp. Available at: https://sref.info/resources/publications/.

Nelson, N. D., Maeglin, R. R. and Walgren, H. E. (1969). Relationship of black walnut wood color to soil properties and site. *Wood Fiber* 1(1), 29–37.

O'Hara, K. L. (2007). Pruning wounds and occlusion: a long-standing conundrum in forestry. *J. For.* 105(3), 131–8. doi:10.1093/jof/105.3.131.

Olcott-Reid, B. and Reid, W. (2007). *Fruit and Nut Production*. Stipes Publishing, L.L.C., Champaign, IL, 597pp.

Pang, K., Van Sambeek, J. W., Navarette-Tindall, N. E., Lin, C.-H., Jose, S. and Garrett, H. E. (2019a). Responses of legumes and grasses to non-, moderate, and dense shade in Missouri, USA. I. Forage yield and its species-level plasticity. *Agrofor. Syst.* 93 (1), 11–24. doi:10.1007s10457-017-0067-8.

Pang, K., Van Sambeek, J. W., Navarette-Tindall, N. E., Lin, C.-H., Jose, S. and Garrett, H. E. (2019b). Responses of legumes and grasses to non-, moderate, and dense shade in Missouri, USA. II. Forage quality and its species-level plasticity. *Agrofor. Syst.* 93 (1), 25–38. doi:10.1007s10457-017-0068-7.

Perkey, A. W., Wilkins, B. L. and Smith, H. C. (1994). *Crop Tree Management in Eastern Hardwoods*. NA-TP-19-93. USDA For. Serv. Northeast. Area State and Private For., Morgantown, WV.

Phelps, J. E. (1989). How management practices influence black walnut wood properties. In: *The Continuing Quest for Quality* (Eds. Phelps, J. E. and McCurdy, D. R.), pp. 21–9. Walnut Council, Indianapolis, IN.

Phelps, J. E. and Chen, P. Y. S. (1989). Lumber and wood properties of plantation-grown and naturally grown black walnut. *For. Prod. J.* 39(2), 58–60.

Ponder Jr., F. (1986). Effect of three weeds on the growth and mycorrhizal infection of black walnut seedlings. *Can. J. Bot.* 64, 1888–92.

Preece, J. E., Navarrete, N. E., Van Sambeek, J. W. and Gaffney, G. R. (1991). An *in vitro* microplant bioassay using clonal white ash to test for tall fescue allelopathy. *Plant Cell Tiss. Org. Cult.* 27, 203–10.

Rast, E. D. 1982. *Photographic Guide to Selected External Defect Indicators and Associate Internal Defects in Northern Red Oak*. Res. Paper NE-511. USDA For. Serv. Northeast. For. Exp. Stn., 20pp. Available at: https://www.fs.usda.gov/treesearch/pubs/15063.

Rast, E. D. and Beaton, J. A. 1985. *Photographic Guide to Selected External Defect Indicators and Associate Internal Defects in Black Cherry*. Res. Paper NE-562. USDA For. Serv. Northeast. For. Exp. Stn., 24pp. Available at: https://www.fs.usda.gov/treesearch/pubs/21736.

Rast, E. D., Beaton, J. A. and Sonderman, D. L. 1988. *Photographic Guide to Selected External Defect Indicators and Associate Internal Defects in Black Walnut*. Res. Paper NE-617. USDA For. Serv. Northeast. For. Exp. Stn., 24pp. Available at: http://www.fs.usda.gov/treesearch/pubs/21802.

Rast, E. D., Beaton, J. A. and Sonderman, D. L. 1989. *Photographic Guide to Selected External Defect Indicators and Associate Internal Defects in White Oak*. Res. Paper NE-628. USDA For. Serv. Northeast. For. Exp. Stn., 24pp. Available at: https://www.fs.usda.gov/treesearch/pubs/21819.

Rast, E. D., Beaton, J. A. and Sonderman, D. L. 1990a. *Photographic Guide to Selected External Defect Indicators and Associate Internal Defects in Yellow-Poplar*. Res. Paper NE-646. USDA For. Serv. Northeast. For. Exp. Stn., 29pp. Available at: https://www.fs.usda.gov/treesearch/pubs/21838.

Rast, E. D., Beaton, J. A. and Sonderman, D. L. 1990b. *Photographic Guide to Selected External Defect Indicators and Associate Internal Defects in Sugar Maple*. Res. Paper NE-647. USDA For. Serv. Northeast. For. Exp. Stn., 35pp. Available at: https://www.fs.usda.gov/treesearch/pubs/21839.

Rast, E. D., Beaton, J. A. and Sonderman, D. L. 1990c. *Photographic Guide to Selected External Defect Indicators and Associate Internal Defects in Yellow Birch*. Res. Paper NE-648. USDA For. Serv. Northeast. For. Exp. Stn., 25pp. Available at: https://www.fs.usda.gov/treesearch/pubs/21840.

Reid, W., Coggeshall, M., Garrett, H. E. and Van Sambeek, J. (2009). *Growing Black Walnut for Nut Production*. Agroforestry in Action AF1011-2009. Univ. Missouri Center for Agrofor., Columbia, MO, 16pp. Available at: http://www.fs.usda.gov/treesearch/pubs/44215.

Rietveld, W. J. (1982). The significance of allelopathy in black walnut cultural systems. In: *Black Walnut for the Future*, pp. 73–86. Gen. Tech. Rept. NC-74. USDA For. Serv. N. Cent. For. Exp. Sta., St. Paul, MN. Available at: https://www.fs.usda.gov/treesearch/pubs/10138.

Rietveld, W. J. (1983). Allelopathic effects of juglone on germination and growth of several herbaceous and woody species. *J. Chem. Ecol.* 9, 295–308. Available at: https://www.fs.usda.gov/treesearch/pubs/19624.

Rink, G. and Van Sambeek, J. W. (1985). Variation among black walnut seedling families in resistance to competition and allelopathy. *Plant and Soil* 88(1), 3–10. Available at: https://www.fs.usda.gov/treesearch/pubs/19625.

Rink, G. and Van Sambeek, J. W. (1987). Variation among four white ash families in response to competition and allelopathy. *For. Ecol. Manage.* 18, 127–34.

Rivest, D. and Vezina, A. (2015). Maize yield patterns on the leeward side of tree windbreaks are site-specific and depend on rainfall conditions in eastern Canada. *Agrofor. Syst.* 89, 237–46.

Rizvi, S. J. H., Tahir, M., Rizvi, V., Kohli, R. K. and Ansari, A. (1999). Interactions in agroforestry systems. *Crit. Rev. Plant Sci.* 18(6), 773–96.

Schlesinger, R. C. (1997). The effects of crowding on black walnut tree growth. In: *Knowledge for the Future of Black Walnut* (Ed. Van Sambeek, J. W.), pp. 139–45. Gen. Tech. Rep. NC-191. USDA For. Serv. North Central For. Exp. Sta., St. Paul, MN. Available at: https://www.fs.usda.gov/treesearch/pubs/10255.

Schlesinger, R. C. (1982). Pruning for quality. In: *Black Walnut for the Future*, pp. 87–91. Gen. Tech. Rept. NC-74. USDA For. Serv. N. Cent. For. Exp. Sta., St. Paul, MN. Available at: https://www.fs.usda.gov/treesearch/pubs/10138.

Schlesinger, R. C. (1986). Pruning Appalachian hardwoods. In: *Proceedings: Guidelines for Managing Immature Appalachian Hardwood Stands* (Eds. Smith, H. C. and Eye, M. E.), pp. 221-7. SAF Publ. 86-02. West Virginia Univ. Books, Morgantown, WV. Available at: https://www.fs.usda.gov/treesearch/pubs/50049.

Schlesinger, R. C. (1988a). Note 3.02 Lateral pruning. In: *Walnut Notes* (Ed. Burde, L.). USDA For. Serv. North Central For. Exp. Sta., 2p. Available at: https://www.fs.usda.gov/treesearch/pubs/11766.

Schlesinger, R. C. (1988b). Note 3.0 First thinning. In: *Walnut Notes* (Ed. Burde, L.). USDA For. Serv. North Central For. Exp. Sta., 2p. Available at: https://www.fs.usda.gov/treesearch/pubs/11767.

Schlesinger, R. C. and Bey, C. F. (1978). Natural improvement in black walnut stem form. In: *Proceedings of the 2nd Central Hardwood Forest Conference* (Ed. Pope, P. E.), pp. 389-400. Purdue University, West Lafayette, IN. Available at: https://www.fs.usda.gov/treesearch/pubs/.

Schlesinger, R. C. and Shigo, A. L. (1989). Pruning central hardwoods. In: *Central Hardwood Notes* (Ed. Hutchinson, J. G.). Note 6.09. USDA For. Serv. North Central For. Exp. Sta., St. Paul, MN. Available at: https://www.fs.usda.gov/treesearch/pubs/11653.

Schlesinger, R. C. and Van Sambeek, J. W. (1986). Ground cover management can revitalize black walnut trees. *North. J. Appl. For.* 3(2), 49-51.

Schlesinger, R. C. and Williams, R. D. (1984). Growth response of black walnut to interplanted trees. *For. Ecol. Manage.* 9, 235-43. Available at: https://www.fs.usda.gov/treesearch/pubs/19623.

Semchenko, M., Lepik, M., Götzenberger, L. and Zobel, K. (2012). Positive effect of shade on plant growth: amelioration of stress or active regulation of growth rate? *J. Ecol.* 100, 459-66. doi:10.1111/j.1365-2745.2011.01936.x.

Shigo, A. L. (1963). *Ring Shake Associated with Sapsucker Injury*. USDA For. Serv. Northeastern For. Exp. Stn., Upper Darby, PA, 10pp. Available at: http://www.fs.usda.gov/treesearch/pubs/3825.

Shigo, A. L. (1984). Compartmentalization: a conceptual framework for understanding how trees grow and defend themselves. *Annu. Rev. Phytopath.* 22(1), 189-214. doi:10.1146/annurev.py.22.090184.001201. Available at: http://www.fs.usda.gov/treesearch/pubs/53382.

Shigo, A. L. (1985). How tree branches are attached to trunks. *Can. J. Bot.* 63, 1391-401.

Shigo, A. L. (1989). *Tree Pruning: a Worldwide Photo Guide*. Shigo and Trees, Durham, NH, 192pp.

Shigo, A. L. (1996). *Tree Basics*. Shigo and Trees, Durham, NH, 40pp.

Shigo, A. L. and Shortle, W. C. (1984). Wound dressings: results of studies over thirteen years. *Arbor. J.* 8, 193-210.

Shigo, A. L., McGinnes Jr., E. A., Funk, D. T. and Rogers, N. (1979). *Internal Defects Associated with Pruned and Nonpruned Branch Stubs in Black Walnut*. Res. Pap. NE-440. USDA For. Serv., Broomall, PA, 27pp. Available at: http://www.treesearch.fs.fed.us/pubs/14974.

Smith, D. E. (1980). Abnormal wood formation following fall and spring injuries in black walnut. *Wood Sci.* 12, 243-51.

Stamps, W. T., Woods, T. L., Linit, M. J. and Garrett, H. E. (2002). Arthropod diversity in alley cropped black walnut (*Juglans nigra* L.) stands in eastern Missouri, USA. *Agrofor. Syst.* 56, 167-75.

Stewart, N. and Dawson, N. (2013). *Forest Thinning - A Landowner's Tool for Healthy Woods*. Ext. Bull. EB-407-2013. University of Maryland, 10pp. Available at: http://extension_umd_edu/woodland.

Stubbs, J. (1986). Hardwood epicormic branching - small knots but large losses. *South. J. Appl. For.* 10, 217-20.

Todhunter, M. N. and Beineke, W. F. (1979). Effect of fescue on black walnut growth. *Tree Planters' Notes* 30(3), 20-3. Available at: https://rngr.net/publications/tpn/30-3/30_3_20_23.pdf/at_download/file.

USDA-FS. (1981). *Quick Reference for Thinning Black Walnut*. USDA For. Serv. North Cent. For. Exp. Stn. and Northeast. Area St. & Private For., St. Paul, MN, 32pp.

USDA-NRCS. (2005). *Land and Water Management Tips for Missouri Landowners*. USDA Natural Resource Conservation Service, Columbia, MO. Available at: https://www.nrcs.usda.gov/wps/portal/nrcs/mo/newsroom/factsheets/.

Van Sambeek, J. W. (2010). Database for estimating tree responses of walnut and other hardwoods to ground cover management practices. *Acta Hortic.* 861, 245-52. Available at: https://www.fs.usda.gov/treesearch/pubs/45876.

Van Sambeek, J. W. (2017). Orchard management using cover crops to improve soil health and pollinator habitat in the Midwestern United States. *The Nutshell (North. Nut Growers Assoc.)* 72(3), 32-45. Available at: https://www.fs.usda.gov/treesearch/pubs/57902.

Van Sambeek, J. W. and Garrett, H. E. (2004). Ground cover management in walnut and other hardwood plantings. In: *Black Walnut in a New Century* (Eds. Michler, C. H., Pijut, P. M., Van Sambeek, J. W., Coggeshall, M. V., Siefert, J., Woeste, K., Overton, R. and Ponder Jr., F.), pp. 85-100. Gen. Tech. Rep. NC-243. USDA For. Serv. North Central Res. Sta., St. Paul, MN. doi:10.2737/NC-GTR-243. Available at: https://www.fs.usda.gov/treesearch/pubs/14714.

Van Sambeek, J. W. and McBride, F. D. (1991). Grass control improves early growth of black walnut more than deep ripping or irrigation. In: *Proceedings of the 2nd Conference on Agroforestry in North America* (Ed. Garrett, H. E.), pp. 42-57. Association for Temperate Agroforestry. https://www.fs.usda.gov/treesearch/pubs/39681.

Van Sambeek, J. W. and Reid, W. (2017). A double row alley-cropping system for establishing nut orchards. *MNGA (Missouri Nut Growers Association) Newsletter* 17(4), 11-14. Available at: https://www.fs.usda.gov/treesearch/pubs/54825.

Van Sambeek, J. W., Ponder Jr., F. and Rietveld, W. J. (1986). Legumes increase growth and alter foliar nutrient levels of black walnut saplings. *For. Ecol. Manage.* 17, 159-67. Available at: https://www.fs.usda.gov/treesearch/pubs/19626.

Van Sambeek, J. W., Schlesinger, R. C., Roth, P. L. and Bocoum, I. (1989). Revitalizing slow-growth black walnut plantings. In: *Proceedings of the 7th Central Hardwood Forest Conference*, pp. 108-14. Gen. Tech. Rep. NC-132. USDA For. Serv. North Central For. Exp. Sta., St. Paul, MN. Available at: https://www.fs.usda.gov/treesearch/pubs/10196.

Von Kiparski, G. R., Lee, L. S. and Gillespie, A. R. (2007). Occurrence and fate of the phytotoxin juglone in alley soils under black walnut trees. *J. Environ. Qual.* 36, 707-17.

White Jr., A. W., Beady, E. R. and Teddars, W. L. (1981). Legumes for supplying nitrogen and studies on legumes in pecan orchards. *Pecan South* 8(4), 24-31.

Willis, R. J. (2000). *Juglans* spp., juglone, and allelopathy. *Allelopathy* 7, 1-55.

Chapter 13

Agroforestry for the cultivation of nuts

Michael A. Gold, University of Missouri, USA

1 Introduction

Row crop agriculture, maize (*Zea mays*) and soybean (*Glycine max*) in particular, covers over 1.28 billion ha of land globally (FAO, 2017). Though extremely productive in terms of yield, these annual cropping systems are heavily dependent on external inputs such as energy, nutrients and pesticides, leading to a suite of ecological consequences (Wolz et al., 2017). Agroforestry includes an integrated set of designed land use practices that can provide economic production while simultaneously enhancing environmental services. As a form of multifunctional agriculture, agroforestry is unique in that it is tree based, adding strategic diversity at various scales in ways that can reduce threats and build resiliency under changing conditions.

Although not defined as such, agroforestry has been practiced worldwide for millennia throughout both the tropical and temperate regions (Smith, 1950; Zhaohua et al., 1991; Nair, 1993; Lelle and Gold, 1994; Lepofsky, 2009; Rossier and Lake, 2014). In the United States, temperate zone agroforestry as currently practiced is relatively recent (Gold and Hanover, 1987). At present, US

http://dx.doi.org/10.19103/AS.2018.0041.17

temperate zone agroforestry is defined as 'intensive land-use management that optimizes the benefits (physical, biological, ecological, economic, and social) from biophysical interactions created when trees and/or shrubs are deliberately combined with crops and/or livestock' (Gold and Garrett, 2009).

Agroforestry design is derived from the scientific discipline of agroecology, which draws upon ecological principles, functions and processes within perennial-based agricultural systems to create sustainable agricultural systems. Applying science to agricultural practices, agroecology and agroforestry focuses on both economic production and ecosystem enhancement (Krebs and Bach, 2018). Agroforestry design incorporates agroecological principles that have been under development for decades (Reijntjes et al., 1992; Vandermeer, 1995; Malézieux, 2012). Primary agroforestry and alley cropping design criteria include imitation of the structure and function of natural ecosystems with species that generate an economic yield to support the farmer (Krebs and Bach, 2018). Recommended principles for designing the cropping system based on this approach include selecting species for complementary functional traits, developing complex trophic levels and reproducing ecological succession (Malézieux, 2012).

Trait-based agroecology has emerged as one of the dominant paradigms in terrestrial ecology (Martin and Isaac, 2018). Studies that evaluate the complex linkages between functional traits, functional diversity and ecosystem functioning have recently begun to explore concepts surrounding 'multifunctionality', the idea that multiple ecosystem functions should be considered as a management goal. Multifunctionality has clear implications for trait-based agroecology and for agroforestry, where land managers manipulate on-farm diversity to enhance ecosystem services beyond yield alone (Finney and Kaye, 2017).

US agroforestry consists of five recognized practices: alley cropping, windbreaks, riparian and upland forest buffers, silvopasture and forest farming. Alley cropping, as one of the five recognized agroforestry practices, is defined as 'trees or shrubs planted in sets of single or multiple rows with agronomic crops, horticultural crops, or forages produced in the alleys between the trees that can also produce additional products' (USDA-FS, 2017). This chapter will primarily focus on the opportunities and challenges associated with alley cropping practices involving overstorey nut crops as one element of the solution to address global needs for economically viable food production while enhancing environmental services (Wolz and DeLucia, 2018).

2 Nut-based agroforestry systems

Alley cropping can be viewed as recreating the structure of a savanna in an agroforestry system with multiple canopy layers consisting of a design that contains an overstorey of nut trees, mid-layer of fruiting small trees and shrubs, and a groundcover of annual crops or perennial grass groundcover. In nut-based agroforestry systems, crops grown in the alleys between tree rows provide annual income while the longer-term nut crop matures. When speciality crops such as herbs, fruits, vegetables, nursery stock or flowers are grown in the alleys, the microclimate created by the trees enhances the economic production of these sensitive high-value crops in stressed environments (Garrett et al., 2009).

It will typically take 8–15 years from the time of establishment of a cultivar-based nut orchard until reliable commercial yields, that is, there are enough nuts to warrant mechanical harvesting. This long time lag is a serious issue when it comes to cash flow and profit generation. For this reason nut orchard growers are turning to alley cropping, that is, establishing intercrops between tree rows until the overstorey trees generate positive cash flow. Many agronomic crops can be successfully intercropped between young chestnut, pecan or black walnut trees including corn, soybeans, winter wheat and milo. Once commercial nut tree yields occur, a shift to hay crops (i.e. clover/grass mixes) can be used to facilitate mechanical nut harvest while continuing to generate income from hay sales.

High-value nut or acorn-producing hardwoods such as walnut (*Juglans* spp.) (Grant et al., 2006; Scott and Sullivan, 2007; Reid et al., 2009; Coggeshall, 2011b; Geyer and Fick, 2014; Cardinael et al., 2015), chestnut (*Castanea* spp.) (Hunt et al., 2012; Wilson et al., 2018), pecan (*Carya illinoensis* (Wangenh.) K. Koch.) (Bugg et al., 1991; Kremer and Kussman, 2011; Fletcher et al., 2012; Van Sambeek and Reid, 2017), hazelnut (*Corylus* spp.) (Molnar et al., 2005) and oak (*Quercus* spp.) (Eichhorn et al., 2006) are the favoured species in alley cropping practices (Molnar et al., 2013; Wilson et al., 2018).

Primary economic benefits and uses for nut-based agroforestry systems include the sequential and integrated production of annual and higher value long-term crops and enhancement of microclimate conditions to improve crop or forage quality and quantity. Environmental benefits include reductions of surface water run-off and erosion; decreasing offsite movement of nutrients or chemicals improvement of soil quality by increasing utilization and cycling of nutrients; and enhancement of habitat for wildlife and beneficial insects (USDA-FS, 2017).

A key advantage of food-producing agroforestry systems is that they are more resilient to climate stressors than are annual cropping systems. Compared to temperate zone annual crop monoculture production practices, integrated tree and crop practices, that is, alley cropping, can increase land use efficiency through improved overall yields compared to separate production of trees and crops in monocultures (Wolz et al., 2017; USDA-FS, 2017). Agroforestry systems can diversify farm portfolios, spread risk and increase both the economic and environmental resilience of farms (USDA-FS, 2017; Krebs and Bach, 2018).

In addition, alley cropping can increase ecosystem resiliency by buffering against weather extremes, diversifying income as a hedge against financial risk, increasing biodiversity, reducing soil erosion and improving nutrient- and water-use efficiency (USDA-FS, 2017; Krebs and Bach, 2018).

The production of nuts and other speciality crops grown in agroforestry systems is increasing in direct response to increasing demand for locally sourced foods (Low and Vogel, 2011; Johnson, 2014; Jose et al., 2018; USDA-ERS, 2015). Fruit, nut and vegetable farms are eight times more likely to sell locally than are other farms (Low and Vogel, 2011). According to statistics collected by the USDA Agricultural Marketing Service, the number of farmers' markets has increased from 1755 to more than 8687, reflecting a 5-fold increase since 1994 (USDA-ERS, 2018). Fruit, nut and vegetable crops produced within agroforestry systems can become a major contributor to increasingly popular direct-to-consumer food systems.

3 Key challenges facing nut-based agroforestry systems

In spite of the strong economic and ecological case justifying the increased adoption of agroforestry and nut tree alley cropping (Jose, 2009; Rigueiro-Rodriguez et al., 2009; Udawatta and Jose, 2012), these systems have not been widely adopted in many regions of the United States where annual monoculture crops dominate the landscape (Valdivia et al., 2012; Moreno et al., 2018). A number of key issues and challenges must be addressed for the promise to become a reality including:

- industry grower support and networks;
- long-term research to select and improve nut tree cultivars across geographic locations;
- detailed production and management information about alley cropping configurations, including efficient mechanical harvest options;
- lack of solid market and financial information;
- policy support.

A knowledge gap exists in terms of what is known regarding the scientific underpinnings and associated practical implementation information for both some of the individual nut tree species in the eastern United States (e.g. chestnut, eastern black walnut, hazelnut) and orchard speciality crops as well as knowledge about design and management of integrated nut tree-based alley cropping systems (Kahn et al., 2011).

There is a need for guidance from extension services in land grant universities. This knowledge network also needs support from other sectors, for example, equipment dealers, private consultants, government agencies,

NGOs, private sector firms that will invest in nut tree alley cropping ventures, as well as farmer-to-farmer networks.

The appeal and viability of agroforestry is substantially undervalued when the productive potential of nut tree crops is not fully explored and exploited. The integration of nut tree species in alley cropping systems, especially those nut trees that have been genetically selected for precocity and high yields across locations, offer the opportunity to improve agroforestry systems' revenue potential and rate of return, enhancing their appeal for prospective adopters (Lovell et al., 2018). There is a need for continued genetic improvement of specific nut tree species and nut tree and companion crop interaction studies.

Production challenges include what to intercrop, how to manage tree and crop competition above and below ground (spatially) and through time (temporally) as the tree canopies develop and eventually dominate the site, how to manage the nut tree crop differently from the annual crops (irrigation, fertility, weed control, pest and disease controls) and how to efficiently mechanically harvest the tree crops within an alley cropping configuration.

To reduce financial risk and enhance adoption there is a need for accurate financial decision support tools and market and consumer information to provide solid growth, yield and price information to use with lenders for bank loans. Finally, there is a need for policy support, including the availability of crop insurance against losses and grants, incentive payments and low interest loans to help support innovative agroforestry production systems.

4 Genetic improvement of nut trees

Long-term research supporting the integration of nut trees into agroforestry systems is underdeveloped. Recommended actions include the development of multiple research trial locations using Land Grant research centres as sites for stable, long-term comparison trials (Lovell et al., 2018) and an increase in USDA funding for public plant breeding (this type of funding existed up to the 1980s) (Shelton and Tracy, 2017). The potential for nut trees to produce a wide range of commercial products and materials has grown recently due to germplasm collection and nascent genetic improvement efforts of edible nut species that have been historically underutilized in the Midwest United States. In this region, none of the important tree nut species, including chestnut, black walnut, northern pecan, hazelnut and oak, have achieved their full potential for increased productivity, disease resistance and broader environmental adaptability (Lovell et al., 2018).

To date, limited numbers of improved selections of chestnut, black walnut, northern pecan, hazelnut and oak germplasm have or soon will be transitioned into replicated performance trials in the Midwest region. This is in response to the need to assess the adaptability of cultivars to different regions (Smith, 1950;

Molnar et al., 2013). Initial selection and breeding work has been formative in developing first-generation nut trees, (Reid et al., 2004; Hunt et al., 2005; Molnar et al., 2005; Coggeshall, 2011a; Anagnostakis, 2012; Molnar and Capik, 2012; Wood et al., 2016) allowing a broader collection of commercially suited nut species to be considered when conceptualizing alley cropping plant assemblages. Improved nut tree cultivars, including further improvement of rootstocks (Grauke and Thompson, 2003), will boost the overall performance of the system.

5 Management of temporal and spatial tree and crop interactions

Alley cropping is designed to provide an increase in total productivity (i.e. considering both crop and tree production) (Muschler, 2015). Complementary functional traits of alley cropping tree/crop components are used to ensure both improved production and resilience. Different use of resources by different species, occupying different ecological niches, forms the basis of this concept of complementarity. Complementary phenology, coupled with strategic management activities, both temporal and spatial, are effective ways to optimize positive interactions between trees and crops for key resources such as light, water and nutrients while minimizing competition (Malézieux, 2012; Lovell et al., 2018; Nicholls et al., 2016; Wells et al., 2018; Wilson et al., 2018).

Agroforestry field experiments are designed to quantify competitive (negative) and facilitative (positive) relationship interactions between trees and crops and/or animals (Jose et al., 2004). Decisions about the experimental design include specific species/varieties and arrangement of species within treatments, necessary size of plots, number of replications to provide statistical significance and arrangement of plots in the landscape (Malézieux, 2012). At the farm level, the practical implementation of these principles differs and designs are specific to each specific location and situation.

Agroforestry systems are dynamic and interactions among system components (i.e. trees and crops), temporal changes, can be expected over the life of the system and will change over time. Although initial complementarity in sharing light, water and nutrient resources can be expected, over time, there will be an overlap between the needs and physical structures of overstorey nut trees and associated annual crop species. In agroforestry and alley cropping systems, this is to be expected. Complementarity can initially result in overyielding, but competition intensifies with time. Strictly in terms of annual crop yield, alley cropping usually results in a decrease in crop yield over time compared to the pure crop because as the trees mature, above and below ground, competition for light, water and nutrients increases between the crop and tree. Thus, agroforestry systems experience a complex series of inter- and intra-specific interactions guided by modification and utilization of light, water and nutrients (Gillespie et al., 2000; Jose et al., 2004, 2009; Thevathasan et al., 2004; Lacombe et al., 2009; Fletcher et al., 2012; Bainard et al., 2013; Cardinael et al., 2015; Inurreta-Aguirre et al., 2018a,b).

Long-term negative yield effects of temporal and spatial tree and crop interactions can be managed not only through complementary seasonal phenology between trees and crops, but also mitigated through annual site manipulation. To prevent tree and crop interaction below ground, when nut trees are planted to establish an alley cropping system, it is feasible to impose annual deep 'ripping' down to 12–18" (put in cm) (1–2 m) from the base of the trees.

A research trial of two Midwestern US alley cropping systems showed no yield decrease in maize (a C4 species) in response to shading (Gillespie et al., 2000). In general, the eastern-most row of maize in a black walnut (*Juglans nigra*) alley cropping system received 11% lower photosynthetically active radiation (PAR) than the middle row. Irrespective of shading, no apparent yield reduction was observed when below-ground competition for nutrients and water was eliminated through trenching and polyethylene barriers. In contrast to black walnut, shading was greater in a red oak (*Quercus rubra*) alley cropping system due to higher canopy leaf area and temporally longer seasonal duration in full leaf during which a 41% reduction in PAR was observed for the eastern row. Similarly, western rows were receiving 17% and 41% lower PAR in comparison to the middle rows in the black walnut Jose et al. (2004).

Double tree-row alley cropping systems have also been designed for pecan to create a viable compromise between trees and crops both in terms of crop management and yield and long-term tree management and yield (Van Sambeek and Reid, 2017).

Lovell et al. (2018) have described long-term agroforestry systems research at two long-term experimental farms, Domaine de Restinclières, located near Montpellier, France, and the University of Missouri Horticulture and Agroforestry Research Center (HARC), located in New Franklin, Missouri, USA. The experiment at Domaine de Restinclières provided important spatial and temporal findings. Tree row orientation resulted in differences in radiation to the alley crop with north–south tree rows providing more homogeneous PAR to the grains. Using a winter wheat alley crop demonstrated temporal and phenological complementarity as the trees and crops did not directly compete for light during much of the year. Winter wheat grows from late fall into early summer, whereas walnut trees bud out in spring/early summer and nut harvest occurs in the fall prior to sowing the winter wheat. Over 20 years, on average, crop yields on agroforestry sites were only reduced 2% compared with monoculture controls. All alleys at Restinclières were managed with commercial-scale farm equipment both for ease of management and directly representative of large-scale application.

Results from long-term chestnut, black walnut and pecan germplasm trials at HARC in Missouri also offer specific information on species selection and management. For example, results of specific trials on chestnut pollination, for both variety compatibility and spacing/density requirements, provide guidance for future alley cropping designs. Higher yields and nut sizes in orchards under trickle irrigation have demonstrated the need for supplemental moisture during periods of summer drought. Research findings from both Restinclières and HARC have informed the development of the recently established Agroforestry for Food field research trial at the University of Illinois designed to study multifunctional woody polyculture systems. The Agroforestry for Food field trial seeks to address several key themes for the future of agroforestry research: food security, climate change, multifunctionality and practical applications.

Augmenting traditional alley cropping through multifunctional woody polyculture systems that include tree crops for food and fodder enhances the potential of alley cropping as a transformative solution to the problem of agriculture across the temperate zone. These systems provide many economic and ecological advantages over conventional row crop agriculture. Key economic drivers of these multifunctional woody polyculture systems of alley cropping include overyielding, utilization of crop analogues compatible with existing staple crops and resilience via crop diversification. Key ecological benefits include enhanced carbon sequestration, soil and nutrient stabilization, biodiversity and resilience to ecological pressures (Wolz et al., 2017).

Effective integration of multifunctional woody polycultures and tree crops in temperate alley cropping requires strategic implementation on marginal lands, an emphasis on highly productive tree crops, practical and optimized multispecies designs and complementary crop combinations for early productivity and management efficiency (Wolz et al., 2017).

6 Orchard design and management

One of the first problems that growers face in planning a new orchard/alley cropping system is deciding on the spacing for the trees - a decision with long-term implications. Growers planning to farm between tree rows need to think about the width of their alleyways. Between-row spacing is primarily determined by machinery width, mature size of nut tree species, with secondary considerations for sunlight requirements of the alley crop. With closer spacing, the intercrop grown in the alley will encounter light competition at an earlier age, requiring a more rapid change to shade-tolerant intercrops (Walter et al., 2015; Wilson et al., 2018). Wider spacing makes intercropping easier but delays the fullest realization of economically viable nut crop yields. Spacings ranging from ~20′ by 20′ to ~60′ by 60′ are recommended depending on both the choice of nut tree species (Coggeshall, 2011b; Hunt et al., 2012), the width of the farm equipment needed for planting and combining the intercrop between the tree rows and the willingness of the landowner to thin out overstorey trees within or between rows over time.

Alley cropping, as with other forms of agroforestry, requires both intensive technical management skill and marketing knowledge (Table 1). Among the many practical considerations, alley cropping with nut trees will require countermeasures against deer and rabbits in the form of either individual tree cages and/or establishment of permanent 8–10′ high fencing around the entire perimeter of the orchard in the long term. Within tree rows, weed control is essential during the establishment years either through the use of herbicides or mulching.

In the Midwest United States, where extended periods without summer rainfall are increasingly common due to more intense climate fluctuations, including an irrigation or fertigation system in the overall design and management is essential for optimizing long-term nut tree productivity. This is true for Chinese chestnut (*Castanea mollissima*) in Missouri where adequate supply of soil moisture greatly improves overall yield and nut size, and may (untested) also improve yield of other tree nut species of interest.

7 Pest management in nut tree alley cropping

There is evidence that biodiverse plant communities, characteristic of natural ecosystems, can be used in agroforestry systems to improve pest management (Bugg et al., 1991; Shapiro-Ilan et al., 2012; Stamps et al., 2009). Stamps et al. (2009) investigated insect pest dynamics, crop yields and small farm economics

Table 1 Considerations for the landowner prior to establishment of nut tree alley cropping

Alley cropping, as with other forms of agroforestry, requires both intensive technical management skill and marketing knowledge.

The following limitations should be considered:

- Removes land from annual crop production and may not provide a financial return from the trees for several years
- Requires a more intensive management system including specialized equipment for the tree management and additional managerial skills and training to manage multiple crops on a given site
- Requires a marketing infrastructure for forest products that may not be presently established
- Trees may be an obstacle to crop cultivation if not carefully planned and managed
- Trees compete with companion crops for light, moisture and nutrients
- Companion crops may compete with trees for moisture and nutrients
- Herbicide drift from crops may damage trees

Source: adapted from Walter et al. (2015).

in an alley cropping practice of alfalfa and black walnut compared to conventionally grown alfalfa. Specifically they examined alfalfa weevil, *Hypera postica* (Gyllenhaal), and its natural enemies in forage grown in an eastern black walnut (*Juglans nigra* L.) alley cropped system and in a traditional monoculture system. The presence of walnut trees increased natural enemy numbers and significantly increased parasitism of alfalfa weevil. Their research demonstrated that alfalfa intercropped with walnut supported significantly more parasitic hymenoptera and/or predators than did traditionally grown alfalfa weevil. Alfalfa weevil mortality from parasitoids and/or fungal infection was generally higher for weevil larvae from alley cropped alfalfa compared to weevils from monocropped alfalfa.

Pecan (*Carya illinoinensis* (Wangenheim) K. Koch) is an economically important nut crop in North America. The pecan weevil, *Curculio caryae* (Horn), is a key pecan pest throughout the southeastern United States stretching into Texas, Oklahoma, Kansas and Missouri. Control recommendations consist mainly of above-ground applications of chemical insecticides (e.g. carbaryl, pyrethroids) to suppress adults. Application is recommended every 7–10 days during peak emergence (generally up to at least a 6-week period). Fungi such as *Beauveria bassiana* (Balsamo) constitute potential alternatives for *C. caryae* control. Shapiro-Ilan et al. (2012) found that persistence and efficacy of endemic *B. bassiana* can be enhanced by the addition of a clover cover crop within the pecan orchard.

8 Financial decision support tools

There is a need to provide solid growth, yield and price information in the form of financial decision support tools and market and consumer information to reduce financial risk for producers and, in turn, to use this detailed information when approaching lenders for bank loans. The USDA National Agriculture Census, providing a wealth of data on all commodity and most speciality crops, added chestnut to the survey for the first time in 2007 and at present no data are collected on eastern black walnut as a nut crop (in contrast to the well-established 'California walnut' industry which is included in the National Agriculture Census). In order to provide some baseline information, Gold et al. (2006) and Aguilar et al. (2009, 2010) have conducted a series of market and consumer studies on the emergent US chestnut industry to supply reliable market trend and consumer preference information. Miller et al. (2016) interviewed owners, managers, consultants, board members and affiliates of five Midwestern businesses engaged in regional nut processing to provide industry leaders with information gleaned from the experience of established Midwestern nut processors and marketers.

The application of agroforestry can be greatly enhanced through the use of financial decision support tools to help integrate diverse sets of information. A variety of decision support tools addressing biophysical, economic and social considerations are available for applying agroforestry (USDA-FS, 2017). Godsey and others have created financial decision support tools for chestnut and elderberry which are designed to assist with establishment and management decisions. The tools also enable landowners to develop enterprise budgets and combine these budgets into annual cash flow plans for evaluation (http://www.centerforagroforestry.org/profit/#budget).

9 Policy support

Additional policy support for nut tree alley cropping, including the availability of crop insurance against losses and grants, incentive payments and low interest loans, is needed to help support innovative agroforestry production systems. Valdivia et al. (2012) examined the barriers preventing landowner adoption of agroforestry practices. Two overarching barriers were identified; one related to transaction costs in establishment of trees, and a second related to profitability concerns. Transaction costs exist due to the weakness of institutions that facilitate access to information, in this case regarding agroforestry practices and policies that can support implementation. Access to information and cost-share programmes would reduce the barriers to adoption of practices that incorporate trees on the landscape, by reducing transaction costs of establishing the practice. Incentives need to include reduction of costs and increased profitability to be appealing to those more engaged in farming.

Cartwright et al. (2017) describe an innovative, pilot-scale cost-share funding opportunity offered through the USDA NRCS Environmental Quality Incentives Program in Missouri. This special funding pool targets landowners interested in establishing agroforestry and/or woody crops on their property to encourage agroforestry adoption. In addition, the USDA Sustainable Agriculture Research and Education Program offers a variety of competitive grant opportunities to support both research and landowners who wish to engage in sustainable agriculture including support for agroforestry, speciality crops and nut tree alley cropping (https://www.sare.org/Grants).

10 Case studies

Pantera et al. (2018) describe the components, structure, ecosystem services and economic value of ten case studies where high value trees (i.e. fruits, nuts, timber) are integrated with understorey crops and/or livestock in Europe. Examples include agroforestry systems with olive, walnut, chestnut, apple and pear. Overall, the value derived from overstorey trees in these systems (olives, apples, nuts, chestnuts) were the primary focus of the farmers with grazing or intercropping of secondary importance. In addition, chestnut woodlands are identified by local stakeholders as one providing good habitat for the commercial production of edible mushroom. Trees grown on a regular grid pattern permitted use of machinery that enabled high outputs for commercial production, while systems with low inputs and levels of mechanization, such as chestnuts in Spain, also delivered high outputs.

This section focuses on two case studies of nut tree alley cropping, starting with Shepherd Farm (Flaim, 2005). Shepherd Farms is in north central Missouri on the Chariton River. The family first began farming there in the late 1960s, growing corn, beans and wheat on 1900 acres. In the early 1980s the farm shifted production to alternative crops. The philosophy behind the switch to a sustainable farming venture that included alley cropping with nut trees is embodied in the following quote from Dan Shepherd's father 'We stand to make a little money doing what others are already doing, or we can make a lot of money doing things others won't'.

According to the Shepherd's marketing of conventional commodities and livestock production means that someone else sets the price. Shepherd Farms switched production to include 270 acres of improved grafted cultivars of alley cropped pecan, pastured buffalo and eastern gamagrass seed production – three unique enterprises for which they could set the price and sell directly to the consumer via their on-farm store. For the first 20 years, positive cash flow in pecan orchards was obtained via alley cropping. They grew wheat and soybeans for small returns that basically paid for their expenses. As the trees matured the Shepherds seeded the alleyways with blue grass. The grass offered a great feed for the Shepherds' buffalo while creating an excellent surface for harvesting over 200 000 pounds of pecans.

The second case study is Hopeton Farms which includes 2100 acres of almonds and 65 acres of organically grown walnuts, located in California's Central Valley. The farm started in the mid-1980s as a partnership of three non-farming landowner families. The owners' interest in creating a more environmentally sound enterprise, involved a 28-acre block of the farm's almond trees in a 'BIOS' (Biologically Integrated Orchard Systems) pilot project initiated by the California Alliance with Family Farmers (CAFF), a California nonprofit organization to develop more sustainable orchard systems. Initial soil tests revealed nutrient-depleted soils and as a result they applied composted dairy manure over the whole farm (Wechsler, 2005).

California almond orchards typically have little-to-no plant cover on the soil, and what is there is mowed very close for ease of harvest. The Hopeton owners decided to try cover crops for soil improvement. In the first year of the project, they experimented with three different cover crop mixes to find the best combination. A low-growing mix, made up of annual clovers and vetches with a small amount of low-growing grasses, worked best. They settled on this, plus an insectary mix every tenth row. Following the success of the original test plot of 28 acres, the owners have added 300–400 acres of almonds under cover crops each year. In addition, instead of removing and burning tree prunings, they now chip up the prunings and put the wood back into the soil. Hopeton

Farms saved US$375 000 in pesticide and fertilizer input costs by cutting back on pesticides and fertilizers in the first year (Wechsler, 2005).

With the dedicated use of compost and ground covers and reduced use of soil chemicals, soil organic matter levels improved, with the return of earthworms throughout the orchard as evidence of improved soil health. With a more diverse habitat, raptors, owls and hawks also returned to the farm (Wechsler, 2005).

11 Conclusion and future trends

Agroforestry and nut tree alley cropping provide transformative solutions beyond monoculture grain production as the only primary food product. Agroforestry systems offer options for mitigation by reducing GHG emissions compared with other agroecosystems, and for adaptation through greater resilience under variable environmental conditions. Multifunctionality seeks to bring ecological and cultural functions into a production system, offering a new alternative for growers in temperate zones. Practical applications are needed to move agroforestry and alley cropping with nut trees beyond basic sciences, to study practical options for growers (Mori et al., 2017; Lovell et al., 2018).

In spite of encouraging initiatives related to soil health, cover crops, precision agriculture and genetic modification of annual crops to increase yield, transformative solutions that address the fundamental issues associated with vast monocultures of annual crops will be necessary for robust and resilient agricultural land use, especially in the face of climate change (Wolz et al., 2017). Agroforestry, including nut tree alley cropping, is one of the transformative approaches in need of further study in the coming decades. If nut tree alley cropping practices are to become one of the successful transformative solutions, they must become proven practices that are ecologically sustainable, economically viable and culturally acceptable. Ecological sustainability requires robust functioning of regulating and supporting ecosystem services alongside the provisioning services at the core of agriculture.

Luedeling et al. (2011a) projected climate change effects upon agroforestry systems that include temperate fruit and nut trees. Concerns exist about early bud break followed by a freeze that could potentially kill the developing buds or flowers. These concerns call for more research on fruit and nut trees' adaptations to climate change and also on the potential development of new or existing cultivars more resilient to these stressors. Other manifestations of climate change are likely to affect fruit and nut production including changes in rainfall, changes to summer and spring heat or increases in pest pressure due to faster reproduction of ectothermic pest organisms. Increasing temperatures are likely to change pest dynamics in California. For example, generation numbers of codling moth, navel orangeworm, web-spinning spider mites and

European red mites are likely to increase, generating new challenges for pest control. Additional research into agroecological food webs is needed in order to project the implications of climate change for pest pressure with greater certainty (Luedeling et al., 2011b).

Future research must continue to address issues of production, regional adaptability, profitability, cultural and consumer acceptability (Lovell et al., 2018). Economic viability means profitability for farmers and prosperity for rural communities. Cultural acceptability entails meeting the aesthetic, ethical and practical needs of rural communities. Access to improved cultivars, solid production guidelines, reliable product supply, growing consumer demand and sound financial decision support tools are assisting the growth of the speciality crop industry and having positive impacts up and down the supply chain (Mori et al., 2017). Increasing production of tree nuts within alley cropping will help to produce the carbohydrates, proteins and oils that are the basic components of food systems and market supply chains (Kahn et al., 2011; USDA-FS, 2017). Major research frontiers for alley cropping with nut trees include within-system tree diversity, tree crops for food, perennial alley crops and trees for crop facilitation (Wolz and DeLucia, 2018). Future work evaluating the long-term yields and biogeochemical consequences of alley cropping with nut trees are critical to widespread adoption of this transformative agricultural alternative (Wolz et al., 2018).

12 Where to look for further information

A number of organizations provide additional resources pertaining to agroforestry, alley cropping and nut trees. Rich sources of information include the USDA National Agroforestry Center, the MU Center for Agroforestry and the Savanna Institute (Table 2).

Table 2 Alley cropping: additional resources

Organization/publication	Location of Information
USDA National Agroforestry Center	https://www.fs.usda.gov/nac/practices/alleycropping.shtml
The Center for Agroforestry at the University of Missouri	http://www.centerforagroforestry.org Alley Cropping: http://www.centerforagroforestry.org/practices/ac.php
Savanna Institute	http://www.savannainstitute.org/resources.html
Alley Cropping video	http://www.youtube.com/watch?v=b8Kwb5ylnPM
United Kingdom and Europe	http://www.agforward.eu/index.php/en/silvoarable-agroforestry-in-the-uk.html
USDA Natural Resources Conservation Service	www.nrcs.usda.gov/wps/portal/nrcs/detailfull/national/landuse/forestry/sustain/guidance
Association for Temperate Agroforestry (AFTA)	http://www.aftaweb.org/about/what-is-agroforestry/alley-croping.html

13 References

Aguilar, F. X., Cernusca, M. M. and Gold, M. A. 2009. Exploratory assessment of consumer preferences for chestnut attributes in Missouri. *HortTechnology* 19(1), 216–23. doi:10.21273/HORTSCI.19.1.216.

Aguilar, F. X., Cernusca, M. M., Gold, M. A. and Barbieri, C. E. 2010. Frequency of consumption, familiarity and preferences for chestnuts in Missouri. *Agroforestry Systems* 79(1), 19–29. doi:10.1007/s10457-009-9266-2.

Anagnostakis, S. L. 2012. Chestnut breeding in the United States for disease and insect resistance. *Plant Disease* 96(10), 1392–403. doi:10.1094/PDIS-04-12-0350-FE.

Bainard, L. D., Koch, A. M., Gordon, A. M. and Klironomos, J. N. 2013. Growth response of crops to soil microbial communities from conventional monocropping and tree-based intercropping systems. *Plant and Soil* 363(1–2), 345–56. doi:10.1007/s11104-012-1321-5.

Bugg, R. L., Sarrantonio, M., Dutcher, J. D. and Phatak, S. C. 1991. Understory cover crops in pecan orchards: possible management systems. *American Journal of Alternative Agriculture* 6(2), 50–62. doi:10.1017/S0889189300003854.

Cardinael, R., Mao, Z., Prieto, I., Stokes, A., Dupraz, C., Kim, J. H. and Jourdan, C. 2015. Competition with winter crops induces deeper rooting of walnut trees in a Mediterranean alley cropping agroforestry system. *Plant and Soil* 391(1–2), 219–35. doi:10.1007/s11104-015-2422-8.

Cartwright, L., Goodrich, N., Cai, Z. and Gold, M. 2017. Using NRCS technical and financial assistance for agroforestry and woody crop establishment through the environmental quality incentives program (EQIP). Agroforestry in Action AF 1016, 4pp. Available at: http://www.centerforagroforestry.org/pubs/NRCS_AgroforestryandWoodyCrop.pdf.

Coggeshall, M. V. 2011a. Use of microsatellite markers to develop new eastern black walnut cultivars in Missouri, USA. *28th International Horticultural Congress*, Lisboa, Portugal. *Acta Horticulturae* 918(1), 221–6. doi:10.17660/ActaHortic.2011.918.27.

Coggeshall, M. V. 2011b. Black walnut: a nut crop for the Midwestern United States. *HortScience* 46(3), 340–2. doi:10.21273/HORTSCI.46.3.340.

Eichhorn, M. P., Paris, P., Herzog, F., Incoll, L. D., Liagre, F., Mantzanas, K., Mayus, M., Moreno, G., Papanastasis, V. P., Pilbeam, D. J., et al. 2006. Silvoarable systems in Europe – past, present and future prospects. *Agroforestry Systems* 67(1), 29–50. doi:10.1007/s10457-005-1111-7.

FAO. 2017. Food and Agriculture Organization of the United Nations, Statistics Division. FAOSTAT. Last Updated: February 2017.

Finney, D. M. and Kaye, J. P. 2017. Functional diversity in cover crop polycultures increases multifunctionality of an agricultural system. *Journal of Applied Ecology* 54(2), 509–17. doi:10.1111/1365-2664.12765.

Flaim, J. 2005. Shepherd farms. In: *The New American Farmer, Profiles of Agricultural Innovation* (2nd edn.). Sustainable Agriculture Network, Beltsville, MD. Available at: https://maawg.files.wordpress.com/2012/07/shepherd-case-study-1.pdf.

Fletcher, E. H., Thetford, M., Sharma, J. and Jose, S. 2012. Effect of root competition and shade on survival and growth of nine woody plant taxa within a pecan (*Carya illinoinensis* K. Koch.) alley cropping system. *Agroforestry Systems* 86(1), 49–60. doi:10.1007/s10457-012-9507-7.

Garrett, H. E., McGraw, R. L. and Walter, W. D. 2009. Alley cropping practices. In: Garrett, H. E. (Ed.), *North American Agroforestry: an Integrated Science and Practice* (2nd edn.). American Society of Agronomy, Madison, WI, pp. 133–62.

Geyer, W. A. and Fick, W. H. 2014. Yield and forage quality of smooth brome in a black walnut alley-cropping practice. *Agroforestry Systems* 89, 107–12.

Gillespie, A. R., Jose, S., Mengel, D. B., Hoover, W. L., Pope, P. E., Seifert, J. R., Biehle, D. J., Stall, T. and Benjamin, T. J. 2000. Defining competition vectors in a temperate alley cropping system in the Midwestern USA: 1. Production physiology. *Agroforestry Systems* 48(1), 25–40. doi:10.1023/A:1006285205553.

Gold, M. A. and Garrett, H.E. 2009. Agroforestry nomenclature, concepts, and practices. In: Garrett, H. E. (Ed.), *North American Agroforestry: an Integrated Science and Practice* (2nd edn). Agronomy Society of America, Madison, WI, pp. 45–56.

Gold, M. A. and Hanover, J. W. 1987. Agroforestry systems for the temperate zone. *Agroforestry Systems* 5(2), 109–21. doi:10.1007/BF00047516.

Gold, M. A., Cernusca, M. M. and Godsey, L. D. 2006. Competitive market analysis: chestnut producers. *HortTechnology* 16(2), 360–9. doi:10.21273/HORTTECH.16.2.0360.

Grant, J., Anderson, K. K., Prichard, T., Hasey, J., Bugg, R. L., Thomas, F. and Johnson, T. 2006. Cover crops for walnut orchards. University of California Agriculture and Natural Resources Publication 21627. Available at: https://anrcatalog.ucanr.edu/pdf/21627e.pdf.

Grauke, L. J. and Thompson, T. E. 2003. Rootstock development in temperate nut crops. In: Janick, J. (Ed.), *Proceedings of the XXVI IHC – Genetics and Breeding of Tree Fruits and Nuts. Acta Horticulturae* 622(622), pp. 553–66. doi:10.17660/ActaHortic.2003.622.59.

Hunt, K. L., Gold, M. A. and Warmund, M. R. 2005. Chinese chestnut cultivar performance in Missouri. In: Abreu, C. G., Rosa, E. and Monteiro, A. A. (Eds), *Proceedings of the Third International Chestnut Congress. Acta Horticulturae* 693(693), pp. 145–8. doi:10.17660/ActaHortic.2005.693.15.

Hunt, K., Gold, M., Reid, W. and Warmund, M. 2012. Growing Chinese chestnut in Missouri. UMCA Agroforestry in Action Guide AF1007-2012, 16pp. Available at: http://extension.missouri.edu/explorepdf/agguides/agroforestry/af1007.pdf.

Inurreta-Aguirre, H. D., Lauri, P. E., Dupraz, C. and Gosme, M. 2018a. Yield components and phenology of durum wheat in a Mediterranean alley-cropping system. *Agroforestry Systems* 92(4), 961–74. doi:10.1007/s10457-018-0201-2.

Inurreta-Aguirre, H. D., Lauri, P. E., Dupraz, C. and Gosme, M. 2018b. Impact of tree root pruning on yield of durum wheat and barley in a Mediterranean alley-cropping system. *Proceedings Fourth European Agroforestry Conference*, Nijmejen, Therapeutic Netherlands. Available at: https://www.researchgate.net/publication/325597457.

Johnson, R. 2014. Fruits, vegetables, and other specialty crops: selected farm bill and other federal programs. Rep. No. R42771. Congressional Research Service, Washington DC, 51pp. Available at: http://www.nationalaglawcenter.org/wp-content/uploads/assets/crs/R42771.pdf.

Jose, S. 2009. Agroforestry for ecosystem services and environmental benefits: an overview. *Agroforestry Systems* 76(1), 1–10. doi:10.1007/s10457-009-9229-7.

Jose, S., Gillespie, A. R. and Pallardy, S. G. 2004. Interspecific interactions in temperate agroforestry. *Agroforestry Systems* 61(1), 237–55.

Jose, S., Holzmueller, E. J. and Gillespie, A. R. 2009. Tree-Crop Interactions in temperate agroforestry. In: Garrett, H. E. (Ed.), *North American Agroforestry: an Integrated Science and Practice* (2nd edn.). American Society of Agronomy, Madison, WI, pp. 57–74.

Jose, S., Gold, M. A. and Garrett, H. E. 2018. Temperate agroforestry in the United States: current trends and future directions. In: Gordon, A. M., Newman, S. M. and Coleman, B. (Eds), *Temperate Agroforestry Systems* (2nd edn.). CAB International, Boston, MA, pp. 51–71. Chapter 3.

Kahn, P. C., Molnar, T., Zhang, G. C. and Funk, C. R. 2011. Investing in perennial crops to sustainably feed the world. *Issues in Science and Technology* 27(4), 75–81.

Krebs, J. and Bach, S. 2018. Permaculture–scientific evidence of principles for the agroecological design of farming systems. *Sustainability* 10(9), 3218–42. doi:10.3390/su10093218.

Kremer, R. J. and Kussman, R. D. 2011. Soil quality in a pecan-kura clover alley cropping system in the Midwestern USA. *Agroforestry Systems* 83(2), 213–23. doi:10.1007/s10457-011-9370-y.

Lacombe, S., Bradley, R. L., Hamel, C. and Beaulieu, C. 2009. Do tree-based intercropping systems increase the diversity and stability of soil microbial communities? *Agriculture, Ecosystems and Environment* 131(1), 25–31.

Lelle, M. A. and Gold, M. A. 1994. Agroforestry systems for temperate climates: lessons from Roman Italy. *Forest and Conservation History* 38(3), 118–26. doi:10.2307/3983919.

Lepofsky, D. 2009. The past, present, and future of traditional resource and environmental management. *Journal of Ethnobiology* 29(2), 161–6. doi:10.2993/0278-0771-29.2.161.

Lovell, S. T., Dupraz, C., Gold, M., Jose, S., Revord, R., Stanek, E. and Wolz, K. J. 2018. Temperate agroforestry research - considering multifunctional woody polycultures and the design of long-term field trials. *Agroforestry Systems* 92(5), 1397–415. doi:10.1007/s10457-017-0087-4.

Low, S. A. and Vogel, S. J. 2011. Direct and intermediated marketing of local foods in the United States. ERR-128. United States Department of Agriculture, Economic Research Service. Available at: https://www.ers.usda.gov/webdocs/publications/44924/8276_err128_2_.pdf?v=41056.

Luedeling, E., Girvetz, E. H., Semenov, M. A. and Brown, P. H. 2011a. Climate change affects winter chill for temperate fruit and nut trees. *PLoS ONE* 6(5), e20155. doi:10.1371/journal.pone.0020155.

Luedeling, E., Steinmann, K. P., Zhang, M., Brown, P. H., Grant, J. and Girvetz, E. H. 2011b. Climate change effects on walnut pests in California. *Global Change Biology* 17(1), 228–38. doi:10.1111/j.1365-2486.2010.02227.x.

Malézieux, E. 2012. Designing cropping systems from nature. *Agronomy for Sustainable Development* 32(1), 15–29. doi:10.1007/s13593-011-0027-z.

Martin, A. R. and Isaac, M. E. 2018. Functional traits in agroecology: advancing description and prediction in agroecosystems. *Journal of Applied Ecology* 55(1), 5–11. doi:10.1111/1365-2664.13039.

Miller, M., Williams, B., Raboin, M., Carusi, C., McNair, R. and Bauer, L. 2016. Growing Midwestern tree nut businesses: five case studies. CIAS, Madison, WI. Available at: https://www.cias.wisc.edu/wp-content/uploads/2016/11/ciasgrowingmidwesttreenutfinallowres.pdf.

Molnar, T. J. and Capik, J. M. 2012. Advances in hazelnut research in North America. *Acta Horticulturae* 940(940), 57–65. doi:10.17660/ActaHortic.2012.940.6.

Molnar, T. J., Goffread, J. C. and Funk, C. R. 2005. Developing hazelnuts for the Eastern United States. In: Tous, J., Rovira, M. and Romero, A. (Eds), *Proceedings of the VIth*

International Conferences on Hazelnut. Acta Horticulturae 686, 609–18. Illinois State Historical Society.

Molnar, T. J., Khan, P. C., Ford, T. M., Funk, C. J. and Funk, C. R. 2013. Tree crops, a permanent agriculture: concepts for the past for a sustainable future. *Resources* 2(4), 457–88. doi:10.3390/resources2040457.

Moreno, G., Aviron, S., Berg, S., Crous-Duran, J., Franca, A., García de Jalón, S., Hartel, T., Mirck, J., Pantera, A., Palma, J. H. N., et al. 2018. Agroforestry systems of high nature and cultural value in Europe: provision of commercial goods and other ecosystem services. *Agroforestry Systems* 92(4), 877–91. doi:10.1007/s10457-017-0126-1.

Mori, G. O., Gold, M. A. and Jose, S. 2017. Specialty crops in temperate agroforestry systems: sustainable management, marketing and promotion for the Midwest region of the U.S.A. In: Montagnini, F. (Ed.), *Integrating Landscapes: Agroforestry for Biodiversity Conservation and Food Sovereignty. Advances in Agroforestry* 12, 331–66. doi:10.1007/978-3-319-69371-2_14.

Muschler, R. G. 2015. Agroforestry: essential for sustainable and climate-smart land use? In: Pancel, L. and Kohl, M. (Eds), *Tropical Forestry Handbook* (2nd edn.). Springer-Verlag, Berlin Heidelberg, pp. 2013–116.

Nair, P. K. R. 1993. *An Introduction to Agroforestry*. Kluwer Academic Publishers, Dordrecht, Netherlands, 505pp.

Nicholls, C. I., Altieri, M. A. and Vazquez, L. 2016. Agroecology: principles for the conversion and redesign of farming systems. *Journal Ecosys Ecography* S5, 010. doi:10.4172/2157-7625.S5-010.

Pantera, A., Burgess, P. J., Mosquera Losada, R., Moreno, G., Lopez-Dıaz, M. L., Corroyer, N., McAdam, J., Rosati, A., Papadopoulos, A. M., Graves, A., et al. 2018. Agroforestry for high value tree systems in Europe. *Agroforestry Systems* 92(4), 945–59. doi:10.1007/s10457-017-0181-7.

Reid, W., Coggeshall, M. V. and Hunt, K. L. 2004. Cultivar evaluation and development for black walnut orchards. In: Michler, C. H., Pijut, P. M., Van Sambeek, J. W., Coggeshall, M. V., Seifert, J., Woeste, K., Overton, R. and Ponder Jr., F. (Eds), *Black Walnut in a New Century. Proceedings of the 6th Walnut Council Research Symposium*, 25–28 July 2004, Lafayette, IN. Gen. Tech. Rep. NC-243. United States Department of Agriculture, Forest Service, North Central Research Station, St. Paul, MN, 188pp.

Reid, W., Coggeshall, M. V., Garrett, H. E. and Van Sambeek, J. W. 2009. Growing black walnut for nut production. 'Agroforestry in Action' publication AF1011-2009. University of Missouri Center for Agroforestry, 16pp. Available at: http://extension.missouri.edu/explorepdf/agguides/agroforestry/af1011.pdf.

Reijntjes, C., Haverkort, B. and Waters-Bayer, A. 1992. *Farming for the Future: an Introduction to Low-External-Input and Sustainable Agriculture*. Macmillan, London, UK.

Rigueiro-Ródriguez, A., Fernandez-Nunez, E., Gonzalez-Hernandez, P., McAdam, J. H. and Mosquera-Losada, M. R. 2009. Agroforestry systems in Europe: productive, ecological and social perspectives. In: Rigueiro-Rodróguez, A., McAdam, J. and Mosquera-Losada, M. R. (Eds), *Agroforestry in Europe: Current Status and Future Prospects, vol. 6*. Springer, Dordrecht, pp. 43–65.

Rossier, C. and Lake, F. 2014. Indigenous traditional ecological knowledge in agroforestry. Agroforestry Notes 44. United States Department of Agriculture, National Agroforestry Center, Lincoln, NE, 8pp.

Scott, R. and Sullivan, W. C. 2007. A review of suitable companion crops for black walnut. *Agroforestry Systems* 71, 185–93. doi:10.1007/s10457-007-9071-8.

Shapiro-Ilan, D. I., Gardner, W. A., Wells, L. and Wood, B. W. 2012. Cumulative impact of a clover cover crop on the persistence and efficacy of *Beauveria bassiana* in suppressing the pecan weevil (Coleoptera: Curculionidae). *Environmental Entomology* 41(2), 298–307. doi:10.1603/EN11229.

Shelton, A. C. and Tracy, W. F. 2017. Cultivar development in the U.S. public sector. *Crop Science* 57(4), 1823–35. doi:10.2135/cropsci2016.11.0961.

Smith, J. R. 1950. *Tree Crops: A Permanent Agriculture*. Devin-Adair, New York, NY.

Stamps, W. T., McGraw, R. L., Godsey, L. and Woods, T. L. 2009. The ecology and economics of insect pest management in nut tree alley cropping systems in the Midwestern United States. *Agriculture, Ecosystems and Environment* 131(1–2), 4–8. doi:10.1016/j.agee.2008.06.012.

Thevathasan, N. V., Gordon, A. M., Simpson, J. A., Reynolds, P. E., Price, G. and Zhang, P. 2004. Biophysical and ecological interactions in a temperate tree-based intercropping system. *Journal of Crop Improvement* 12(1-2), 339–63.

Udawatta, R. P. and Jose, S. 2012. Agroforestry strategies to sequester carbon in temperate North America. *Agroforestry Systems* 86(2), 225–42. doi:10.1007/s10457-012-9561-1.

USDA-ERS. 2015. Trends in U.S. local and regional food systems: report to Congress. Administration Publique No. 068. United States Department of Agriculture, Economic Research Service, Washington DC, 47pp.

USDA-ERS. 2018. National count of farmers market directory listings: Graph 1994–2017. Available at: https://www.ams.usda.gov/sites/default/files/media/NationalCountofFMDirectory17.JPG.

USDA-FS. 2017. Agroforestry: enhancing resiliency in U.S. agricultural landscapes under changing conditions. Schoeneberger, M. M., Bentrup, G. and Patel-Weynand, T. (Eds). Gen. Tech. Report WO-96. United States Department of Agriculture, Forest Service, Washington DC. doi:10.2737/WO-GTR-96.

Valdivia, C., Barbieri, C. and Gold, M. A. 2012. Between forestry and farming: policy and environmental implications of the barriers to agroforestry adoption. *Canadian Journal of Agricultural Economics/Revue Canadienne d'Agroeconomie* 60(2), 155–75. doi:10.1111/j.1744-7976.2012.01248.x.

Vandermeer, J. 1995. The ecological basis of alternative agriculture. *Annual Review of Ecology and Systematics* 26(1), 201–24. doi:10.1146/annurev.es.26.110195.001221.

Van Sambeek, J. and Reid, W. 2017. A double row alley-cropping system for establishing nut orchards. *MNGA (Missouri Nutrition Growers Association) Newsletter* 17(4), 11–4.

Walter, W. D., Jose, S. and Zamora, D. 2015. Alley cropping. In: Gold, M. A., Cernusca, M. M. and Hall, M. (Eds), *Training Manual for Applied Agroforestry Practices – 2015 Edition. MU Center for Agroforestry*, pp. 31–49. Available at: http://www.centerforagroforestry.org/pubs/training/index.php.

Wechsler, D. 2005. Hopeton farms. In: *The New American Farmer, Profiles of Agricultural Innovation* (2nd edn.). Sustainable Agriculture Network, Beltsville, MD. Available at: https://www.sare.org/Learning-Center/Books/The-New-American-Farmer-2nd-Edition.

Wells, L., Conner, P., Brock, J. and Hudson, W. 2018. Organic pecan production. UGA Coop Extension Bulletin 1493. Available at: https://secure.caes.uga.edu/extension/publications/files/pdf/B%201493_1.PDF.

Wilson, M. H., Lovell, S. T. and Carter, T. 2018. *Perennial Pathways: Planting Tree Crops. Designing and Installing Farm-Scale Edible Agroforestry*. Savanna Institute, Madison, WI.

Wolz, K. J. and DeLucia, E. H. 2018. Alley cropping: global patterns of species composition and function. *Agriculture, Ecosystems and Environment* 252, 61–8. doi:10.1016/j.agee.2017.10.005.

Wolz, K. J., Lovell, S. T., Branham, B. E., Eddy, W. C., Keeley, K., Revord, R. S., Wander, M. W., Yang, W. H. and DeLucia, E. H. 2017. Frontiers in alley cropping: transformative solutions for temperate agriculture. *Global Change Biology* 24, 883–94.

Wolz, K. J., Branham, B. E. and DeLucia, E. H. 2018. Reduced nitrogen losses after conversion of row crop agriculture to alley cropping with mixed fruit and nut trees. *Agriculture, Ecosystems and Environment* 258, 172–81. doi:10.1016/j.agee.2018.02.024.

Wood, B. W., Grauke, L. J. and Bock, C. H. 2016. The pecan provenance collection at Byron Ga - a unique resource for the long-term survival of the US industry. *Pecan Grower* 27, 28–46.

Zhaohua, Z., Maoyi, F. and Stastry, C. B. 1991. Agroforestry in China - an overview. In: Zhaohua, Z., Mantang, C., Shiji, W. and Youxu, J. (Eds), *Agroforestry Systems in China*. Chinese Academy of Forestry and International Development Research Center, Canada.

Chapter 14

Agroforestry for fruit trees in Europe and Mediterranean North Africa

Pierre-Éric Lauri, INRA, France; Karim Barkaoui, CIRAD, France; Mohammed Ater, Abdelmalek Essaadi University, Morocco; and Adolfo Rosati, CREA, Italy

1 Introduction

In his reflection about agroforestry design, Oldeman (1992) considers that the 'main criteria distinguishing agroforestry systems from other types of agrosystems rest upon the former's forest properties'. This forest legacy of agroforestry has deeply affected all initial definitions such as the one proposed by ICRAF in the 'first agroforestry paradigm' that is combining 'the protective characteristics of forestry with the productive attributes of both forestry and agriculture' (Van Noordwijk et al., 2016). More recently, agroforestry was redefined as a type of 'land use that combines aspects of agriculture and forestry, including the agricultural use of trees' (Van Noordwijk et al., 2016). This broader definition does not limit the term 'tree' to its use in forestry, timber production for example, and extends it to the contribution of trees to the agricultural production (e.g. fruit production). This shift in the definition of agroforestry, especially about the purpose of trees, is well integrated in the newly defined concept of 'multifunctional woody polycultures' that emphasizes the various agricultural uses of trees and shrubs and their crop-oriented management within farming systems (Mosquera-Losada et al., 2017; Lovell et al., 2017).

http://dx.doi.org/10.19103/AS.2018.0041.18

Fruit trees are considered as high value for agroforestry (Den Herder et al., 2017; Pantera et al., 2018) and are the primary drivers of agroforestry adoption worldwide (Wolz and Delucia, 2018). Fruit trees are still the main components of agroforestry systems (AFS) in many tropical regions where they contribute to domestic consumption and support short supply chain (Nair, 1991; Parrotta et al., 2015). However, this is not the case in Europe, and to a lesser extent in North Africa, where fruit trees are nowadays mostly grown in dedicated and often intensive monospecific orchards, hereafter referred to as 'conventional' orchards (Herzog, 1998).

Although conventional orchards may reach high fruit production and high economic profit, they also have an increasing dependence on external inputs such as irrigation water, fertilizers and pesticides (Simon et al., 2017). They may therefore generate environmental pollution causing human and animal health issues and raising questions about their sustainability regarding both access to resources and quality of the environment (IPES-Food, 2016). For instance, fruit trees in conventional orchards usually have high planting densities, for example around 3000 trees ha^{-1} and more rarely above 4000 trees ha^{-1} in apple (Close and Bound, 2017) or up to 2254 trees ha^{-1} in so-called 'super high-density' olive orchards (Diez et al., 2016), making the control of pests and diseases challenging without an intensive use of pesticides. Thanks to genetic improvement, and sometimes the use of low-vigour rootstocks for species such as apple, peach or cherry, trees in conventional orchards are generally early bearing (e.g. for apple, mature yield around year 6 or 7; Robinson, 2003) and well adapted to full or assisted mechanization but require high inputs. They may have mowed grass alleys or even bare soil between tree rows to limit the competition for resources with other plant species and facilitate the technical interventions within the fields. However, soil tilling has led to increased soil erosion, degraded soil fertility and accelerated leaching of pollutants into water bodies. Soil erosion is worse in the case of orchards planted on slopes, which occurs particularly often with olive trees for example. Moreover, frequent tilling and the abandoning of manuring practices reduced soil organic matter.

In contrast, modern agroforestry systems (AFS) that combine fruit trees with viable intercrops can represent a solution to the negative externalities of conventional orchards, and help preserve fruit tree landscapes. Olive landscapes are good examples of such a possible alternative in the Mediterranean area, both in Europe and North Africa. To prevent further soil and fertility loss in conventional olive orchards, the use of cover crop has been increasingly recommended and adopted in the last decades. Mulches or cover crops provide barriers to evaporation, radiation and heat fluxes, and stimulate soil biological activity with positive effects on soil structure and water retention (Neilsen et al., 2017). While cover crop has become common, the use of economically viable cover crops has not been considered much by either farmers or researchers, probably because intercropping and grazing appear as old, risky and thus unproductive practices. On the contrary, given that covered soils are less exposed to erosion and fertility loss, there is no reason why marketable cash or cover crops should not be intercropped, as long as their cultivation is technically compatible with modern orchard management and farmers' objectives (Rosati et al., 2009, 2011). For instance, intercropping in olive orchards can help increase farmers' income in a context where olive orchards have high harvesting costs but olive oil generally achieves low prices. Additionally, in some countries, olive trees are protected, due to their cultural value or legacy laws, and cannot be removed by farmers. This implies large risks of abandonment, since while olive cultivation is not economically viable, trees cannot be removed in order to change crop or trees. Intercropping, therefore, becomes the only way to diversify the production and increase the farm's income. Similarly, in Europe, cider and fresh apple combined with intercrops and possibly animals, such as sheep or poultry, attract growers' interest due to the greater productivity potential of these systems. Combining fruit-crops with other crop or non-crop (e.g. flower strips) plants, herbaceous or perennials, may directly contribute to the farmers' income and may also promote the biological regulation of pests and diseases (Albert et al., 2017; Rosati et al., 2009, 2012b; Pantera et al., 2018).

In this chapter, our objectives are first to illustrate the large diversity of fruit tree-based agroforestry in Europe and the Mediterranean North Africa, that is excluding the Mediterranean South Africa but including the oases regions on the edge of the Saharan desert, with a focus on the two most represented species-based (apple, olive) systems, and on two place-based (arganery, oasis) systems. Secondly, we detail some biological and agronomical specific characteristics of fruit trees, and present updated work on how current trends in both fruit tree cultivation and agroforestry may converge into the design of agronomically and ecologically sound, and economically viable fruit tree-based AFS, hereafter referred as FT-AFS.

2 Extent and classification of fruit tree-based agroforestry systems (FT-AFS) in Europe and in Mediterranean North Africa

We quantified and classified the extent and type of FT-AFS in Europe and in Mediterranean North Africa. For Europe, we used the LUCAS (Land Use and Coverage Area frame Survey) database, which covers the 28 countries of the European Union. We followed the methodology used by den Herder et al. (2017), but used a more recent version of the LUCAS database (EUROSTAT, 2015). Additionally, we provide specific details about European FT-AFS, which were not specified in den Herder et al. (2017), where they were included in a broader category, called 'high-value tree agroforestry'. Unfortunately, similar statistics were not available for the Mediterranean North Africa and therefore it was difficult to provide a similar comprehensive analysis of FT-AFS across the entire Euro-Mediterranean space. However, a literature survey combined with multiple interviews with experts and field trips in North Africa allowed us to describe and compare the main existing FT-AFS.

Overall, FT-AFS have a very small extent compared to conventional orchards, especially in Europe where they represent only a maximum of 0.24% of the total UAA (Utilized Agricultural Area; recalculated from den Herder et al., 2017) and 5.98% of the total area planted with fruit trees (conventional orchards and FT-AFS). FT-AFS have a small extent also compared to timber or fodder tree agroforestry intended for wood and energy production: only 11.26% of the total European agroforestry areas are allocated to fruit trees. However, FT-AFS are still an important part of the traditional agricultural systems, as they remain a typical component of the agricultural identity of Europe. In Mediterranean North Africa, FT-AFS probably have a greater extent than in Europe, with different forms, from the coast to the mountains and desert inlands, but they are rapidly declining, being replaced by conventional orchards.

FT-AFS encompass a high structural diversity, differing in their composition and diversity in fruit tree species, their associated understory crops and their spatial arrangement. They also differ in their associated farming management and in the role of fruit trees within the system. Amongst the 30 fruit tree species (or group of species) recorded in the LUCAS database in Europe, 17 fruit tree species (or group of species) are currently cultivated in FT-AFS. Olive (60.48%) and apple (15.20%) are by far the most widespread fruit tree species in FT-AFS covering more than 75% of the total European FT-AFS area (Fig. 1).

Other fruit tree species such as pear (5.12%), grapevine (4.32%), plums (3.68%) and cherry (3.52%) are also significant in European FT-AFS. More generally, apart from olive, the typical fruit tree species of European FT-AFS belong to the Rosaceae family (e.g. apple, apricot, cherry, nectarine, peach, pear, plum), which, together, cover 29.92% of the total FT-AFS area, reflecting the high diversity of Rosaceae spp. in European fruit production. In contrast, *Citrus* spp. (e.g. lemon and orange) only cover 1.12% of total European FT-AFS.

A wide range of fruit tree species is also cultivated in FT-AFS in the Mediterranean North Africa. Actually, most of the fruit tree species found in European FT-AFS are also widespread in FT-AFS in Mediterranean North Africa. In particular, olive and apple are by far amongst the most widespread fruit species used in Mediterranean North African FT-AFS. On the other hand, *Citrus* spp. are rarely cultivated in both Mediterranean North Africa and European FT-AFS. However, species like fig, pomegranate or quince that are currently marginal in Europe are more widely cultivated in FT-AFS in North of Africa. Furthermore, other important species like oleaster (including feral olive trees and *Olea europaea var. sylvestris*), prickly pear, date palm and argan, absent or rare in Europe, are also typically cultivated in FT-AFS.

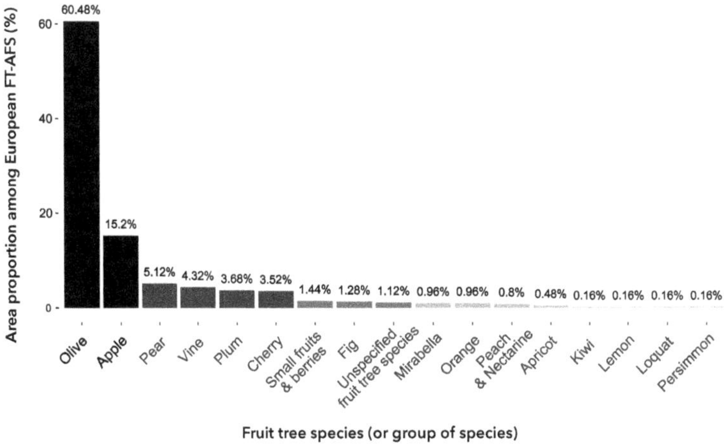

Figure 1 Diversity and area proportion of the 17 fruit tree species (or group of species) cultivated in European FT-AFS. Olive and apple are the main two species, covering more than 75% of the total European FT-AFS. The importance of olive in European FT-AFS reveals the dominance of Mediterranean agroforestry in Europe. Data from LUCAS database (EUROSTAT, 2015).

We identified three main but non-exclusive types of FT-AFS that differ from each other and from conventional orchards based on structural criteria (Fig. 2):

1 *Fruit-livestock agroforestry* is the integration of livestock production with fruit trees and usually considered as a traditional silvopastoral practice. All types of AFS from grazed orchards to more extensive grasslands with sparse fruit trees are included in fruit-livestock agroforestry. In Europe, it is the most important type of FT-AFS which covers 67.84% of the total FT-AFS area (Fig. 2), including grazed apple and olive orchards. In the Mediterranean North Africa, fruit-livestock agroforestry is also an emblematic and widespread form of agriculture (e.g. grazed olive orchards, grazed argan forests etc.). Fruit-livestock agroforestry is part of the category 'Grazed high value tree agroforestry' from den Herder et al. (2017).

2 *Woody polycultures* include any type of association of (1) a fruit tree species with another fruit tree species, including nuts, or (2) a fruit tree species with at least one forest tree species cultivated for timber, energy or fodder or without delivering regular production (e.g. hedgerows). Mixed orchards or complex rural fruit forests therefore may be assimilated to woody polycultures. In Europe, woody polycultures cover 20% of the total FT-AFS area (Fig. 2). Associations of fruit tree species (e.g. apple-pear) are the most usual form of woody polycultures (19.20% of the total

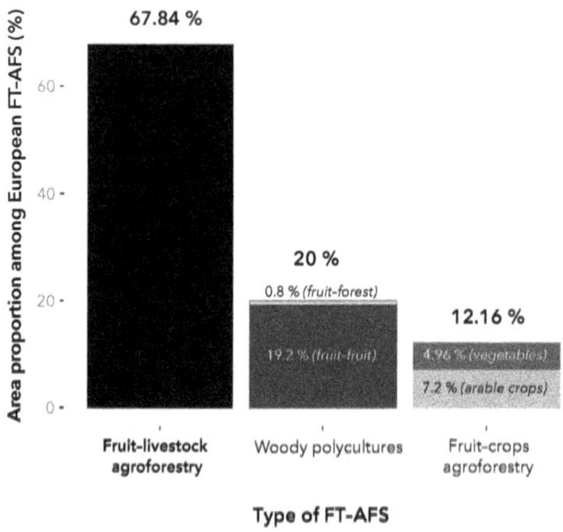

Figure 2 Area proportion of the five different types of FT-AFS identified in Europe. Fruit-animal agroforestry is the main type of FT-AFS, with 67.84% of the total European FT-AFS area, followed by woody polycultures (grouping fruit-fruit and fruit-wood polycultures, totalling 20%). Fruit-crops agroforestry (grouping arable crops and vegetables) is the least widespread in Europe with only 12.16% of the total European FT-AFS area; most of them are found in Southern Europe where Mediterranean conditions prevail. Data from LUCAS database (EUROSTAT, 2015).

FT-AFS area) while associations of fruit trees with forest trees are rare (0.8% of the total FT-AFS area). In Mediterranean North Africa, woody polycultures are part of traditional agro-ecosystems (e.g. combinations of olive-fig-almond-carob trees in rural forests, date palms overtopping several other fruit tree species in the oases). This type of FT-AFS was not classified in den Herder et al. (2017) probably because it is not usually considered agroforestry; however, it was considered as such by Mosquera-Losada et al. (2017).

3 *Fruit-crops agroforestry* is the integration of arable crops, harvested fodder, managed cover crops or high-value vegetables with fruit trees and is usually associated to silvoarable practices. All types of systems from intercropped orchards to extensive arable lands or vegetable farming with sparse fruit trees may be included in fruit-crops agroforestry. In Europe, fruit-crops agroforestry covers only 12.16% of the total FT-AFS area (Fig. 2), including 7.20% of arable agroforestry (e.g. apple-wheat association) and 4.96% of vegetable agroforestry (e.g. olive-tomato association). In contrast, fruit-crops agroforestry is likely to be more important in Mediterranean North Africa, especially because the traditional associations of fruit trees with arable crops (e.g. olive-cereals-dry pulses association) and with vegetables (e.g. olive-onion or olive-fresh faba bean association) are still

widespread in many regions. Fruit-crops agroforestry is part of the category 'Intercropped high value tree agroforestry' from den Herder et al. (2017).

3 Emblematic FT-AFS in Europe and in Mediterranean North Africa

It is only around the middle of the twentieth century that fruit tree growing was isolated from traditional multispecies systems. Until this period, fruit trees were cultivated in agrosilvopastoral systems combining crops, pastures and trees across Europe and in the Mediterranean (Herzog, 1998). These systems were known as 'Joualles', 'Verger haute tige' or 'Pré-verger' in France (Laugénie, 1965; Bérion et Pourcelot, 2007; Dupraz and Liagre 2011), 'Dehesa', 'Pomadaras' or 'Bocage fruit landscapes' in Asturias, Spain (Guichard, 1975), 'Piantate' and 'Seminativi arborati' (arable land with trees) in Italy (Dupraz et al., 2018) or 'Streuobst' in Germany, Austria and Switzerland (Herzog, 1998). More intensive multispecies systems under favourable water conditions were identified as 'Coltura promiscua' in Italy, 'Huertas' in Spain or 'Oasis' in North Africa and in the Eastern part of the Mediterranean basin. Products from these traditional systems were mostly oriented towards family consumption and the surplus of fruit was sold to neighbouring cities (Herzog, 1998).

Despite their structural diversity, these traditional fruit tree systems generally consisted of tall trees planted at low density (less than 100 trees ha^{-1}) leaving a large space for crops and/or forages (hayed or grazed). They were composed of one or more fruit species with pip-fruit species mixed with stone-fruit species, and possibly grapevines depending on the site (e.g. 'Hautains' mixing grapevine and fruit trees where the latter were the structural support of grapevines (Pointereau, 2006; Dupraz and Liagre, 2011)). These traditional AFS formed the so-called 'Bocages', typically composed of hedgerows and scattered trees within arable lands, placed next to open landscapes or 'open fields'. Historians interpret this juxtaposition as the result of the evolution of landscapes according to the specific history of territories. Bocage landscapes corresponded to marginal polyculture systems in scattered settlements related to the poor and illiterate populations whereas open landscapes corresponded to a central agrarian monoculture system community grouped around habitats (Maury, 1992; Flatrès and Flatrès, 1997). These two systems differed in their purpose: bocages were usually part of the food-production system whereas open fields were strictly oriented towards cash crops only.

The four case studies presented hereafter are emblematic of existing FT-AFS inherited and adapted from traditional multispecies systems. In two of them, apple agroforestry in Europe and olive agroforestry in both Europe and Mediterranean North Africa, apple or olive trees are intercropped with forages, annual arable crops or vegetables possibly combined with grazing depending on the growing conditions. Two other systems are presented as typical traditional FT-AFS in North Africa, which are particularly well suited to harsh growing conditions: arganery and oasis. These two systems are illustrated for Morocco which is in the southernmost part of the Mediterranean basin.

3.1 Apple-based AFS in Europe

Although most of the apple (*Malus domestica* Borkh.) production in Europe comes from high-density monospecific orchards, some traditional intercropped systems combining apples, as the main or even the only fruit tree, with other plants and possibly animals are now regaining popularity (Pantera et al., 2018). Environmental concerns are often one of the main motivations to develop these systems, for example to reduce soil erosion and nitrate leaching while increasing carbon sequestration (Kuster et al., 2012), or to reduce pest and disease using ecological regulations (Simon et al., 2017; Smith et al., 2016). The economic interest in apple-based AFS also derives from the fact that growing various agricultural products and animals on the same plot is a way to ensure more stable economic returns over the years. Apple-based AFS, including cereals, field vegetables and potatoes (Smith et al., 2016) or field vegetables and poultry, and possibly sheep (GRAB, 2017; Timmermans and Bestman, 2016), are now well documented (see Section 4.3) (Fig. 3).

Whether or not apple-based AFS are combined with animals, a general characteristic of these systems is the low planting density, with large distances between tree rows and/or within the row. This design has strong implications in terms of the selection of appropriate apple plant material. Although the use of own-rooted apple trees with high vigour (Maguylo and Lauri, 2007) may be convenient for apple-based AFS practices and systems, the cultivated apple tree is most often a grafted tree combining two components: the rootstock and the scion. The former ensures that the vegetative growth potential of the whole tree is positively related to root development and also to the precocity of bearing (the lower the vigour of the rootstock the earlier the bearing). The scion (i.e. the grafted cultivar) is the part of the tree that produces the fruit.

The apple tree in FT-AFS has to be vigorous in order to be able to compete with the other plants in the system, and to be self-supporting, that is without the need of a support system. This implies that the low-vigour rootstocks (e.g. type M9) that are used in current commercial apple orchards may not be convenient for FT-AFS. Taking advantage of the large variation in rootstock vigour available for apple (Ferree, 1988), high-vigour rootstocks, like those from the Malling Merton series (e.g. MM111; Smith et al., 2016) or from the Cornell-Geneva series (Fazio, 2017; Lordan et al., 2017) for example, are recommended to improve anchoring of the fruit tree into the soil as

Figure 3 Grazed apple orchard (©Nathalie Corroyer, CRA Normandie, France).

well as nutrients and water absorption. Ideally, the selection for high-vigour rootstocks also needs to consider the resistance or tolerance to some major pests and diseases such as fire blight caused by *Erwinia amylovora*, crown and root rot caused by *Phytophthora cactorum*, and also replant disease complex (Fazio, 2017).

The second component of the apple tree, the cultivar, is primarily chosen for fruit characteristics and market preferences. However, as for the rootstock, the success of the apple-based AFS also depends on the selection of cultivars that are resistant or tolerant to the main pests and diseases prevalent in the region (Smith et al., 2016).

Once the plant material is determined, the physical design of the system is an effective, although partial, means to reduce the spread of pathogens through both changes in the tree and whole-system microclimate. For instance, a large distance between apple trees favours air circulation and thus decreases leaf wetness (Simon et al., 2006). Plant arrangement may also act to dilute, or create physical barriers against, pathogens and insects (Simon et al., 2010; Smith et al., 2016).

3.2 Olive-based AFS in Europe and Morocco

Olive (*Olea europaea* L.) cultivation is among the oldest tree cultivations dating back millennia (Zohary and Hopf, 1994; Terral et al., 2004; Breton et al., 2006; Besnard et al., 2013; Besnard, 2016). This cultivation affected both the culture and the economy of the Mediterranean area (Loumou and Giourga, 2003;

Kaniewski et al., 2012). Since the beginning, olive trees, like other fruit trees, were intercropped with other crops, like legumes and grains, vineyards and forages (Sestini, 1963; Lelle and Gold, 1994) thus forming AFS. When crops were not present, the olive groves provided grazing land for animals, while the latter weeded and fertilized the orchards (Vannucci, 2009). As an evergreen species, olive pruning materials also provided forage during the cold or the dry season, unlike other deciduous fruit trees.

In the last century, specialized olive orchards increased their extent in Europe. In order to increase production, trees were planted closer to each other (Brugnoli and Varanini, 2005), but intercropping and grazing remained common until well after the Second World War, when multifunctional agriculture disappeared for most fruit tree species. Olive orchards, however, are still mostly managed with large trees and traditional cultivars, allowing intercropping and grazing, which still occurs in a relatively small but significant fraction of the olive cultivation area. While it is hard to estimate this area precisely, given the large area of olive cultivation, that is more than 5 million ha in Europe and more than 10 million ha worldwide (EUROSTAT, 2015), even a relatively small fraction of the total olive area represents a substantial amount of olive-based AFS. Additionally, even super-high-density olive orchards present opportunities for modern intercropping.

In Morocco, as in the rest of the Mediterranean North Africa, rural landscapes are strongly linked to olive growing, which has undergone its first development in the northern areas of Morocco since the Roman period (Fig. 4; Akerraz and Lenoir, 1981; Moukhli et al., 2013). Olive systems are very diversified and characterized by a strong regional identity. In the most traditional olive orchards in the northern region of Morocco, around Chefchaouen in the 'Jbala zones', the olive tree is still usually associated with other fruit species, especially fig, carob and grapevine that play an important role in the domestic economy. The extensive and scattered type of plantation has favoured the practice of polycultures, where, in addition to other fruit trees, farmers grow annual arable crops, mainly cereals (e.g. durum wheat, bread wheat, barley), legumes (faba beans, lentils, chickpeas) and forage. In addition, olive orchards are grazed by extensive livestock, mainly goats. This traditional type of olive management is very old and was described by travellers in the eleventh century (Mouliéras, 1899) and by the first agronomists of the Spanish colonial administration (Torrés, 1930; Alta Comisarìa de España en Marruecos, 1948). However, intercropping tends to decrease following the modernization and mechanization of olive orchards or when conditions are too challenging for it, either because of dense shade under tall trees or more often because of steep slopes.

The quality and interests of these traditional olive-based AFS have been recognized by Moroccan agricultural development organizations and are currently the subject of several studies (Daoui and Fatemi, 2014). Another

Figure 4 Olive landscape in spring in the region of Moulay Idriss Zerhoun, Northern Morocco (©Karim Barkaoui).

characteristic of traditional olive orchards is the presence of spontaneous olive trees represented by the oleaster trees, locally called 'Berri'. These can be wild (*Olea europea* ssp. *europea* var. *sylvestris*) as well as feral, corresponding to crop escapements, generated as seedlings of cultivated olive. The recognition of both cultivated and spontaneous forms is very old and well described in the agronomic manuals (Ibn Al Awam, 2000) and by Andalusian botanists (Al Khattabi, 1995). These abundant spontaneous forms in the oleaster and mastic (*Pistacia lentiscus*) forest formations of which they are a characteristic species (Emberger, 1939; Benabid, 1984) are commonly exploited by local people. In most of the traditional orchards oleaster is used as rootstock for olive varieties (e.g. in the 'Jbala zones', 100% of farmers use oleaster as a rootstock; Ater et al., 2016).

3.3 Arganery and palm-based oasis as traditional FT-AFS in Morocco

Arganery (Fig. 5) is an AFS structured around a keystone species, the argan tree (*Argania spinosa* L. Skeuls) that is an undomesticated fruit species (Emberger, 1925, 1938; Boudy, 1951; Benabid, 2000; Msanda et al., 2004; Peltier, 1982).

Arganery represents a good example where traditional practices constitute an agro-biodiversity conservation model with argan included in multispecies orchards (mostly olive, grapevine and fig tree), valorizing the practice of extensive breeding and the coexistence of cultivated and spontaneous compartments. The argan forest covers some 828 000 ha in so-called pre-forest

Figure 5 Arganery in Morocco (©Mohammed Ater).

stands and occurs in essentially arid or semi-arid bioclimates spread over an estimated area of between 4 000 000 ha (M'Hirit et al., 1998) and 5 038 400 ha (ANDZOA, 2016). The argan tree has a strong identity and symbolic value for the populations of southwestern Morocco (Nouaim, 2005). Thus, human life in the argan tree areas is closely linked to argan cultivation with strong cultural and heritage landscapes. 'Argan, practices and know-how related to the Arganery' has been included in the 'Representative list of the intangible cultural heritage of humanity' of UNESCO in 2014. Previously, an Arganery Biosphere Reserve was recognized in 1998 by UNESCO (El Fasskaoui, 2009).

Traditional arganery is composed of two types of formations: the arganery forest, mostly in hilly areas, and the arganery orchard characterized by intercropping argan trees with cereals (Fig. 6) and more recently with market gardening.

Recent studies show that arganery is mostly an agrosilvopastoral system developed in the last millennium contrary to the dominant idea that places arganery as a forest system (Ballouche, 2015). This system is based on three productions: argan (for oil, wood, grazing), livestock (mainly goats; El Aïch et al., 2005) and intercrops (mainly barley) (Benchekroun and Buttoud, 1989; Al Ifriqui, 2004; Lefhaili, 2014; Bourbouze and El Aïch, 2005; Tarrier and Benzyane, 2003). Indeed, more than two million people benefit from the exploitation of traditional argan tree-based AFS, which supports about 90% of the local rural economy. Indigenous communities hold important ecological knowledge on the functioning of the ecosystem resulting in an efficient management of arganery landscape (El Harousse et al., 2012) as demonstrated by a complex legal status of the land including private ('Melk') or collective ('Mouchaa') status, and privatized seasonal fencing ('Agdal') (De Ponteves et al., 1990).

Figure 6 Argan trees intercropped with cereals in Morocco (©Mohammed Ater).

The oasis agro-ecosystem can be considered as a good example of intensive FT-AFS well adapted to arid climates. Knowledge about water and soil management is the founding element of this type of FT-AFS. In Morocco, oases are scattered over an area of 115 563 km² (15% of total area) associated with 1.7 million people (5% of the total Moroccan population; Renevot et al., 2009). These ingenious systems face strong threats and disruptions that have led to initiatives for their conservation, particularly with the creation of the biosphere reserve of southern Moroccan oases (RBOSM, 'Réserve de Biosphère des oasis du Sud du Maroc') in 2000.

The palm-based oasis FT-AFS is characterized by its vertical structure in three strata: (1) the upper stratum with the date palm (*Phoenix dactylifera* L.) as a keystone species of the oasis system, (2) the intermediate stratum with fruit trees (olive, fig, grapevine, almond, pomegranate, apricot) and (3) the lower stratum with herbaceous crops (legumes, cereals, fodder and vegetable crops). It is an intensive system based on polyculture and livestock farming. However, this organization scheme may vary depending on geography, relief or water sources. There are, therefore, various types of oasis that can be quantified through diverse variables such as the relative importance of the main crops and more generally of the agro-biodiversity. Although the date palm is still a main species of the agro-ecosystem, it is partly replaced by other fruit species such as almond or other Rosaceae species at high altitude. The typical example of this type of Oasis in the High Atlas, in a region called 'Imilchil', integrates vegetables, cereals and livestock in rotation under fruit trees. This system has been recognized as a 'Globally Important Agricultural Heritage Systems' by FAO in 2011.

4 Reinventing FT-AFS

4.1 Biological specificities of fruit trees and consequences for fruit tree management in agroforestry

In a given FT-AFS, the growth and production of the fruit tree depend on both its intrinsic specificities and especially how they differ from those of the timber tree, and the way the fruit tree is managed taking into account a more competitive aboveground and belowground environment.

The evolution of the concept of agroforestry, from forest tree-based systems (Oldeman, 1992) to 'multifunctional woody polycultures' (Lovell et al., 2017) underlines the diversity of tree uses and perceptions, from trees that are grown for vegetative products only (fodder for livestock, fuel, shelter or timber) and are often designated by the term 'forest trees', to those grown for fruit. In the former case, the aim of all cultural operations is to stimulate vegetative growth avoiding as much as possible reproductive growth. For example, for timber trees, the main objective is to build a trunk with a defined size and shape, and high wood quality. For fruit trees, on the contrary, biomass allocation to the reproductive organs is of primary interest and breeding programmes have focused on enhancing carbon allocation to fruits, encompassing both aspects of fruit quality and regular bearing across consecutive years. These objectives are met through genetic improvement (e.g. apple, Brown and Maloney, 2003; olive, Bellini et al., 2002) and cultural operations (e.g. apple, Lauri and Corelli-Grappadelli, 2014; peach, Plenet et al., 2009). These two biomass allocation strategies, towards either vegetative or reproductive growth, result in very different biomass patterns and tree architecture. For example, forest trees from Fagaceae (genera: *Fagus*, *Quercus*) and Pinaceae (genera: *Pinus*, *Picea*, *Abies*, *Pseudotsuga*) species allocate 0.5–3% of photosynthates into fruit formation (Fernandez-Martinez et al., 2017), whereas fruit trees may allocate from 7% (e.g. cherry; Kappel 1991) to 60–70% (e.g. apple, Getachew 2000, in Wünsche and Ferguson, 2005) of photosynthates into fruit formation depending on the species.

These two biomass allocation strategies characterizing timber and fruit trees are likely subtended by a strong genetic determinism. There is also a genetic variability within each fruit trees species in the vegetative-reproductive interplay as shown in several species, including apple (Lespinasse, 1992; Lauri and Laurens, 2005) and olive (Rosati et al., 2013, 2018). Cultural operations carried out all along the tree lifespan are therefore of utmost importance to balance vegetative growth and fruiting as shown in apple for example (Lauri and Laurens, 2005; Lauri et al., 2011; Lauri and Corelli-Grappadelli, 2014; Musacchi and Greene, 2017). In past decades, a great number of articles and books have been published on fruit tree management revealing a common agreement on two main points (Lauri, 2018). Firstly, there is no unique good training (i.e. giving the tree a given shape and controlling scaffold branches orientation) and pruning (i.e. cutting away branch parts) management for fruit trees. This is well illustrated for apple trees where more than 30 'training systems' (sets of training and pruning rules) have been formalized all over the world (Robinson, 2003). This highlights that, depending on the geographical region and even the grower, the initial choices of planting distance, tree height, support system and cultural operations during the orchard lifespan depend not only on the plant material but also on the cultivation and socio-economic contexts. Secondly, a main objective of all training and pruning procedures is to increase the canopy porosity to light, avoiding strong vegetative reactions, and to partition photosynthates to fruit-bearing shoots and not to vegetative shoots (e.g. water sprout) unless the structural framework of the tree has to be rebuilt. Such methods have been recently developed especially for apple trees (Lauri and Corelli-Grappadelli, 2014; Breen et al., 2016) and have to be tightly adapted to the endogenous potential of each species and variety (Lauri et al., 2011). Therefore, the biologically-based fruit tree management is a main component of the sustainability of fruit tree cultivation (e.g. apple; Lauri et al., 2009) and has to be considered when implementing a FT-AFS in relation with the choice of the plant material (variety and possibly rootstock; e.g. for apple, see Section 3.1) and the design of the system (e.g. planting distances, use of a support system) (Lauri, 2018).

The question is then how to choose the fruit tree plant material most suitable for FT-AFS and how to adapt the conventional 'training system' initially designed and formalized for conventional orchards to fruit trees grown in FT-AFS. Three main aspects have to be taken into account to limit detrimental above- and belowground competition that characterizes all multispecies systems. Firstly, concerning the plant material, a high plant vigour is needed, implying that the variety should be either grafted on a vigorous rootstock (see for apple Section 3.1) or even used as own-rooted tree (as usual in traditional olive AFS, and possibly in apple, cf Section 3.1). Secondly, the spatial and temporal design of the FT-AFS, that is the layout and the possible removal of trees at certain times along the FT-AFS lifespan, respectively, has to include a good evaluation of minimal distances between the fruit trees and the other plants, whether perennials or herbaceous (see Section 6.2). Thirdly, training and pruning has to be adapted to both the plant material and distances and even though the main concepts driving current tree management are still valid (e.g. high canopy porosity to light to improve leaf functioning and fruit quality), the management of the whole tree has to consider the possible presence of animals, avoiding low bent branches for example (cf Section 4.3).

4.2 Diversifying the ecosystem services provided by FT-AFS

Taking advantage of concepts developed in the field of ecology, recent research aimed at analysing 'ecosystem services' (Millennium Ecosystem

Assessment, 2005; Haines-Young and Potschin, 2018) in orchards, especially with apple (Demestihas et al., 2017). More precisely, Demestihas et al. (2017) considered that orchards can be optimized through the choice of plant material and agricultural practices to provide the following six ecosystem services to a satisfying level: fruit production, climate regulation, soil nitrogen availability, water regulation, pest and disease control and pollination.

The management of the orchard to provide these various ecosystem services, especially when it has to mitigate trade-offs between some services (e.g. production vs regulation services), may take advantage of the empirical and scientific knowledge existing on biodiversity, and the structural and temporal management of existing FT-AFS (Simon et al., 2017). For example, among the various principles and practices promoting a more ecological production in fruit tree cultivation, two main standards co-exist in current fruit tree cultivation which may be synergistic with the renewed interests in FT-AFS, 'Integrated Fruit Production' (IFP) and 'Organic Farming' (OF).

IFP is defined by the IOBC (International Organization for Biological and Integrated Control of Noxious Animals and Plants) as 'an economical and high quality production framework giving priority to ecologically safer methods, minimizing the undesirable side effects and use of agrochemicals, and enhancing the safeguard of the environment and human health' (Damos et al., 2015). Organic Farming is defined by the IFOAM (International Federation of Organic Agriculture Movements) as 'an integrated farming system that strives for sustainability, the enhancement of soil fertility and biological diversity whilst, with rare exceptions, prohibiting synthetic pesticides, antibiotics, synthetic fertilizers, genetically modified organisms, and growth hormones' (Eventbrite, 2018). Although the area of fruit production in OF has been increasing in recent years (Granatstein et al., 2016), OF is still currently less developed than IFP. Basically, a main difference between IFP and OF is that the OF approach aims to manage the agro-ecosystem as a whole insisting on the restricted use of synthetic inputs and their derivatives (Weibel and Häseli, 2003) while IFP guidelines do not have strict regulations on the use of synthetic products (Granatstein and Peck, 2017). However, IFP and OF both share in common that agroforestry principles are important levers to improve the overall functioning of orchards and the provisioning of ecosystem services. Indeed, both IFP and OF orchards generally include other annual and/or perennial herbaceous species between the fruit tree rows, as well as bushes and/or trees in hedgerows around the orchard. In addition, it should be noticed that the adoption of either IFP or OF is independent of the level of external inputs, since both IFP and OF systems can be implemented under high or low external input levels (Parr et al., 1990; Parisi et al., 2014), and therefore correspond to different economic strategies.

Whether managed with IFP or OF, orchards can be oriented towards the production of one fruit and are therefore typical monocropping orchards, or are diversified to secure the food-production service. In this latter case they combine one species of fruit tree and of vegetables (e.g. apple tree combined with soybean or peanut, Gao et al., 2013) or various fruit tree species and various annual crops such as vegetables and cereals (e.g. 'verger maraîcher' in France, GRAB, 2017; Smith et al., 2016) as shown in Fig. 7. These latter systems are no longer 'conventional' orchards but are developed in the framework of novel land use systems, typically FT-AFS.

4.3 Benefits of integrating animals in FT-AFS

As shown above, fruit-livestock agroforestry is the most widespread form of FT-AFS in Europe and in Mediterranean North Africa (Mosquera-Losada et al., 2016). Besides the socio-economic reasons that have encouraged farmers to integrate livestock to fruit tree cultivation in the traditional systems, there are several agro-ecological reasons to reconsider the integration of animals into orchards and to strengthen the positive interactions between trees and animals. For instance, sheep are being studied as a means to weed orchards (e.g. cider apple and to a lesser extent fresh apple) and even vineyards, with low environmental impact while also increasing product diversification and overall productivity (Shropshire Sheep Breeders Association, 2008). In addition, by eating fallen apple leaves with scab or other infections, sheep might decrease the incidence of some diseases (Corroyer and Upson, 2015). However, sheep do eat leaves on the tree as well. Consequently, the integration of sheep into

Figure 7 Peach with cabbage in a 'verger maraîcher', GRAB experiment, Avignon, France (©Pierre-Éric Lauri).

apple orchards has to be optimized by an appropriate training and pruning strategy, for example promoting high stem trees, as found in traditional cider apple orchards where sheep grazing is still practised (Burrough et al., 2010), or by restricting the presence of sheep from fall to spring before trees begin to grow. Sheep breed is important as different breeds are more or less attracted by tree leaves, and have different tendencies to stand on hind legs to browse higher up on the tree. Sheep management in apple orchards is currently being studied in several places, including the United Kingdom (McAdam and Ward, 2017; Burgess et al., 2017a,b) and France (Corroyer and Upson, 2015). In vineyards, sheep can be trained to not eat grapevines, and thus be used as effective means of weed control (Meadows, 2008). Although not investigated yet, the same training could probably be used to avoid damage to trees in low stem fruit tree orchards.

Weeding with sheep is easier in taller orchards, such as walnut and chestnut orchards, and some research work on this topic is being carried out in Spain and Greece, among other places (Pantera et al., 2018). To avoid damage to the trees, weeding can be done with small animals. Chickens and hens are being studied to weed orchards, or alternatively orchards are planted in pasture areas of free-range chickens, in several places including the Netherlands (Timmermans and Bestman, 2016) and the United Kingdom (Woodland Trust, 2014). Geese are grass eaters and thus good weeding agents in both orchards and vineyards, as well as in many herbaceous crops (Geiger and Biellier, 1993). Interestingly, the benefits of the grazed agroforestry is also for the animals and, as shown by Timmermans and Bestman (2016), chickens kept in apple-based AFS exhibit signs of reduced stress, with reduced pecking damage.

Recent research has focused on free-range chickens grazing the olive orchard, thus providing weeding and fertilization (Rosati et al., 2009, 2012a,b,c, 2014) (Fig. 8).

Chickens feel protected from predators under the trees, and thus venture further away from their shelter. This enhances their foraging activity, making them better able to exploit the forage and feed (i.e. microfauna) available in the pasture (Dal Bosco et al., 2014). The enhanced foraging improves chicken meat quality (Dal Bosco et al., 2016). Paolotti et al. (2016) showed that raising free-range chicken in olive orchards, rather than in areas dedicated only to the chickens, has environmental benefits for both the animal and the olive sectors. In fact, the animals can provide both the required fertilization and weeding in the orchard, which otherwise represent the greatest environmental impact in the orchard. The latter, instead, provides grazing land, thus eliminating the land use impact due to grazing.

Chickens can weed and fertilize not only the orchard, but also understory crops grown in the orchard. For instance, recent studies on the cultivation of a perennial wild asparagus (*Asparagus acutifolius*) suggests that this is a

Figure 8 Chicken reared in an olive orchard. Chicken shelters are situated at the margin of the fenced orchard (©Adolfo Rosati).

suitable understory crop in olive orchards (Rosati, 2001, 2011, 2014; Rosati and Falavigna, 2000; Rosati et al., 2005, 2009, 2012a,b,c, 2014; Mantovani et al., 2016) (Fig. 9). Chickens can weed and fertilize the olives as well as the wild

Figure 9 Wild asparagus (*Asparagus acutifolius*) cultivated along tree row in a high-density olive orchard (©Adolfo Rosati).

asparagus, since its mature plants are prickly and chickens do not damage them (Rosati et al., 2009, 2012a,b,c; Rosati, 2014).

5 The possible interest of multistrata FT-AFS

A great majority of FT-AFS found in Europe and in Mediterranean North Africa have two strata (e.g. fruit trees and arable crops or vegetables), except the palm-based oases that are typically multistrata FT-AFS where palm trees overtop fruit trees in the upper strata and annual or perennial herbaceous crops in the lower strata. The concept of 'multifunctional woody polycultures' developed in some temperate areas (Lovell at al., 2017) emphasizes the interest of multiple tree crops, among which fruit and/or nut trees are cultivated in multistrata systems. This concept takes inspiration from tropical AFS that are characterized by both a high number of woody and non-woody species and a multistrata vertical stratification (Lauri et al., 2016). In multistrata FT-AFS, the various species share the vertical space, with timber trees at the highest stratum generally overtopping fruit trees (e.g. mango) and cocoa trees in the intermediate stratum, and staple food at the lowest stratum (Parrotta et al., 2015; Saj et al., 2017). Apart from the well-documented production, and regulation and maintenance service categories, cultural services can also be developed by such systems combining for example leisure and pick your own activities.

A main aspect to consider in multistrata FT-AFS is the endogenous ability of the fruit tree to grow under shade. In general, temperate fruit tree species perform well under full sun conditions with significant negative effect of shade on carbohydrate photosynthates and fruit growth and colouring (Correlli-Grappadelli, 2003). However, in climatic regions with hot and dry summer, that is typical of Mediterranean North Africa, fruit trees may be subjected to an excess of solar radiation during the summer months with known deleterious effects on fruit quality (sunburn) and leaf functioning (photooxidative damage) (e.g. apple; Correlli-Grappadelli and Lakso, 2007; Racsko and Schrader, 2012). This issue is commonly handled using shade nets that are also used against hail. Therefore, the provision of multiple ecosystem services including disease and pest regulation that are delivered by ground covers and hedgerows in IFP and OF orchards for example (e.g. against *Dysaphis plantaginea* in cider apple (Albert et al., 2017) and against *Aphis cidricola* in fresh apple (Song et al., 2013)) could be enhanced by an increasingly targeted climate extremes (excess of light, heat) attenuation service. In this case, overstory trees, grown for food (nuts) or other goods (timber, fodder) or even for providing supporting services (e.g. legume trees improving soil fertility through nitrogen fixation and recycling of nutrients; Ong et al., 1991), may provide convenient shade to the fruit trees, provided that overstory trees are carefully pruned to avoid excessive shading (see Section 6.2).

It should be noticed that, apart from the light and temperature issues, multistrata systems are also known to reduce wind speed reducing evaporative stress which may have positive effects on leaf functioning (Jose et al., 2004) but could also increase certain diseases whose development is enhanced by air and surface relative humidity (e.g. apple scab). Otherwise, AFS have known effects in decreasing high temperatures during the day and increasing low temperatures during the night (Gosme et al., 2016). Although these effects are of small amplitude, the cumulative effects over the whole season may have consequences on plant development and especially on fruit maturation which, as documented in apple for example, is improved by alternating temperatures during night and day (Saure, 1990).

6 Remaining challenges

When growing fruit trees in AFS two strongly related issues that are common to IFP and OF FT-AFS and also specific to multistrata AFS must be addressed.

6.1 Biodiversity and genetics of fruit trees

There is a two-faceted issue regarding biodiversity in FT-AFS: (i) the selection of fruit tree diversity and the associated crops to be assembled and (ii) the conservation of fruit tree genetics and associated (cultivated or not) plant diversity. Apart from the choice of the species that are assembled in the system, directly related to the targeted ecosystem services (belonging to one of the three categories, production, regulation and maintenance, cultural), a main aspect shared by all sustainable production systems is the optimal use of the genetic variability within each species. In a review of apple ideotypes adapted to low-input fruit production agro-ecosystems, Parisi et al. (2014) establish three main characteristics: regular income for the grower that implies regularity of fruiting and good fruit quality and also easy-to-manage tree architecture to decrease tree management workload; tolerance and durable resistance to the most important pests and diseases, which is often encountered in old genotypes and that should be better integrated for breeding and screening new cultivars; hardiness, that is the ability to produce under limiting environmental conditions. In the case of multistrata FT-AFS the ability to grow and to fruit in the shade could also be an objective for breeding (Dupraz and Liagre, 2011; Ehret et al., 2015).

Growing several species together in FT-AFS often results in a higher tree and plant density compared to conventional orchards. This may increase competition for belowground water and nutrients, and may eventually result in decreased growth and yield of fruit trees, especially during the first years (e.g. apple trees intercropped with non-nitrogen-fixing annuals; Robinson and

O'Kennedy, 1978). In this context, the choice of either the fruit tree genotype (e.g. olive) or the rootstock (e.g. apple) is crucial to optimize the response of fruit trees to competition. For example, the intensity and impacts of competition will differ depending on whether the roots of the different species are developing in the same soil layers or in complementary soil layers. However, this choice has to be made also considering planting distances, high-vigour fruit trees demanding larger space than low-vigour fruit trees.

In traditional agro-ecosystems, some fruit tree species can have several coexisting forms (i.e. from domesticated to spontaneous) depending on their level of domestication and management. The distinction between fully domesticated cultivated forms and spontaneous wild forms is made by the mode of reproduction of trees, which consists mainly in vegetative propagation for cultivated trees and spontaneous sowing for wild trees. The coexistence of these two modes of reproduction creates favourable dynamics for the conservation of intraspecific genetic diversity of the various fruit tree species by maintaining a connected gene flow between cultivated and wild trees (e.g. date palm, *Phoenix dactilifera* L., INRA, 2011; apricot, *Prunus armeniaca* L., Bourguiba et al., 2012; Mamouni et al., 2014). For example, in the region of Beni Boufrah in Morocco, the observed dynamics of genetic diversification of almond trees was explained by the importance of the recruitment of spontaneous almond tree seedlings within traditional almond AFS associated with cereals (Delplancke et al., 2013; Delplancke and Aumeeruddy-Thomas, 2017). Apart from conservation issues, the increase of genetic diversity following such diversification pathway promotes tree stands that best adapt to the local environmental conditions and heterogeneity. Indeed, in Northern Morocco forests are cleared and oleaster is preserved by farmers (Aumeeruddy-Thomas, 2010). This practice contributes to the creation of a forest-based olive agro-ecosystem and favours olive intraspecific diversity that is highly adapted to heterogeneous mountain environments. In the traditional olive AFS of Northern Morocco, grafting olive cultivars on wild spontaneous oleaster used as rootstocks is a traditional practice that enables olive to be highly resilient to chronic drought, thanks to well-adapted root systems. In addition, the conservation and integration of wild genetic diversity into cultivated systems favours positive interactions between trees and other species of the ecosystem. For example, oleaster rootstocks allow grafted olive cultivars to benefit from mycorrhizal symbiosis (Sghir et al., 2013) and better use of soil water and nutrients (Caravaca et al., 2003).

6.2 Design of the FT-AFS and tree management

The spatial layout includes both the horizontal and the vertical planes. In almost all the documented multispecies systems the design is based on arrangement in rows, whether straight or curved, which generally offers the best convenience in terms of ergonomics (Gliessman, 1985; Gallaher, 2009; Dupraz and Liagre, 2011). In the typical two-strata systems, forage and/or vegetables in 8–10 m or even 12 m (Smith et al., 2016) wide alleys separate fruit tree rows. In some novel experiments with fruit trees intercropped with vegetables, vegetable alleys alternate with forage alleys to permit convenient access to fruit trees (GRAB, 2017). In the multistrata systems, both maximum tree height and distances between trees along the row have to be tightly managed depending on the species (trees, shrubs and annuals) to avoid excessive shading on the understory plants. According to Lovell et al. (2017), tree rows (whether timber or fruit trees) in a South–North orientation should be set at a minimum of 9 m to avoid a constantly shaded side of trees. Moreover, the spatial pattern needs to be designed together with the temporal scheduling to optimize the production service in the first years of the system. Options such as the co-plantation or alternatively the sequential planting of the perennial species can also be considered depending on the species but also on the ergonomics (access to trees for care and harvest). Similarly, during the system lifespan, overstory and understory tree removal can be done to increase within-row and between-row distances, improving the light microclimate within the system (concept of 'dynamic agroforestry' developed on cocoa AFS; Andres et al., 2016).

Once fruit trees are planted, the way they are individually managed through cultural operations is of paramount importance. Concerning overstory trees, removing branches is a way to efficiently manage the below-tree microclimate, especially light and temperature, provided a good balance is found between the timber or fodder production service and the regulation service (here, microclimate mitigation for the understory plants) of the overstory tree (e.g. pollarding; Le Bec et al., 2016). Training and pruning of the fruit trees have to be planned in relation to the spatial layout of the system. During the tree lifespan, the objective is to improve both the shoot placement in space through training and pruning to optimize light interception and the growth of the fruit-bearing shoots to maintain a regular and high-quality fruiting. In the case of multistrata systems where fruit trees are understory, this latter aspect is crucial.

7 Future trends and conclusion

Fruit tree-based AFS in Europe and Mediterranean North Africa are at the meeting point of at least two converging motivations:

- the 'ecosystem service motivation' including the cultural and societal services, and associated functions, related to the multispecies and multistrata structure (e.g. pick your own farms, leisure).

- the 'resilience motivation' that is especially important in the context of climate change, to harness food security with various plant productions including fruit, but also timber, aromatic plants and animals.

Although FT-AFS are often based on traditional practices with significant empirical knowledge, more fundamental and applied research is needed to improve the provisioning of fruit and other goods in a sustainable way. In this framework, research priorities should focus on how to optimize both the inter-species and within-species (among genotypes) functional biodiversity, and the temporal (e.g. delayed vs simultaneous plantation of the various species comprising the system) and spatial (occupation into the horizontal and vertical planes) design of the FT-AFS.

8 Where to look for further information

Further information on fruit trees in temperate agroforestry systems can be found in the recently published second edition of the book *Temperate Agroforestry Systems* (Gordon, A.M., Newman, S.M. and Coleman, B., Eds), CABI, 2018.

The European Agroforestry Federation website (EURAF; http://www.eurafagroforestry.eu/welcome) provides access to a lot of information, some of it related to fruit trees, including agroforestry projects and results, documents, presentations at conferences, including the four European agroforestry conferences organized by EURAF.

The AGFORWARD (AGroFORestry that Will Advance Rural Development) is the largest European project on agroforestry. The project website (https://www.agforward.eu) contains leaflets on different innovative and best agroforestry practices developed within the project, some of them on fruit trees especially apple and olive (e.g. https://www.agforward.eu/index.php/en/grazed-orchards-in-france.html).

The journal *Agroforestry Systems* (https://link.springer.com/journal/10457) contains original and review articles, some of them discussing the interest and specificities of fruit tree-based agroforestry systems. A special issue of this journal collects 24 papers on the main results of the AGFORWARD project: https://link.springer.com/journal/10457/92/4/page/1.

9 References

Akerraz, A. and Lenoir, M. (1981), 'Les huileries de Volubilis', *Bulletin d'Archéologie Marocaine* 14, 69–119.

Albert, L., Franck, P., Gilles, Y. and Plantegenest, M. (2017), 'Impact of agroecological infrastructures on the dynamics of *Dysaphis plantaginea* (Hemiptera: Aphididae) and its natural enemies in apple orchards in Northwestern France', *Environmental Entomology* 46, 528–37.

Alifriqui, M. (2004), *L'écosystème de l'arganier*, PNUD, Paris, France.

Al Khattabi, M. L. (1995), 'Umdat al Ṭabīb fī Ma'rifat al Nabāt par Abū al-Khayr Al-Ishbīlī, Edition annotée (Guide des plantes à l'usage du médecin)', Dar Al Gharb Al Islami (Ed.), Rabat, Morocco.

Alta Comisarìa di España en Marruecos (1948), *La obra material: Tétouan y Madrid*, Talleres del Instituto Geographico y Catastral, España.

Andres, C., Comoé, H., Beerli, A., Schneider, M., Rist, S. and Jacobi, J. (2016), 'Cocoa in monoculture and dynamic agroforestry', in Lichtfouse, E. (Ed.), *Sustainable Agriculture Reviews*, Sustainable Agriculture Reviews Volume 19, Springer, Cham, pp. 121-53.

ANDZOA (Agence Nationale pour le Développement des Zones Oasiennes et de l'Arganier) (2016), *Données générales*, http://andzoa.ma/fr/andzoa/zone-dintervention/zone-oasienne/donnees-generales/zone-dintervention-de-landzoa/ (Last accessed on 15 January 2018).

Ater, M., Essalouh, L., Ilbert, H., Moukhli, A. and Khadari, B. (2016), *L'Oléiculture au Maroc de la préhistoire à nos jours: Pratiques, diversité, adaptation, usages, commerce et politiques*, CIHEAM, Montpellier, Options Méditerranéennes: Série A. Séminaires Méditerranéens, n° 118, http://om.ciheam.org/option.php?IDOM = 1026

Aumeeruddy-Thomas, Y. (2010), 'Des clones aux semis: Domestication des arbres en Méditerranée, un continuum entre nature et culture. Exemple de *Ficus carica* L., *Olea europaea* L. et *Castanea sativa* L', in Delhon, C., Théry-Parisot, I. and Thiébault, S. (Eds), *Des Hommes et des plantes: Exploitation du milieu et gestion des ressources végétales de la préhistoire d'Antibes*, APDCA, pp. 379-90.

Ballouche, A. (2015), 'Le palimpseste du paysage ou comment lire dans l'espace. L'exemple de l'arganeraie du sud-ouest marocain', in Trivisani-Moreau, I., Taïbi, A. D. and Oghina-Pavie, C. (Eds), *Traces du végétal, nouvelles recherches sur l'imaginaire*, PUR, Rennes, France, pp. 23-42.

Bellini, E., Giordani, E., Parlati, M. V. and Pandolfi, S. (2002), 'Olive genetic improvement: Thirty years of research', *Acta Horticulturae* 586, 105-8.

Benabid, A. (1984), 'Étude phytosociologique des peuplements forestiers et préforestiers du Rif centro-occidental (Maroc)', *Travaux de l'Institut Scientifique, Série Botanique*, Rabat, Morocco, number 34.

Benabid, A. (2000), *Flore et écosytèmes du Maroc: évaluation et préservation de la biodiversité*, Ibis Press, Paris, France.

Benchekroun, F. and Buttoud, G. (1989), 'L'arganeraie dans l'économie rurale du sud-ouest marocain', *Forêt Méditerranéenne* 11(2), 127-36.

Bérion, P. and Pourcelot, J. (2007), 'Approche géographique du terroir du Kirsh de Fougerolles', *Territoire* 36, 6-9.

Besnard, G. (2016), 'Origin and domestication', in Rugini, E., Baldoni, L., Muleo, R. and Sebastiani, L. (Eds), *The Olive Tree Genome, Compendium of Plants Genomes*, Springer International Publishing AG, Dordrecht.

Besnard, G., Khadari, B., Navascués, M., Fernández-Mazuecos, M., El Bakkali, A., Arrigo, N., Baali-Cherif, D., Brunini-Bronzini de Caraffa, V., Santoni, S., Vargas, P. and Savolainen, V. (2013), 'The complex history of the olive tree: From Late Quaternary diversification of Mediterranean lineages to primary domestication in the northern Levant', *Proceedings of the Royal Society* B 280, 20122833.

Boudy, P. (1951), *Description forestière du Maroc*, Publication de la station forestière du Maroc, Rabat, Morocco.

Bourbouze, A. and El Aïch, A. (2005), 'L'élevage caprin dans l'arganeraie: l'utilisation conflictuelle d'un espace', *Cahiers Agricultures* 14(5), 447-53.

Bourguiba, H., Audergond, J. M., Krichena, L., Trifi-Faraha, N., Mamouni, A., Trabelsi, S. and Khadari, B. (2012), 'Genetic diversity and differentiation of grafted and seed propagated apricot (*Prunus armeniaca* L.) in the Maghreb region', *Scientia Horticulturae* 142, 7-13.

Breen, K. C., Palmer, J., Tustin, D. S. and Close, D. C. (2016), 'Artificial spur extinction alters light interception by 'Royal Gala' apple trees', *Acta Horticulturae* 1130, 265-72.

Breton, C., Médail, F., Pinatel, C. and Bervillé, A. (2006), 'De l'olivier à l'oléastre: Origine et domestication de l'*Olea europea* L. dans le bassin méditerranéen', *Cahiers de l'Agriculture* 15(4), 329-36.

Brown, S. K. and Maloney, K. E. (2003), 'Genetic improvement of apple: Breeding, markers, mapping and biotechnology', in Ferree D. C. and Warrington I. J. (Eds), *Apples - Botany, Production and Uses*, CABI Publishing, Wallingford, UK, pp. 31-60.

Brugnoli, A. and Varanini, G. M. (2005), *Olivi e olio nel medioevo italiano*, Volume 29, CLUEB, Bologna, Italy.

Burgess, P. J., Chinery, F., Eriksson, G., Pershagen, E., Pérez-Casenave, C., Lopez Bernal, A., Upson, A., Garcia de Jalon, S., Giannitsopoulos, M. and Graves, A. (2017a), 'Grazed orchards in England and Wales', http://www.agforward.eu/index.php/en/Grazed_Orchards.html/ (Last accessed on 22 December 2017).

Burgess, P. J., Chinery, F., Eriksson, G., Pershagen, E., Pérez-Casenave, C., Lopez Bernal, A., Upson, A., Garcia de Jalon, S., Giannitsopoulos, M. and Graves, A. (2017b), 'Lessons learnt - Grazed orchards in England and Wales', AGFORWARD project, http://www.agforward.eu/index.php/en/Grazed_Orchards.html/ (Last accessed on 12 January 2017).

Burrough, A. E., Oines, C. M., Oram, S. P. and Robertson, H. J. (2010), *Traditional Orchard Project in England - The Creation of an Inventory to Support the UK Habitat Action Plan*, Natural England, London, UK.

Caravaca, F., Díaz, E., Barea, J. M., Azcón-Aguilar, C. and Roldán, A. (2003), 'Photosynthetic and transpiration rates of *Olea europaea* subsp. *sylvestris* and *Rhamnus lycioides* as affected by water deficit and mycorrhiza', *Biologia Plantarum* 46(4), 637-9.

Close, D. C. and Bound, S. A. (2017), 'Advances in understanding apple tree growth: The manipulation of tree growth and development', in Evans, K. (Ed.), *Achieving Sustainable Cultivation of Apples*, Burleigh Dodds Series in Agricultural Science, number 018, Burleigh Dodds Science Publishing Limited, Cambridge, UK, pp. 53-84.

Correlli-Grappadelli, L. (2003), 'Light relations', in Ferree, D. C. and Warrington, I. J. (Eds), *Apples - Botany, Production and Uses*, CABI Publishing, Wallingford, UK, pp. 195-216.

Correlli-Grappadelli, L. and Lakso, A. N. (2007), 'Is maximizing orchard light interception always the best choice?' *Acta Horticulturae* 732, 507-18.

Corroyer, N. and Upson, M. (2015), 'Research and development protocol for grazed orchards in France', http://www.agforward.eu/index.php/en/grazed-orchards-in-france.html/ (Last accessed on 22 December 2017).

Dal Bosco, A., Mugnai, C., Rosati, A., Paoletti, A., Caporali, S. and Castellini, C. (2014), 'Effect of range enrichment on performance, behaviour and forage intake of free-range chickens', *Journal of Applied Poultry Research* 23, 137-45.

Dal Bosco, A., Mugnai, C., Mattioli, S., Rosati, A., Ruggeri, S., Ranucci, D. and Castellini, C. (2016), 'Transfer of bioactive compounds from pasture to meat in organic free-range chickens', *Poultry Science* 95, 2464–71.

Damos, P., Escudero Colomar, L. A. and Ioriatti, C. (2015), 'Integrated fruit production and pest management in Europe: The apple case study and how far we are from the original concept?' *Insects* 6(3), 626–57.

Daoui, K. and Fatemi, Z. (2014), 'Agroforestry systems in Morocco: The case of olive tree and annual crops association in Saïs region', in Behnassi, M., Shahid, S. A. and Habib-Mintz, N. (Eds), *Science, Policy and Politics of Modern Agricultural Systems*, Springer Science and Business Media, Dordrecht. doi:10.1007/978-94-007-7957-0_19.

Delplancke, M. and Aumeeruddy-Thomas Y. (2017), 'Des semis et des clones: Domestication de l'amandier (*Prunus dulcis*) à la frontière entre beldi (ici) et romi (ailleurs), Bni-Boufrah, Rif, Maroc', *Revue d'ethnoécologie*, http://ethnoecologie.revues.org/3166.

Delplancke, M., Alvarez, N., Benoit, L., Espíndola, A., Joly, H. I., Neuenschwander, S. and Arrigo, N. (2013), 'Evolutionary history of almond tree domestication in the Mediterranean basin', *Molecular Ecology* 22(4), 1092–104.

Demestihas, C., Plénet, D., Génard, M., Raynal, C. and Lescourret, F. (2017), 'Ecosystem services in orchards. A review', *Agronomy for Sustainable Development* 37, 12. doi:10.1007/s13593-017-0422-1.

den Herder, M., Moreno, G., Mosquera-Losada, R. M., Palma, J. H. N., Sidiropoulou, A., Santiago Freijanes, J. J., Crous-Durand, J., Paulo, J. A., Tomé, M., Pantera, A., Papanastasis, V. P., Mantzanase, K., Pachana, P., Papadopoulos, A., Plieninger, T. and Burgess, P. J. (2017), 'Current extent and stratification of agroforestry in the European Union', *Agriculture, Ecosystems and Environment* 241, 121–32.

De Ponteves, E., Bourbouze, A. and Narjisse, H. (1990), 'Occupation de l'espace, droit coutumier et législation forestière dans l'arganeraie marocaine', *Cahiers de la Recherche Développement* 26, 28–43.

Díez, C. M., Moral, J., Cabello, D., Morello, P., Rallo, L. and Barranco, D. (2016), 'Cultivar and tree density as key factors in the long-term performance of super high-density olive orchards', *Frontiers in Plant Sciences* 7, 1226. doi:10.3389/fpls.2016.01226.

Dupraz, C. and Liagre, F. (2011), *Agroforesterie, des arbres et des cultures*, Éditions France-Agricole, Paris, France.

Dupraz, C., Lawson, G. J., Lamersdorf, N., Papanastasis, V. P., Rosati, A. and Ruiz-Mirazo, J. (2018), 'Temperate agroforestry: The European way', in Gordon, A., Newman, S. and Coleman, B. (Eds), *Temperate Agroforestry Systems*, CABI, Wallingford, UK, https://www.cabi.org/bookshop/book/.

Ehret, M., Grab, R. and Wachendorf, M. (2015), 'The effect of shade and shade material on white clover/perennial ryegrass mixtures for temperate agroforestry systems', *Agroforestry Systems* 89, 557–70.

El Aïch, A., Bourbouze, A. and Morand-Fehr, P. (2005), *La chèvre dans l'arganeraie*, Institut Agronomique et Vétérinaire, Actes Editions, Rabat, Morocco.

El Fasskaoui, B. (2009), 'Fonctions, défis et enjeux de la gestion et du développement durables dans la Réserve de Biosphère de l'Arganeraie (Maroc)', *Études caribéennes*, http://journals.openedition.org/etudescaribeennes/3711. doi:10.4000/etudescaribeennes.3711 (Last accessed on 15 December 2017).

El Harousse L., Aziz, L., Bellefontaine, R. and El Amrani, M. (2012), 'Le savoir écologique de deux populations habitant l'arganerie (Essaouira)', *Sécheresse* 23(2), 67–77.

Emberger, L. (1925), 'Le domaine naturel de l'arganier', *Bulletin de la Société Botanique de France* 72, 770–4.

Emberger, L. (1938), *Les arbres du Maroc*, Larose, Paris, France, Volume 1.

Emberger, L. (1939), 'Aperçu sur la végétation du Maroc', *Veröff. Geobot. Inst. Rübel Zürich* 14, 40–157.

EUROSTAT (2015), http://ec.europa.eu/eurostat/web/lucas/data (Last accessed on 15 December 2017).

Eventbrite (2018), https://www.eventbrite.com/e/international-conference-on-organic-farming-biodynamics-biological-treatment-cse-as-tickets-45675620009 (Last accessed on 22 June 2018).

Fazio, G. (2017), 'Evaluating and improving rootstocks for apple cultivation', in Evans, K. (Ed.), *Achieving Sustainable Cultivation of Apples*, Burleigh Dodds Series in Agricultural Science, number 018, Burleigh Dodds Science Publishing Limited, Cambridge, UK, pp. 135–64.

Fernandez-Martinez, M., Vicca, S., Janssens, I. A., Maria Espelta, J. and Peñuelas, J. (2017), 'The role of nutrients, productivity and climate in determining tree fruit production in European forests', *New Phytologist* 213, 669–79.

Ferree, D. C. (1988), 'Role of rootstocks and spur-type scions for controlling vegetative growth of apple and peach trees', *HortScience* 23, 464–7.

Flatrès H. and Flatrès P. (1997), 'Mutations agricoles et transformations des paysages en Europe', *Norois*, 173, 173–93.

Gallaher, R. N. (2009), *'Multiple cropping systems', in Management of Agricultural, Forestry and Fisheries Enterprises – Volume I, Multiple cropping systems, Encyclopedia of Life Support Systems*, http://eolss.net/eolss-sampleallchapter.aspx (Last accessed on 15 December 2017).

Gao, L., Xu, H., Bi, H., XI, W., Bao, B., Wang, X., Bi, C. and Chang, Y. (2013), 'Intercropping competition between apple trees and crops in agroforestry systems on the Loess Plateau of China', *PLoS ONE*, 8(7): e70739. doi:10.1371/journal.pone.0070739.

Geiger, G. and Biellier, H. (1993), 'Weeding with geese', Extension publications, University of Missouri, https://extension2.missouri.edu/g8922 (Last accessed on 15 December 2017).

Gliessman, S. R. (1985), 'Chapter 5 – Multiple cropping systems: A basis for developing an alternative agriculture', in *Innovative Biological Technologies for Lesser Developed Countries – Workshop proceedings*, OTA, Washington DC, pp. 69–83.

Gosme, M., Dufour, L., Inurreta Aguirre, H. D. and Dupraz, C. (2016), Microclimatic effect of agroforestry on diurnal temperature cycle, in Gosme, M. (Ed.), *Proceedings of the 3rd European Agroforestry Conference*, Montpellier, France, May 2016, pp. 183–86, ISBN: 978-2-87614-717-1, EAN: 9782876147171.

GRAB (Groupement de Recherche en Agriculture Biologique) (2017), http://www.grab.fr/wp-content/uploads/2017/09/guide_verger-maraicher_smart_GRAB_web-1.pdf (Last accessed on 22 December 2017).

Granatstein, D. and Peck, G. (2017), 'Assessing the environmental impact and sustainability of apple cultivation', in Evans, K. (Ed.), *Achieving Sustainable Cultivation of Apple*, Burleigh Dodds Science Publishing Limited, Cambridge, UK, pp. 524–49.

Granatstein, D., Kirby, E., Ostenson, H. and Willer, H. (2016), 'Global situation for organic tree fruits', *Scientia Horticulturae* 208, 3–12.

Guichard, M. F. (1975), 'La région du VinhoVerde (Portugal)', *Revue géographique des Pyrénées et du Sud-Ouest* 46(4), 403–29.

Haines-Young, R. and Potschin M. B. (2018), 'Common International Classification of Ecosystem Services (CICES) V5.1 and Guidance on the Application of the Revised Structure', Available from www.cices.eu

Herzog, F. (1998), 'Streuobst: A traditional agroforestry system as a model for agroforestry development in temperate Europe', *Agroforestry Systems* 42, 61–80.

Ibn Al Awwam – Kitab Al Filaha, (2000), *Le livre de l'agriculture*, Edition revue et corrigée avec introduction de El Faïz. *Actes Sud* and *Sinbad*.

INRA (2011), *Atlas du palmier dattier au Maroc*, Rabat, Morocco.

IPES-Food (2016), 'From uniformity to diversity: A paradigm shift from industrial agriculture to diversified agroecological systems', *International Panel of Experts on Sustainable Food systems*, http://www.ipes-food.org (Last accessed on 22 December 2017).

Jose, S., Gillespie, A. R. and Pallardy, S. G. (2004), 'Interspecific interactions in temperate agroforestry', *Agroforestry Systems* 61, 237–55.

Kaniewski, D., Van Campo, E., Boiy, T., Terral, J. F., Khadari, B. and Besnard, G. (2012), 'Primary domestication and early uses of the emblematic olive tree: Palaeobotanical, historical and molecular evidences from the Middle East', *Biological Reviews* 87, 885–99.

Kappel, F. (1991), 'Partitioning of above-ground dry matter in 'Lambert' sweet cherry trees with or without fruit', *Journal of the American Society for Horticultural Science* 116(2), 201–5.

Kuster, M., Herzog, F., Rehnus, M. and Sorg, J. P. (2012), 'Innovative agroforstsysteme – on farm monitoring von Chancen und Grenzen/systèmes agroforestiers novateurs – monitoring des opportunités et limites', *Agrarforschung Schweiz/Recherche Agronomique Suisse* 3(10), 470–7.

Laugénie C. (1965), 'La culture du Noyer au Perigord', *Revue géographique des Pyrénées et du Sud-Ouest* 36(2), 135–58.

Lauri, P. É. (2018), 'Apple tree architecture and cultivation – A tree in a system', *Acta Horticulturae* (in press).

Lauri, P. É. and Corelli-Grappadelli, L. (2014), 'Tree architecture, flowering and fruiting. Thoughts on training, pruning and ecophysiology', *Acta Horticulturae* 1058, 291–8.

Lauri, P. É. and Laurens, F. (2005), 'Architectural types in apple (*Malus Xdomestica* Borkh.)', in Ramdane Dris (Ed.), *Crops: Growth, Quality and Biotechnology*, World Food Limited, Helsinki, Finland, pp. 1300–14.

Lauri, P. É., Costes, E., Regnard, J. L., Brun, L., Simon, S., Monney, P. and Sinoquet H. (2009), 'Does knowledge on fruit tree architecture and its implications for orchard management improve horticultural sustainability? An overview of recent advances in the apple', *Acta Horticulturae* 817, 243–9.

Lauri, P. É., Hucbourg, B., Ramonguilhem, M. and Méry, D. (2011), 'An architectural-based tree training and pruning – identification of key features in the apple', *Acta Horticulturae* 903, 589–96.

Lauri, P. É., Mézière, D., Dufour, L., Gosme, M., Simon, S., Gary, C., Jagoret, P., Wery, J. and Dupraz, C. (2016), 'Fruit trees in agroforestry systems – Review and prospects for the temperate and Mediterranean zones', in Gosme, M. (Ed.), *Proceedings of the 3rd European Agroforestry Conference*, Montpellier, France, May 2016, pp. 106–9, ISBN: 978-2-87614-717-1, EAN: 9782876147171.

Le Bec, J, Bailly, A., Brossier, B. and Dupraz, C. (2016), 'Trunk-crown growth trade-off in pollarded trees: Influence on wood production', in Gosme, M. (Ed.), *Proceedings of the 3rd European Agroforestry Conference*, Montpellier, France, pp. 111–13, May 2016, ISBN: 978-2-87614-717-1. EAN: 9782876147171.

Lefhaili, A. (2014), *Forest Resources Assessment: Morocco Country Report*, FAO, Rome, Italy.

Lelle, M. A. and Gold, M. A. (1994), 'Agroforestry systems for temperate climates: Lessons from Roman Italy', *Forest & Conservation History* 38, 118–26

Lespinasse, Y. (1992), 'Breeding apple tree: Aims and methods', in Rousselle-Bourgeois, F. and Rousselle, P. (Eds), *Proceedings of the Joint Conference of the EAPR Breeding and Varietal Assessment Section and the EUCARPIA Potato Section*, January 1992, Landerneau, France, INRA, Ploudaniel, France, pp. 103–10.

Lordan, J., Fazio, G., Francescatto, P. and Robinson, T. (2017), 'Effects of apple (*Malus ×domestica*) rootstocks on scion performance and hormone concentration', *Scientia Horticulturae* 225, 96–105.

Loumou A. and Giourga C. (2003), 'Olive groves: The life and identity of the Mediterranean', *Agriculture and Human Values* 20(1), 87–95.

Lovell, S. T., Dupraz, C., Gold, M., Jose, S., Revord, R., Stanek, E. and Wolz, K. (2017), 'Temperate agroforestry research: Considering multifunctional woody polycultures and the design of long-term field trials', *Agroforestry Systems*. doi:10.1007/s10457-017-0087-4.

Maguylo, K. and Lauri, P. É. (2007), 'Growth and fruiting characteristics of eight untrained genotypes assessed on both M.9 and own-rooted trees in Southern France', *Acta Horticulturae* 732, 93–9.

Mamouni, A., El Bakkali, A., Lambert, P., Krichen, L., Oukabli, A., Audergon, J. M., Chatelet, P., El Modafar, C. and Khadari, B. (2014), 'Bottleneck and gene flow effects impact the genetic structure of seed-propagated apricot populations in Moroccan oasis agroecosystems', *Plant Genetic Resources* 12(2), 215–25.

Mantovani, D., Benincasa, P. and Rosati, A. (2016), 'Olive (*Olea europaea* L.) and wild asparagus (*Asparagus acutifolius* L.) agroforestry system: Asparagus performance and its best positioning in the olive orchard', in: Gosme, M. (Ed.), *Proceedings of the 3rd European Agroforestry Conference*, Montpellier, France, May 2016, pp. 229–31, ISBN: 978-2-87614-717-1, EAN: 9782876147171.

Maury, R. (1992), 'La genèse des paysages agraires en France. La Touraine en est-elle une synthèse?', *Norois* 153, 5–27.

McAdam, J. and Ward, F. (2017), 'System report: Grazed orchards in Northern Ireland', http://www.agforward.eu/index.php/en/grazed-orchards-in-northern-ireland-uk.html/ (Last accessed on 22 December 2017).

Meadows, R. (2008), 'Trained ovines chomp on weeds, avoid vines', *California Agriculture* 62(1), 10–10.

M'hirit, O., Benzyane, M., Benchekroun, F., El Yousfi, S. M. and Bendaanoun, M. (1998), *L'arganier, une espèce fruitière-forestière à usages multiples*, Mardaga, Sprimont, Belgium.

Millennium Ecosystem Assessment (2005), *Ecosystems and Human Well-Being: Synthesis*, Island Press. https://www.millenniumassessment.org/documents/document.356.aspx.pdf (Last accessed on 22 December 2017).

Mosquera-Losada, M. R., Santiago Freijanes, J. J., Pisanelli, A., Rois, M., Smith, J., den Herder, M., Moreno, G., Malignier, N., Mirazo, J. R., Lamersdorf, N., Ferreiro Domínguez, N., Balaguer, F., Pantera, A., Rigueiro-Rodríguez, A., Gonzalez-Hernández, P., Fernández-Lorenzo J. L., Romero-Franco, R., Chalmin, A., Garcia de Jalon, S., Garnett, K., Graves, A. and Burgess, P. J. (2016), Extent and success of current policy measures to promote agroforestry across Europe, Deliverable

8.23 for EU FP7 Research Project: AGFORWARD 613520 (8 December 2016), 95pp, Available in: https://www.agforward.eu/index.php/es/extent-and-success-of-current-policy-measures-to-promote-agroforestry-across-europe.html?file=files/agforward/documents/Deliverable%208.23%20Extent%20and%20Success%20of%20Current%20Policy%20Measures%208%20Dec%202016.pdf

Mosquera-Losada, M. R., Santiago Freijanes, J. J., Pisanelli, A., Rois, M., Smith, J., den Herder, M., Moreno, G., Lamersdorf, N., Ferreiro Domínguez, N., Balaguer, F., Pantera, A., Papanastasis, V., Rigueiro-Rodríguez, A., Aldrey, J. A., Gonzalez-Hernández, P., Fernández-Lorenzo, J. L., Romero-Franco, R., Lampkin, N. and Burgess, P. J. (2017), Deliverable 8.24: How can policy support the appropriate development and uptake of agroforestry in Europe? 7 September 2017, 21pp. Available at: https://www.agforward.eu/index.php/en/how-can-policy-support-the-uptake-of-agroforestry-in-europe.html?file=files/agforward/documents/Deliverable%208_24%20How%20can%20policy%20support%20agroforestry%281%29.pdf

Moukhli, A., Haouane, H., El Modafar, C. and Khadari, B. (2013), 'Histoire de l'introduction et de la diffusion de l'oléiculture au Maroc', in: Ilbert H., Tekelioglu, Y., Cagatay, S. and Tozanli, S. (Eds), *Indications Géographiques, Dynamiques Socio-économiques et Patrimoine Bio-culturel en Turquie et dans les pays Méditerranéens*, CIHEAM-IAMM, Montpellier, France, pp. 169–96 (Options Méditerranéennes: Série A. Séminaires Méditerranéens, n. 104). 2. Séminaire International d'Antalya, 16–19 December 2010, Antalya, Turkey.

Mouliéras A. (1899), Le Maroc inconnu: étude géographique et sociologique. Deuxième partie: Exploration des Djebala (Maroc septentrional), Augustin Challamel, Paris, France.

Msanda, F., El Aboudi, A. and Peltier, J. P. (2004), 'Biodiversité et biogéographie de l'arganeraie marocaine', *Cahiers Agricultures* 4(14), 357–64.

Musacchi, S. and Greene, D. (2017), 'Innovations in apple tree cultivation to manage crop load and ripening', in Evans, K. (Ed.), *Achieving Sustainable Cultivation of Apples*, Burleigh Dodds Series in Agricultural Science, number 018, Burleigh Dodds Science Publishing Limited, Cambridge, UK, pp. 195–237.

Nair, P. K. R. (1991), 'State-of-the-art of agroforestry systems', *Forest Ecology and Management* 45, 5-29.

Neilsen, G. H., Neilsen, D. and Forge, T. (2017), 'Advances in soil and nutrient management in apple cultivation', in Kate E. (Ed.), *Achieving Sustainable Cultivation of Apples*, Burleigh Dodds Series in Agricultural Science, number 018, Burleigh Dodds Science Publishing Limited, Cambridge, UK, pp. 239–78.

Nouaim R. (2005), *L'arganier au Maroc entre mythes et réalités: une civilisation née d'un arbre*, L'Harmattan, Paris, France.

Oldeman, R. A. A. (1992), 'Architectural models, fractals and agroforestry design', *Agriculture, Ecosystems and Environment* 41, 179–88.

Ong, C. K., Corlett, J. E., Singh, R. P. and Black, C. R. (1991), 'Above and below ground interactions in agroforestry systems', *Forest Ecology and Management* 45, 45-57.

Pantera A., Burgess, P. J., Mosquera Losada, R., Moreno, G., Lopez-Díaz, M. L., Corroyer, N., McAdam, J., Rosati, A., Papadopoulos, A. M., Graves, A., Rigueiro Rodríguez, A., Ferreiro-Domínguez, N., Fernandez Lorenzo, J. L., Gonzalez-Hernandez, M. P., Papanastasis, V. P., Mantzanas, K., Van Lerberghe, P. and Malignier, N. (2018), 'Agroforestry for high value tree systems in Europe', *Agroforestry Systems* (in press). doi:10.1007/s10457-017-0181-7.

Paolotti, L., Boggia, A., Castellini, C., Rocchi, L. and Rosati, A. (2016), 'Combining livestock and tree crops to improve sustainability in agriculture: A case study using the Life Cycle Assessment (LCA) approach', *Journal of Cleaner Production* 131, 351–63.

Parisi, L., Jamar, L., Lateur, M., Laurens, F. and Lauri, P. É. (2014), 'Adapting apple ideotypes for organic and low-input, competitive systems', in Bellon, S. and Penvern, S. (Eds), *Organic Farming, Prototype for Sustainable Agricultures*, Springer, Dordrecht, the Netherlands, pp. 131–48.

Parr, J. F., Papendick, R. I., Youngberg, I. G. and Meyer, R. E. (1990), 'Sustainable agriculture in the United States', in Edwards C. A., Lal, R., Madden, P., Miller, R. H. and House, G. (Eds), *Sustainable Agricultural Systems*, Soil and Water Conservation Society, Ankeny, IA.

Parrotta, J. A., Dey de Pryck, J., Darko, O. B., Padoch, C., Powell, B. and Sandbrook, C. (2015), 'The historical, environmental and socio-economic context of forests and tree-based systems for food security and nutrition', in Vira, B., Wildburger, C. and Mansourian, S. (Eds), *Forests, Trees and Landscapes for Food Security and Nutrition. A Global Assessment Report*, pp. 51–85, https://www.researchgate.net/publication/276950253 (Last accessed on 22 December 2017).

Peltier, J. P. (1982), 'Les séries de l'Arganeraie steppique dans le Souss (Maroc)', *Ecologia Mediterranea*, number 10.

Plénet, D., Giauque, P., Navarro, É., Millan, M., Hilaire, C., Hostalnou, É., Lyoussoufi, A. and Samie, J. J. (2009), 'Using on-field data to develop the EFI information system to characterize agronomic productivity and labour efficiency in peach (*Prunus persica* L. Batsch) orchards in France', *Agricultural Systems* 100, 1–10.

Pointereau, P. (2006). *La diversité des systèmes arborés et des pratiques de gestion dans le sud de l'Europe: les dehesas ibériques et les hautains méditerranéens*, Maison Botanique Boursay, Paris, France.

Racsko, J. and Schrader, L. E. (2012), 'Sunburn of apple fruit: Historical background, recent advances and future perspectives', *Critical Reviews in Plant Sciences* 31, 455–504.

Renevot G., Bouaziz A., Ruf, T. and Raki M. (2009), 'Pratiques d'irrigation du palmier dattier dans les systèmes oasiens du Tafilalt, Maroc', *Proceedings of the International Symposium 'Agriculture durable en région Méditerranéenne (AGDUMED)'.* Rabat, Morocco, 14–16 mai 2009, pp. 196–211.

Robinson, T. L. (2003), 'Apple-orchard planting systems', in Ferree, D. C. and Warrington, I. J. (Eds), *Apples – Botany, Production and Uses*, CABI Publishing, Wallingford, UK, pp. 345–408.

Robinson, D. W. and O'Kennedy, N. D. (1978), 'The effect of the overall herbicide systems of soil management on the growth and yield of apple trees "Golden Delicious"', *Scientia Horticulturae* 9, 127–36.

Rosati, A. (2001), 'Un possibile futuro per l'asparago selvatico', *L'Informatore Agrario* 7, 89–92.

Rosati, A. (2011), 'Asparagi nell'oliveto e la produzione raddoppia', *Olivo e Olio* 7/8, 20–4.

Rosati, A. (2014), 'Coltivare asparagi selvatici e allevare polli in un piccolo oliveto', *Vita in Campagna* 12, 44–8.

Rosati, A. and Falavigna, A. (2000), 'Germinazione dei semi di asparago selvatico', *L'Informatore Agrario* 56/46, 53–5.

Rosati, A., Pepe, R., Senatore, A., Perrone, D. and Falavigna, A. (2005), 'Produttività dell'asparago selvatico', *L'Informatore Agrario* 8, 75–7.

Rosati A., Caporali, S. and Paoletti, A. (2009), 'Olive, asparagus and animals: An agroforestry model for temperate climate in developed countries', *Proceedings of the III OLIVEBIOTEQ -For a Renovated, Profitable and Competitive Mediterranean Olive Growing Sector*, Sfax, Tunisia, 15-19 December 2009, pp. 229-33, ISBN: 978-9938-9513-0-1.

Rosati, A., Castellini, C., Dal Bosco, A., Mugnai, C. and Paoletti, A. (2012a), 'Manuale per la coltivazione consociata olivo, asparago selvatico, pollo rustico', *Edizioni 3A-PTA*. ISBN: 88-88417-06-0, 87.

Rosati, A., Castellini, C., Dal Bosco, A., Mugnai, C., Paoletti, A. and Caporali, S. (2012b), 'Olive agroforestry: An inverse approach to agroforestry', in Mosquera-Losada, M. R., Pantera, A., Rosati, A., Amaral, J., Smith, J. and Dupraz, C. (Eds), *What Priorities for European Agroforestry*, Book of abstracts of the 1st European agroforestry conference, 9-10 October 2012, Brussels, Belgium, p. 24, ISBN: 978-84-96351-79-0.

Rosati, A., Concezzi, L., Dal Bosco, A., Mugnai, C. and Paoletti A. (2012c), 'Olivo Asparago selvatico, Pollo rustico. Video on the olive, wild asparagus and free-range chicken system', https://www.youtube.com/watch?v=ALw73WwUr2o (Last accessed on 22 December 2017).

Rosati, A., Paoletti, A., Caporali, S. and Perri, E. (2013), 'The role of tree architecture in super high density olive orchards', *Scientia Horticulturae* 161, 24-9.

Rosati, A., Piottoli, L., Cartoni, A., Dal Bosco, A. and Castellini, C. (2014), 'Polli al pascolo nell'oliveto Risparmio a tutto campo', *Olivo e Olio* 6, 4-7.

Rosati, A., Paoletti, A., Al Hariri, R. and Famiani, F. (2018), 'Fruit production and branching density affect shoot and whole-tree wood to leaf biomass ratio in olive', *Tree Physiology* (*in press*). doi:10.1093/treephys/tpy009.

Saj, S., Durot, C., Mvondo Sakouma, K., Tayo Gamo, K. and Avana-Tientcheu, M. L. (2017), 'Contribution of associated trees to long-term species conservation, carbon storage and sustainability: A functional analysis of tree communities in cacao plantations of Central Cameroon', *International Journal of Agricultural Sustainability* 15(3), 282-302.

Saure, M. C. (1990), 'External control of anthocyanin formation in apple', *Scientia Horticulturae* 42, 181-218.

Sestini, A. (1963), *Il Paesaggio*, Touring Club Italiano, Italy.

Sghir, F., Chliyeh, M., Kachkouch, W., Khouader, M., Ouazzani Touhami, A., Benkirane, R. and Douira, A. (2013), 'Mycorrhizal status of *Olea europaea* spp. *oleaster* in Morocco', *Journal of Applied Biosciences* 61, 4478-89.

Shropshire Sheep Breeders' Association (2008), *Two Crops from One Acre: A Comprehensive Guide to using Shropshire Sheep for Grazing Tree Plantations*, http://www.shropshire-sheep.co.uk/publications/SSBA%20P16%20Two%20Crops%20Booklet.pdf (Last accessed on 15 January 2018).

Simon, S., Lauri, P.É., Brun, L., Defrance, H. and Sauphanor, B. (2006), 'Does fruit tree architecture manipulation affect the development of pests and pathogens? - a case study in apple orchard', *Journal of Horticultural Science & Biotechnology* 81(4), 765-73.

Simon, S., Bouvier, J. C., Debras, J. F. and Sauphanor, B. (2010), 'Biodiversity and pest management in orchard systems. A review', *Agronomy for Sustainable Development* 30, 139-52.

Simon, S., Lesueur-Jannoyer, M., Plénet, D., Lauri, P.É. and Le Bellec, F. (2017), 'Methodology to design agroecological orchards: Learnings from on-station and on-farm experiences', *European Journal of Agronomy* 82, 320-30.

Smith, J., Crossland, E. M. and Wolfe, M. S. (2016), 'Silvoarable agroforestry: An alternative approach to apple production?', in *Proceedings of the International Farming Systems Association (IFSA), Europe Group for 2016*, 12–15 July 2016, Newport, UK, http://archive.harper-adams.ac.uk/events/ifsa-conference/papers/3/3.3%20Smith.pdf (Last accessed on 22 December 2017).

Song, B., Tang, G., Sang, X., Zhang, J., Yao, Y. and Wiggins, N. (2013), 'Intercropping with aromatic plants hindered the occurrence of *Aphis citricola* in an apple orchard system by shifting predator-prey abundances', *Biocontrol Science and Technology* 23(4), 381–95.

Tarrier, M. R. and Benzyane, M. (2003), 'L'arganeraie marocaine se meurt: problématique et bio-indication', *Sécheresse* 1(1), http://www.secheresse.info/spip.php?article228 (Last accessed on 22 December 2017).

Terral, J. F., Alonso, N., Buxoi Capdevilla, R., Chatti, N., Fabre L., Fiorentino, G., Marinval, P, Perz Jordan, G, Pradat, B., Rovira, N. and Alibert, P. (2004), 'Historical biogeography of olive domestication (*Olea europea* L.) as revealed by geometrical morphometry applied to biological and archeological material', *Journal of Biogeography* 31, 63–77.

Timmermans, B. and Bestman, M. (2016), 'Quality of apple trees and apples in poultry free range areas', in Gosme, M. (Ed.), *Proceedings of the 3rd European Agroforestry Conference*, Montpellier, France, May 2016, pp. 420–3, ISBN: 978-2-87614-717-1, EAN: 9782876147171.

Torrés, A. M. (1930), 'La riquez olivarera de la Regiòn Occidental de la Zona de Protectorado de España en Marruecos', *Maroc: Direction de colonizacion*.

van Noordwijk, M., Coe, R. and Sinclair, F. (2016), 'Central hypotheses for the third agroforestry paradigm within a common definition', Working paper 233, World Agroforestry Centre (ICRAF) Southeast Asia Regional Program, Bogor, Indonesia. doi:10.5716/WP16079.PDF (Last accessed on 22 December 2017).

Vannucci, S. (2009), *L'ulivo e l'olio, Storia dell'olivo*, Diritti di sfruttamento economico: Bayer Crop Science S.r.l., Milano, http://www.colturaecultura.it (Last accessed on 22 December 2017).

Weibel, F. and Häseli, A. (2003), 'Organic apple production – with emphasis on European experiences', in Ferree, D. C. and Warrington, I. J. (Eds), *Apples – Botany, Production and Uses*, CABI Publishing, Wallingford, UK, pp. 551–83.

Wolz, K. J. and DeLucia, E. H. (2018), 'Alley cropping: Global patterns of species composition and function', *Agriculture, Ecosystems and Environment* 252, 61–8.

Woodland Trust (2014), 'Woodland Trust guide to tree planting for free range poultry', https://www.woodlandtrust.org.uk/publications/2014/04/tree-planting-for-free-range-poultry/ (Last accessed on 22 December 2017).

Wünsche, J. N. and Ferguson, I. B. (2005), 'Crop load interactions in apple', *Horticultural Reviews* 31, 231–92.

Zohary, D. and Hopf, M. (1994), *Domestication of Plants in the Old World*, 2nd edition, Clarendon Press, Oxford, UK.

Part 4

Tropical agroforestry

Chapter 15

Moving up the scale: challenges in tropical agroforestry

John Lynam,[1] Independent Consultant

1 Introduction

The science, practice and investment in agroforestry in the developing world have increased significantly over the last couple decades. From its early beginnings, agroforestry has moved from a 'technology in search of a problem' to a principal solution to critical global agendas, including climate-smart agriculture, agroecological intensification, land rehabilitation and payment for ecosystem services (Leakey, 2012). Agroforestry has been in a position to respond to a broadening of the global development agenda as reflected in the Sustainable Development Goals (SDGs), particularly in the agricultural and forestry sector and by extension into natural resources, ecosystem services and sustainable landscapes in conjunction with the more traditional focus on smallholder productivity and rural poverty. The increasing role of agroforestry in this expanding development agenda builds on the multifunctionality of agroforestry, where agroforestry can achieve different objectives at farm, landscape and policy levels.

In many respects, the challenges facing agroforestry in the developing world stem from being able to align agroforestry practice and the associated science to the very high expectations created by growing investment in

1 The author would like to thank Ric Coe for highly constructive comments on an earlier draft, without implicating him in the final result.

http://dx.doi.org/10.19103/AS.2018.0041.21

realizing the SDGs. While multifunctionality of agroforestry has been key to this alignment, the development objectives implicit in multifunctionality that are of importance to the policymaker and the scientist may differ from those of the farmer or communities deciding on land use. It is the farmers and communities that must apply agroforestry technologies and management practices, and the challenge is in finding congruence in objectives between the policymakers who manage investment in agroforestry, the researchers who design the technologies and increasingly investigate the processes in applying agroforestry, and farmers and other land users who adopt the practice. For investors and policymakers, the intent is to achieve adoption rates at a relevant scale that justifies the investment in carbon sequestration or land rehabilitation; for farmers it is improving livelihood objectives.[2]

Given that current application of agroforestry is supported by external finance (to be discussed in more detail further in the chapter), the principal challenges for agroforestry achieving its potential in the tropics lie in how to bridge between the objectives of investors and the uptake of agroforestry by farmers and communities. These challenges can be framed at different scales of application, and this chapter is organized at four such scales, namely agroecological intensification at farm scale, managing sustainable landscapes, developing supportive agroforestry policies and how these three scales come together in scaling of agroforestry practices. Optimally, interventions at each higher scale will reinforce application at lower scales. The challenges discussed in each section in many ways reflect an R&D agenda for future development of agroforestry in the tropics.

2 The challenge of agroforestry in agroecological intensification

The multifunctionality of agroforestry at farm scale integrates farmer income with provision of ecosystem services. The positive benefits of agroforestry on carbon sequestration, improved nutrient retention and cycling, increased biodiversity, enhanced soil organic matter and improvements in water balance and hydrological function reduce the footprint of agricultural cultivation and provide ecological services to both the farmer and external to the farm to the broader ecosystem. However, for the farmer, these must be matched by either commensurate improvements in productivity of crops and livestock produced on the farm or by generation of marketable products from the trees themselves, either wood products or fruits, nuts and other raw materials.[3]

2 There are, of course, policymakers and investors who are motivated by livelihood improvement, especially for the poor, but agroforestry is not often seen as a principal means in achieving these objectives, apart from high-value tree crops, which are discussed later. In bridging that gap see van Noordwijk (2017).

3 'Commensurate' here also incorporates farmer time preferences, given there is a lag between farmer investment and the benefit stream from agroforestry.

Agroecological approaches are often contrasted with input-based approaches to increasing productivity. In the context of smallholder systems, these two 'paradigms' are more complementary in sub-Saharan Africa, the uplands of Southeast Asia and the forest margins of all three continents. The two are more competing approaches in other parts of the developing world. In the former regions, systems are nutrient limited and fertilizer is relatively expensive compared to output prices. Limited fertilizer applications, particularly phosphorus, are essential for root growth and nitrogen fixation. Limited fertilizer use in agroforestry in such contexts improves nutrient use efficiency in the whole farming system and enhances the yield gains from the utilization of improved crop varieties. In higher input systems (which usually predominate in contexts of high agroecological potential and well-developed markets), there is a larger trade-off between profitability of crop production and the integration of trees into the system, unless these are high-value trees or there are niches for trees within the farming system. In practice, however, this complementarity in low-input systems is not achieved because there is investment in only one 'paradigm',[4] usually the input-based approaches and primarily because of the perceived facility in scaling the production and income impact of the approach, where many argue at the detriment of impact on ecosystem services and longer-term sustainability.

Farmer adoption depends on how agroforestry meets the objectives of the farmer and thus how (s)he understands multifunctionality. Agroforestry can meet a range of objectives or it can involve trade-offs between objectives. Assessing trade-offs can apply to the choice among agroforestry systems, the choice between an agroforestry system and annual crop production, and the period between investment and initiation of return. The farmer will value differently the various service flows from agroforestry – often involving trade-offs, for example using leaves as fodder versus incorporation in the soil. The choice problem in adoption involves an understanding of the farmer weights on objectives and an understanding of the trade-offs in ecological service flows (counting productivity itself as a service flow) that optimally can be valued and thus compared (Klapwijk et al., 2014). Farmers also value flexibility in land use which requires a premium in moving to a more perennial system.

The choice problem introduces a critical challenge of asymmetries of knowledge between farmers and agroforestry providers and in many cases a lack of sufficient knowledge across both. On the one hand, there is usually a lack of understanding by providers of weights on objectives, even if farmers have a knowledge of ecosystem services. Extension approaches thus tended to incorporate farmer participatory research, where those implicit objectives

4 Agroforestry, itself, is not immune to this single intervention approach, as for example the wholesale distribution of tree seedlings of a particular species. A systems approach often in some cases does not extend from the science to the practice.

of farmers can be made explicit. Productivity and income, complemented by subsistence and nutrition, are expected to be the dominant farmer objective, with the question of how the farmer assesses trade-offs in these objectives with the provision of other ecosystem services. Ecosystem services such as water and soil organic matter/nutrient cycling can also enhance productivity, but there is usually a time lag in these effects and a choice problem between tree systems that enhance these services and trees that provide marketable products. In an Asian or Latin American context, the portfolio of trees tends to focus on fruit or high-value product trees, while in a smallholder context in sub-Saharan Africa there is a broadening to include 'fertilizer' trees such as *Tephrosia* ssp and *Gliricidia sepium*[5] and trees that provide fodder for livestock.

The tendency for farmers to focus on trees that provide products and incomes, much like a focus on grain legumes in legume species for soil replenishment, produces another challenge which is the development of stable markets for these products. For higher-value tree crops with well-developed markets, such as rubber, oil palm, cocoa, tea and coffee, there is a tendency to plant in monoculture where more biodiverse systems are found to be more sustainable (Boreux, 2016; van Noordwijk et al., 2017). These tree crops tend to have a limited agroecological range and many are dominated by production in large plantations. To expand both the ecological range and access to smallholders requires tree crop products where markets are not well developed and where development of these markets is a major challenge. Some products such as fruits tend to be perishable, others require processing and still others have limited or seasonal markets where price depression occurs with significant increases in production. Some of these products are new markets principally in developed countries, for example allanblackia or moringa, and the difficulty is developing sufficient and continuous supply to foster investment in end-market processing and market infrastructure, particularly when supply is dependent on smallholder production, where aggregation costs and scale economies in processing often lead to plantation production systems. For many agroforestry programmes, various tree products are planted which leads to a rather more specific challenge for agroforestry of developing and accessing multiple markets, similar to horticulture but without the geographical concentration found in that sector.

An assessment of farmer objectives and trade-offs among alternative agroforestry is an indicator of potential adoption and frames the choice of agroforestry alternatives in a dissemination programme. Facilitating the adoption process has its own challenges. Agroforestry is an investment by the farmer, as there is often a significant time lag in generating either product or ecosystem services. Programmes for payment for ecosystem services (PES)

5 For an interesting exchange on assessing the benefit of fertilizer trees on farm in Malawi see: Coe et al. (2017) and Silechi and Akinnifesi (2017).

have been piloted to provide incentives for his investment, where credits are provided, for example in carbon trading (Foster and Neufeldt, 2014). In regions such as Africa where market failures are often the rule, the PES concept has been extended to co-investment (Namirembe et al., 2014). In many respects, the larger challenge is the provision of management options and tree seedlings, which are provided locally through nurseries. These can be privately run or community based, and often involve specialized techniques such as grafting.[6] Sustainability of nurseries is an issue, as local demand for seedlings can be easily saturated. A recent innovation to solve this problem is that of relay organizations (Degrande et al., 2014). These issues beg the question of how most cost effectively to organize extension approaches for agroforestry. A range of approaches have been developed and piloted, but the limited research on dissemination methods, particularly at the World Agroforestry Centre, has tended to focus on improving farmer-to-farmer approaches using volunteer farmer trainers.[7] There have not been studies that compare and evaluate different methods, nor when linked to other support services such as credit or PES.

Moreover, there have been only a limited number of adoption studies and far less impact studies for agroforestry. Adoption studies are difficult for agroforestry because of the wide choice in tree and management options. Since agroforestry is essentially extended within a project framework, there is a critical issue of defining what constitutes farmer adoption and how the intensity of adoption (how extensively agroforestry is adopted within the farming system) is defined.[8] This extends beyond just tree adoption – which is often the focus in adoption studies – to such management practices as farmer-managed natural regeneration (World Agroforestry Centre, 2013) and sustainable cocoa and oil palm systems. This 'plasticity' raises the issue of how wide a choice of options are offered to farmers and how management options are extended to farmers, for example tree establishment techniques, population, spacing, fertilization and pruning.[9] An hypothesis is that management (particularly given the lag between

6 The concept of community based rural resource centres provides both seedlings and management information. See Takoutsing et al. (2014).

7 See http://www.worldagroforestry.org/ras/farmer-farmer-extension

8 Agroforestry is particularly problematic for economic adoption studies, comparable to conservation agriculture where both involve a number of practices. Econometric analysis is based on adoption of a single technique compared to an alternative, usually existing farmer practice. This limits the understanding of farmer adoption of more system-based technologies. Thus, in examples of such studies, conservation agriculture was defined as minimum tillage and agroforestry as a fertilizer tree (among a number of species of fertilizer trees). See Coulibaly et al. (2016) and Arslan et al. (2013).

9 The concept of plasticity allows substantial scope for farmer adaptation of agroforestry systems. An argument can be made that agroforestry should be seen as the application of certain core principles rather than rigorously defined practices, which changes significantly how evaluation of agroforestry projects is designed. Ric Coe (pers. comm.) provides the following example: 'In a project in central Tanzania the "technologies" being prompted were rotational woodlots, fodder banks and improved ngitiris (grazing reserve areas). Researchers worried about management of each, such as how long to let a woodlot grow before returning it to cropland, and how to crop it before the next tree crop. But farmers saw all these as the same, and on a daily or seasonal basis made decisions such as: cut a tree and let it resprout, cut it and crop, graze among the trees, protect emergent trees from grazing. ... It was impossible to say what had been "adopted" but would have been easy to see the extent to which a range of principles were being understood and exploited'.

tree planting and benefits) is especially critical in achieving the benefits of agroforestry and therefore in eventual adoption or disadoption of the practice.[10] Management, however, is very difficult to monitor and is usually done in surveys after the end of the project. Moreover, it is mediated by the effectiveness of the extension method employed. There is a clear challenge in understanding farmer choice and adoption (and disadoption) in agroforestry projects.

Perhaps the larger challenge is reducing dependence on projects in the dissemination of agroforestry. Autonomous adoption dependent on more classical diffusion processes is key to such cost reduction (and to scaling, which is discussed in the last section). Such processes are dependent on access to tree seedlings and information (usually based on farmer to farmer exchange) but most importantly on farmer 'learning by doing'. For agroforestry systems, this is not a straightforward process, involving as it does such a potential range of management 'dimensions'.[11] It might be argued that there is a certain robustness in adaptability of agroforestry systems, that such farmer experimentation leads to not an optimal system but one sufficiently fit to the farmer context. Diffusion studies generally focus on single technological components such as improved varieties or inorganic fertilizer and there is a question of how more knowledge-based, system technologies are transferred from farmer to farmer, through social networks, or through communication media and experimented with by farmers. The first, most binding constraint is access to tree germplasm, which experience would suggest requires an outside facilitation. Much of this work is community based and the question is when does the market for tree seedlings become large enough to motivate private nurseries, such as are found in principal urban areas.

Agroforestry is considered to be one of the principal entry points into agroecological intensification since it brings a perennial component that recycles nutrients, reduces run-off and enhances soil moisture, and sequesters carbon in the system and leads to improvements of overall system productivity while reducing the agricultural 'footprint'. Not unexpectedly, agroecological intensification faces all the challenges summarized above for agroforestry systems. These are not insuperable challenges but they do require innovations in the delivery and financing of agroforestry in smallholder agriculture. In the health sector, such monitoring of process innovations - sometimes involving large scale random controlled trial designs - is known as operations research and in the agricultural sector, this is increasingly known as research

10 Understanding disadoption is in many ways as important as understanding adoption. Again an example from conservation agriculture in Zambia suggests rates of 95% of farmers disadopting the technology (Arslan et al., 2013, p. v). A study of improved fallows and biomass transfer in Western Kenya found 10% disadoption but just within the project period [and this with relatively low rates of adoption and even lower intensity of adoption. See Place et al. (2003)].

11 This should not overlook an increasing 'natural' trend in tree cover on farms. 'Remote sensing data show that in 2010, 43% of all agricultural land globally had at least 10% tree cover and that this has increased by 2% over the previous ten years' (Zomer et al., 2016).

in development (Coe et al., 2014). This is the subject of the scaling challenge discussed in the last section.

3 The challenge of agroforestry in sustainable landscapes

Landscapes have within the last decade become an organizing framework for both research and dissemination strategies for agroforestry. In practice, it has become a framework that merges the interests and programmes of both the agricultural and forestry sectors. The landscape approach builds on its use in ecology and allows a more targeted approach to multifunctionality through support to alternative land uses. The landscape approach has its origins in the area of conservation exploring alternative land uses in the buffer areas around natural reserves. By design, this entailed a particular focus on community engagement in these buffer areas. The more recent extension has been to focus on landscapes as a basis for understanding and delivering the provision of ecosystem services. Landscapes provide a more appropriate scale for both monitoring ecosystem services and organizing programmes around PES, including adaptation and mitigation programmes for climate-smart agriculture.[12]

Landscapes tend to be defined as a mosaic of different land uses and the tendency is to characterize landscapes in terms of the predominant land use, such as agriculture, savannas or drylands. Programmes using the landscape approach with agroforestry predominate in agricultural or forest margin landscapes, although there is an increasing emphasis on defining landscapes in terms of the degree of land degradation, particularly in relation to the SDG on land degradation neutrality (Gray et al., 2016).

Two reviews of the experience with integrated landscape initiatives (ILI) point to a number of early conclusions about the approach in Africa and Latin America (Hart et al., 2015; Milder et al., 2014). Only within the last couple of decades or less have projects started to be organized around landscapes, although there were precursors such as integrated watershed management. The approach is still developing standards of practice. Conservation projects were the leading organizational rationale in both Africa and Latin America and remain dominant in Latin America. The growth in the approach in Africa involves agricultural projects and increasingly land degradation, which focuses on either farming or pastoralism. The projects were assessed in terms of work in four domains, namely agriculture, conservation, livelihoods and institutional planning and coordination. Most projects involved investments in a combination of these, demonstrating the multifunctionality of the approach. However, the area of institutional planning and coordination dominated on both continents. The approach indeed has at its core the requirement for

12 For an excellent compendium on the theory and approach to landscapes, see Minang et al. (2015).

collective decision-making, more often than not involving negotiation over trade-offs in objectives, coordination and capacity building, in effect a focus on governance mechanisms.

Governance is one of the key challenges in ILI. Most projects focus on governance at a local level and these involve stakeholder groups organized around a coordination platform. Over 80% of the groups on both continents were farmer groups, supported through the platform by NGOs, extension services and in some cases local government bodies. Hart et al. argue that, 'initiative leaders in LAC may perceive platforms for coordination as an important foundation for achieving specific outcomes related to their primary objectives. African initiatives that invested in the creation or strengthening of coordination bodies reported significantly more outcomes than those that did not invest in coordination bodies across all domains' (2014, p. 97). Projects are time limited and the survey did not assess the sustainability of these platforms. Moreover, capacity strengthening was a major investment area. As noted by Hart et al., 'These results suggest that capacity building activities are centrally important to the work of landscape initiatives and foundational to the achievement of other objectives' (2014, p. 95).

Given the large early investment in coordination and capacity and the lag in investment in ecosystem restoration and impact on productivity, time frames are lengthy in order to ensure returns to landscape investments. An assessment of watershed management programmes in Ethiopia found that it required seven years for projects to start to realize a positive return (Schmidt and Tadess, 2012). However, returns increased at an increasing rate after that as soil, water and ecological health had been restored.

Nevertheless, there are relatively few such benefit cost evaluations of ILI. Barron and Noel in a meta-analysis of agricultural watershed management projects found 'very few cases with benefit-to-cost evaluation at full project level, or separate costing of hard and soft components. The synthesis suggests higher development success rates in communities with an initial level of social capital, where projects were implemented with cost- and knowledge-sharing between involved stakeholders, and where one or more "agents of change" were present to facilitate leadership and communications' (Barron and Noel, 2011). Thus, coordination and capacity costs are essential to project success, but there are as yet little data to suggest that those costs yield a return over investment costs.[13]

The surveys also found that agroforestry is utilized in over half the agricultural ILIs in both Latin America and Africa (there was no data on agroforestry in conservation projects), as agroforestry and social forestry are a principal element

13 There is an inherent conundrum in projects investing in a landscape approach requiring clear returns on investment when landscapes by their very nature introduce measurement issues that are flexible, vague and even disputed that make the estimation process inconclusive (R. Coe, pers. comm.).

in the implementation of landscape programmes. Implementing landscape initiatives with agroforestry as a dominant element encompasses the challenges in farm-level extension programmes in agroforestry, namely markets, nurseries and the lag between investment and return, and adds a few others. The stakeholder formation, the capacity building and the coordination/negotiation platform, discussed previously, are preconditions for planning and implementation at a landscape scale. The project will usually have an overall objective such as climate change mitigation and adaptation, land degradation, hydrological management or biodiversity conservation, that is ecosystem services that can best be improved at a landscape level. These are not usually land user or farmer objectives and must be balanced against productivity and income objectives. More critically, landscape level interventions usually entail differentiated interventions across the landscape and as close to universal participation as possible. Both are difficult and the latter usually entails an extended time frame for the project. Most projects follow the principle of 'common entry point',[14] which argues for the joint identification of targets that allow collective action (Sayer et al., 2013). Project time frames, however, usually do not allow getting beyond the common entry point to more differentiated landscape planning.

Even with a common entry point, there is the additional challenge of scaling up agroforestry interventions within the project landscape. At one level compared to standard extension programmes, this should be facilitated by coordinated planning of nurseries, market access and credit (and/or PES). The challenge here is how to distribute project resources within a sufficiently large landscape to be cost-effective, given that there are strong arguments for continuing project costs over an extended time frame. Such considerations argue for an operational scale that maximizes the effectiveness of project resources. This challenge is particularly important in the relatively recent development of impact funds as a principal means of financing livelihood, climate mitigation or land degradation objectives using agroforestry (impact funds are discussed in more detail in Section 5). Such financing is based on generating a return over project costs, which puts a significant priority on efficiency in the use of such overhead costs. Such efficiency gains come from generating spillovers from project focal points and in redistributing project resources from those focal points when more autonomous diffusion processes and sustainability of stakeholder platforms are indicated. Minang et al. capture the problem this way, 'Domains of similarity that include modes of decision-making need to be recognized to facilitate cross site learning; existing ecological stratification needs social counterparts' (2015, p. xxxiii). It is fair to say that landscape projects have not reached this level of project management, as it is rare that hard and soft costs are monitored in such programmes.

14 Such entry points are usually defined at an outcome level, for example hydrological response or improved livelihoods.

As with all integrated approaches to development (e.g. the integrated rural development approaches of the 1970s), the complexity of the theory often overwhelms the practice and implementation. Landscape approaches are still in an experimentation and development stage, but there is enough evidence to support continued investment in ILIs but with the need that the practice has to continue to evolve. In essence, there is a research agenda in the realm of operations research on implementation of ILI and the role of agroforestry within the approach. Mbow et al. lay out a series of research questions for improving practice in project implementation for climate-smart landscapes (2014). These primarily involve research on better targeting of agroforestry systems to specific conditions.

4 The challenge of policy in support of agroforestry

Most articles surveying the potential of either agroforestry or landscape approaches mention policy as a critical challenge in realizing that potential. Policy, however, is a very broad rubric with a plethora of different instruments. Policy can run from broad government strategies in a sector to price policies, investment policies, regulatory controls and the organization and administration of government entities. Moreover, policies can be developed at various administrative levels, from local ordinances to district/state level to national policies, and including international policies such as REDD+ and investment funds. The district/state level has become particularly important with the trend over the last couple of decades to decentralization and subsidiarity. Compatibility and subsidiarity across these different levels are an issue, as suggested by the relatively recent national agroforestry policy in India, which attempted to harmonize regulations at the state level (National Agroforestry Policy, 2014).

The challenge with the policy environment for agroforestry derives from the advantages coming from its multifunctionality. Administratively, agroforestry cuts across ministries of agriculture, forestry, natural resources and land, with no particular organizational responsibility in any – although the Indian Agroforestry Policy resides in the Ministry of Agriculture, which has given agroforestry both priority and direct implementation capacity. Agroforestry, because it is a relatively recent policy focus, has struggled to achieve a 'level playing field' in terms of regulatory, commodity, input and land policies. These policies are set at the national level and below. Examples of policy constraints on the expansion of agroforestry would include the effect of fertilizer input subsidies on adoption of fertilizer trees in Zambia and Malawi (where fertilizer trees can contribute to an exit strategy from the budgetary dependence on fertilizer subsidies, where fertilizers continue to be necessary but agroforestry improves the nutrient efficiency of fertilizer use and thus the profitability). Also is the example of the reinvestment in the parkland system in Niger where farmer-managed natural regeneration leading to increased tree density was promoted by 'change in policy and institutional focus from protection by the state of remaining trees and woodlands, to enabling regeneration and harvesting of locally managed tree stocks to directly benefit local communities investing in protection and management of trees on farms and the reform of the Forestry Code in 2004 to officially sanction local ownership and management rights for trees on farms' (Gray et al., 2016, p. 109).

These examples highlight the nature of the challenge facing policy change in support of agroforestry, namely that such changes need to be context specific, depending on the particular agroforestry system, the existing laws, regulations and policies currently in place, and the administration unit and level which enforce the policy. Such context specificity in turn implies national policy research capacity to undertake the evaluation of the costs and benefits of policy changes that are critical input into the policy reform process. National capacity to undertake policy research is constrained in tropical, developing countries, usually existing in policy research institutes or universities. Given the scope of potential research in the policy arena, the complexity of the agroforestry policy agenda and the lack of clear administrative channels for agroforestry policies, developing capacity to undertake policy research on agroforestry requires external facilitation, and such capacity is also highly constrained in international institutions undertaking agroforestry research. Thus, policy is recognized as a central constraint on the widespread uptake of agroforestry, but clear avenues to developing such policies remain unclear.

There have been two principal lines of attack in addressing this challenge. The first, building on the success of the India Agroforestry Policy, is to develop an overall policy framework to support agroforestry in the country. This is the approach suggested by a 2013 FAO paper on the subject, which argues that policy change is necessary (1) to eliminate legal and institutional constraints on agroforestry, (2) to support positive outcomes of agroforestry and (3) to compensate farmers for the delay in returns (FAO, 2013, pp. 11–13). The World Agroforestry Centre, which facilitated the development of the national policy in India, is pursuing this approach in a number of countries.[15] Such an overall framework together with the identification of the implementing agency, for example the Natural Resource Authority in the case of Rwanda, then allows and potentially prioritizes policy research on existing policies and regulations that constrain the development of the approach.

The other approach being pursued by the Forestry, Trees and Agroforestry CGIAR Research Program and by the Center for International Forestry Research is to support development of international policy agreements and encourage country buy-in to them. REDD+ is in many ways the best example of this, but there

15 See for example, World Agroforestry Centre (2015) for the case of Rwanda.

is also work on certification and standards of sustainability, usually associated with production and international trade of high-value tree commodities (Tayleur et al., 2017). These approaches require long-term engagement and in the end primarily target large-scale production systems where costs of meeting standards are a critical consideration. In general, 'public policy, due to perverse incentives and implementation failures, often is ineffective in dealing with negative environmental impacts. Sustainability standards and associated certification schemes have made contributions to ameliorating some of these impacts, but these schemes show mixed results with regard to environmental performance and the promotion of better inclusion of smallholders and rural communities in global value chains' (CGIAR Research Program on Forests, Trees and Agroforestry, 2016, p. 116). In summary, policy research in support of agroforestry has a toe hold in the policy research community but needs a far more dedicated effort to get agroforestry on the agenda of national policy makers.

5 The challenge of developing agroforestry at scale

Achieving impact at scale has become a particular focus of investments in agriculture and natural resource management, in many ways linked to the entry of the Gates Foundation into the field of agricultural development. There was a sense that there was a need to move from piloting projects to large investments that would generate a change process in the sector, what has recently entered the development lexicon as an agricultural transformation. The scaling problem has resulted in the development of two quite different strategies – although with some degree of complementarity which is often not appreciated – and with a particular focus on sub-Saharan Africa. Each focuses on increasing productivity and each targets the smallholder farmer. The 'Green Revolution' strategy focuses on seed and fertilizer technologies and investment in the upstream market capacities for both delivery of these inputs and the efficient marketing of the increased output. The transformative component is the creation of an environment that incentivizes private sector investment which creates an autonomous growth process or transformation.

Agroecological intensification using agroforestry is a more holistic strategy which targets project investments primarily at farm and landscape scales working in a systems framework to improve livelihoods while reducing ecological externalities (Milder et al., 2014). This is a more incremental approach that invests in longer-term sustainability of the productivity gains and the natural resource base and contributes to mitigation of climate change. As with other system approaches, it relies on participatory, bottom-up methods that require initial facilitation. The overarching challenge for agroforestry is scaling, particularly as current expansion of agroforestry depends on public funding through a project modality. The issue is reaching a tipping point where the time and cost of project investments in agroforestry and landscape approaches build over time to the point where there is widespread understanding of the benefits (as with farmer-managed natural regeneration in Niger), leading to autonomous investment and uptake – in many ways similar to what van Noordwijk (2017) terms type III knowledge into action pathways. These initial public investments then 'crowd in' farmer and private sector investment into agroforestry. The increasing finance of development through impact funds is both an opportunity and a challenge to increasing the financial base for agroforestry.

The scaling issue encompasses all the other challenges laid out in the previous three sections of the chapter and asks the question of how to integrate extension, nurseries, markets and policy at a sufficient scale when each of those areas is context specific. Moreover, if these essential elements are integrated through landscape approaches and projects, the question then becomes how to scale these ILIs. As with any new approach that requires some sophistication in the implementation, scaling requires the development of capacity in the methods, more particularly in national agencies. Given that landscape approaches are cross sectoral, the question arises of how and where to develop such capacity at a national level. As with the agroforestry policy in India, this requires identifying an agency that can lead in the approach but with the potential of linking to other ministries and/or state capacities. ILIs in turn require that there is potential to develop a field implementation capacity. Field capacity has been aided by the policy in many countries to decentralize functions and administrative control, but the human resource capacity at district or state level is in most cases quite limited, as for example in agricultural extension.

These capacity issues define in part the challenges that will face the increasing financial flows that are coming through impact funds, such as the Moringa Fund which finances only agroforestry investments.[16] The Moringa Fund understands the capacity constraint and has established 'the Agroforestry Technical Assistance Facility (ATAF) (which) was created by Moringa Partnership to provide technical assistance in relation to investments of the Fund with the goal to amplify and upscale positive environmental and social impacts triggered through Moringa investments. ATAF is a grant based mechanism parallel to the investment of the Moringa Fund'.[17] However, impact funds in addition to achieving social and environmental goals must also deliver a competitive return on the investment. The Moringa Fund does this by taking an equity stake in agroforestry firms, for example those involved in producing cashews, tea and coffee. Two livelihood funds managed by Livelihood Ventures focus on smallholder commodity production principally based on trees and generate returns based either on carbon credits or commodity supply guarantees.[18] To

16 See https://www.moringapartnership.com/
17 https://www.moringapartnership.com/agroforestry-technical-assistance-facility/
18 For the Livelihoods Fund for Family Farming see http://www.livelihoods.eu/l3f/

do well, these investment projects have to cover organizational, facilitation and administrative costs at the same time as guaranteeing outcome performance either in terms of carbon stocks or smallholder productivity and production, those are the two key challenges in scaling of capacity and cost-effectiveness in project implementation.

6 Future trends and conclusion

There has been increased investment in agroforestry R&D in the tropics in the last two decades. Because of those investments, the nature of the challenges in the development of agroforestry has shifted to a higher scale which focuses on mainstreaming agroforestry practice in the development agenda of countries in the tropics. As this chapter has outlined, reaching this goal has a number of challenges. The concluding question is whether those challenges are insurmountable. In this author's assessment, the answer is no, as long as agroforestry can maintain the investment flows of the last two decades. As the last section argues, the prospects for that are positive but with a significant shift in the nature of that finance from international development aid to sources of private finance. Maintaining those flows in turn requires an increased level of accountability and clear evidence of a positive return on those investments.

The challenges above in part lay out a research agenda in both the area of technical change and the area of knowledge into action (van Noordwijk, 2017), or what has been termed research *in* development (RinD) (Coe et al., 2014). This research agenda reflects the inherent challenge in the effective organization of systems research, namely how to integrate a number of very different components to achieve the multifunctionality in outcomes. Such systems research has been further complicated by defining the system boundary as a landscape and then the more fundamental challenge of how to scale both within and across such landscape systems. These challenges are at the frontier of applied systems research and have an interplay of how to organize a range of different stakeholders to achieve a required level of coordination across those stakeholders and to facilitate the sharing in financial flows and decision-making. RinD entails research designs overlaid on development programmes such as ILIs and involve experimental or quasi-experimental research designs. There is an explosion of work in application of these methods in agricultural development projects and scaling of agroforestry will build on this work.

Along with its obvious opportunities, multifunctionality also creates challenges in the area of accountability in that it requires measuring the benefit streams across multiple outcomes. Investment in agroforestry in the past has rested on an agreed consensus around its potential, as there have been very few impact studies and these have been done in relatively short time frames after the project interventions and have primarily concentrated on the productivity and income benefit flows. Valuing the improvement in flows of environmental services, particularly at a landscape level, has essentially not been done.[19] Such valuation of the non-productivity benefits will be essential in generating the returns for financing from impact funds, organized through payments for environmental services. However, these are next steps, but it is important that the project development side does not get ahead of the ability to assess actual costs and benefit flows in these projects.

In many ways this is probably the overriding challenge, namely in achieving the potential of agroforestry in the sustainable development agenda that research has to stay ahead of practice on the ground. Moreover, as with any systems-based approach, research, often in an RinD framework, must work on multiple fronts at the same time and a lack of progress in any one may set back the whole enterprise. In particular, a shift in research to more of a focus on downstream process based research should not come at the expense of more upstream research that feeds the research pipeline and thus the reasons for investing in agroforestry. This pressure to meet these multiple challenges ensures that those working in the field are committed, creative and focused on the end goal, seeing agroforestry enhance livelihood and landscapes in the developing world.

7 Where to look for further information

Agroforestry is an empirical science in that there is a very close feedback loop between research and practice. As agroforestry R&D has moved up in scale, the research agenda, which feeds into implementation methodology, has significantly broadened and at the same time, there has been a closer integration of the science and the practice. Looking forward, much of the research agenda is focused at the landscape level with its implications for scaling. Plot- and farm-level research is integrated into this larger scale. Maintaining an effective interaction between science and practice will be critical to resolving many of the challenges laid out above. The landscape scale will in turn lead to a closer integration of environmental services, the evaluation of which will support many of the newer financing mechanisms, particularly the impact funds. Similarly, a multiple market methodology will need to be developed to support projects that rely on generating a clear financial return, often requiring work on new market development. Possibly the more innovative research will be on understanding context specificity with the increasing requirement to achieve scale with agroforestry interventions.

19 There is an increasing amount of research on the measurement of physical flows of environmental services, usually focusing on a single environmental service. There is far less work on integrating these at a landscape level and virtually none on then valuing these services. See Jose (2009).

A locus for much of this work will be the World Agroforestry Centre based in Nairobi, Kenya, but with offices throughout the tropics and subtropics. The Centre led much of the research on tropical agroforestry, including the work carried out under the Alternatives to Slash and Burn program. CATIE also carries out significant research on agroforestry but focused almost wholly on Central America. There still are relatively few national agroforestry research programmes, in many ways the largest being the Central Agroforestry Research Institute, founded in 1988, in Jhansi, India. Agroforestry still has few dedicated research centres or university departments that carry out a well-conceived research agenda on tropical agroforestry, but there are an increasing number of individual scientists undertaking such research. The 4th World Congress on Agroforestry will be held in 2019 as a platform for assessing the state of both research and practice and as a venue for networking individual scientists. In many ways the most widespread capacity in agroforestry practice is the multiplicity of NGOs in the developing world that have implemented agroforestry projects.

8 References

Arslan, A., N. McCarthy, L. Lipper, S. Asfaw and A. Cattaneo (2013). Adoption and intensity of adoption of conservation farming practices in Zambia. ESA Working Paper No. 13-01. Rome, Italy: FAO.

Barron, J. and S. Noel (2011). Valuing soft components in agricultural water management interventions in meso-scale watersheds: A review and synthesis. *Water Alternatives* 42(2):145–54.

Boreux, V., Vaast, P., Madappa, L. P., Cheppudira, K. G., Garcia, C. and Ghazoul, J. (2016). Agroforestry coffee production increased by native shade trees, irrigation, and liming. *Agronomy for Sustainable Development* 36:42.

CGIAR Research Program on Forests, Trees and Agroforestry (2016). *Revised Phase II Full Proposal 2017-2022*. Bogor, Indonesia: CIFOR, p. 116.

Coe, R., F. Sinclair and E. Barrios (2014). Scaling up agroforestry requires research 'in' rather than 'for' development. *Current Opinion in Environmental Sustainability* 6:73–7.

Coe, R., J. Njoloma and F. Sinclair (2017). To control or not to control: How do we learn more about how agronomic innovations perform on farms. *Experimental Agriculture* (in press), 7p.

Coulibaly J. Y., B. Chiputwa, T. Nakelse and G. Kundhlande (2016). Adoption of agroforestry and its impact on household food security among farmers in Malawi. ICRAF Working Paper No. 223. Nairobi, Kenya: World Agroforestry Centre (ICRAF).

Degrande A., Y. Siohdjie Yeptiep, S. Franzel, E. Asaah, B. Takoutsing, A. Tsobeng and Z. Tchoundjeu (2014). Disseminating agroforestry innovations in Cameroon: Are relay organisations effective? In: Van Lauwe B., Van Asten P. and Blomme G. (Eds), *Agro-ecological Intensification of Agricultural Systems in the African Highlands*. New York, NY: Routledge, pp. 221–30.

FAO (2013). Advancing agroforestry on the policy agenda: A guide for decision-makers, by G. Buttoud, in collaboration with O. Ajayi, G. Detlefsen, F. Place & E. Torquebiau. Agroforestry Working Paper No. 1. Rome, Italy: FAO, 37pp.

Foster, K. and H. Neufeldt (2014). Biocarbon projects in agroforestry: Lessons from the past for future development. *Current Opinion in Environmental Sustainability* 6:148–54.

Gray, E., N. Henninger, C. Reij, R. Winterbottom and P. Agostini (2016). Integrated landscape approaches for Africa's drylands. World Bank Studies. Conference Edition. Washington DC: World Bank.

Hart, A. K., J. C. Milder, N. Estrada-Carmona, F. A. J. DeClerck, C. A. Harvey and P. Dobie (2015). Integrated landscape initiatives in practice: assessing experiences from 191 landscapes in Africa and Latin America. In: Minang, P. A., van Noordwijk, M., Freeman, O. E., Mbow, C., de Leeuw, J. and Catacutan, D. (Eds), *Climate-Smart Landscapes: Multifunctionality in Practice*. Nairobi, Kenya: World Agroforestry Centre (ICRAF), pp. 89–101.

Jose, S. (2009). Agroforestry for ecosystem services and environmental benefits: An overview. *Agroforestry Systems* 76:1–10.

Klapwijk, C. J., M. T. van Wijk, T. S. Rosenstock, P. J. A. van Asten, P. K. Thornton and K. E. Giller (2014). Analysis of trade-offs in agricultural systems: Current status and way forward. *Current Opinion in Environmental Sustainability* 6: 110–15.

Leakey, R. (2012). *Living with the Trees of Life – Towards the Transformation of Tropical Agriculture*. Wallingford, UK: CABI.

Mbow, C., M. Van Noordwijk, E. Luedeling, H. Neufeldt, P. A. Minang and G. Kowero (2014). Agroforestry solutions to address food security and climate change challenges in Africa. *Current Opinion in Environmental Sustainability* 6:61–7.

Milder, J. C., A. K. Hart, P. Dobie, J. Minai and C. Zaleski (2014). Integrated landscape initiatives for African agriculture, development, and conservation: A region-wide assessment. *World Development* 54:68–80.

Minang, P. A., M. van Noordwijk, O. E. Freeman, C. Mbow, J. de Leeuw and D. Catacutan (Eds) (2015). *Climate-Smart Landscapes: Multifunctionality in Practice*. Nairobi, Kenya: World Agroforestry Centre (ICRAF).

Namirembe, S., B. Leimona and M. van Noordwijk, F. Bernard and K. E. Bacwayo (2014). Co-investment paradigms as alternatives to payments for tree-based ecosystem services in Africa. *Current Opinion in Environmental Sustainability* 6:89–97.

National Agroforestry Policy (2014). New Delhi, India: Government of India, Department of Agriculture and Cooperation, Ministry of Agriculture.

Place, F., M. Adato, P. Hebinck and M. Omosa (2003). The impact of agroforestry-based soil fertility replenishment practices on the poor in western Kenya. FCND Discussion Paper 160. Washington DC: IFPRI, p. 14.

Sayer, J., T. Sunderland, J. Ghazoul, J. L. Pfund, D. Sheil, E. Meijaard, M. Venter, A. Klintuni Boedhihartono, M. Day, C. Garcia, C. van Oosten and L. E. Buck (2013). Ten principles for a landscape approach to reconciling agriculture, conservation, and other competing land uses. *Proceedings of the National Academy of Science* 110(21):8349–56.

Schmidt, E. and F. Tadess (2012). Household and plot level impact of sustainable land and watershed management practices in the Blue Nile. ESSP Research Note 18. Ethiopian Development Research Institute and International Food Policy Research Institute.

Silechi, G. W. and F. K. Akinnifesi (2017). Comments on Coe et al. (2016)–'Loading the Dice in Favour of the Farmer…'. *Experimental Agriculture* (in press), 6p.
Takoutsing, B., Z. Tchoundjeu, A. Degrande, E. Asaah and A. Tsobeng (2014). Scaling-up sustainable land management practices through the rural resource centre: Reconciling farmers' interests with research agendas. *Journal of Agricultural Education and Extension* 20(5):463–83.
Tayleur, C., A. Balmford, G. M. Buchanan, S. H. M. Butchart, H. Ducharme, R. E. Green, J. C. Milder, F. J. Sanderson, D. H. L. Thomas, J. Vickery and B. Phalan (2017). Global coverage of agricultural sustainability standards, and their role in conserving biodiversity. *Conservation Letters* 10(5):610–18.
van Noordwijk, M. (2017). Integrated natural resource management as pathway to poverty reduction: Innovating practices, institutions and policies. *Agricultural Systems* (in press), 12p.
van Noordwijk M., P. Pacheco, M. Slingerland, S. Dewi and N. Khasanah (2017). Palm oil expansion in tropical forest margins or sustainability of production? Focal issues of regulations and private standards. Working Paper 274. Bogor, Indonesia: World Agroforestry Centre (ICRAF) Southeast Asia Regional Program.
World Agroforestry Centre (2013). Farmer-managed natural regeneration: How to regenerate pasture and farmland on a low budget. Technical Brief. Nairobi, Kenya: World Agroforestry Centre (ICRAF).
World Agroforestry Centre (2015). Harnessing the potential of agroforestry to boost yields and strengthen food security in Rwanda. Policy Brief 31. Nairobi, Kenya: ICRAF.
Zomer, R. J., H. Neufeldt, J. Xu, A. Ahrends, D. A. Bossio, A. Trabucco, M. van Noordwijk and M. Wang (2016). Global Tree Cover and Biomass Carbon on Agricultural Land: The contribution of agroforestry to global and national carbon budgets. *Scientific Reports* 6:29987.

Chapter 16

Tropical tree domestication in agroforestry

Damase Khasa and Alain Atangana, Université Laval, Canada

1 Introduction

Trees, whether occurring in wild stands or in agricultural landscapes, play a key role in the livelihoods of smallholders in the tropics. They provide agricultural products such as fruit and nuts, fodder for livestock, firewood, timber and medicinal substances (Leakey et al. 2005).

Trees also provide services such as shade for crops and livestock as well as a means to retain moisture and nutrients in soil and prevent soil erosion. More broadly, trees provide ecosystem services such as carbon sequestration and conservation of biodiversity.

People have long exploited inter- and intra-species variation to select, propagate and cultivate superior types to meet one or more of these needs. Advances in knowledge now allow scientists to work in collaboration with farmers to better understand and optimize the benefits of domesticating particular tree species within agroforestry systems (Atangana et al. 2014a). The need for collaboration has become particularly urgent given rates of deforestation in the tropics. The World Agroforestry Centre (ICRAF) was created in 1977 to promote agroforestry as an alternative to poor land-use systems including slash-and-burn practices in the tropics. Continuing rates of deforestation also led to the World Commission on Forests and Sustainable Development (WCFSD) to

http://dx.doi.org/10.19103/AS.2018.0041.22

recommend that more extensive support be provided to community-based agroforestry projects, with the aim of providing sustainable livelihoods for poor farmers in the tropics without the need for further deforestation for crop and livestock farming (WCFSD 1999).

This chapter discusses examples of tree species that have undergone domestication in Latin America and in the Congo Basin, and discusses methods that could be used to address key issues in participatory tree domestication, based on our studies on *Allanblackia floribunda* Oliv. (Clusiaceae). Other examples are drawn from our studies on *Irvingia* species (Irvingiaceae), *Dacryodes edulis* (G.Don) H.J. Lam (Burseraceae), *Annickia chlorantha* (Oliv.) Setten and Maas (Annonaceae) and *Baillonella toxisperma* Pierre (Sapotaceae).

2 Tropical tree domestication principles and techniques

Domestication of woody plants may be understood as the selection and management of these plants to increase their benefits to human kind (Simons and Leakey 2004; Leakey et al. 2012). A major objective of tree domestication in agroforestry is the creation or enrichment of complex, multi-strata and species-rich agro-ecosystems that mimic natural forests. The process of domestication helps to reduce deforestation through the planting and management of selected germplasm from high-value multipurpose trees which can both help to preserve ecosystems and provide sustainable livelihoods for farmers. Those promoting tree domestication promoters argue that increasing the availability of timber and non-timber forestry products (NTFPs) with potential dietary and commercial value to local farmers would not only reduce poverty and poor nutrition in tropical rural areas but would also reduce pressure on natural resources in the wild, thereby protecting forests. This win-win strategy has been used as the main hypothesis behind tree domestication activities by ICRAF and its local partners in the tropics (Leakey et al. 2005).

Key steps in tree domestication as implemented by ICRAF and its local partners include (Tchoundjeu et al. 2006):

- the identification of priority species in each region using ethnobotanical, biophysical and market studies,
- the conservation of genetic diversity of species undergoing tree domestication (Atangana et al. 2010),
- the development of methods for phenotypic selection of superior individuals (e.g. phenotypic characterization of traits of importance for domestication and selection of 'plus trees' for breeding),
- germplasm management,

- the development or fine-tuning of protocols for the vegetative propagation of priority species for domestication (Atangana et al. 2006; Atangana and Khasa 2008),
- distribution to farmers,
- management to ensure healthy trees (e.g. prevention/control of pests),
- the development of post-harvest techniques to add value to both timber and non-wood products (NWP),
- processing of agroforestry products (Tchoundjeu et al. 2005),
- the characterization of value chains for each product,
- market development of agroforestry products and smallholder cultivation (Jamnadass et al. 2010),
- the development of farmer's entrepreneurship skills so as to increase farmers' benefits from the sale of NWP.

Some of these issues are discussed in more detail below.

2.1 Identification of priority species for domestication

Farmers in the tropics use a range of wood products and NWP from trees for various purposes throughout the year. Table 1, for instance, shows the range of forest products commonly used in the humid lowlands of West and Central Africa. This highlights the potential pitfalls of establishing too narrow a list of top priority species for domestication in such a region. Another key consideration is the importance of consulting farmers themselves. The priority-setting process of top species for domestication should take into account the use of NTFPs by farmers and market needs. For example, in the humid tropics of Africa, farmer surveys identify priority species with a wide range of uses, from spices, traditional medicine products, fruit production, soup thickeners and vegetables to fuelwood, timber for construction and raw materials used in furniture making such as rattan. Priorities may differ within a region and must reflect the various uses of forest/savannah products by local farmers.

2.2 Selection of 'plus trees'

Once priority species for domestication are identified, the next step is developing methods for the selection and propagation of trees/shrubs with desirable characteristics (or 'plus trees'). The selection is facilitated by a thorough characterization of phenotypic and genetic variation in the species of interest in their respective natural distribution zone. Phenotypic characterization studies should target the traits of interest, for example, fruit characteristics such as size and taste (Atangana et al. 2001; Leakey et al. 2002) or fatty acid profile of seeds (Atangana et al. 2011b). Prior to the planning of the characterization

Table 1 Examples of forest products that are commonly used in the humid lowlands of West and Central Africa, but that were not ranked as top priority species for domestication globally

Species	Product(s)	Use
Afrostyrax lepidophyllus Midbr. (Huaceae) or bush onion	Bark and seeds	Spice
Allanblackia floribunda Oliv. (Clusiaceae)	Fat extracted from the seeds	Cash generation
	Bark	Traditional medicine
Alstonia boonei De Wild. (Apocynaceae)	Bark	Traditional medicine (bark extracts are used to cure malaria)
Annickia chlorantha (Oliv.) Setten & Maas (Sapotaceae)	Bark	Traditional medicine (used to cure malaria)
	Sap	Yellow dye
Baillonella toxisperma Pierre (Sapotaceae)	Edible oil extracted from the seeds	Food
	Bark	Traditional medicine
	Wood	Timber
Monodora myristica (Gaertn.) Dunal (Annonaceae)	Seeds	Spice
	Roots and bark	Traditional medicine (used to cure headache, diarrhoea, toothache)
Pentaclethra macrophylla Benth. (Fabaceae) or Mubala	Leaf extracts	Used to cure diarrhoea
	Oil extracted from the seeds	Used in cooking, and for soap making
Scorodophloeus zenkeri Harms (Fabaceae)	Bark	Spice
Tetrapleura tetraptera (Schumach. & Thonn.) Taub. (Fabaceae)	Fruit	Spice

All products listed in the table are sold in local markets.

of phenotypic variation, quantitative descriptors for variation in the traits of interest should be identified (Leakey et al. 2000).

The assessment of phenotypic characterization of variation in wild stands allows the selection of 'plus trees' that form first-generation breeding populations (Atangana et al. 2002, 2011b). In the Congo Basin for instance, the tree domestication strategy that has been implemented so far consists in the selection of 'plus trees' and the vegetative propagation of selected individuals in rural nurseries using simple, robust methods that local farmers can employ (Leakey et al. 1990). Though facilitating an increase in genetic gains through broad-sense heritability, this approach can lead to a potential increase of consanguinity in breeding and production populations if the genetic base is narrow.

2.3 Development of vegetative propagation methods for selected trees

The vegetative propagation of species identified for domestication requires using simple, inexpensive (i.e. affordable to smallholders) and robust methods such as non-mist propagators in rural nurseries (Leakey et al. 1990). This has been made possible by the experimental development of practical rooting protocols for leafy unimodal juvenile stems of the species. The factors that have been reported to influence the rooting of leafy stem cuttings include the propagation environment, auxin application, leaf area supported on the cutting, cutting length and diameter, genetic origin of the cutting and stock plant management (Leakey et al. 1990; Mesén et al. 1997; Tchoundjeu and Leakey 1996; Atangana et al. 2006, 2011a; Ngo Mpeck and Atangana 2007).

While exogenous auxin application at the cutting base prior to insertion into the poly-propagators was reported as the most important factor that influences rooting, it is important to note that some species such as *Allanblackia floribunda* are insensitive to auxin application (Atangana et al. 2006; Atangana and Khasa 2008). For such species, a combination of factors that are reported to influence rooting in leafy stem cuttings would improve rooting percentage (Atangana et al. 2006). Ngo Mpeck and Atangana (2007) found that 85% of *B. toxisperma* cuttings with 75 cm^2 leaf area that have indole-3-butyric acid (IBA) applied on their base develop roots 14 weeks after their insertion in sand as a rooting medium in non-mist poly-propagators.

The factors listed above as influencing rooting ability in leafy stem cuttings may not yield high rooting rates on some species that are difficult to root. To increase rooting percentage in the so-called 'difficult-to-root' species, it is first desirable to identify the combination of factors that provides the best results (Atangana et al. 2006; Ngo Mpeck and Atangana 2007) and identify the clones that provide better rooting percentages, as clonal variation has been reported to influence rooting in leafy stem cutting (Atangana and Khasa 2008).

One study used factorial designs to test the simultaneous effects of cutting stem morphology and genetic origin on rooting of *A. floribunda*, a reputed 'difficult-to-root' species (Atangana et al. 2006). The study found no relation between cutting stem morphology and genetic origin on the proportion of cuttings that rooted (Atangana and Khasa 2008). The study was initiated to test whether clonal variation in rooting percentage is due to variation in cutting morphology, as postulated by Hoad and Leakey (1994, 1996) on *Eucalyptus grandis* W. Hill. ex Maiden (Myrtaceae) and Dick et al. (2004) on *Triplochiton scleroxylon* K. Schum (Malvaceae). Indeed, Hoad and Leakey (1994, 1996) observed that manipulating stockplant light environment influenced dry matter partitioning and significantly modified the morphology and physiology of the cuttings, thus enhancing rooting in reputed 'difficult-to-root' clones. Dick et al. (2004) found that genetic variation in rooting potential was minimized by optimizing nutrient application, length of new shoots, total leaf area and node position.

2.4 Integration of trees into agroforestry systems

There are various ways of integrating trees with crop cultivation. Many tropical agroforestry systems include alley cropping where hedgerows of trees or shrubs with high potential for biological nitrogen fixation are intercropped with alleyways of cultivated plants to maintain soil fertility or to replace nutrients exported by crops. By protecting soil health, alley cropping improves the long-term sustainability of crop production and enhances security by diversifying output, for example, by providing fodder and fuelwood to smallholders. Legumes typically used in alley cropping are most often fodder species such as *Leucaena leucocephala* (Lam.) de Wit (Fabaceae: Mimosoideae) and *Calliandra calothyrsus* Meisn. (Fabaceae: Caesalpinioideae). Other legumes that are used in alley cropping include *Acacia* species, *Gliricidia sepium* (Jacq.) Kunth ex Walp. (Fabaceae), *Sesbania sesban* and *S. grandiflora* (L.) Poiret (Fabaceae), whereas actinorhizal plants that are used in alley cropping include *Alnus nepalensis* D. Don, *A. acuminata* (Betulaceae) and *Casuarina equisetifolia* L. (Casuarinaceae). A non-legume and non-actinorhizal plant, *Tithonia diversifolia* (Hemsl.) A. Gray (Asteraceae) is also used in alley cropping systems. Legumes are often cultivated in rotational woodlots which are stands of trees that are used as fuelwood sources by smallholders in tropical Africa (Kwesiga et al. 1999, 2003; Nyadzi et al. 2003; Kimaro et al. 2007; Akinnifesi et al. 2008).

3 Domesticated tree species

Tree domestication has been implemented by tropical farmers for centuries, as reported by Galluzzi et al. (2015) for *Bactris gasipaes* in Southern America, Parker et al. (2010) for caimito (*Chrysophyllum cainito* L., Sapotaceae) in the New World tropics and by Leakey et al. (2004a) for *Dacryodes edulis* and *Irvingia gabonenis* Aubry-Lecomte ex O'Rorke (Irvingiaceae) in Cameroon and Nigeria. Tree domestication has been carried out by ICRAF and local National Agriculture Research Services (NARS), for example, in the Peruvian Amazon (Weber et al. 2001), in the Congo Basin (Tchoundjeu et al. 2006), in the Sahel, in Southern Africa, East Africa, Asia and Oceania since the early 1990s (Leakey et al. 2012). Examples of domesticated tree species in agroforestry systems include:

- Capirona (*Calycophyllum spruceanum* (Benth.) Hook. f. ex K. Schum (Rubiaceae)) in the Amazon Basin

- *Irvingia gabonensis* Oliv. (Clusiaceae) in the Congo Basin
- *Vitellaria paradoxa* C.F. Gaertn (Sapotaceae), *Parkia biglobosa* (Jacq.) R.Br. ex G.Don. (Fabaceae) and *Ziziphus mauritiana* Lam. (Rhamnaceae) in the Sahel
- *Sclerocarya birrea* (A. Rich.) Hochst. (Anacardiaceae), *Strychnos cocculoides* Baker (Loganiaceae), *Uapaca kirkiana* Müll. Arg. (Phyllanthaceae), *Vangueria infausta* Burch. (Rubiaceae) and *Syzygium cordatum* (Hochst.; Myrtaceae) in Southern Africa
- *Vitex pubescens* Vahl (Verbenaceae) and *Gmelina arborea* Roxb ex Sm. (Verbenaceae) in Southeast Asia
- *Artocarpus altilis* (Parkinson) Fosberg (Moraceae), *Barringtonia edulis* Seem (Lecythidaceae), *Canarium indicum* L., *C. harveyi* Seem (Burseraceae) and *Inocarpus fagifer* (Parkinson ex Zollinger) Fosberg (Fabaceae) in Oceania.

Tropical perennial crop-based agroforestry systems have been well-documented in Atangana et al. (2014a) and agroforestry tree species are listed in the Agroforestree (AFT) Database 4.0 (www.worldagroforestry.org/treedb/).

Tree species that are found in perennial crop-based agroforestry in the tropics also consist of cultivated crops such as cocoa (*Theobroma cacao* L., Sterculiaceae), coffee (*Coffea arabica* L. or *C. canephora* L., Rubiaceae), rubber (*Hevea brasiliensis* (Willd. Ex A. Juss.) Müll. Arg., Euphorbiaceae), *Camellia sinensis* (L.) Kuntze (Theaceae) and coconut (*Cocos nucifera* L., Arecaceae). These woody species constitute the basis of cocoa, coffee, rubber or coconut-based agroforestry systems. Seedlings of these tree crops often benefit from the shade provided by multipurpose trees that are retained during farm establishment. For instance, in cocoa agroforests of the humid lowlands of tropical Africa, multipurpose tree species such as *B. toxisperma* which produce edible fat and timber are used to provide shade to *T. cacao* seedlings. Fruit and medicinal trees also constitute a major component of home gardens, which are common in tropical landscapes. Some examples of fruit trees found in tropical home gardens include *Citrus sinensis* (L.) Osbeck (Rutaceae), *Mangifera indica* L. (Anacardiaceae), *D. edulis*. (Burseraceae) or *Psidium guajava* L. (Myrtaceae; Atangana et al. 2014a). The following sections provide examples of key species that have been targeted for domestication in the Amazon and the Congo Basins.

3.1 Tree species for domestication in the Amazon Basin

3.1.1 Bactris gasipaes *Kunth (Arecaceae) or Pejibaye palm or peach palm*

Peach palm, along with *Capirona*, bolaina blanca (*Guazuma crinita* Mart., Malvaceae) and ice-cream-bean (*Inga edulis* Mart., Fabaceae), is one of the top priority species that have first been subjected to farmer-driven domestication activities by ICRAF in the Peruvian Amazon (Labarta Chávarri and Weber 1998; Weber et al. 2001). Peach palm is a multipurpose perennial crop that is widely spread in Central and South America. It produces heart-of-palm (palmito) and edible fruits which can be used, for example, for flour and meal, oil production and as animal rations (Clement and Mora Urpi 1987). The species is semi-domesticated, is also found in wild populations (e.g. var. *chichagui*; Galluzzi et al. 2015), and prefers lowland well-drained soils with annual rainfall ranging between 2000 mm and 5000 mm (Mora-Urpi et al. 1997).

Peach palm has been the target of improvement programmes by the University of Costa Rica, the Colombian Scientific Research Institute for the Cauca Valley and the Brazilian National Research Institute for Amazonia since the 1970s (Clement and Mora Urpi 1987). Pejibaye palm was also identified by ICRAF and the Peruvian National Institute for Agricultural Research and Extension as one of the top 23 priority species for domestication in the Peruvian Amazon (Sotelo-Montes and Weber 1997). ICRAF's domestication activities in the Peruvian Amazon include participatory and farmer-driven plant selection, testing adaptation of seed sources and accelerating the delivery of high-value germplasm to farmers. As an example of farmer-driven tree selection in peach palm, domestication activities targeted the selection and development of oily fruit tree varieties for the production of cooking oil, and starchy fruit tree varieties for the production of flour, on the basis of farmers' preferences (Weber et al. 2001). An important key issue for fruit tree domestication in Latin America is the reduction of genetic diversity in domesticated populations of priority species. Indeed, local farmers in Latin America most often collect germplasm from a few trees with desirable fruit characteristics for cultivation, increasing the risk of inbreeding in cultivated populations of the species (O'Neill et al. 2001).

3.1.2 Calycophyllum spruceanum *(Benth.) Hook. f. ex K. Schum (Rubiaceae) or Capirona*

Capirona has been targeted for domestication in the Peruvian Amazon for its use as timber and as fuelwood (Weber et al. 2001). The bark of the species also has medicinal properties. Domestication of the species has focussed on identifying genetic variation using a range of techniques. These include:

- amplified length polymorphism analysis (Russel et al. 1999)
- phenotypic variation in stem growth and branch wood traits among provenances in the Peruvian Amazon (Sotelo-Montes et al. 2003)
- analysing coppice-shoot growth among provenances (Boivin-Chabot et al. 2004)
- studying stem growth and wood traits (Weber and Sotelo-Montes 2005)

- analysing wood shrinkage and its correlations with tree growth and wood density at an early age (Sotelo-Montes et al. 2007a)
- testing mechanical properties (Sotelo-Montes et al. 2007b)
- correlating wood density with growth at an early age (Sotelo-Montes et al. 2007c)
- correlating wood colour with tree growth and wood density at an early age (Sotelo-Montes et al. 2008).

Genetic variation studies using molecular markers are also important to understand population genetic diversity and structure of the species and consequently to help domestication strategies. Phenotypic variation studies have been followed by low-intensity phenotypic selection of best germplasm to maintain genetic diversity throughout breeding cycles (Weber et al. 2009). However, a low intensive selection strategy results in low genetic gains. Next steps towards domestication of the species may involve genetic selection of elite trees while increasing genetic gain and maintaining enough genetic variation in breeding and production populations so as to avoid the risk of inbreeding (Atangana et al. 2010).

3.1.3 Guazuma crinita *Mart. (Malvaceae) or bolaina blanca*

Bolaina blanca is a fast-growing tree species which is cultivated in Latin America for construction timber and fuelwood. The species has been included in a tree domestication programme undertaken by ICRAF and its national agricultural research partners in the Peruvian Amazon (Sotelo-Montes and Weber 1997; Weber et al. 2001). Domestication activities have focussed on genetic variation in tree growth and wood density (Weber and Sotelo-Montes 2008), on tree growth at an early age (Rochon et al. 2007) as well as stem form and mortality (Weber et al. 2011) of the species for selection of provenances (Weber et al. 2001). Seeds were collected from on-farm provenance trials so that farmers would rapidly select best provenances for charcoal production and timber (Weber et al. 2011). Next steps towards domestication and cultivation of improved germplasm of the species include the selection of elite trees with an increase in genetic gain while implementing strategies to maintain sufficient genetic diversity in breeding and production populations.

3.1.4 Inga edulis *Mart. (Fabaceae) or ice-cream-bean, Guama or Sucrin*

Inga edulis is a fruit tree that is widely grown in Latin America for various purposes including timber, shade, biological nitrogen fixation (Nichols and Carpentier 2006) and the production of local alcoholic beverages. Extracts from leaves of the species are rich in polyphenolic compounds with medicinal properties (Silva et al. 2007). Seeds from the pods of the species are surrounded with a waxy coat tasting like vanilla, hence the name ice-cream-bean (Larbi et al. 2000). Farmers in the Peruvian Amazon have provided researchers with their preferred fruit and tree characteristics as well as potential varieties for development (Potters 1997; Weber et al. 2001). As farmers have been planting the species in Latin America for decades, a key issue was conservation of genetic diversity in cultivated populations. An analysis of planted and unplanted populations of *I. edulis* revealed that farmers had reduced genetic diversity when domesticating the species in Amazonia (Hollingsworth et al. 2005). One of the strategies to conserve genetic diversity is to identify and design *in situ* conservation areas (O'Neill et al. 2001). Another option is the facilitation of gene flow between domesticated and wild populations of the species throughout breeding cycles.

3.2 Tree species for domestication in the Congo Basin

3.2.1 Irvingia gabonensis *and* I. wombolu *Vermoesen (Irvingiaceae), or bush mango tree*

Irvingia gabonensis produces edible fruits, while its kernels and those of *I. wombolu* are used as soup thickeners and are extensively traded in the humid lowlands of West and Central Africa (Leakey 1999). Consequently, *I. gabonensis* and *I. wombolu* were ranked as top priority species for domestication in a household survey undertaken in Gabon, Cameroon, Nigeria and Ghana by ICRAF and the local NARS. Trees of these species are most often retained when clearing land for farming, and they are widely distributed in farms growing food crops, fallow areas, home gardens and cocoa or coffee farms in this area. Both species were domesticated centuries ago by local farmers in the humid tropics of Nigeria and Cameroon (Leakey et al. 2004a). Participatory domestication of both species has been implemented by ICRAF and its local partners and consists of:

- characterizing phenotypic variation in fruit and kernel traits (Leakey et al. 2000; Atangana et al. 2001; Anegbeh et al. 2003)
- selecting 'plus trees' with desirable fruit and nut characteristics (Atangana et al. 2002; Leakey et al. 2004b)
- a comprehensive germplasm collection to establish genebanks in Cameroon and Nigeria (Ladipo et al. 1996; Tchoundjeu et al. 1998, 2006)
- developing affordable, simple and robust vegetative propagation protocols for mass propagation of selected germplasm (Shiembo et al. 1996)

- integrating and managing selected germplasm in existing agroforests, borrowing from local fruit tree growing strategies (Degrande et al. 2006; Tchoundjeu et al. 2006).

Domestication also includes developing farmer entrepreneurship to optimize the marketing of *Irvingia* nuts (Tchoundjeu et al. 2006) and assessing traditional methods of nut preservation (Tchoundjeu et al. 2005).

An important objective of domesticating indigenous fruit trees in the humid lowlands of West and Central Africa is to reduce the time until fruit is produced to 3–4 years, as most farmers in the humid forest zones of Cameroon, Gabon and the Democratic Republic of the Congo are discouraged from planting forest trees because of the long waiting period before the trees fruit. However, some farmers in Nigeria and Southwest Cameroon plant *I. wombolu* trees in gardens or as shade plants for cocoa. Vegetative propagation of selected trees by marcotting or air-layering seems to be an option that addresses this issue, and promising results have been obtained at the ICRAF stations in Onne, Nigeria, and Nkolbisson Yaoundé, Cameroon (Tchoundjeu et al. 2006). The next step should involve establishing mother blocks of germplasm from marcotts, and propagating this germplasm on a large scale using leafy stem cuttings. Germplasm distribution pathways should then be developed in rural areas, as should village nurseries where farmers are trained in tree selection, vegetative propagation and other methods of tree domestication, which seem to be pivotal at this stage (Tchoundjeu et al. 2006). ICRAF and its partners have made significant progress domesticating *Irvingia* spp. trees in the humid lowlands of West and Central Africa, but mass propagation of selected germplasm is still limited, and domestication activities have also not been widely adopted by farmers, though they are being scaled up by ICRAF. Farmers also benefit more from selling *Irvingia* kernels and from the paste obtained from processing the kernels, which is another challenge.

3.2.2 Dacryodes edulis *(G. Don) H.J. Lam. (Burseraceae) or Safou*

Boiled or roasted safou fruits are widely consumed in the humid forest zone of West and Central Africa, and they are also traded regionally and internationally (Awono et al. 2002). Safou trees are extensively planted in their natural distribution zone in Cameroon, such as in Makenene, where safou gardens are common (Tchoundjeu et al. 2002). They are also common in home gardens, on food crop and cocoa farms, and in fallows in this zone (Tchoundjeu et al. 2002). Domestication of safou by ICRAF and local NARS in the humid lowlands of this region has been well-described elsewhere (Tchoundjeu et al. 2002, 2006; Leakey et al. 2012) and consists of:

- phenotypic characterization of fruit and nut traits (Leakey et al. 2002)

- provenance trials using 20 accessions in Barombi-Kang and Minkoameyos, Cameroon
- collecting germplasm by air-layering in Central and Western Cameroon (Tchoundjeu et al. 2002)
- creating range-wide seed collection in Cameroon, Equatorial Guinea, Gabon and Nigeria to establish genebanks in these countries (Tchoundjeu et al. 2002)
- establishing 'village nurseries' for mass propagation and distribution of 'plus tree' germplasm (Tchoundjeu et al. 2002, 2006).

Domestication has also included research into germplasm management in local agroforests and fruit processing post-harvest (Tchoundjeu et al. 2006; Leakey et al. 2012). Although considerable progress has been made in domesticating safou trees for fruit production, pest management is still a concern, as syrphid flies (Syrphidae) cause the flowers of this species to abort, hampering fruit production. Preserving safou fruits is a major issue for poor farmers, although some producers who can afford ovens to dry the fruit to extend preservation. Safou fruits are rich in fat (Omoti and Okiy 1987; Leakey 1999) and can be processed for fat production. Safou trees have been cultivated in Cameroon for decades but information on the heritability of desirable fruit characteristics to calculate genetic gains in safou breeding is still lacking. The optimum spacing between safou germplasm and other crops and trees in safou-based agroforestry systems is also unknown.

3.2.3 Ricinodendron heudelotii *(Baill.) Pierre ex Pax (Euphorbiaceae), ezezang or ndjanssang*

This is a fast-growing tree which produces spherical, plum-like, indehiscent, two to three seeded, lobed capsules, and its kernels are widely used as a thickening agent for soups (Leakey 1999) and traded in tropical Africa where the species is endemic from Senegal to Angola (Ngono and Ndoye 2004). The species also occurs in the Central African Republic, Democratic Republic of the Congo, Uganda, Kenya, Tanzania, Zambia and Madagascar (Tchoundjeu and Atangana 2006). Phenotypic variation in fruit and seed traits has been assessed (Ngo Mpeck et al. 2003) and protocols have been developed for vegetative propagation of the species using leafy stem cuttings (Shiembo et al. 1997) as part of a domestication programme. Little is known about the ecological interaction between ndjanssang and other crops in agroforestry systems, or the phenology of flowering or mechanisms of pollination, although agroforestry opportunities and constraints of the species have been well documented (Djeugap et al. 2013). Ndjanssang genetic resources also need to be characterized for germplasm collections and living genebanks (Atangana et al. 2014a).

3.2.4 *Cola acuminata* Schott & Endl. (Malvaceae), *C. anomala* K. Schum. and *C. nitida* (Vent.) Schott & Endl. (Sterculiaceae) or kola nuts

Kola nuts are widely traded in West and Central Africa (Ndoye et al. 1997). They are also consumed locally as stimulants and have been used as additives for soft drinks in Europe and America. Domestication of the genus has focussed on developing protocols for marcotting trees, establishing genebanks in Nigeria, and characterization and morphological assessment of germplasm (Adebola and Morakinyo 2006). The next steps in domesticating *Cola* spp. may involve fine-scale genetic characterization of the species in its natural distribution zone to select germplasm that could be used in multi-location trials. This characterization should be combined with the study of interspecific hybridization, as the species is enormously variable (Adebola and Morakinyo 2006). Methods should be developed to minimize pest infestation of kola nuts, as *Balanogastris kolae* and *Saphrorhinus* spp. attack the nuts before harvest and during storage.

3.2.5 *Prunus africana* (Hook.f.) Kalkman (Rosaceae) or African cherry

African cherry bark extracts are used to treat prostate gland hypertrophy and benign prostatic hyperplasia (Cristoni et al. 2000), which explains the over-exploitation of the bark (Stewart 2003) in its natural distribution zone, reducing genetic diversity in natural populations of the species (Farwig et al. 2008). Trade in African cherry bark can generate over US$150 million in Cameroon, one of the top exporters of the bark. Natural populations of this evergreen tree species can be found at elevations between 700 and 3000 in Cameroon, Equatorial Guinea, São Tomé e Príncipe, Angola, the Democratic Republic of the Congo, Burundi, Rwanda, Uganda, South Sudan, Ethiopia, Tanzania, Zambia, Zimbabwe, Mozambique, Swaziland, South Africa and Madagascar. Protocols for vegetative propagation of African cherry using leafy stem cuttings have been developed by Tchoundjeu et al. (2002), and Avana et al. (2004) assessed the genetic diversity in a few samples of the species in Cameroon using amplified fragment length polymorphism and microsatellite markers. *P. africana* trees have been planted near Dschang in Cameroon by the National Office of Forest Development and in Kenya by the Forest Department of Kenya for timber production (Dawson et al. 2000). The phylogeography of the species in East and West Africa investigated by Kadu et al. (2011) clearly revealed a former migration corridor between East and West African highlands. Kadu et al. (2012) also assessed the geographic variation in the bioactive constituents of African cherry bark extracts throughout Africa. The next step in domesticating the African cherry may involve assessing genetic control of the bioactive constituents of the bark extracts and the genetic diversity of wild and cultivated populations, using genetic markers, such as microsatellites and single-nucleotide polymorphisms (SNPs) (Nadeem et al. 2018).

4 Case study: participatory domestication of *Allanblackia floribunda*, a high-value agroforestry tree species in Central Africa

The genus *Allanblackia,* or tallow tree, consists of forest trees that occur in wild stands and fallows in the evergreen lowland, deciduous forests and mountains of tropical Africa. Ten species have been recorded to date: *A. gabonensis* (Pellegr.) Bamps, *A. floribunda*, *A. kimbiliensis* Spirl., *A. kisonghi* Vermoesen, *A. marienii* Staner, *A. monticola* Mildbr., *A. parviflora* A. Chevalier, *A. stanerana* Exell & Mendonça, *A. stuhlmanni* (Engl.) Engl. and *A. ulugurensis* Engl. Tallow trees bear many seeded berry-like fruits (Fig. 1). Species that have been traded for the edible white fat that is extracted from their seeds include *A. stuhlmannii* in Tanzania, *A. floribunda* in Nigeria and *A. parviflora* in Ghana. The seeds are rich in unsaturated fatty acids, mostly stearic and oleic acids (Crockett 2015; Atangana et al. 2011b), and the fat is used in the production of margarine with health-promoting properties. Unilever PLC has guaranteed a market for *Allanblackia* seeds, purchasing them from farmers in Ghana, Nigeria and Tanzania as part of a sustainable *Allanblackia* oil business developed from a public-private partnership between Unilever, ICRAF, the International Union for Conservation of Nature (IUCN), NARS, farming communities and market traders (Jamnadass et al. 2010).

To ensure a sustainable supply of fat from *Allanblackia* seeds, ICRAF is promoting *Allanblackia* cultivation, and the species have been enrolled in the ICRAF participatory tree domestication programme. This programme targets vegetative propagation of 'plus trees' with desirable fruit and seed characteristics and introduction of germplasm into farms. Practical rooting protocols for *A. floribunda* leafy stem cuttings using simple, low-cost and robust methods have been developed to start *Allanblackia* domestication (Atangana

Figure 1 Fruits (a, b) and seeds (c) of *Allanblackia* species. Source: Alain Tsobeng (ICRAF).

et al. 2006, Atangana and Khasa 2008), as seeds of the species are recalcitrant (35% germination rate for *A. parviflora* seeds; Ofori et al. 2011). Optimizing factors reputed to affect cutting rooting, such as genetic origin (clone) and rooting medium, or clone and rooting medium and leaf area, increases the rooting percentage in *A. floribunda* leafy stem cuttings (Atangana et al. 2006). The best rooting percentage identified for *A. floribunda* was 68.7% (Atangana et al. 2006) and methods of grafting this species have been developed with a success rate of up to 80% (Asaah et al. 2011).

Methods of participatory *Allanblackia* tree domestication have included phenotypic characterization of fruit and seed traits, and 20 *A. floribunda* 'plus trees' have been selected to form a first-generation breeding population by independent culling of the 70 fruiting trees that were assessed in wild stands (Atangana et al. 2011b). Levels of stearic and oleic acid in seed fat were also measured by gas chromatography. The study found extensive among-tree variation in the traits measured, but these characteristics are low to moderately heritable in these species (Atangana et al. 2011b). Fruit mass was identified as a good predictor of total seed mass per fruit and mean seed mass per fruit. However, none of the characteristics measured, including fruit mass, length and width, seed mass and number of seeds per fruit, were found to be related to the stearic or oleic acid content of the seed fat (Atangana et al. 2011b). The study recommended optimizing sylvicultural practices to increase fruit production throughout the *Allanblackia* breeding cycle (Atangana et al. 2010). A larger sample of fruiting trees from across the natural distribution zone of the genus may also allow higher selection intensity to achieve greater genetic gains.

Sampling sufficient fruiting trees in wild stands is an issue when domesticating forest trees in the tropics for fruit and seed production. Indeed, the number of forest trees of a particular species fruiting in the humid lowlands of tropical Africa may hardly reach the numbers required to assess phenotypic variation, which is usually around 200–300. This is why first-generation breeding populations for fruit and seed production in this region are usually small, which is a concern for the genetic diversity of source populations and inbreeding, both in breeding and production populations. For example, we sampled 20 *A. floribunda* 'plus trees' for breeding from wild stands in Cameroon (Atangana et al. 2011b) and have investigated whether establishing a breeding programme with such a small sample affects neutral genetic diversity and inbreeding in both the breeding and production populations (Atangana et al. 2010). Our study found that this sampling could preserve wild genetic diversity in breeding populations, although there is a slight increase in inbreeding in the first-generation breeding population (Atangana et al. 2010). Our results are congruent with those of Ingvarsson and Dahlberg (2018), who showed that 20 unique clonal progenitors are needed to effectively capture 95% of the genetic diversity present in the base population. Measures have been recommended to control inbreeding in *A. floribunda*, including increasing the size of the breeding population, assessing gene flow between natural and breeding populations and implementing appropriate breeding strategies, such as subgroup breeding using many breeding populations (Atangana et al. 2010).

Prior to this research, we developed ten microsatellite DNA markers (simple sequence repeats or SSRs) for *A. floribunda* and tested their transferability to *A. gabonensis* and *A. stanerana* (Anderson et al. 2010). Nuclear SSR markers are polymorphic and abundant in the genome, hence their suitability for genetic population studies. The SSR markers that we developed were used to investigate genetic structure and gene flow in *A. floribunda* (Atangana et al. 2010). These markers are also valuable for population genetics studies in other species in the genus, as eight of these exhibited polymorphisms with *A. gabonensis,* and seven were polymorphic with *A. stanerana* (Anderson et al. 2010). The next step would be to assess the transferability of these markers to *A. stuhlmannii*, *A. parviflora*, *A. kimbiliensis* and *A. ulugurensis*.

With the advent of next-generation DNA sequencing (NGS) technologies, the recently developed genotyping by sequencing (GBS) technique offers a highly cost-effective alternative for simultaneous SNP discovery and genotyping (Deschamps et al. 2012; Nadeem et al. 2018). NGS that enables the mass sequencing of genomes and transcriptomes, producing a vast array of genomic information through bioinformatics, facilitates analysis of the genotype and its relationship with the phenotype, especially for the complex traits that are desirable in genetic improvement of agroforestry species. These molecular breeding tools and methods of improving varieties by genomic selection function much more rapidly than conventional techniques and are now available and well-documented in plant genomics (Varshney and Dubey 2009; Resende Jr. et al. 2011; Pérez-de-Castro et al. 2012; Isik et al. 2015).

Tallow tree domestication also included practical protocols to root leafy stem cuttings from this genus. We have investigated vegetative propagation of *A. floribunda* using this method and found that the species is difficult to root, even when cuttings were treated with IBA before placing in non-mist poly-propagators (Atangana et al. 2006; Atangana and Khasa 2008). A species can be labelled difficult to root because of the time it would take to obtain high leafy stem cutting rooting rates, and the rooting period may last as long as 2 months when the protocol is fully developed in a species (e.g. *Khaya ivorensis* A.Chev., Meliaceae; Tchoundjeu and Leakey 1996). It can last over 30 weeks when the species is reputedly difficult to root (e.g. *A. floribunda*; Atangana et al. 2006; Atangana and Khasa 2008). Therefore, the effect of factors that are known to accelerate rooting should be investigated in any previously unstudied species. Auxin application is the main factor that has been identified as accelerating rooting of leafy stem cuttings in tree species (Leakey 2004; Atangana et al.

2011a), so investigating the effects of auxin application over time and any other factors that can influence rooting may enable the optimum rooting period for a species to be identified (Ngo Mpeck et al. 2009). Indeed, Ngo Mpeck et al. (2009) found that the propagation period for *Annickia chloranta* can be longer than 13 weeks in a study that investigated the simultaneous effects of exogenous auxins (indole-3-acetic acid (IAA), IBA and 1-naphthaleneacetic acid (NAA)) and substrate type (sand, 50:50 mix of sand and sawdust, and sawdust). The study also found that temporal effects of exogenous auxin on the rooting of *A. chlorantha* leafy stem cuttings were amplified by the composition of the substrate (Ngo Mpeck et al. 2009). The study suggested that the effects of within-species variation in endogenous rooting inhibitors and rooting co-factors should be assessed. It also suggested that the simultaneous effects of rooting inhibitors/co-factors and substrate composition on *A. chlorantha* leafy stem cutting rooting should be investigated (Ngo Mpeck et al. 2009).

5 Conclusion

The species that have been enrolled to date in the participatory tree domestication programme led by ICRAF and its partners in the tropics are valued for products such as their edible fruits and nuts that are traded locally and internationally (Leakey 1999; Awono et al. 2002; Ndoye et al. 1997). Research into participatory tree domestication can contribute to improved and sustainable crop production and improve smallholder livelihoods in the tropics in several ways, including by breeding cultivars of high-value multipurpose tree species that have been identified as a top priority species for domestication and by making germplasm available to farmers. After over two decades of participatory tree domestication for high-value NTFPs in the tropics, more needs to be done to determine whether widespread cultivation of germplasm with desirable characteristics for consumption and sale has reduced pressure on local forests. Such a study should include the biophysical and socio-economic dimensions of tree domestication and identify key issues that should be addressed to deliver on the promise of a win-win situation where farmers' income is increased and pressure on natural forests is reduced. Questions that need to be answered include: to what extent have tree domestication activities reduced slash-and-burn practices in a particular biome over the last two decades? Do tree domestication activities correlate with forested areas locally? Is local biodiversity related to tree domestication activities? What are the key issues to address to boost tree domestication while increasing the biodiversity in a particular biome? To what extent has participatory domestication of indigenous trees improved farmers' livelihoods and improved the environment? Answering these questions would refine the initial win-win hypothesis and eventually develop other hypotheses to guide tree domestication activities in the tropics and help develop cultivars that would meet farmers' requirements while still preserving biodiversity. For example, developing cultivars of *I. gabonensis* for fruit or nut production based on the fruit and nut ideotypes characterized by Leakey et al. (2000) would help diversify crop production and sources of income for smallholders in the humid tropics of West and Central Africa. Developing cultivars of various fruit tree species with different fruiting periods throughout the year would also increase crop production and smallholder livelihoods in the tropics. The characteristics that could guide tree selection for cultivar development include regular and abundant fruit production every year for sustainable crop production.

Developing multi-strata and complex agroforests with the germplasm of top priority species for domestication would also contribute to improved and sustainable crop production for smallholders in the tropics. Trees help to form the humus layer of the soil and, therefore, maintain the organic matter in the soil and adequate coverage of the soil surface. Trees also contribute to root turnover and nutrient cycling, and to erosion control and soil stabilization by the roots. Therefore, species-rich and complex agroforestry systems increase soil productivity, sustaining and increasing crop production. The home gardens that are popular in the tropics are an example of a complex agroforest system. Home gardens are primarily intended to provide households with edible products, and they consist of multilayered agroforestry systems that are made up of trees, shrubs and herbaceous plants that are managed by the households surrounding the compound. As many as 60 species have been recorded in a home garden (Kumar and Nair 2004), and this diversity and multilayered structure make home gardens a model of productive and sustainable agroforestry. Tropical home gardens provide several agroforestry products that have a food, social, medicinal and economic value to local farmers (Atangana et al. 2014a). For example, multipurpose indigenous agroforestry species such as *Garcinia kola* Heckel (Clusiaceae), *Dacryodes edulis*, *Irvingia gabonensis*, *Ricinodendron heudelotii* and *Cola* species can be found in tropical home gardens. This diversity of food and other NWP ensures the economic viability of home gardens in the tropics. Agroforestry systems are resilient landscapes in which the germplasm of important multipurpose indigenous trees and shrubs can be managed to ensure sustainable production and preservation of genetic resources. Germplasm of high-value tree and shrub species can be planted in these agroforestry systems in a way that prevents negative ecological interactions (e.g. competition for soil nutrients and light) between the germplasm and surrounding plants. Research into these systems should therefore identify the appropriate spacing between the germplasm of domesticated priority species and the species found in local agroforests (Atangana et al. 2014c).

A new agroforestry project (2018–21) led by Université Laval and funded by the Quebec Government Green Fund as part of its 2013-20 Action Plan on Climate Change aims to develop climate-change resilient agroforestry systems to improve the livelihoods of smallholders in Cameroon. The project will focus on building local capacity and be carried out in close collaboration with the University of Dschang, an agricultural and environmental training centre, and ABIOGeT, a Cameroonian non-governmental organization. The project will contribute to Cameroon's commitment to COP-21 (Paris) to reduce its GHG emissions by 32% by 2035 and improve socio-economic and environmental resilience in the regions of Adamawa and northern Cameroon. The government will also use the project's scientific data in Cameroon's National REDD+ (Reducing Emissions from Deforestation and Forest Degradation) Strategy. Improved home gardens and other innovative practices will be promoted, as home garden agroforestry systems have been proven to preserve biodiversity in Bangladesh (Bardhan et al. 2008).

Agroforestry systems also provide opportunities to generate cash while creating resilient landscapes. Increasing the incomes of tropical farmers by increasing farmers' benefits within agroforestry tree product (AFT) market chains would also contribute to improved and sustainable crop production (Facheux et al. 2012). ICRAF scientists in the humid lowlands of West and Central Africa have promoted group product sales, developed post-harvest processing and storage methods at a village level and facilitated a village fund to allow for off-season sales to improve incomes (Gyau et al. 2011; Facheux et al. 2012; Degrande et al. 2014). Other initiatives, such as including agroforestry systems in the Clean Development Mechanism of the Kyoto protocol would surely increase the incomes of tropical farmers through the sale of carbon on the carbon market under the REDD+ scheme (Atangana et al. 2014e), contributing to improved and sustainable crop production by smallholders. Multi-strata agroforests are a type of ecoagriculture (Scherr and McNeely 2007) that has a high potential for carbon sequestration, so they deserve to be included in carbon markets (Albrecht and Kandji 2003; Montagnini and Nair 2004; Schoeneberger 2009; Nair et al. 2009, 2010). The sale of carbon as a payment for ecosystem services would therefore constitute another benefit of agroforestry systems for tropical smallholders who will be motivated to maintain multipurpose trees on farms, sustaining food production and improving their livelihoods.

6 Acknowledgements

This work was supported by a Natural Science and Engineering Research Council, Canada (NSERC) Discovery Grant to Damase Khasa. We are grateful to the anonymous reviewers who made constructive comments which help improve the manuscript. We are also grateful to Professor Roger Leakey

(Vice Chairman of the International Tree Foundation, Edinburgh, United Kingdom), Dr. Marie Louise Avana (University of Dschang, Cameroon) and Dr. Zacharie Tchoundjeu (Former Regional Coordinator of ICRAF West and Central Africa) who have shared their wealth of experience in participatory tree domestication.

7 Where to look for further information

This chapter attempts to summarize the authors' contribution to the development of new tree crops in the tropics using agroforestry techniques. More information on tropical tree crops and agroforestry can be found in the ICRAF Agroforestree Database 4.0 (www.worldagroforestry.org/treedb/). Other relevant information on tree domestication in the tropics can be found in the papers published by Professor Roger Leakey, and in the ICRAF tree domestication programme.

8 References

Adebola, P. O. and Morakinyo, J. A. 2006. Evaluation of morpho-agronomic variability of wild and cultivated kola in south western Nigeria. *Genetic Resources and Crop Evolution* 53(4), 687–94. doi:10.1007/s10722-004-3558-1.

Akinnifesi, F. K., Sileshi, G., Ajay, O. C., Chirwa, P. W., Kwesiga, F. R. and Harawa, R. 2008. Contributions of agroforestry research and development to livelihood of smallholder farmers in Southern Africa: 2. Fruit, medicinal, fuelwood and fodder tree systems. *Agricultural Journal* 3(1), 76–88.

Albrecht, A. and Kandji, S. T. 2003. Carbon sequestration in tropical agroforestry systems. *Agriculture, Ecosystems and Environment* 99(1–3), 15–27. doi:10.1016/S0167-8809(03)00138-5.

Anderson, C. M., Aparicio, G. J., Atangana, A. R., Beaulieu, J., Bruford, M. W., Cain, F., Campos, T., Cariani, A., Carvalho, M. A., Chen, N., et al. 2010. Permanent genetic resources added to Molecular Ecology Resources database 1 December 2009 - 31 January 2010. *Molecular Ecology Resources* 10, 576–9.

Anegbeh, P. O., Usoro, C., Ukafor, V., Tchoundjeu, Z., Leakey, R. R. B. and Schreckenberg, K. 2003. Domestication of *Irvingia gabonensis*: 3. Phenotypic variation of fruits and kernels in Nigeria. *Agroforestry Systems* 58(3), 213–8. doi:10.1023/A:1026094021408.

Asaah, E. K., Tchoundjeu, Z., Wanduku, T. N. and Van Damme, P. 2010. Understanding structural roots system of 5-year-old African plum tree (*D. edulis* (G. Don) H. J. Lam) of seed and vegetative origins. *Trees* 24(5), 789–96. doi:10.1007/s00468-010-0449-2.

Asaah, E. K., Tchoundjeu, Z., Ngahane, W., Tsobeng, A., Kouodiekong, L., Jamnadass, R. and Simons, A. 2011. *Allanblackia floribunda*: a new oil tree crop for Africa: amenability to grafting. *New Forests* 41(3), 389–98. doi:10.1007/s11056-010-9230-z.

Atangana, A. R. and Khasa, D. P. 2008. Preliminary survey of clonal variation in rooting of *Allanblakia floribunda* leafy stem cuttings. *Canadian Journal of Forest Research* 38(1), 10–5. doi:10.1139/X07-151.

Atangana, A. R., Tchoundjeu, Z., Fondoun, J.-M., Asaah, E., Ndoumbe, M.. and Leakey, R. R. B. 2001. Domestication of *Irvingia gabonensis*: 1. Phenotypic variation in fruits and kernels in two populations from Cameroon. *Agroforestry Systems* 53(1), 55-64. doi: 10.1023/A:1012293700774.

Atangana, A. R., Ukafor, V., Anegbeh, P., Asaah, E., Tchoundjeu, Z., Fondoun, J.-M., Ndoumbe, M. and Leakey, R. R. B. 2002. Domestication of *Irvingia gabonensis*: 2. The selection of multiple traits for potential cultivars from Cameroon and Nigeria. *Agroforestry Systems* 55(3), 221-9. doi:10.1023/A:1020584823505.

Atangana, A. R., Tchoundjeu, Z., Asaah, E. K., Simons, A. J. and Khasa, D. P. 2006. Domestication of *Allanblackia floribunda*; amenability to vegetative propagation. *Forest Ecology and Management* 237(1-3), 246-51. doi:10.1016/j.foreco.2006.09.081.

Atangana, A. R., Beaulieu, J. and Khasa, D. P. 2010. Wild genetic diversity preservation in a small-sized first-generation breeding population of *Allanblakia floribunda* (Clusiaceae). *Tree Genetics and Genomes* 6(1), 127-36. doi:10.1007/s11295-009-0234-9.

Atangana, A. R., Mpeck-Nyemeck, M. L., Chang, S. X. and Khasa, D. P. 2011a. Auxin regulation and function: insights from studies on rooted leafy stem cuttings of tropical tree species. In: Keller, A. H. and Fallon, M. D. (Eds), *Auxins: Structure, Biosynthesis and Functions*, pp. 53-66.

Atangana, A. R., van der Vlis, E., Khasa, D. P., van Houten, D., Beaulieu, J. and Hendrickx, H. 2011b. Tree-to-tree variation in stearic and oleic acid content in seed fat from *Allanblackia floribunda* from wild stands: potential for tree breeding. *Food Chemistry* 126(4), 1579-85. doi:10.1016/j.foodchem.2010.12.023.

Atangana, A. R., Khasa, D. P., Chang, S. and Degrande, A. 2014a. *Tropical Agroforestry*. Springer+Business Media, Dordrecht, Heidelberg, London and New York, 380pp.

Atangana, A. R., Khasa, D. P., Chang, S. and Degrande, A. 2014b. Biological nitrogen fixation and mycorrhizal associations in agroforestry. In: Atangana, A. R., Khasa, D., Chang, S. and Degrande, A. (Eds), *Tropical Agroforestry*. Springer, pp. 173-202.

Atangana, A. R., Khasa, D., Chang, S. and Degrande, A. 2014c. Ecological interactions and productivity in agroforestry systems. In: Atangana A. R., Khasa, D., Chang, S. and Degrande, A. (Eds), *Tropical Agroforestry*. Springer, pp. 151-72.

Atangana, A. R., Khasa, D., Chang, S. and Degrande, A. 2014d. Phytoremediation in tropical agroforestry. In: Atangana A. R., Khasa, D., Chang, S. and Degrande, A. (Eds), *Tropical Agroforestry*. Springer, pp. 343-51.

Atangana, A. R., Khasa, D., Chang, S. and Degrande, A. 2014e. Agroforestry and the carbon market in the tropic. In: Atangana A. R., Khasa, D., Chang, S. and Degrande, A. (Eds), *Tropical Agroforestry*. Springer, pp. 353-65.

Avana, T. M. L., Tchoundjeu, Z., Bell, J. M., Vaillant, A. and Chevallier, M.-H. 2004. Diversité génétique du *Prunus africana*(Hook. F.) Kalkam au Cameroun. *Bois et Forêts des Tropiques* 282(4), 41-51.

Awono, A., Ndoye, O., Schreckenberg, K., Tabuna, H., Isseri, F. and Temple, L. 2002. Production and marketing of Safou (*Dacryodes edulis*) in Cameroon and internationally: market development issues. *Forests, Trees and Livelihoods* 12(1-2), 125-47. doi:10.1080/14728028.2002.9752416.

Bardhan, S., Jose, S., Biswas, S., Kabir, K. and Rogers, W. 2008. Homegarden agroforestry systems: an intermediary for biodiversity conservation in Bangladesh. *Agroforestry Systems* 85(1), 29-34. doi:10.1007/s10457-012-9515-7.

Boivin-Chabot, S., Margolis, H. A. and Weber, J. C. 2004. Variation in coppice-shoot growth among provenances of *Calycophyllum spruceanum* Benth. in the Amazon Basin. *Forest Ecology and Management* 198(1-3), 249-60, doi:10.1016/j.foreco.2004.04.009.

Clement, C. R. and Mora Urpi, J. E. 1987. Pejibaye palm (*Bactris gasipaes*, Arecaceae): multi-use potential for the lowlands humid tropics. *Economic Botany* 41(2), 302-11.

Cristoni, A., Di Pierro, F. and Bombardelli, E. 2000. Botanical derivatives for the prostate. *Fitoterapia* 71(1), S21-8. doi:10.1016/S0367-326X(00)00174-X.

Crockett, S. L. 2015. Allanblackia oil: phytochemistry and use as a functional food. *International Journal of Molecular Sciences* 16(9), 22333-49. doi:10.3390/ijms160922333.

Dawson, I., Were, J. and Lengkeek, A. 2000. Conservation of *Prunus africana*, an over-exploited African medicinal tree. *Forest Genetic Resources* 28, 27-33.

Degrande, A., Schreckenberg, K., Mbosso, C., Anegbeh, P., Okafor, V. and Kanmegne, J. 2006. 'Farmers' fruit tree growing strategies in the humid forest zone of Cameroon and Nigeria. *Agroforestry Systems* 67(2), 159-75. doi:10.1007/s10457-005-2649-0.

Degrande, A., Gyau, A., Foundjem-Tita, D. and Tollens, E. 2014. Improving smallholders' participation in tree product value chains: experiences from the Congo basin. *Forests, Trees Livelihoods* 23(1-2), 102-15. doi:10.1080/14728028.2014.886867.

Deschamps, S., Llaca, V., May, G. D. 2012. Genotyping-by-sequencing in plants. *Biology* 1(3), 460-83; doi:10.3390/biology1030460.

Dick, J. McP., Leakey, R. R. B., McBeath, C., Harvey, F., Smith, R. I. and Woods, C. 2004. Influence of nutrient application rate on growth and rooting potential of the West African hardwood *Triplochiton scleroxylon*. *Tree Physiology* 24(1), 35-44. doi:10.1093/treephys/24.1.35.

Djeugap, F. J., Bernier, L., Dostaler, D., Khasa, D. P., Afontem, D. and Nwaga, D. 2013. Opportunités et contraintes agroforestières de *Ricinodendron heudelotii* au Cameroun. *International Journal of Biological and Chemical Sciences* 7, 344-55.

Facheux, C., Gyau, A., Russell, D., Foundjem-Tita, D., Mbosso, C., Franzel, F. and Tchoundjeu, Z. 2012. Comparison of three modes of improving benefits to farmers within agroforestry product market chains in Cameroon. *African Journal of Agricultural Research* 7(15), 2336-43. doi:10.5897/AJAR11.577.

Farwig, N., Braun, C. and Böhning-Gaese, K. 2008. Human disturbance reduces genetic diversity of an endangered tropical tree, *Prunus africana* (Rosaceae). *Conservation Genetics* 9(2), 317-26. doi:10.1007/s10592-007-9343-x.

Galluzzi, G., Dufour, D., Thomas, E., van Zonneveld, M., Escobar Salamanca, A. F., Giraldo Toro, A., Rivera, A., Duque, H. S., Suárez Baron, H., Gallego, G., et al. 2015. An integrated hypothesis on the domestication of *Bactris gasipaes*. *PLOS ONE* 10(12), e0144644. doi:10.1371/journal.pone.0144644.

Gyau, A., Mbosso, C., Tchoundjeu, C., Foundjem-Tita, D., Asaah, E. and Franzel, S. 2011. Antecedents and effects of group sales on supply chain performance: the case of kola nut production and marketing in Cameroon. *Journal of Development and Agricultural Economics* 3(13), 621-6.

Hoad, S. P. and Leakey, R. R. B. 1994. Effects of light quality on gas exchange and dry matter partitioning in *Eucalyptus grandis* W. Hill ex Maiden. *Forest Ecology and Management* 70(1-3), 265-73. doi:10.1016/0378-1127(94)90092-2.

Hoad, S. P. and Leakey, R. R. B. 1996. Effects of pre-severance light quality on the vegetative propagation of *Eucalyptus grandis* W. Hill ex Maiden: cutting morphology, gas exchange and carbohydrate status during rooting. *Trees* 10(5), 317-24. doi:10.1007/BF02340778.

Hollingsworth, P. M., Dawson, I. K., Goodall-Copestake, W. P., Richardson, J. E., Weber, J. C., Sotelo-Montes, C. and Pennington, R. T. 2005. Do farmers reduce genetic diversity when they domesticate tropical trees? A case study from Amazonia. *Molecular Ecology* 14(2), 497-501. doi:10.1111/j.1365-294X.2005.02431.x.

Ingvarsson, P. K. and Dahlberg, H. 2018. The effects of clonal forestry on genetic diversity in wild and domesticated stands of forest trees. *Scandinavian Journal of Forest Research*, 1-10, doi:10.1080/02827581.2018.1469665.

Isik, F., Kumar, S., Martinez-Garcia, P. J., Iwata, H. and Yamamoto, T. 2015. Acceleration of forest and fruit tree domestication by genomic selection. *Advances in Botanical Research* 74, 93-124. doi:10.1016/bs.abr.2015.05.002.

Jamnadass, R., Dawson, I. K., Anegbeh, P., Asaah, E., Atangana, A., Cordeiro, N. J., Hendrickx, H., Henneh, S., Kadu, C. A., Kattah, C., et al. 2010. *Allanblackia*, a new tree crop in Africa for the global food industry: market development, smallholder cultivation and biodiversity management. *Forests, Trees and Livelihoods* 19(3), 251-68. doi:10.1080/14728028.2010.9752670.

Kadu, C. A. C., Schueler, S., Konrad, H., Muluvi, G. M., Eyog-Matig, O., Muchugi, A., Williams, V. L., Ramamonjisoa, L., Kapinga, C., Foahom, B., et al. 2011. Phylogeography of the Afromontane *Prunus africana* reveals a former migration corridor between East and West African highlands. *Molecular Ecology* 20(1), 165-78. doi:10.1111/j.1365-294X.2010.04931.x.

Kadu, C. A. C., Parich, A., Schueler, S., Konrad, H., Muluvi, G. M., Eyog-Matip, O., Muchugi, A., Williams, V. L., Ramamonjisoa, L., Kapinga, C., et al. 2012. Bioactive constituents in *Prunus africana*: geographical variation throughout Africa and associations with environmental and genetic parameters. *Phytochemistry* 83, 70-8. doi:10.1016/j.phytochem.2012.06.001.

Kimaro, A. A., Timmer, V. R., Mugasha, A. G., Chamshama, S. A. O. and Kimaro, D. A. 2007. Nutrient use efficiency and biomass production of tree species for rotational woodlot systems in semi-arid Morogoro, Tanzania. *Agroforestry Systems* 71(3), 175-84. doi:10.1007/s10457-007-9061-x.

Kumar, M. B. and Nair, P. K. R. 2004. The enigma of tropical homegardens. *Agroforestry Systems* 61, 135-52.

Kwesiga, F. R., Franzel, S., Place, F., Ohiri, D. and Simanza, C. P. 1999. *Sesbania sesban* improved fallows in eastern Zambia: their inception, development and farmer enthusiasm. *Agroforestry Systems* 47(1/3), 49-66. doi:10.1023/A:1006256323647.

Kwesiga, F., Akinnifesi, F. K., Mafongoya, P. L., McDermott, M. H. and Agumya, A. 2003. Agroforestry research and development in southern Africa during the 1990s: review and challenges ahead. *Agroforestry Systems* 59(3), 173-86. doi:10.1023/B:AGFO.0000005222.68054.38.

Labarta Chávarri, R. and Weber, J. C. 1998. Valorización económica de bienes tangibles de cinco species arbóreas agroforestales en la Cuenca Amazónica Peruana. *Revista Forestal Centroamericana* 23, 12-21.

Ladipo, D. O., Fondoun, J.-M.. and Ganga, N. 1996. Domestication of the bush mango (*Irvingia* spp.) some exploitable intraspecific variations in west and central Africa. In: Leakey, R. R. B., Temu, A. B., Melnyk, M. and Vantomme, P. (Eds), *Domestication and Commercialization of Non-Timber Forest Products in Agroforestry Systems*.

Non-Wood Forest Products, No. 9. Food and Agriculture Organization, Rome, Italy, pp. 193–205.

Larbi, A., Awojide, A. A., Adekunle, I. O., Ladipo, D. O. and Akinlade, J. A. 2000. Fodder production responses to pruning height and fodder quality of some trees and shrubs in a forest-savanna transition zone in southwestern Nigeria. *Agroforestry Systems* 48(2), 157–68. doi:10.1023/A:1006291413670.

Leakey, R. R. B. 1999. Potential for novel food products from agroforestry trees: a review. *Food Chemistry* 66(1), 1–14. doi:10.1016/S0308-8146(98)00072-7.

Leakey, R. R. B. 2001. Win-win land-use strategies for Africa: 2. Capturing economic and environmental benefits with multistrata agroforests. *The International Forestry Review* 3(1), 11–8.

Leakey, R. R. B. 2004. Physiology of vegetative reproduction. In: Burley, J., Evans, E. and Younqist, J. A. (Eds), *Encyclopaedia of Forest Sciences*. Academic Press, London, pp. 1655–68.

Leakey, R. R. B., Mesén, J. F., Tchoundjeu, Z., Longman, K. A., Dick, J. McP., Newton, A., Matin, A., Grace, J., Munro, R. C. and Muthoka, P. N. 1990. Low-technology techniques for the vegetative propagation of tropical trees. *Commonwealth Forestry Review* 69, 247–57.

Leakey, R. R. B., Fondoun, J. -M., Atangana, A. and Tchoundjeu, Z. 2000. Quantitative descriptors of variation in the fruits and seeds of *Irvingia gabonensis*. *Agroforestry Systems* 50(1), 47–58. doi:10.1023/A:1006434902691.

Leakey, R. R. B., Atangana, A. R., Kengni, E., Waruhiu, A. N., Usoro, C., Anegbeh, P. O. and Tchoundjeu, Z. 2002. Domestication of *Dacryodes edulis* in Western and Central Africa: characterization of genetic variation. *Forests, Trees and Livelihoods* 12(1–2), 57–71. doi:10.1080/14728028.2002.9752410.

Leakey, R. R. B., Tchoundjeu, Z., Smith, R. I., Munro, R. C., Fondoun, J.-M., Kengue, J., Anegbeh, P. O., Atangana, A. R., Waruhiu, A. N., Asaah, E., et al. 2004a. Evidence that subsistence farmers have domesticated indigenous fruits (*Dacryodes edulis* and *Irvingia gabonensis*) in Cameroon and Nigeria. *Agroforestry Systems* 60(2), 101–11. doi:10.1023/B: AGFO.0000013259.95628.22.

Leakey, R. R. B., Greenwell, P., Hall, M. N., Atangana, A. R., Usoro, C., Anegbeh, P. O., Fondoun, J.-M.. and Tchoundjeu, Z. 2004b. Domestication of *Irvingia gabonensis*: 4. Tree-to-tree variation in food thickening properties and in fat and protein contents of dika nut. *Food Chemistry* 90(3), 365–78. doi:10.1016/j.foodchem.2004.04.012.

Leakey, R. R. B., Tchoundjeu, Z., Schreckenberg, K., Shackleton, S. E. and Shackleton, C. M. 2005. Agroforestry tree products (AFTPs): targeting poverty reduction and enhanced livelihoods. *International Journal of Agricultural Sustainability* 3(1), 1–23. doi:10.1080/14735903.2005.9684741.

Leakey, R. R. B., Weber, J. C., Page, T., Cornelius, J. P., Akinnifesi, F. K., Roshetko, J. M., Tchoundjeu, Z. and Jamnadass, R. 2012. Tree domestication in agroforestry: progress in the second decade (2003–2012). In: *Agroforestry - The Future of Global land Use*. Springer, pp. 145–73. doi:10.1007/978-94-007-4676-3_11.

Mesén, F., Newton, A. C. and Leakey, R. R. B. 1997. The effects of propagation environment and foliar area on the rooting physiology of *Cordia alliodora* (Ruiz & Pavon) Oken cuttings. *Trees* 11(7), 404–11. doi:10.1007/PL00009683.

Montagnini, F. and Nair, P. K. R. 2004. Carbon sequestration: an underexploited environmental benefit of agroforestry systems. *Agroforestry Systems* 61, 281–95.

Mora-Urpi, J., Weber, J. C. and Clement, C. R. 1997. Peach palm - Bactris gasipaes Kunth. IPGRI, Rome, Italy. Available at: http://www.worldagroforestry.org.downloads/publications (Accessed on 16 December 2017).

Nadeem, M. A., Nawaz, M. A., Shahid, M. Q., Doğan, Y., Comertpay, G., Yıldız, M., Hatipoğlu, R., Ahmad, F., Alsaleh, A., Labhane, N., et al. 2018. DNA molecular markers in plant breeding: current status and recent advancements in genomic selection and genome editing. *Biotechnology and Biotechnological Equipment* 32(2), 261-85, doi:10.1080/13102818.2017.1400401.

Nair, P. K. R., Mohan Kumar, B. and Nair, V. D. 2009. Agroforestry as a strategy for carbon sequestration. *Journal of Plant Nutrition and Soil Science* 172(1), 10-23. doi:10.1002/jpln.200800030.

Nair, P. K. R., Nair, V. D., Mohan Kumar, B. and Showalter, J. M. 2010. Carbon sequestration in agroforestry systems. *Advances in Agronomy* 108, 237-307. doi:10.1016/S0065-2113(10)08005-3.

Ndoye, O., Pérez, M. R. and Eyebe, A. 1997. The markets of non-timber forest products in the humid forest zone of Cameroon. Rural Development Forestry Network Paper, 22c. ODI, London.

Nichols, J. D. and Carpentier, F. L. 2006. Interplanting *Inga edulis* yields nitrogen fixation benefits to *Terminalia Amazonia*. *Forest Ecology and Management* 233(2-3), 344-51. doi:10.1016/j.foreco.2006.05.031.

Ngo Mpeck, M.-L. and Atangana, A. R. 2007. Rooting of leafy stem cuttings of *Baillonella toxisperma*. *Forest Science* 53(5), 571-9.

Ngo Mpeck, M.-L., Asaah, E., Tchoundjeu, Z. and Atangana, A. R. 2003. Strategies for the domestication of *Ricinodendron heudelotii*: evaluation of variability in natural populations from Cameroon. *Journal of Food, Agriculture and Environment* 1, 257-62.

Ngo Mpeck, M.-L., Atangana, A. R. and Omgba, A. M. 2009. Temporal variation in rooting of *Annickia chloranta* leafy stem cuttings. *New Forests* 38(3), 273-83. doi:10.1007/s11056-009-9146-7.

Ngono, L. D. and Ndoye, O. 2004. Njansang and bush mango: Cameroonian seeds in national and international markets. In: Lopez, C. and Shanley, P. (Eds), *Riches of the Forests: for Health Life and Spirit in Africa*, pp. 21-4.

Nyadzi, G. I., Otsyina, R. M., Banzi, F. M., Bakengesa, S. S., Gama, B. M., Mbwambo, L. and Asenga, D. 2003. Rotational woodlot technology in northwestern Tanzania: tree species and crop performance. *Agroforestry Systems* 59(3), 253-63. doi:10.1023/B:AGFO.0000005226.62766.05.

Ofori, D. A., Peprah, T., Cobbinah, J. R., Atchwerebour, H. A., Osabutey, F., Tchoundjeu, Z., Simons, A. J. and Jamnadass, R. 2011. Germination requirements of *Allanblackia parviflora* seeds and early growth of seedlings. *New Forests* 41(3), 337-48. doi:10.1007/s11056-011-9252-1.

Omoti, U. and Okiy, A. 1987. Characteristics and composition of the pulp oil and cake of the African pear, *Dacryodes edulis* (G.Don) H. J. Lam (Burseraceae). *Journal of the Science of Food and Agriculture* 38, 67-72.

O'Neill, G. A., Dawson, I., Sotelo-Montes, C., Guarino, L., Guariguata, M., Current, D. and Weber, J. C. 2001. Strategies for genetic conservation of trees in the Peruvian Amazon basin. *Biodiversity and Conservation* 10(6), 837-50. doi:10.1023/A:1016644706237.

Parker, I. M., López, I., Petersen, J. J., Anaya, N., Cubilla-Rios, L. and Potter, D. 2010. Domestication syndrome in Caimito (*Chrysophyllum cainito* L.): fruit and seed characteristics. *Economic Botany* 64(2), 161-75. doi:10.1007/s12231-010-9121-4.

Pérez-de-Castro, A. M., Vilanova, S., Cañizares, J., Pascual, L., Blanca, J. M., Díez, M. J., Prohens, J. and Picó, B. 2012. Application of genomic tools in plant breeding. *Current Genomics* 13(3), 179–95. doi:10.2174/138920212800543084.

Potters, J. 1997. Farmers' knowledge and perceptions about tree use and management: the case of Trancayacu, a Peruvian community in the Amazon. MSc thesis dissertation. Wageningen Agricultural University, Wageningen, The Netherlands.

Resende Jr., M. F. R., Muñoz, P., Acosta, J. J., Peter, G. F., Davis, J. M., Grattapaglia, D., Resende, M. D. V. and Kirst, M. 2011. Accelerating the domestication of trees using genomic selection: accuracy of prediction models across ages and environments. *The New Phytologist* 193(3), 617–24. doi:10.1111/j.1469-8137.2011.03895.x.

Rochon, C., Margolis, H. A. and Weber, J. C. 2007. Genetic variation in growth of *Guazuma crinita* (Mart.) trees at an early age in the Peruvian Amazon. *Forest Ecology and Management* 243(2-3), 291–8. doi:10.1016/j.foreco.2007.03.025.

Russel, J. R., Weber, J. C., Booth, A. and Powell, W. 1999. Genetic variation of *Calycophyllum spruceanum* in the Peruvian Amazon Basin, revealed by amplified fragment length polymorphism (AFLP) analysis. *Molecular Ecology* 8(2), 1999–204.

Scherr, S. J. and McNeely, J. A. 2007. Ecoagriculture strategies for poverty reduction and biodiversity conservation. Paper presented at the *International Workshop on 'Reconciling Rural Poverty and Resource Conservation: Identifying Relationships and Remedies'*, 2-3 May 2003, Cornell University, Ithaca, NY.

Schoeneberger, M. M. 2009. Agroforestry: working trees for sequestering carbon on agricultural lands. *Agroforestry Systems* 75(1), 27–37. doi:10.1007/s10457-008-9123-8.

Shiembo, P. N., Newton, A. C. and Leakey, R. R. B. 1996. Vegetative propagation of *Irvingia gabonensis*, a West African fruit tree. *Forest Ecology and Management* 87(1–3), 185–92. doi:10.1016/S0378-1127(96)03781-4.

Shiembo, P. N., Newton, A. C. and Leakey, R. R. B. 1997. Vegetative propagation of *Ricinodendron heudelotii*, a West African fruit tree. *Journal of Tropical Forest Science* 9, 514–25.

Silva, E. M., Rogez, H. and Larondelle, Y. 2007. Optimization of extraction of phenolics from *Inga edulis* leaves using response surface methodology. *Separation and Purification Technology* 55(3), 381–7. doi:10.1016/j.seppur.2007.01.008.

Simons, A. J. and Leakey, R. R. B. 2004. Tree domestication in tropical agroforestry. *Agroforestry Systems* 61, 167–81.

Sotelo-Montes, C. and Weber, J. C. 1997. Priorización de species arbóreas para sistemas agroforestales en la selva baja Del Perú. *Agroforestería en las Américas* 4(14), 12–7.

Sotelo-Montes, C., Vidaurre, H. and Weber, J. C. 2003. Variation in stem-growth and branch-wood traits among provenances of *Calycophyllum spruceanum* Benth. from the Peruvian Amazon. *New Forests* 26(1), 1–16. doi:10.1023/A:1024425519480.

Sotelo-Montes, C., Beaulieu, J. and Hernández, R. E. 2007a. Genetic variation in wood shrinkage and its correlations with tree growth and wood density of *Calycophyllum spruceanum* at an early age in the Peruvian Amazon. *Canadian Journal of Forest Research* 37(5), 966–76. doi:10.1139/X06-288.

Sotelo-Montes, C., Beaulieu, J. and Hernández, R. E. 2007b. Genetic variation in wood and mechanical properties of *Calycophyllum spruceanum* at an early age in the Peruvian Amazon. *Wood and Fiber Science* 38, 578–90.

Sotelo-Montes, C., Beaulieu, J. and Hernández, R. E. 2007c. Radial variation in wood density and correlations with growth of *Calycophyllum spruceanum* at an early age in the Peruvian Amazon. *Wood and Fiber Science* 39(3), 377–87.

Sotelo-Montes, C., Hernández, R. E., Beaulieu, J. and Weber, J. C. 2008. Genetic variation in wood color and its correlations with tree growth and wood density of *Calycophyllum spruceanum* at an early age in the Peruvian Amazon. *New Forests* 35(1), 57–73. doi:10.1007/s11056-007-9060-9.

Stewart, K. M. 2003. The African cherry (*Prunus africana*): can lessons be learned from an over-exploited medicinal tree? *Journal of Ethnopharmacology* 89(1), 3–13. doi:10.1016/j.jep.2003.08.002.

Tchoundjeu, Z. and Atangana, A. 2006. *Ndjanssang: Ricinodendron Heudelotti*. University of Southampton Centre for Underutilised Crops, Southampton, UK.

Tchoundjeu, Z. and Leakey, R. R. B. 1996. Vegetative propagation of African mahogany: effects of auxin, node position, leaf area and cutting length. *New Forests* 11(2), 125–36. doi:10.1007/BF00033408.

Tchoundjeu, Z., Duguma, B., Foudoun, J.-M. and Kengue, J. 1998. Strategy for the domestication of indigenous fruit trees of West Africa: case of *Irvingia gabonensis* in southern Cameroon. In: *Proceedings of 5th Annual Conference of the Cameroon Committee of Biosciences*, 17–19 December 1997, Yaoundé, Cameroon.

Tchoundjeu, Z., Kengue, J. and Leakey, R. R. B. 2002. Domestication of *Dacryodes edulis*: state-of-the-art. *Forests, Trees and Livelihoods* 12(1-2), 3–13. doi:10.1080/14728028.2002.9752407.

Tchoundjeu, Z., Atangana, A. R. and Degrande, A. 2005. Indigenous methods for preserving bush mango kernels in Cameroon. *American Journal of Applied Sciences* 2(9), 1337–42. doi:10.3844/ajassp.2005.1337.1342.

Tchoundjeu, Z., Asaah, E. K., Anegbeh, P. O., Degrande, A., Mbile, P., Facheux, C., Tsobeng, A., Atangana, A. R., Ngo Mpeck, M.-L. and Simons, A. J. 2006. Putting participatory domestication into practice in West and Central Africa. *Forests, Trees and Livelihoods* 16(1), 53–69. doi:10.1080/14728028.2006.9752545.

Varshney, R. K. and Dubey, A. 2009. Novel genomic tools and modern genetic and breeding approaches for crop improvement. *Journal of Plant Biochemistry and Biotechnology* 18(2), 127–38. doi:10.1007/BF03263311.

WCFSD. 1999. Our forests-our future. Summary report. World Commission on Forests and Sustainable Development (WCFSD), Winnipeg, p. 38.

Weber, J. C. and Sotelo-Montes, C. S. 2005. Variation and correlations among stem growth and wood traits of *Calycophyllum spruceanum* Benth. from the Peruvian Amazon. *Silvae Genetica* 54(1-6), 31–41. doi:10.1515/sg-2005-0005.

Weber, J. C. and Sotelo-Montes, C. 2008. Geographic variation in tree growth and wood density of *Guauma crinita* Mart. in the Peruvian Amazon. *New Forests* 36(1), 29–52. doi:10.1007/s11056-007-9080-5.

Weber, J. C., Sotelo-Montes, C., Vidaurre, H., Dawson, I. K. and Simons, A. J. 2001. Participatory domestication of agroforestry trees: an example from the Peruvian Amazon. *Development in Practice* 11(4), 425–33. doi:10.1080/09614520120066710.

Weber, J. C., Sotelo-Montes, C., Ugarte, J. and Simons, T. 2009. Phenotypic selection of *Calycophyllum spruceanum* on farms in the Peruvian Amazon: evaluating a low-intensity selection strategy. *Silvae Genetica* 58(1-6), 172–9. doi:10.1515/sg-2009-0023.

Weber, J. C., Sotelo-Montes, C., Cornelius, J. and Ugarte, J. 2011. Genetic variation in tree growth, stem form and mortality of *Guazuma crinite* in slower- and faster-growing plantations in the Peruvian Amazon. *Silvae Genetica* 60(2), 70–8.

Wyant, J. 1996. Agroforestry - an ecological perspective. *Agroforestry Today* 8, 1.

Chapter 17

Tropical agroforestry and ecosystem services: trade-off analysis for better design strategies

Rolando Cerda, CATIE (Centro Agronómico Tropical de Investigación y Enseñanza), Costa Rica; Luis Orozco-Aguilar, The University of Melbourne, Australia; Norvin Sepúlveda, CATIE (Centro Agronómico Tropical de Investigación y Enseñanza), Costa Rica; Jenny Ordoñez and Geovana Carreño-Rocabado, CATIE (Centro Agronómico Tropical de Investigación y Enseñanza) and World Agroforestry Centre (ICRAF), Costa Rica; and Freddy Amores, Willan Caicedo, Samuel Oblitas and Eduardo Somarriba, CATIE (Centro Agronómico Tropical de Investigación y Enseñanza), Costa Rica

1 Introduction

Farms within a given agricultural landscape play a dual role in providing farmers livelihoods while delivering crucial ecosystem services over time (Tscharntke et al., 2011; Harvey et al., 2014; Laurance et al., 2014; Zomer et al., 2014; Vlek et al., 2017; Zheng et al., 2019). Agroforestry covers between 200 and 357 million ha in Latin America (Nair, 2012; Somarriba et al., 2012), thus providing various products for family consumption and contributing to a set of ecosystem services such as soil protection, carbon sequestration, biodiversity conservation, and regulation of water and nutrient cycles (Nair, 2012; Schroth and da Mota, 2014; Zomer et al., 2016). The most prominent agroforestry examples in the region are fallow-based annual cropping systems,

http://dx.doi.org/10.19103/AS.2018.0041.24

silvopastoral systems, and shaded tree-crop systems involving coffee (*Coffea* spp.) and cocoa (*Theobroma cacao* L.). Agroforestry represents an opportunity to achieve the millennium development goals while curbing tree cover loss, restoring degraded land, and reducing pressure on remaining forests (Perfecto and Vandermeer, 2010; Power, 2010; Kragt and Robertson, 2014; Rockstrom et al., 2017).

A large body of research has documented a wide list of provisioning and regulating services from tropical agroforestry systems (AFS). So far, carbon sequestration, nutrient cycling, maintenance of water quality/infiltration, and enhanced landscape connectivity are the most common and studied ecosystem services (Jose, 2009; Lin, 2010; Rousseau et al., 2012; Somarriba et al., 2013; Harvey et al., 2014; Lorenz and Lal, 2014). Likewise, coffee and cocoa AFS provide timber, building materials, fruits, firewood, fibers, honey, and medicinal plants (Lopez-Sampson and Somarriba, 2005; Dahlquist et al., 2007; Salgado-Mora et al., 2007; Deheuvels et al., 2012; Cerda et al., 2014; de Sousa et al., 2015). Silvopastoral systems including live fences and riparian forests aid landscape connectivity and provide shelter for migratory birds, while dispersed trees in pastures improve cattle behavior and milk and meat yield, cast shade that enriches grass nutritional value, and improve water quality (Esquivel et al., 2007; Montagnini et al., 2013; Benegas et al., 2014; Bussoni et al., 2017). In tropical Latin America, there are many types of AFS with varying design features, plant richness, structure, and management regimes, and therefore their performance in providing ecosystem services differs widely in space and over time (Rodriguez et al., 2006; Boreux et al., 2013; Rapidel et al., 2015; Rahman et al., 2016; Mortimer et al., 2017). Furthermore, AFS can exhibit trade-offs between ecosystem services that need to be balanced and reduced to simultaneously offer several benefits for farmers in a sustainable way (Nelson et al., 2009; Mckechnie et al., 2011; Howe et al., 2014; Vaast and Somarriba, 2014; Grossman, 2015; Rapidel et al., 2015; Cordingley et al., 2016; Mora et al., 2016; Rahman et al., 2016; Tamburini et al., 2016; Dade et al., 2018).

This chapter offers an overview of ecosystem services delivered by tropical AFS, and presents practical approaches for trade-off analysis between ecosystem services and plant biodiversity for better design (or redesign) and management of AFS. It is structured as follows: Section 2 highlights the main provisioning and regulating services provided by tropical AFS; Section 3 provides an overview of practical approaches to assess trade-offs; and Section 4 presents a case study of trade-off analysis, using data of a so-called Sentinel Landscape–El Tuma-La Dalia in Nicaragua, useful to derive recommendations to enhance the delivery of ecosystem services from five AFS (cocoa AFS, coffee AFS, pastures–silvopastoral, grain fields, and homegardens).

2 Overview on ecosystem services provided by tropical agroforestry

AFS deliver key provisioning and regulating ecosystem services at both farm and landscape scale (Kuyah et al., 2016). The quality and extent of those ecosystem services depend on the structural complexity, species richness, management intensity, and the spatial scale of the AFS on the land (Beer et al., 2003; Harvey et al., 2014; Johnson et al., 2014; Laurance et al., 2014; Kennedy et al., 2016). An overview of notable ecosystem services provided by the most common tropical AFS in different continents is given below.

2.1 Provisioning services for domestic consumption, savings, and sales

The economic contribution of AFS to family income varies according to tree species richness and density, and the proportion of basal area occupied by managed trees (Jagoret et al., 2014; Somarriba et al., 2014). Recent research argues that species richness increases the income of small farmers of eastern Amazonia and Ivory Coast (Gyau et al., 2014). Associated trees in coffee AFS, cocoa AFS, pastures, and grain fields offer a significant contribution of timber to annual income (use or sale of timber) and family savings (timber in standing, harvestable trees); yet timber harvest rates are sometimes unsustainable, thus compromising a steady supply of timber over time (Fischer and Vasseur, 2002; Bentley et al., 2004; Jaimez et al., 2013; Somarriba et al., 2014). In general terms and despite low timber market prices, the net value from timber sales could represent 11–49% of the total revenue from these AFS (de Sousa et al., 2015). This amount could be 58% higher if farmers were able to improve management practices (Schroth and da Mota, 2014; de Sousa et al., 2015). With adequate management, it is possible to obtain up to 3 m^3 ha^{-1} $year^{-1}$ of commercial timber from AFS (Orozco-Aguilar and Somarriba, 2005). In Nicaragua, consumption and sale of coffee, fruits, timber, and banana from coffee AFS equaled the economic returns from high-yielding full-sun coffee plantations (Pinoargote et al., 2017). Both in cocoa and coffee AFS, the production of timber, firewood, and fruits accounted for up to 30% of total annual income for small farmers in Central America (Cerda et al., 2014; Pinoargote et al., 2017), and similar important valuation of fruits from either simple or complex coffee AFS was reported in Guatemala and Peru (Rice, 2011). Apart from providing incomes and savings, fruits are a source of vitamins, minerals, and traditional medicines, particularly important in areas with low health care facilities/access (Johns et al., 2013). Fruit tree population in coffee AFS, cocoa AFS, and homegardens function as a complementary option for food and nutrition for rural families (Mahapatra and Panda, 2012); yet, the lack of market channels or added value facilities in rural areas lower the chances for farmers to fully realize the benefits of local trade (Pfund et al., 2011; Rice, 2011).

In silvopastoral systems, the benefits of managing trees in live fences or dispersed in the pastures contribute to an increase in the production from livestock and also offer products and goods. The shade provided by trees reduces air temperatures that dry out pastures, as well as improving animal well-being, which in tuns improve milk and meat yields (Montagnini et al., 2013). Cattle ranchers keep important tree species such as *Acacia pennatula* because it produces fence posts, forage, and firewood either for in-farm use or for sales (Somarriba, 2012).

A set of studies highlighted the multifunctional role of homegardens in providing income, food, timber, and ecosystem services while decreasing pressure on natural forests and hence saving and storing carbon (Albuquerque et al., 2005; Salinas-Peba and Parra-Tabla, 2007). For instance, homegardens in Uganda are important for household nutrition and health since they contain a variety of useful crops which constitute year-round supply of food (Whitney et al., 2017). In addition, Mendez et al. (2001) reported that Nicaraguan homegardens provide 40 different plant products for rural families, as well as the benefit of space for working and socializing. In Sri Lanka, homegardens cover nearly 15% of the land area, offering climate regulation, protection against natural hazards, enhanced land productivity, and increased crop diversity and food security for rural poor (Mattsson et al., 2013). Mohri et al. (2013) provided a comprehensive assessment of ecosystem services in homegardens in Indonesia, Sri Lanka, and Vietnam and found that traditional homegardens maintain high provision of ecosystem services.

2.2 Regulation of pests and diseases in coffee and cocoa AFS

Controlling pests and diseases is an ecosystem service that the biodiversity in agroecosystems can provide (Pumariño et al., 2015; Mortimer et al., 2017). In the case of AFS, associated plant biodiversity could have either positive or negative effects on the development of pests. On one hand, shade trees can hamper pathogens by favoring their natural predators, by forming barriers to the movement of infectious propagules or pests, and by changing the microclimate of the system (Beer et al., 1998; Staver et al., 2001; Charbonnier et al., 2017). Trees also contribute to improve soil fertility, thus plants can be vigorous and improve their resistance to pests and diseases (Avelino et al., 2012; Tully et al., 2012). On the other hand, high and dense shade could host or favor the development of pathogens including fungi and insects (López-Bravo et al., 2012; Soto-Pinto et al., 2012; Cerda et al., 2017a).

Tree species diversity, spatial arrangement, and phenology of associated trees in AFS may play a role in favoring or limiting biological control of pest and disease outbreaks (Schroth et al., 2000; Sonwa et al., 2002; Perfecto and Vandermeer, 2008). For instance, clustered rather than regular or random spatial disposition of shade trees favored the development of diseases in Costa Rican cocoa AFS and reduced the crop yield (Ngo Bieng et al., 2013). High diversity and size of shade trees in traditional AFS resulted in a considerable heterogeneity of plot light conditions causing that populations of mirids (*Sahlbergella singularis*) were patching distributed coinciding with light spots (Babin et al., 2010). Moreover, Opoku et al. (2002) reported that some species such as *Funtumia elastica, Sterculia tragacantha, Dracaena mannii*, and *Ricinodendron heudelotii* may host *Phytophthora megakarya*, a devastating cocoa pathogen. Whereas, in other studies desirable effects of plant biodiversity were documented. For instance, Bisseleua et al. (2013) documented that greater species richness of native shade tree species, with well-managed shade cover, was associated with a higher number of beneficial arthropods such as wasp nests and spider webs, which can predate pests. Another study reported that the number of parasitoid families was enhanced by tree species richness and density (Sperber et al., 2004). In coffee AFS, the presence of legume trees contributed to nitrogen-fixing and the presence of both resident and migratory birds, which are capable of predating pests (Narango et al., 2019). Apart from measurements of population, incidence, and severity of pests, several researchers suggest that primary yield losses (those caused in the current year) and secondary yield losses (the reduction of yielding capacity due to negative effects of pests in the previous year) must be assessed as indicators of regulation of pests and diseases (Wang et al., 2015; Cerda et al., 2017b; Andreotti et al., 2018; Dade et al., 2018). A complete review regarding supporting and regulating ecosystem services in cocoa and coffee AFS is presented by Martin et al. (2017), Mortimer et al. (2017), and Sauvadet et al. (2019).

2.3 Nutrient cycling and soil quality

Soil fertility and nutrient cycling is a well-researched feature of several AFS in both tropical and temperate regions (Schroth et al., 2001; Pinho et al., 2012; Lorenz and Lal, 2014; Tamburini et al., 2016). In the broadest sense, it is believed that nutrient cycling is more efficient in AFS than in monoculture because of complementary nutrients uptake, greater litter deposition, reduced nutrient leaching, and improved soil quality (van Noordwijk and Purnomosidhi, 1995; Nair, 2001; Defrenet et al., 2016). A complete review on nutrient stocks, nutrient cycling, and soil changes in cocoa AFS is presented elsewhere (Hartemink, 2005). Moreover, Seneviratne et al. (2010), Mattsson et al. (2013), Goenster et al. (2014), and Mohri et al. (2013) provided a compilation of the potential, limitations, and the effects of plant structure, species richness, spatial arrangement, and management practices on nutrient cycling in homegardens.

Important studies focused in cocoa AFS reflect the service that agroforestry can provide to improve soil quality. Isaac et al. (2007) reported important amounts of both biomass in cocoa trees and N, P, K uptake by cocoa trees under shade (43–80%, 22–45%, and 96–100% for N, P, and K; respectively). Large amounts of plant litter and roots of shade trees deposited in AFS play a vital role in nutrient cycling and soil quality (Monroe et al., 2016). Overall increase of soil organic matter and nutrients in mixed cocoa AFS was better than in traditional cocoa plantations (Arevalo-Gardini et al., 2015). Rousseau et al. (2012), by using soil bulk density, sum of bases, pH, and carbon, were able to separate cocoa AFS plots and forests into five distinct clusters along a low-to-high "soil quality" gradient; the study concluded that cocoa AFS can conserve soil and provide a high level of soil-related ecological services. Silva Moço et al. (2008) reported a significant litter layer on the soil of old cocoa AFS and cabrucas (rustic AFS) in Bahia, Brazil, which resulted in higher abundance and diversity of soil fauna. Finally, Pérez-Flores et al. (2017) reported that 35-year-old cocoa produced more litter than 55-year-old cocoa (2042 vs 1570 kg DM ha^{-1} $year^{-1}$), but both of them are capable of recovering the nutrients extracted by the cocoa main harvest, indicating that leaf fall constitute the most important nutrient recycling processes in AFS. Other authors stated that trees in AFS are useful but not completely enough to supply all nutrients needed. For instance, Wartenberga et al. (2017) reported that increasing tree diversity in cocoa AFS did not increase neither soil fertility parameters nor cocoa yields; the study concluded that tree diversification alone may not be an effective solution to mitigate soil degradation. The competition among roots in the system should be carefully managed, because it may limit phosphorus and potassium uptake (Abdulai et al., 2017).

The mitigation of nutrient leaching losses is another important ecosystem service, which is affected by small-scale management decisions. For instance, similar N and P losses at 100 cm of depth were reported between organically and conventionally managed AFS (119 kg N ha^{-1} $year^{-1}$ and 1.5 kg P ha^{-1} $year^{-1}$, respectively) (Tully et al., 2012); however, N losses declined linearly with increasing shade tree biomass, which is determined by farmers' management. Furthermore, Defrenet et al. (2016) reported no significant effect of trees on coffee fine root biomass (coffee root systems represented 49% of the total plant biomass in the 1.5 m soil depth), suggesting that coffee root systems are very competitive in the topsoil.

2.4 Carbon sequestration

The potential of AFS to store atmospheric carbon has been widely assessed in tropical regions (Albrecht and Kandji, 2003; Montagnini and Nair, 2004; Nair et al., 2009; Zomer et al., 2016). In general terms, AFS (at an average

age of 14 years) are capable of sequestering 7.2 Mg ha^{-1} year^{-1} (Kim et al., 2016). Carbon stored in aboveground biomass has been calculated in coffee and cocoa AFS, silvopastoral systems, homegardens, live fences, and riparian forests (Nair, 2012; Montagnini et al., 2013; Schroth et al., 2015; Jha et al., 2014). Nevertheless, belowground biomass and soil carbon has been evaluated to a less extent due to presumably high variability in soil properties and the difficulty of capturing mining full data from soil profiles. A review on soil organic carbon sequestration in AFS is presented elsewhere (Lorenz and Lal, 2014). In the below sections, we present relevant examples of stored carbon in AFS, mainly in aboveground biomass.

Somarriba et al. (2013) reported that total carbon stock of Central American cocoa AFS (soil + biomass + dead biomass) was 117 ± 47 Mg ha^{-1}, with 51 Mg ha^{-1} in the soil and 49 Mg ha^{-1} (42% of total carbon) in aboveground biomass (cocoa and canopy trees). The study concluded that stored carbon and sequestration rates in cocoa AFS are significant and similar to those in other cocoa-growing regions around the world. Similarly, successional AFS in Alto Beni, Bolivia, had a mean aboveground carbon stock of 143.7 Mg ha^{-1} (±5.3) with notable differences among plots depending on the tree diversity and shade canopy complexity (Jacobi et al., 2014). Jadan et al. (2015) reported that total carbon stocks (aboveground and belowground C) of diversified cocoa AFS in Ecuador were in the range of 85.2–141.4 Mg ha^{-1}. These registered values were equivalent to half the stored carbon in near-primary forests and three times higher than the full-sun cocoa. Rustic cocoa or "cabrucas" in Bahia, Brazil, store carbon estimated in the range of 87–187 Mg ha^{-1} (Schroth et al., 2015). In other studies, it was found that cabrucas with 55% shade level can sustain yield levels up to of 585 kg ha^{-1}, twice the current regional average of 285 kg ha^{-1}, and are compatible with an aboveground C stock in the large shade trees (>30 cm of trunk diameter) (Schroth et al., 2014).

In Central Sulawesi, Indonesia, cocoa AFS allow high carbon storage and sequestration without yield losses. This was demonstrated by Abou Rajab et al. (2016) who reported that total aboveground and belowground carbon in biomass was five times larger in multi-shade tree systems than in cacao monoculture. With 11, 18, and 57 Mg C ha^{-1} in monoculture, Cacao-*Gliricidia*, and multi-shade tree system, respectively. In another study, Kessler et al. (2012) reported that carbon stocks were about 60% of those of natural forest, suggesting that 1.6 ha of optimally managed agroforest can contribute to the conservation of carbon stocks as much as 1 ha of natural forests.

Bolivian and Peruvian cacao and coffee AFS also store significant amount of aboveground carbon when managed properly (Jacobi et al., 2014; Vebrova et al., 2014; Ehrenbergerová et al., 2016). For instance, polyculture-shade organic coffee plots store on average 184.2 (±16.5) Mg C ha^{-1} in living and dead biomass, and soil organic matter (Soto-Pinto et al., 2010). Moreover, carbon storage of coffee AFS with different dominant shading trees in Villa Rica, Peru, were in the range of 119.9-177.5 Mg ha^{-1}. Most carbon was fixed in the soil (57-99%), followed by aboveground tree biomass (23-32%), tree belowground biomass (8-9%), coffee shrubs (0.2-2%), and litter (1%). Noponen et al. (2013) used data from two long-term coffee experiments in Costa Rica and Nicaragua and reported that coffee AFS sometimes can function as sinks or sources of soil organic carbon depending on management and shade types. In Sumber-Jaya (Lampung, Indonesia), conversion of all full-sun coffee to AFS while protecting the remaining forest could increase average landscape level C stocks by 10 Mg ha^{-1} over 20 years, or 0.5 Mg C ha^{-1} year^{-1} (van Noordwijk et al., 2002).

Homegardens also have a substantial potential to store carbon (Nair, 2001; Kumar and Nair, 2004). Recent research conducted by Saha et al. (2009) and Kumar (2011) reported that the soil C stock was positively related to plant diversity of homegardens in Kerala, India. Overall, within 1 m profile, soil C content ranged from 101 to 127 Mg ha^{-1}. Similarly, Mattsson et al. (2013, 2015) assessed several homegardens in a dry zone area of Moneragala district, Sri Lanka, and reported a range of aboveground biomass stock of 13-56 Mg C ha^{-1} favored by tree density.

2.5 Regulation of the water cycle and water consumption

The potential of AFS to provide quality water and regulate water balance is one of the less researched topics in modern agroforestry in Latin America; however, important findings have been already documented. Trees could modify the water cycle by intercepting rain drops, increasing transpiration rates, and facilitating water infiltration, thus reducing water runoff (Beer et al., 2003; Cannavo et al., 2011; Mueller et al., 2012; Montagnini et al., 2013).

In cocoa and coffee AFS there are relevant studies that highlight hydrological services but also warn of possible drawbacks. Padovan et al. (2018) found that in a suboptimal site coffee plants in AFS were more efficient water users than in full-sun coffee because a greater proportion of rainfall was used by plant transpiration rather than being lost by soil evaporation. A study of coffee AFS in Southern Mexico reported that with 60-80% shade cover, daily soil evaporation rates decreased by 41% compared to the low shade site (10-30% shade), although high levels of soil moisture were maintained in the dry season with only 30-65% shade cover (Lin, 2010). Wu et al. (2016) suggested that interspecific competition for water can enhance water use efficiency of drought-avoidance plants (i.e., rubber trees, *Hevea brasiliensis*) and lead to complementarity between the root distributions of plants in rubber AFS (i.e., rubber with tea, coffee, and cocoa). Furthermore, Cannavo et al. (2011) documented that runoff was lower and infiltration rates higher in coffee AFS than in monoculture (5.4% and 8.4% of total rainfall, respectively). Whereas, Poppenborg and Hölscher (2009) documented that emergent trees in cocoa AFS tended to reduce rain water input, and produced clear spatial patterns in throughfall distribution since estimated rainfall interception loss was 4% and 16% of gross precipitation in cocoa only and cocoa plus tree plots, respectively. These results suggest that a reduced water availability may lead to reduced cocoa bean yields in times of water shortage. Abdulai et al. (2017) documented that soil water content in full sun is higher than in AFS (*Albizia ferruginea* and *Antiaris toxicaria*) suggesting that cocoa mortality in AFS is linked to strong competition for soil water; therefore, promoting cocoa AFS as drought resilient system especially under climate change needs to be carefully analyzed.

Trees also improve hydrologic services in pasture lands, a dominant land use type across Latin America. For instance, tree cover in pastures was negatively correlated with runoff (R = −0.71; P = 0.01) and positively linked to infiltration rates (R = 0.75; P = 0.01) (Rios et al., 2008). Benegas et al. (2014) concluded that trees in pastures reduce surface runoff at the highest observed rainfall intensities (>50 mm h^{-1}).

3 Practical approaches to assess trade-offs between different ecosystem services, and between ecosystem services and biodiversity

Balancing production and conservation requires a better understanding of the multiple trade-offs and synergies among ecosystem services to aid design interventions toward resilient agricultural landscapes (Howe et al., 2014; Cordingley et al., 2016; Rahman et al., 2016). There are several practical approaches to analyze and identify possible trade-offs between indicators of ecosystem services and plant biodiversity. Among quantitative analysis there are exploratory analysis of relationships (such as principal component analysis), correlations, linear and nonlinear relationships, calculation of magnitude of trade-offs, and analysis of services in relation to the allocation of trees in the system. Among analysis which combines qualitative and quantitative information, there are multi-criteria approaches (MCA), multi-criteria decision analysis (MCDA), analytic hierarchy process (AHP), and Bayesian approaches. In the below sections we provide a short overview of these practical approaches to assess trade-offs potentially applicable to AFS.

3.1 Approaches for quantitative analysis of trade-offs

3.1.1 Principal component analysis

Principal component analysis (PCA) among indicators of ecosystem services, and variables of structure and tree diversity of shade canopies, is useful in order to explore the overall relationships among shade canopy features and ecosystem services. The position of indicators and variables across different axis of variation of PCA can reveal whether relationships are positive or negative, which are most commonly assessed in the first and second axis, but could also be assessed in more dimensions. This latter then could be denoting possible trade-offs, which deserve more attention.

3.1.2 Analysis of Spearman's correlations

Spearman's correlation analysis between pairs of variables are useful to identify types and levels of relationships (Felipe-Lucia et al., 2014; Tamburini et al., 2016). In the case of AFS, such correlations can be estimated between pairs of indicators of ecosystem services, and between ecosystem services and shade canopy features. This analysis provides a first insight of whether positive or negative relationships between two variables are present. Negative coefficients denote trade-offs, the higher the coefficient the higher the trade-off.

3.1.3 Linear and nonlinear regression analysis

Linear and nonlinear regression analysis can be done between original values of pairs of indicators of ecosystem services, and between ecosystem services and shade canopy features (Rapidel et al., 2015; Cerda et al., 2017a). Significant negative relationships would be denoting trade-offs, with the R^2 as an indicator of the strength of the influence of one ecosystem service on the reduction of the other one.

3.1.4 Estimation of the magnitude of trade-offs

Estimation of the magnitude of trade-offs are indicators of ecosystem services that can be standardized to values between 0 and 1 to analyze relationships between pairs of indicators, and then calculate the root mean square deviation (RMSD) as representatives of the magnitudes of the trade-offs (Bradford and D'Amato, 2012; Lu et al., 2014). The higher the RMSD, the higher the trade-off. This approach provides a promising way to visualize trade-offs. Paired ecosystem services plotted closer to the 1:1 line represent the ideal relationship (no trade-offs). Paired ecosystem services plotted far from the 1:1 line indicates otherwise (greater trade-offs) within the agroecosystem (Fig. 1). The standardization (0–1) for each observation of indicators of ecosystem services is done as follows (Lu et al., 2014):

$$ES_{std} = \left(ES_{obs} - ES_{\min}/ES_{\max} - ES_{\min}\right)$$

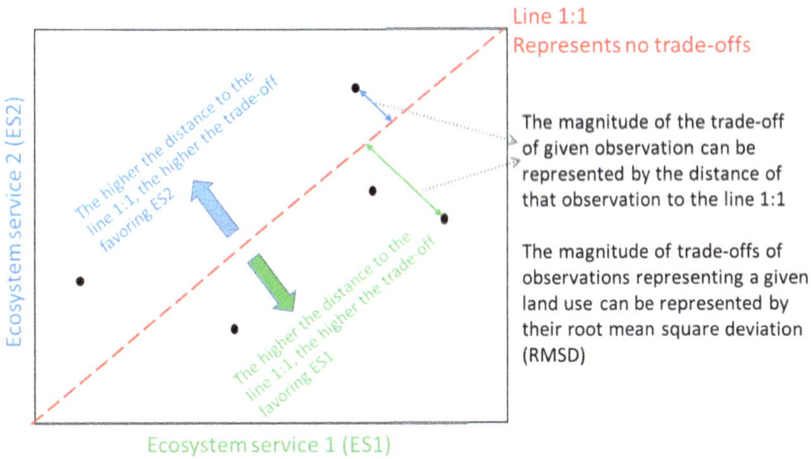

Figure 1 Diagram of trade-off analysis between two indicators of ecosystem services by standardization of ecosystem services values and use of root mean square deviation (RMSD). Source: modified from Bradford and D'Amato (2012) and Lu et al. (2014).

where ES_{std} is the standardized value of any ES, ES_{obs} is an observed value, and ES_{min} and ES_{max} are the minimum and maximum observed values.

The RMSD is calculated as follows (Lu et al., 2014):

$$RMSD = \sqrt{\frac{1}{n-1}\sum_{i=1}^{n}\left(ES_i - \hat{ES}\right)^2}$$

where ES_i is the standardized value of ES_i and $\hat{ES}$ is the expected value of the i number of ESs. RMSD quantifies the average difference between an individual ES_{std} and the mean ES_{std}. In a two-dimensional plot, $\hat{ES}$ is on the 1:1 ESs (Fig. 1).

3.1.5 The competitive allocation of stand basal area

AFS are represented by two, interconnected tree stands: plants of the main crop and shade canopy. Stand basal area (usually expressed in $m^2\ ha^{-1}$) is used as an index of stand density. Basal area is a composite variable that takes into consideration both population density (trees ha^{-1}) and tree size (stem cross-sectional area, a function of tree stem diameter). Three levels of analysis are required. First, maximum, acceptable total basal area may vary according to site conditions (e.g., may differ between humid and dry areas, or between lowlands and highlands). Second, the allocation of basal area to the main crop or to the shade canopy varies according to the AFS typology. For instance, more basal area may be allocated to low-input system under mixed canopies than to high-input system under simpler shade canopies. Third, shade canopy basal area may be competitively allocated to various shade canopy components in line with farmers' production goals. For instance, farmers may allocate more basal area to timber trees at the expense of reduced allocation to fruit trees (Fig. 2). The competitive allocation of tree basal area model herein proposed requires determining the allocation of basal area to all components to optimally achieve multiple farmers' objectives (Somarriba et al., 2018).

3.1.6 Identification of promising individual systems with the least trade-offs

Apart from identifying trade-offs, it is worth to identify the most promising individual systems (those with least trade-offs and achieving simultaneously desired values of ecosystem services), which could serve as successful models to follow (Cerda et al., 2017a). When plotting in two dimensions, for instance, ecosystem service 1 vs. ecosystem service 2, or ecosystem service vs. biodiversity indicator, or ecosystem service vs. tree basal area of the system, it is also possible to define desired values for the given indicators of ecosystem services, and identify which systems surpass such desired value and at the same time reach good value of the other indicator (Fig. 3).

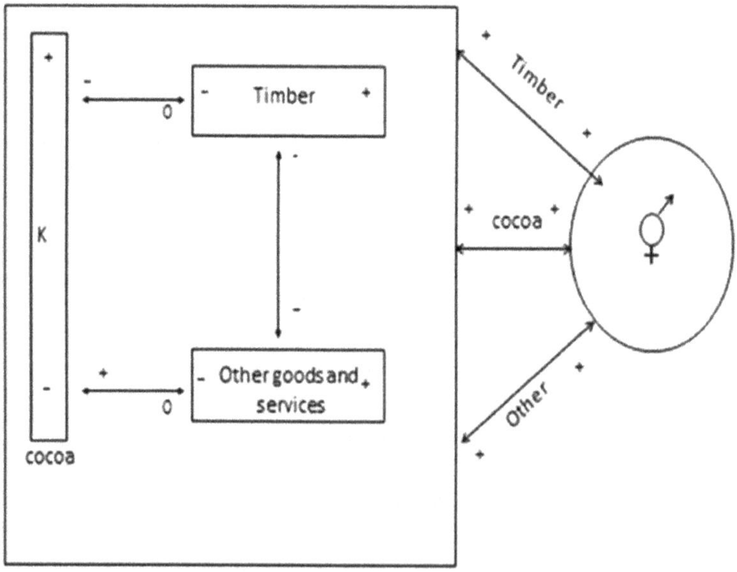

Figure 2 Analysis of competitive allocation of basal areas between tree components of an agroforestry system that may result in trade-offs. Source: Somarriba et al. (2018).

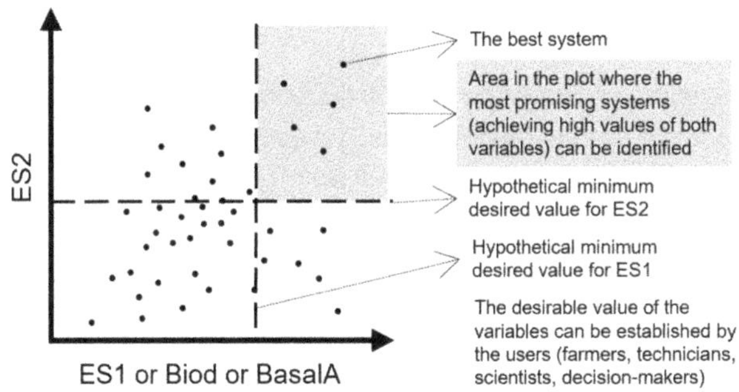

Figure 3 Approach to identify the most promising systems. ES1: ecosystem service 1; ES2: ecosystem service 2; Biod: biodiversity in the system; BasalA: basal area of trees in the system. Source: modified from Rapidel et al. (2015) and Cerda et al. (2017a).

3.2 Approaches combining qualitative and quantitative data for trade-offs analysis

3.2.1 Multi-criteria approaches and multi-criteria decision analysis

These approaches can be used to analyze several types of ecosystem services at the same time. They offer the advantage to compare qualitative information such as data from cultural services with quantitative data of other services or with economic indicators, and they also permit to deal with incomplete or uncertain information (Koschke et al., 2012). During the MCA, different stakeholders are gathered to rank the ecosystem services of interest and/or to assign weight to each service according to its importance to reach a specific goal, and then compare the contributions of the services to such goal, thus trade-offs between services can be evaluated (Sijtsma et al., 2012; Domptail et al., 2013). The MCDA follows similar steps like MCA, and the last one consist of developing of possible scenarios to support important decision-making processes, which include the valuation of possible trade-offs and alternatives to balance them (Saarikoski et al., 2015).

3.2.2 The Analytic Hierarchy Process

The AHP is originally defined as an effective tool to help decision-makers to identify priorities and based on them make the best decisions, by analyzing pairwise comparisons between benefits pursued and then synthesizing results (Saaty, 1980). In the context of ecosystem services, AHP can help to identify trade-offs among several ecosystem services, and then identify priorities to better balance negative relationships. Similar to MCA and MCDA, in the AHP a weight to the ecosystem services or to the factors that affect them can be assigned, which can also help to prioritize the most important ones (Marre et al., 2016).

3.2.3 Bayesian approaches

There are Bayesian network models that incorporate qualitative, categorical, and/or quantitative numerical data to model ecosystem services. These models are especially useful to generate information on ecosystem services when there are uncertainties on their performance, lack of quantitative data, or when the services are intangible or difficult to measure (Landuyt et al., 2013; Xue et al., 2017). They combine the knowledge of experts and stakeholders in the analysis of benefits and decision-making process in participatory manners. In the context of ecosystem services, recent studies utilized Bayesian network models to assess trade-offs among water regulation services and uses of water (including agricultural irrigation), with three steps: identification and assessment of the ecosystem services of interest; analysis of trade-offs and synergies among ecosystem services and their use; and the use of Bayesian models to assess economic compensations needed when a given action should be modified in order to avoid negative effects on ecosystem services (Xue et al., 2016, 2017). Such steps and models could be used also in AFS in order to identify the need of compensations if the balance of trade-offs require a reduction of a given benefit for the farmer.

4 Case study: application of trade-off analysis to derive better design strategies for agroforestry systems

4.1 Study site: The Sentinel Landscape-El Tuma-La Dalia, Nicaragua

El Tuma-La Dalia is a municipality of the department of Matagalpa, located in the north center of Nicaragua. El Tuma-La Dalia is part of the Sentinel Landscape Nicaragua-Honduras, where research and initiatives are promoted in order to achieve better management and/or reintroduction of trees at farm and landscape level to increase ecosystem restoration. The landscape of El Tuma-La Dalia is mountainous, where 60% of the land is in slopes and 60% of the landscape is fragmented with low forest cover. More than 50% of the land is cultivated in farms whose sizes range from ~0.5 to 300 ha, and most commonly smallholders have individual private ownership and good accessibility. The climate is subtropical, semi-humid forest, with annual precipitation between 2000 and 2500 mm. The temperature ranges between 22°C and 24°C. Land use includes mainly cattle ranching, coffee plantations, and staple grains, although homegardens and cocoa plantations are also important. All of these land uses have tree components; therefore, we consider all of them as AFS in the Sentinel Landscape. They are the main source of income for families; however, crop yields are low, due to the impact of diseases (e.g., coffee rust), pests (affecting staple grains), and the alternation of drought and excessive rainfall. Agricultural practices and climate change have taken their toll on the environment: the flow of water sources has decreased and are contaminated, and the soils are less fertile. Soils are clayey and loamy clay, with erosion moderate and mostly with agricultural coverage. Further description on this landscape is also given by Somarriba et al. (2017), and in this link: https://www1.cifor.org/sentinel-landscapes/sentinel-landscapes-sites/regional-sentinel-landscapes/nicaragua-honduras-sentinel-landscape.html.

4.2 Data collection

We used data set from an ongoing project "The CGIAR Research Program on Forests, Trees and Agroforestry (FTA)" that assessed tree cover in AFS of cocoa, coffee, pastures, grains, and homegardens in the Sentinel Landscape. This project has generated information regarding provisioning and regulating services from each AFS. We used yield data of all products produced by the system (not only products from the main crop) to calculate cash flow, value of domestic consumption, and family benefit as indicators of provisioning services. We used data of carbon in the aboveground biomass in the shade canopy as indicator of a regulating service. In addition, we selected data about plant diversity (species richness, plant density, basal area) to describe tree canopy structure and reflect the plant biodiversity in the AFS. We used these data from 31 cocoa AFS, 41 coffee AFS, 63 grain fields, 61 pastures (silvopastoral), and 75 homegardens, for a total of 271 plots for this case study.

4.3 Shade canopy features and indicators of ecosystem services

The structure and plant diversity data were obtained through field inventories. Trees recorded in each AFS were identified at species level, and basic measurements such as trunk diameter, tree height, and local use were taken. These field inputs allowed us to calculate tree density and tree basal area, tree species richness, and the reciprocal of Simpson index ($1/D$) as representative of plant diversity.

We calculated economic indicators (cash flow, value of domestic consumption, and family benefit) as representatives of provisioning services, to reflect the overall contribution of the AFS to family's well-being (Cerda et al., 2014; Pinoargote et al., 2017). Economic indicators comprise incomes provided by the main crops as well as other products such as bananas, other fruits, and timber. Farm gate prices per product, cultural practices, applied inputs, and production costs (labor and inputs) were obtained through farmers' interview. Calculated economic indicators are described below:

- $GI = AS \times MP$
- $CF = GI - CC$

- $VDC = ADC \times MP$
- $FB = CF + VDC$

where GI = gross income from sale of agroforestry products; AS = amount of agroforestry products for sale; MP = market price; CC = cash costs; CF = cash flow; FB = family benefit; VDC = value of domestic consumption; ADC = amount of agroforestry products for domestic consumption. All results were expressed in United States dollars as USD.

Finally, we quantified aboveground carbon stock in the trees of the shade canopy (not including carbon in the main crops) of the AFS as indicator of regulating service. Tree inventory inputs were computed and combined with published allometric equations to calculate biomass and then stored carbon for coffee shrubs, cocoa plants, and shade trees; dispersed trees in pastures; associated trees in grain fields and tree population in homegardens (Somarriba et al., 2013; Andrade et al., 2016; Pinoargote et al., 2017).

4.4 Data analysis

Since we had availability of quantitative data of indicators of ecosystem services and shade canopy features, we applied the qualitative approaches described in Section 3, which permits to identify possible trade-offs between paired ecosystem services, and between ecosystem services and plant diversity; assess relationships between ecosystem services and shade canopy features; and estimate the degree of trade-offs. That way, we were interested in obtaining quantitative results to support recommendations to balance possible trade-offs in each AFS being assessed. Therefore, we applied (1) PCA, performed with data standardized between −1 and 1, and the analysis worked with a correlation matrix; (2) Spearman's correlations and linear relationships, performed with original data; and (3) magnitude of trade-offs, performed with the values of indicators of ecosystem services standardized to a scale 0–1 and the RMSD for each type of AFS with formulas described in the Subsection 3.1.4. Based on the evidence of trade-offs with the set of techniques, we derived recommendations to improve the design of AFS for better provisioning of ecosystem services.

4.5 Results

4.5.1 Structure and plant diversity in the AFS of the Sentinel Landscape

The five AFS within the Sentinel Landscape were contrasting in terms of cultivated area, tree diversity, densities, and basal areas. Pastures with trees was the land use with the most area per system followed by grain fields, cocoa and coffee AFS with similar areas, and homegardens with the lowest area. Yet, homegardens were the most complex systems as expressed by high densities and tree basal areas (especially fruit trees), followed by coffee and cocoa AFS. Grain fields and pastures had similar characteristics in terms of tree density, basal area, and species diversity (Table 1 and Fig. 4).

4.5.2 Ecosystem services provided by agroecosystems

Important differences were found between economic indicators (representatives of provisioning services) and carbon stocks (representative of regulating services) among AFS. Homegardens contributed the most to the overall family benefit, but thanks only to their high domestic consumption. Moreover, cash flow of homegardens was negative on average, which means that most production is consumed on farm. Both coffee and cocoa AFS registered higher contributions to cash flow than the other systems, and higher value of standing timber, which represents an important family provisioning good for future needs. Grain fields, such as homegardens, contributed more to the value of domestic consumption than to the cash flow. Finally, pastures showed the lowest indicators of provisioning in general. Carbon stock was higher in coffee AFS, followed by cocoa AFS, homegardens, and least in pastures and grain fields. (Table 2, Fig. 4).

4.6 Trade-offs between ecosystem services and plant diversity

4.6.1 Principal component analysis

This analysis permitted to visualize, in an exploratory way, both positive and negative relationships (possible trade-offs). Among ecosystem services most relationships were positive or neutral. There were only negative relationships of cash flow and value of standing timber with the value of domestic consumption. Carbon in the shade canopy showed no negative relationships with economic indicators. Regarding other possible trade-offs, plant diversity (1/*D*) showed a negative relationship with value of domestic consumption, and the area of the systems were negatively related to family benefit, value of domestic consumption, and carbon stocks in the shade canopy (Fig. 4).

4.6.2 Spearman's correlations and linear regressions

We found significant negative relationships denoting trade-offs between the value of domestic consumption with cash flow, value of standing timber, and plant biodiversity. Given that coefficients of correlation were low in general, such negative relationships would be considered as slight trade-offs. Based

Table 1 Structural features of five dominant agroforestry systems in the Sentinel Landscape, Nicaragua

	Cocoa AFS		Coffee AFS		Grain fields		Pastures		Homegardens	
Characteristics (Unit)	Mean	SD	Mean	SD	Mean	SD	Mean	SD	Mean	SD
Area of the system (ha)	1.9	2.5	1.8	1.8	2.0	2.2	8.3	13.5	0.2	0.2
Density of service trees (*n*/ha)	33.2	28.9	43.6	24.8	8.5	9.5	14.8	11.9	41.7	41.7
Density of post trees (*n*/ha)	1.5	3.3	4.5	7.8	0.8	2.0	1.0	2.2	1.8	4.6
Density of fruit trees (*n*/ha)	35.2	49.0	51.5	55.2	4.8	7.3	6.7	9.8	109.6	88.2
Density of timber trees (*n*/ha)	38.7	36.1	55.1	45.1	19.6	19.2	22.9	19.2	26.7	30.0
Total density of shade canopy (*n*/ha)	108.6	85.7	154.7	86.4	33.6	30.1	45.4	29.7	180.1	111.8
Basal area of service trees (m^2/ha)	1.7	1.2	2.0	0.9	0.4	0.4	0.7	0.5	1.7	2.2
Basal area of post trees (m^2/ha)	0.2	0.4	1.1	3.3	0.1	0.3	0.1	0.3	0.1	0.6
Basal area of fruit trees (m^2/ha)	1.0	1.6	1.3	1.6	0.2	0.3	0.2	0.3	3.2	3.2
Basal area of timber trees (m^2/ha)	2.7	2.0	3.8	2.8	1.2	1.7	1.2	0.9	1.4	2.1
Total basal area of shade canopy (m^2/ha)	5.5	3.4	8.2	4.1	1.9	2.1	2.1	1.4	6.5	4.7
Total species richness (number)	26.4	15.5	31.5	12.2	12.4	6.8	29.6	20.9	12.9	8.5
1/*D*	9.3	5.5	10.1	4.7	6.1	3.5	8.1	4.9	7.1	4.0

SD: standard deviation; *n*: individuals; AFS: agroforestry systems; 1/*D*: reciprocal of Simpson index.

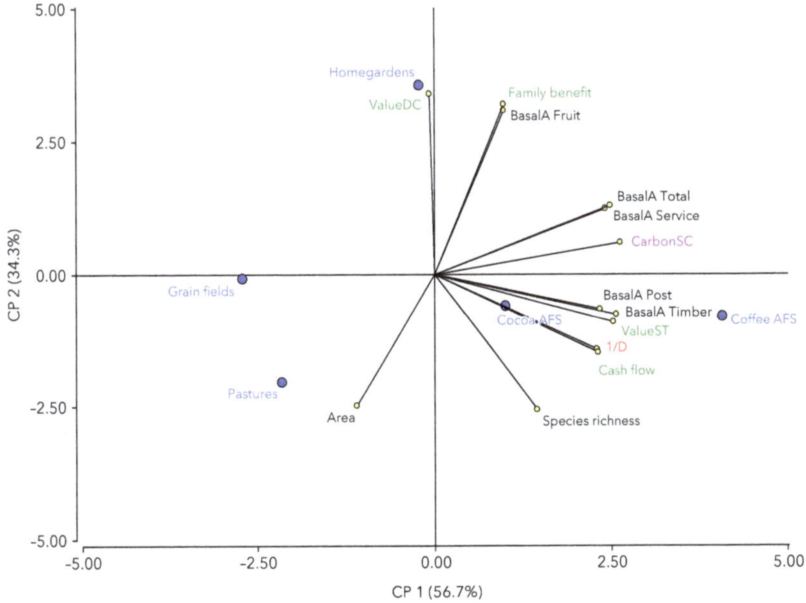

Figure 4 Principal component analysis to explore the relationships among structural features of tree cover and ecosystem services in five agroforestry systems in the Sentinel Landscape, Nicaragua. Agroforestry systems are represented by blue circles. Indicators of provisioning services are represented in green; ValueDC: value of domestic consumption; ValueST: value of standing timber. Carbon in the shade canopy is the indicator of a regulating service and is showed in purple. 1/D is the reciprocal of Simpson index as representative of plant biodiversity in the shade canopies and is displayed in red. Structural features of tree cover are represented in black. BasalA: basal area of trees; Fruit: fruit trees; Service: service trees; Timber: timber trees; Post: trees useful for posts.

on linear regression analysis with paired ecosystem services (similar to Table 3), the same significant negative relationships ($P < 0.05$) were found but with even lower R^2 than the coefficients of Spearman, supporting that such trade-offs would be slight. As well as in the PCA, no negative relationships were found between economic indicators and carbon in the shade canopy (Table 3).

4.6.3 Magnitude of trade-offs

The analysis of relationships using standardized indicators and RMSD revealed trade-offs between economic indicators and plant diversity, and carbon stocks. The trade-offs favoring one indicator over another one, and the magnitude of such trade-offs, were different depending on the type of AFS as it is explained subsequently.

Table 2 Ecosystem services provided by five dominant agroforestry systems in the Sentinel Landscape, Nicaragua

Indicators of ecosystem services	Cocoa AFS		Coffee AFS		Grain fields		Pastures		Homegardens	
	Mean	SD	Mean	SD	Mean	SD	Mean	SD	Mean	SD
Cash flow (USD/ha)	426	776	796	953	126	567	226	400	−753	1521
Value of domestic consumption (USD/ha)	529	753	555	663	745	551	223	245	2978	3042
Family benefit (USD/ha)	955	1035	1351	1123	871	708	449	479	2225	2564
Value of standing timber (USD/ha)	4857	3833	6564	5189	2001	2723	1896	1688	2163	3294
Carbon in the shade canopy (Mg/ha)	20	16	36	32	6	5	9	5	22	17

SD: standard deviation; AFS: agroforestry systems.

Table 3 Spearman's correlations between economic indicators of ecosystem services of five dominant agroecosystems of the Sentinel Landscape, Nicaragua (coefficients/*P*-value)

	ValueDC	Cash flow	Family benefit	ValueST	CarbonSC	1/D
ValueDC	1	*5.20E-06*[a]	0	*0.04*[a]	0.14	*3.60E-05*[a]
Cash flow	*−0.27*[a]	1	1.40E-07	5.60E-09	0.01	1.30E-03
Family benefit	0.74	0.31	1	0.35	5.60E-05	0.12
ValueST	*−0.12*[a]	0.34	0.06	1	0	1.80E-09
CarbonSC	0.09	0.16	0.24	0.52	1	6.30E-11
1/*D*	*−0.25*[a]	0.19	−0.1	0.35	0.38	1

[a] significant negative relationships denoting trade-offs.
1/*D*: reciprocal of Simpson index.

Most AFS favored plant diversity over economic indicators, especially in the case of the relationship with cash flow, denoting the existence of trade-offs. The AFS with higher magnitude of trade-offs favoring plant diversity were coffee AFS, cocoa AFS, and pastures. Homegardens showed lower trade-offs and was the only system exhibiting more observations closer to the 1:1 line (the line of no trade-offs) in the relationships of plant diversity with value of domestic consumption and family benefit (Fig. 5).

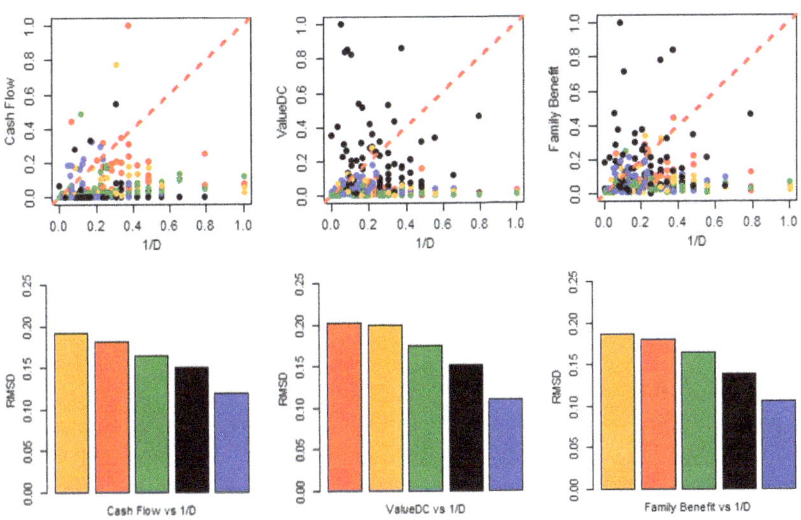

Figure 5 Analysis of trade-offs between economic indicators and plant biodiversity (reciprocal of Simpson index) in the shade canopy of five agroforestry systems in the Sentinel Landscape, Nicaragua. Red line is the 1:1 line, which represents no trade-offs. Points and bars: black: homegardens; blue: grain fields; green: pastures; red: coffee AFS; orange: cocoa AFS. Magnitude of trade-offs are represented by the root mean square deviation (RMSD): the higher the RMSD, the greater the trade-off.

Trade-offs between economic indicators and carbon stocks were identified, and they were different depending on the type of AFS. Most coffee AFS, cocoa AFS, homegardens, and pastures favored carbon stocks over cash flow, while most of grain fields on the contrary, favored cash flow instead. The relationships of the other economic indicators with carbon stocks showed a similar trend, except for homegardens. Approximately half of such systems favored more the value of domestic consumption and family benefit over carbon stocks. In general, the magnitude of trade-offs was higher in coffee AFS, and then in cocoa AFS and homegardens (Fig. 6).

There were also trade-offs between carbons stocks and plant biodiversity. In general, most observations of the types of AFS favored more plant diversity over carbon stocks, except homegardens. Approximately, half of homegardens favored more carbon stocks and the other half plant diversity. Only few observations of coffee AFS clearly favored more the carbon stocks over plant diversity. The magnitude of trade-offs was higher for coffee AFS, pastures, and cocoa AFS (Fig. 7). But in general, the magnitudes of trade-offs of this relationship were lower than in the relationships of economic indicators with carbon and plant biodiversity (Figs. 5 and 6).

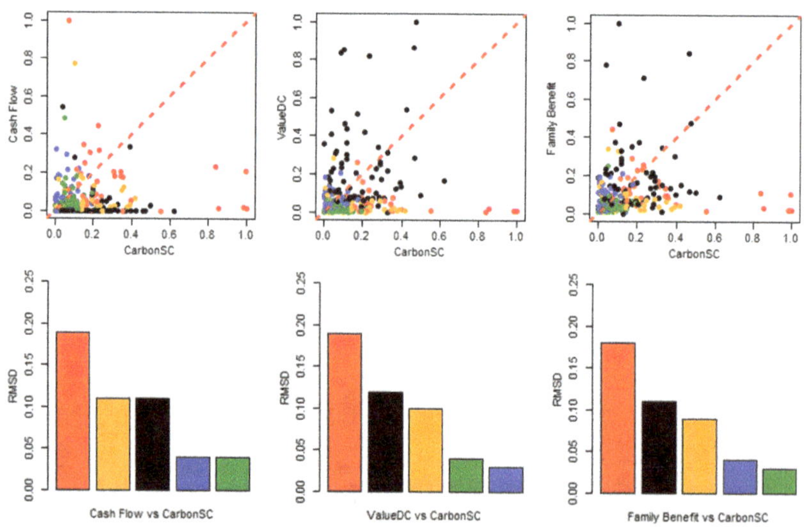

Figure 6 Analysis of trade-offs between economic indicators and carbon in the shade canopy of five agroforestry systems in the Sentinel Landscape, Nicaragua. Red line is the 1:1 line, which represents no trade-offs. Points and bars: black: homegardens; blue: grain fields; green: pastures; red: coffee AFS; orange: cocoa AFS. Magnitude of trade-offs are represented by the root mean square deviation (RMSD): the higher the RMSD, the greater the trade-off. SC: shade canopy.

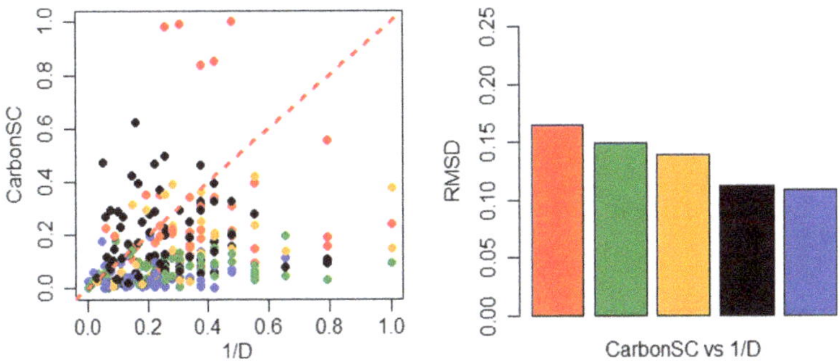

Figure 7 Analysis of trade-offs between carbon and plant diversity (reciprocal of Simpson index) in the shade canopy of five agroforestry systems in the Sentinel Landscape, Nicaragua. Red line is the 1:1 line, which represents no trade-offs. Points and bars: black: homegardens; blue: grain fields; green: pastures; red: coffee AFS; orange: cocoa AFS. Magnitude of trade-offs are represented by the root mean square deviation (RMSD): the higher the RMSD, the greater the trade-off. SC: shade canopy.

4.6.4 Promising individual AFS

By applying the approach of identifying individual promising systems (Fig. 3), and considering that individual systems to be cataloged as promising should be located at least in coordinates ≥0.5 of both paired indicators in Figs. 5–7, practically none of the 271 individual systems evaluated would be cataloged as promising AFS to provide the provisioning and regulating services (and plant diversity) simultaneously. This means that all systems either reported very low values of indicators of ecosystem services or reported only good values (≥0.5) only for one service compared to the other.

4.6.5 Relationships between ecosystem services and shade canopy features

Based on Spearman's correlation analysis, a few significant and negative relationships between provisioning services and shade canopy features were found, denoting trade-offs. It is possible to highlight the negative relationship between cash flow and densities and basal area of fruit trees. There were also negative relationships between value of domestic consumption and family benefit with post trees, but this does not represent a considerable issue given the low density of such trees. Contrary to what was expected, the relationships between value of domestic consumption and family benefit with species richness and plant diversity were negative (Table 4). This latter indicates that plant diversity within the five AFS is not contributing as expected to family benefits. This confirms the findings with the analysis of magnitudes of trade-offs.

Table 4 Spearman's correlations analysis between economic indicators of ecosystem services and shade canopy features of five agroforestry systems in the Sentinel Landscape, Nicaragua

	Cash flow		Value of domestic consumption		Family benefit		Value of standing timber		Carbon SC	
	Coef	*P*-value	Coef	*P*-value	Coef	*P*-value	Coef	*P*-value	Coef	*P*-value
Density of service trees	0.04	0.5198	0.15	0.0131	0.23	0.0001	0.33	<0.0001	0.44	<0.0001
Density of post trees	0.27	<0.0001	*−0.28*	*<0.0001*[a]	*−0.12*	*0.0465*[a]	0.21	0.0005	0.24	0.0001
Density of fruit trees	*−0.2*	*0.0008*[a]	0.46	<0.0001	0.43	<0.0001	0.18	0.0038	0.53	<0.0001
Density of timber trees	0.22	0.0004	−0.03	0.5762	0.09	0.1623	0.79	<0.0001	0.4	<0.0001
Total density of shade canopy	−0.11	0.0674	0.38	<0.0001	0.4	<0.0001	0.39	<0.0001	0.6	<0.0001
Basal area of service trees	0.13	0.0391	0.06	0.3154	0.18	0.0028	0.45	<0.0001	0.58	<0.0001
Basal area of post trees	0.21	0.0004	*−0.2*	*0.0012*[a]	−0.07	0.2377	0.21	0.0004	0.26	<0.0001
Basal area of fruit trees	*−0.16*	*0.0065*[a]	0.43	<0.0001	0.42	<0.0001	0.2	0.0008	0.62	<0.0001
Basal area of timber trees	0.34	<0.0001	−0.1	0.0955	0.07	0.2197	0.98	<0.0001	0.53	<0.0001
Total basal area of shade canopy	0.06	0.3553	0.24	0.0001	0.34	<0.0001	0.65	<0.0001	0.82	<0.0001
Total species richness	0.37	<0.0001	*−0.43*	*<0.0001*[a]	*−0.19*	*0.0021*[a]	0.5	<0.0001	0.45	<0.0001
1/*D*	0.19	<0.0013	*−0.25*	*<0.0001*[a]	−0.10	0.1185	0.35	<0.0001	0.38	<0.0001

[a] significant negative relationships denoting trade-offs.
1/*D*: reciprocal of Simpson index.

The overall performance of provisioning indicators (family benefit) in relation to basal area of different types of trees across AFS is presented in Fig. 8. In the figures it is possible to see which observations (individual systems) surpass a family benefit of 2500 USD ha^{-1} $year^{-1}$, which we consider as a desirable family benefit in the context of the Sentinel Landscape under study. Likewise, it is possible to identify, per each AFS, which observation achieved the maximum value of both family benefit and tree basal area. For instance, the best paired family benefit and total tree basal area values for pastures, grain fields, coffee AFS, cocoa AFS, and homegardens were 3, 5, 9, 12, and 16 m^2 ha^{-1}, respectively (Fig. 8). In summary, these will be the maximum expected values of basal area recommended to manage in each AFS to avoid important trade-offs.

4.7 Discussion

In this case study we assessed trade-offs between provisioning and regulating services, and plant diversity, which are considered the most relevant in order to reflect contributions to human well-being and environment conservation (Vlek et al., 2017; Dade et al., 2018; Zheng et al., 2019). Our analysis can be framed in a landscape-scale approach aimed at minimizing trade-offs, which postulates that landscape composition, configuration, and their dynamics affect biodiversity patterns and ecological processes that generate ecosystem services (Bennett et al., 2009; Tscharntke et al., 2012; Wong et al., 2015; Zheng et al., 2019). Although the approaches followed in this study did consider several dominant features (area covered, tree basal area, tree richness, family benefits) of the AFS, the study did not consider the dynamics of the agricultural landscape, excluding the analysis of drivers that affect ecosystem services (Perfecto and Vandermeer, 2010). Trade-offs among ecosystem services are also linked to complex social dynamics, stakeholders' behavior, and decision-making (Power, 2010; van Vliet et al., 2012; Gaba et al., 2015; Manners and Varela-Ortega, 2018). Thus, the main challenge for researchers is to aid users to simultaneously meet multiple objectives from multiple stakeholders at the landscape level by designing interventions to reduce trade-offs (Diaz et al., 2011; Groot et al., 2012; Rahman et al., 2016; Spake et al., 2017). The use of the ecosystem service trade-off approaches applied in this study contributes to such challenge because allowed us to (1) quantify important ecosystem services provided by dominant land uses across an agricultural landscape in Nicaragua, (2) identify pairs of provisioning and regulating ecosystem services being compromised, and (3) provide important information and actionable knowledge on how to minimize trade-offs. In the following sections, we highlight the main utilities, limitations, and implications of our findings.

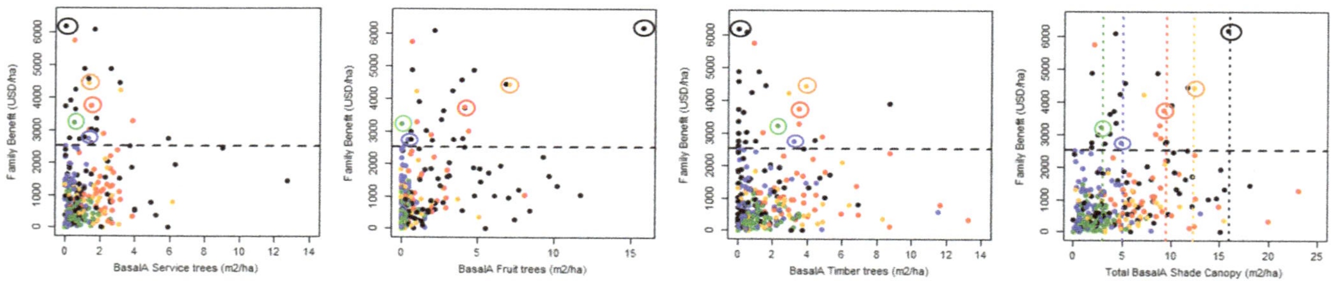

Figure 8 Relationships between family benefit and basal area of different types of trees in the shade canopy of agroforestry systems in the Sentinel Landscape, Nicaragua. Points: black: homegardens; blue: grain fields; green: pastures; red: coffee AFS; orange: cocoa AFS. Horizontal black lines indicate the acceptable value of family benefit considered in this study. Colored circles point to the systems that achieved the highest family benefits of each type of agroforestry system. Vertical colored lines denote the total basal area in the shade canopy of the systems which achieved the highest family benefits of each agroforestry systems.

4.7.1 Advantages and limitations of the approaches used in this case study

Analyses with PCA, Spearman's correlations, and linear regressions were useful to identify trade-offs in a traditional way, in the sense that only negative relationships would be denoting trade-offs. In addition, the approach of using a 1:1 line between pairs of ecosystem services and RMSD turned out even more useful because it permitted to identify more trade-offs not detected in the traditional manner. With this approach it was possible to deduce that there were trade-offs between indicators of provisioning services and carbon stocks, and plant biodiversity, and with different magnitudes according to the type of AFS. Thus, such approach offers the opportunity to discover trade-offs between ecosystems no matter how they are correlated, and also compare quantitatively the degree or magnitude of the trade-offs among different systems (Lu et al., 2014).

The approach of graphing pairs of ecosystem services indicators is also useful to identify individual promising systems (those that achieve high values of both services) as it was suggested (Rapidel et al., 2015) and done in other studies of AFS following the approach shown in Fig. 3 (Cerda et al., 2017a). In this case study, the approach was useful to realize that in the context of the Sentinel Landscape there were no promising systems achieving high values of economic indicators (provisioning services), carbon stocks, and plant diversity at the same time. This situation became an important warning that, in general, all systems should be improved to minimize trade-offs and maximize benefits.

The relationships between some indicators of ecosystem services cannot always be expected to be linear, therefore, the interpretations must have warnings. For instance, relationships between incomes and plant diversity, or between carbon stocks and plant diversity could be linear until some point and then experiment curves. In that sense, the use of the 1:1 line in graphs of pairs of ecosystem services can be used, and it was demonstrated that it is useful as the reference to identify which services is being more benefited over other services (trade-offs); but, it cannot necessarily be expected that the best systems should follow the line up to the maximum values and have scores close to 1 for both services.

Another limitation of the trade-off analysis using RMSD and 1:1 line for identifying trade-offs between pairs of ecosystem services is the lack of analysis of causality between each other. This situation is a drawback as well as in other studies within the landscape-scale approach, which at the present time has been poorly discussed in the literature (Spake et al., 2017; Dade et al., 2018). Other methods are needed to improve the understanding of the complex and dynamic nature of ecosystem service trade-offs in agricultural landscapes, including spatial and temporal analysis and trade-offs between beneficiaries–users of ecosystem services (Johnson et al., 2014; Laurance et al., 2014; Behrman et al., 2015; Cordingley et al., 2016; Zheng et al., 2019).

4.7.2 How to balance trade-offs between provisioning services and plant biodiversity, and regulating services

The majority of observations (individual systems) of all types of AFS in this case study benefited more the plant diversity over the economic indicators (representatives of provisioning services). This finding indicates that in most of the systems such trade-offs occur due to two possible reasons: there are many tree species that do not provide products for the consumption of the families or domestic animals, nor other products for on-farm uses (timber, firewood, materials) as it would be desired; and/or farmers are not taking advantage of the products they could harvest. Similar results were found in cocoa AFS and coffee AFS in other sites (Cerda et al., 2014; Pinoargote et al., 2017). It is not expected to have a perfect positive linear relationship between provisioning and plant diversity, but the trade-offs can be balanced in favor of family benefits with farmers concentrating efforts specially in the management of useful trees that produce tangible goods. It is also desired to assess the relationship of tree species with the diet diversity of families, in order to place more management efforts to those species providing nutritive edible products for domestic consumption in different seasons of the year (Somarriba et al., 2017). In cases with excessive density of trees that do not contribute with tangible goods, it doesn't mean the total elimination of such species because they can provide other services (protection, soil improvement, others), but at least reduction of their populations.

The trade-offs detected with carbon stocks in the shade canopy reflect that regulating services are being more benefited than provisioning services, especially in coffee and cocoa AFS. In these cases, carbon stocks can be positively related to density of trees and/or shade cover, indicating that high carbon stocks involve excessive shade cover that leads to less production of crops (Zuidema et al., 2005; Wade et al., 2010). However, there are manners to balance this trade-off by using timber tree species with morphological characteristics useful to produce timber and store carbon without producing negative effects on the crops, such trees are those with large and tall trunks whose crowns have small leaves and low occlusion (Somarriba et al., 2013).

The magnitude of trade-offs of grain fields and pastures were lower in the major part of relationships between pairs of ecosystem services, but it is not an indicator that these systems perform better. Their economic indicators were low, therefore in these cases efforts must be focused in increasing the quantities of both provisioning and regulating services per hectare.

Given the different amounts of benefits (provisioning and regulating services) and magnitudes of trade-offs among the AFS assessed in this study, the interventions must be tailored to each of them. In this regard, the approach of considering basal areas of trees in relation with family benefits is an important contribution of this case study. Figure 8 provides recommendations on the adequate amount of basal areas to be managed in the shade canopies for each type of tree and each type of AFS. The user (scientist, technician, or farmer) can calculate the density of trees aimed to achieve and not surpass such basal areas. The same approach can be applied to any given AFS in other areas.

4.7.3 Usefulness of results for decision-making processes and optimization

The type of findings encountered in this study case can serve to highlight the importance of AFS and as important inputs for decision-making processes at large scales involving several types of stakeholders. The quantification of ecosystem services and the trade-offs assessed could be utilized in broader approaches such as MCA, MCDA, and AHP. In those, stakeholders like farmers, technicians, scientists, politicians, and others could assign weights to give more importance to given ecosystem services or plant diversity according to the necessities of their landscape or region, analyze the trade-offs identified, combine with information of other ecosystem service, prioritize actions/measures to minimize trade-offs, and project scenarios of impacts on ecosystem services (Koschke et al., 2012; Domptail et al., 2013; Marre et al., 2016). It is imperative that the actual and potential benefits of trees on farms be exposed to high-level decision-makers, given that the contribution of trees in agricultural landscapes are practically invisible in politics (Somarriba et al., 2017).

The approaches of this study as well as in many others deal with trade-offs and recommendations to minimize them. However, for the optimization of AFS in order to maximize synergies, this is not enough and must be complemented with other methods. For instance, analysis with simulations and mathematical models which go deeper on maximum benefits that could be achieved with given intervention to AFS. For instance, research is needed to define and mathematically describe the interactions between all tree components in AFS (Somarriba et al., 2018). Furthermore, additional key indicators should be added, such as pest regulation, hydrologic services, and others. Such methods should include in the analysis of changes in plant composition and farming management, identified as the main factors influencing ecosystem services (Rudel et al., 2002; Villamagna et al., 2013; Rockstrom et al., 2017).

4.8 Conclusion and guidelines for balancing trade-offs in AFS

In the study site, in general, most observations across AFS (coffee AFS, cocoa AFS, grain fields, pastures, and homegardens) favored more plant diversity and a regulating service (carbon sequestration) over provisioning services (reflected by economic indicators). Such trade-offs need to be balanced in order to obtain higher incomes and overall tangible benefits for the farmers' families. By doing so, the exploration of economic indicators in relation to tree basal area will help to better identify appropriate levels of basal areas per type of tree that should set as thresholds to simultaneously obtain family benefits and key ecosystem services at least at plot and farm levels.

Based on the analysis of trade-offs of indicators of ecosystem services and relationships with tree basal area, the following recommendations can be drawn for the studied landscape:

- Across AFS, grain fields and pastures offered the lowest values of provisioning and regulating services. Therefore, these land uses require more tree-based interventions to increase the overall contributions of trees to increase provisioning (family benefits) and regulating services.
- Homegardens provided important values for domestic consumption but not for cash flow, denoting a trade-off between those economic indicators. Thus, efforts should be made to promote the cultivation and sale of plant species that offer marketable products to increase family cash flow year-round.
- Contrary to homegardens, coffee and cocoa AFS need to increase the offer of products for domestic consumption. Such intervention will help create a better balance between family benefit and carbon stocks, and between family benefit and plant biodiversity.
- Data from Fig. 8 can be considered as a technical advice to make informed decisions regarding the most suitable amount of basal areas per type of tree and total basal area for (re-)design of the shade canopy per each type of AFS. Farmers must be encouraged to do specific improvements to each type of AFS managed in the farm, that way the increment of ecosystem services will be noted first at plot scale, then at farm scale, and potentially at landscape scale if all stakeholders agree on the type of interventions.

The approaches applied in this chapter, with their advantages and limitations, are useful for suggesting measures to minimize trade-offs and achieve better balances among ecosystem services and plant diversity. A next step will be the maximization of synergies and overall benefits; for that end, further research is needed with simulation and mathematical models aimed at optimazing the composition and management of AFS. Such research must also consider the multi-objectives, priorities, and decision-making processes of stakeholders in a given landscape.

5 Acknowledgment

The research was carried out by CATIE as part of the CGIAR Research Program on Forests, Trees and Agroforestry (FTA). FTA's research is supported by CGIAR Fund Donors: www.cgiar.org/funders.

6 Where to look for further information

FTA "is the world's largest research for development program to enhance the role of forests, trees and agroforestry in sustainable development and food security and to address climate change. CIFOR leads FTA in partnership with Bioversity International, CATIE, CIRAD, ICRAF, INBAR and TBI." Further information on research, publications, and news of this program can be found at: http://foreststreesagroforestry.org/.

The partners in other projects also develop research on many types of agroforestry systems and ecosystem services they provide. The description and initiatives related to these topics, as well as references of publication can be found in the web pages of each institution:

- Bioversity International: https://www.bioversityinternational.org/
- CATIE: https://catie.ac.cr/en/
- CIFOR: https://www.cifor.org/
- CIRAD: https://www.cirad.fr/en
- ICRAF: http://worldagroforestry.org/
- INBAR: https://www.inbar.int/
- TBI: https://www.tropenbos.org/

CIRAD and CATIE also lead the "Mesoamerican Scientific Platform for Agroforestry (PCP)," which concentrates research efforts on agroforestry with perennial crops (mainly coffee and cocoa). This platform is functioning since 2007 and their themes of research, publications, and materials can be found at: https://www.pcpagroforestry.com/.

7 References

Abdulai, I., Vaast, P., Hoffmann, M. P., Asare, R., Jassogne, L., Van Asten, P., Rotter, R. P. and Graefe, S. 2017. Cocoa agroforestry is less resilient to sub-optimal and extreme climate than cocoa in full sun. *Global Change Biology* 24(1), 273-86. doi:10.1111/gcb.13885.

Abou Rajab, Y., Leuschner, C., Barus, H., Tjoa, A. and Hertel, D. 2016. Cacao cultivation under diverse shade tree cover allows high carbon storage and sequestration without yield losses. *PLoS ONE* 11(2), e0149949. doi:10.1371/journal.pone.0149949.

Albrecht, A. and Kandji, S. T. 2003. Carbon sequestration in tropical agroforestry systems. *Agriculture, Ecosystems and Environment* 99(1-3), 15-27. doi:10.1016/S0167-8809(03)00138-5.

Albuquerque, U. P., Andrade, L. H. C. and Caballero, J. 2005. Structure and floristics of homegardens in Northeastern Brazil. *Journal of Arid Environments* 62(3), 491-506. doi:10.1016/j.jaridenv.2005.01.003.

Andrade, H. J. C., Segura, M. A., Feria, M. and Suárez, W. 2016. Above-ground biomass models for coffee bushes (*Coffea arabica* L.) in Líbano, Tolima, Colombia. *Agroforestry Systems* 92(3), 775-84. doi:10.1007/s10457-016-0047-4.

Andreotti, F., Mao, Z., Jagoret, P., Speelman, E. N., Gary, C. and Saj, S. 2018. Exploring management strategies to enhance the provision of ecosystem services in complex smallholder agroforestry systems. *Ecological Indicators* 94, 257-65. doi:10.1016/j.ecolind.2018.06.048.

Arevalo-Gardini, E., Canto, M., Alegre, J., Loli, O., Juica, A. and Baligar, V. 2015. Changes in soil physical and chemical properties in long term improved natural and traditional agroforestry management systems of cacao genotypes in Peruvian Amazon. *PLOS ONE* 9(7), e101872. doi:10.1371/journal.pone.0101872.

Avelino, J., Romero-Gurdian, A., Cruz-Cuellar, H. F. and Declerck, F. A. 2012. Landscape context and scale differentially impact coffee leaf rust, coffee berry borer, and coffee root-knot nematodes. *Ecological Applications: a Publication of the Ecological Society of America* 22(2), 584-96. doi:10.1890/11-0869.1.

Babin, R., ten Hoopen, G. M., Cilas, C., Enjalric, F., Yede, Gendre, P. and Lumaret, J.-P. 2010. Impact of shade on the spatial distribution of Sahlbergella singularisin traditional cocoa agroforests. *Agricultural and Forest Entomology* 12(1), 69-79. doi:10.1111/j.1461-9563.2009.00453.x.

Beer, J., Muschler, R., Kass, D. and Somarriba, E. 1998. Shade management in coffee and cacao platations. *Agroforestry Systems* 38, 139-64.

Beer, J., Harvey, C. A., Ibrahim, M., Harmand, J. M., Somarriba, E. and Jimenez, F. 2003. Servicios ambientales de los sistemas agroforestales. *Agroforesteria en las Americas* 10, 80-7.

Behrman, K. D., Juenger, T. E., Kiniry, J. R. and Keitt, T. H. 2015. Spatial land use trade-offs for maintenance of biodiversity, biofuel, and agriculture. *Landscape Ecology* 30(10), 1987-99. doi:10.1007/s10980-015-0225-1.

Benegas, L., Ilstedt, U., Roupsard, O., Jones, J. and Malmer, A. 2014. Effects of trees on infiltrability and preferential flow in two contrasting agroecosystems in Central America. *Agriculture, Ecosystems and Environment* 183, 185-96. doi:10.1016/j.agee.2013.10.027.

Bennett, E. M., Peterson, G. D. and Gordon, L. J. 2009. Understanding relationships among multiple ecosystem services. *Ecology Letters* 12(12), 1394-404. doi:10.1111/j.1461-0248.2009.01387.x.

Bentley, J. W., Boa, E. and Stonehouse, J. 2004. Neighbor trees: shade, intercropping, and cacao in Ecuador. *Human Ecology* 32(2), 241-70. doi:10.1023/B:HUEC.0000019759.46526.4d.

Bisseleua Daghela, H. B., Fotio, D., Yede, Missoup, A. D. and Vidal, S. 2013. Shade tree diversity, cocoa pest damage, yield compensating inputs and farmers' net returns in West Africa. *PLOS ONE* 8(3), e56115. doi:10.1371/journal.pone.0056115.

Boreux, V., Kushalappa, C. G., Vaast, P. and Ghazoul, J. 2013. Interactive effects among ecosystem services and management practices on crop production: pollination in coffee agroforestry systems. *Proceedings of the National Academy of Sciences of the United States of America* 110(21), 8387–92. doi:10.1073/pnas.1210590110.

Bradford, J. B. and D'Amato, A. W. 2012. Recognizing trade-offs in multi-objective land management. *Frontiers in Ecology and the Environment* 10(4), 210–6. doi:10.1890/110031.

Bussoni, A., Alvarez, J., Cubbage, F., Ferreira, G. and Picasso, V. 2017. Diverse strategies for integration of forestry and livestock production. *Agroforestry Systems* 93(1), 333–44. doi:10.1007/s10457-017-0092-7.

Cannavo, P., Sansoulet, J., Harmand, J. -M., Siles, P., Dreyer, E. and Vaast, P. 2011. Agroforestry associating coffee and Inga densiflora results in complementarity for water uptake and decreases deep drainage in Costa Rica. *Agriculture, Ecosystems and Environment* 140(1–2), 1–13. doi:10.1016/j.agee.2010.11.005.

Cerda, R., Deheuvels, O., Calvache, D., Niehaus, L., Saenz, Y., Kent, J., Vilchez, S., Villota, A., Martinez, C. and Somarriba, E. 2014. Contribution of cocoa agroforestry systems to family income and domestic consumption: looking toward intensification. *Agroforestry Systems* 88(6), 957–81. doi:10.1007/s10457-014-9691-8.

Cerda, R., Allinne, C., Gary, C., Tixier, P., Harvey, C. A., Krolczyk, L., Mathiot, C., Clément, E., Aubertot, J.-N. and Avelino, J. 2017a. Effects of shade, altitude and management on multiple ecosystem services in coffee agroecosystems. *European Journal of Agronomy* 82, 308–19. doi:10.1016/j.eja.2016.09.019.

Cerda, R., Avelino, J., Gary, C., Tixier, P., Lechevallier, E. and Allinne, C. 2017b. Primary and secondary yield losses caused by pests and diseases: assessment and modeling in coffee. *PLOS ONE* 12(1), e0169133. doi:10.1371/journal.pone.0169133.

Charbonnier, F., Roupsard, O., le Maire, G., Guillemot, J., Casanoves, F., Lacointe, A., Vaast, P., Allinne, C., Audebert, L., Cambou, A., Clement-Vidal, A., Defrenet, E., Duursma, R. A., Jarri, L., Jourdan, C., Khac, E., Leandro, P., Medlyn, B. E., Saint-Andre, L., Thaler, P., Van Den Meersche, K., Barquero Aguilar, A., Lehner, P. and Dreyer, E. 2017. Increased light-use efficiency sustains net primary productivity of shaded coffee plants in agroforestry system. *Plant, Cell and Environment* 40(8), 1592–608. doi:10.1111/pce.12964.

Cordingley, J. E., Newton, A. C., Rose, R. J., Clarke, R. T., Bullock, J. M. and Mori, A. 2016. Can landscape-scale approaches to conservation management resolve biodiversity-ecosystem service trade-offs? *Journal of Applied Ecology* 53(1), 96–105. doi:10.1111/1365-2664.12545.

Dade, M. C., Mitchell, M. G. E., McAlpine, C. A. and Rhodes, J. R. 2018. Assessing ecosystem service trade-offs and synergies: the need for a more mechanistic approach. *AMBIO*. doi:10.1007/s13280-018-1127-7.

Dahlquist, R. M., Whelan, M. P., Winowiecki, L., Polidoro, B., Candela, S., Harvey, C. A., Wulfhorst, J. D., McDaniel, P. A. and Bosque-Perez, N. A. 2007. Incorporating livelihoods in biodiversity conservation: a case study of cacao agroforestry systems in Talamanca, Costa Rica. *Biodiversity and Conservation* 16(8), 2311–33. doi:10.1007/s10531-007-9192-4.

Defrenet, E., Roupsard, O., Van den Meersche, K., Charbonnier, F., Pastor Perez-Molina, J., Khac, E., Prieto, I., Stokes, A., Roumet, C., Rapidel, B., de Melo Virginio Filho, E., Vargas, V. J., Robelo, D., Barquero, A. and Jourdan, C. 2016. Root biomass, turnover and net primary productivity of a coffee agroforestry system in Costa Rica: effects of soil depth, shade trees, distance to row and coffee age. *Annals of Botany*. pii:mcw153.

Deheuvels, O., Avelino, J., Somarriba, E. and Malezieux, E. 2012. Vegetation structure and productivity in cocoa-based agroforestry systems in Talamanca, Costa Rica. *Agriculture, Ecosystems and Environment* 149, 181–8. doi:10.1016/j.agee.2011.03.003.

de Sousa, K. F. D., Detlefsen, G., de Melo Virginio Filho, E., Tobar, D. and Casanoves, F. 2015. Timber yield from smallholder agroforestry systems in Nicaragua and Honduras. *Agroforestry Systems* 90, 207–18.

Diaz, S., Quetier, F., Caceres, D. M., Trainor, S. F., Perez-Harguindeguy, N., Bret-Harte, M. S., Finegan, B., Pena-Claros, M. and Poorter, L. 2011. Linking functional diversity and social actor strategies in a framework for interdisciplinary analysis of nature's benefits to society. *Proceedings of the National Academy of Sciences of the United States of America* 108(3), 895–902. doi:10.1073/pnas.1017993108.

Domptail, S., Nuppenau, E.-A., Azebaze, N. M., Brown, L. D., Falk, T., Finckh, M., Zoch, L., Kowalski, B., Pröpper, M., Stellmes, M. and Overmann, J. 2013. Using trade-offs and synergies in ecosystem services in resource management. *Biodiversity and Ecology* 5, 185–93.

Ehrenbergerová, L., Cienciala, E., Kučera, A., Guy, L. and Habrová, H. 2016. Carbon stock in agroforestry coffee plantations with different shade trees in Villa Rica, Peru. *Agroforestry Systems* 90(3), 433–45. doi:10.1007/s10457-015-9865-z.

Esquivel, M. J., Harvey, C. A., Finegan, B., Casanoves, F. and Skarpe, C. 2007. Effects of pasture management on the natural regeneration of neotropical trees. *Journal of Applied Ecology* 45(1), 371–80. doi:10.1111/j.1365-2664.2007.01411.x.

Felipe-Lucia, M. R., Comín, F. A. and Bennett, E. M. 2014. Interactions among ecosystem services across land uses in a floodplain agroecosystem. *Ecology and Society* 19(1). doi:10.5751/ES-06249-190120.

Fischer, A. and Vasseur, L. 2002. Smallholder perceptions of agroforestry projects in Panama. *Agroforestry Systems* 54(2), 103–13. doi:10.1023/A:1015047404867.

Gaba, S., Lescourret, F., Boudsocq, S., Enjalbert, J., Hinsinger, P., Journet, E.-P., Navas, M.-L., Wery, J., Louarn, G., Malézieux, E., Pelzer, E., Prudent, M., Ozier-Lafontaine, H. 2015. Multiple cropping systems as drivers for providing multiple ecosystem services: from concepts to design. *Agronomy for Sustainable Development* 35(2), 607–23. doi:10.1007/s13593-014-0272-z.

Goenster, S., Wiehle, M., Gebauer, J., Ali, A. M. and Buerkert, A. 2014. Carbon and nutrient fluxes and balances in Nuba Mountains homegardens, Sudan. *Nutrient Cycling in Agroecosystems* 100(1), 35–51. doi:10.1007/s10705-014-9624-y.

Groot, J. C. J., Oomen, G. J. M. and Rossing, W. A. H. 2012. Multi-objective optimization and design of farming systems. *Agricultural Systems* 110, 63–77. doi:10.1016/j.agsy.2012.03.012.

Grossman, J. J. 2015. Ecosystem service trade-offs and land use among smallholder farmers in eastern Paraguay. *Ecology and Society* 20(1). doi:10.5751/ES-06953-200119.

Gyau, A., Smoot, K., Kouame, C., Diby, L., Kahia, J. and Ofori, D. 2014. Farmer attitudes and intentions towards trees in cocoa (*Theobroma cacao* L.) farms in Côte d'Ivoire. *Agroforestry Systems* 88(6), 1035–45. doi:10.1007/s10457-014-9677-6.

Hartemink, A. E. 2005. Nutrient stocks, nutrient cycling, and soil changes in cocoa ecosystems: a review. *Advances in Agronomy* 86, 227–53. doi:10.1016/S0065-2113(05)86005-5.

Harvey, C. A., Chacón, M., Donatti, C. I., Garen, E., Hannah, L., Andrade, A., Bede, L., Brown, D., Calle, A., Chará, J., Clement, C., Gray, E., Hoang, M. H., Minang, P., Rodríguez, A. M., Seeberg-Elverfeldt, C., Semroc, B., Shames, S., Smukler, S., Somarriba, E., Torquebiau, E., van Etten, J. and Wollenberg, E. 2014. Climate-smart landscapes: opportunities and challenges for integrating adaptation and mitigation in tropical agriculture. *Conservation Letters* 7(2), 77–90. doi:10.1111/conl.12066.

Howe, C., Suich, H., Vira, B. and Mace, G. M. 2014. Creating win-wins from trade-offs? Ecosystem services for human well-being: a meta-analysis of ecosystem service trade-offs and synergies in the real world. *Global Environmental Change* 28, 263–75. doi:10.1016/j.gloenvcha.2014.07.005.

Isaac, M. E., Timmer, V. R. and Quashie-Sam, S. J. 2007. Shade tree effects in an 8-year-old cocoa agroforestry system: biomass and nutrient diagnosis of *Theobroma cacao* by vector analysis. *Nutrient Cycling in Agroecosystems* 78(2), 155–65. doi:10.1007/s10705-006-9081-3.

Jacobi, J., Andres, C., Schneider, M., Pillco, M., Calizaya, P. and Rist, S. 2014. Carbon stocks, tree diversity, and the role of organic certification in different cocoa production systems in Alto Beni, Bolivia. *Agroforestry Systems* 88(6), 1117–32. doi:10.1007/s10457-013-9643-8.

Jadan, O., Cifuentes, M., Torres, B., Seleis, D., Veintimilla, D. and Gunter, S. 2015. Influence of tree cover on diversity, carbon sequestration and productivity of cocoa systems in the Ecuadorian Amazon. *Bois et Forets des Tropiques* 325(325), 36–47. doi:10.19182/bft2015.325.a31271.

Jagoret, P., Kwesseu, J., Messie, C., Michel-Dounias, I. and Malézieux, E. 2014. Farmers' assessment of the use value of agrobiodiversity in complex cocoa agroforestry systems in central Cameroon. *Agroforestry Systems* 88(6), 983–1000. doi:10.1007/s10457-014-9698-1.

Jaimez, R. E., Araque, O., Guzman, D., Mora, A., Espinoza, W. and Tezara, W. 2013. Agroforestry systems of timber species and cacao: survival and growth during the early stages. *Journal of Agriculture and Rural Development in the Tropics and Subtropics* 114, 1–11.

Jha, S., Bacon, C. M., Philpott, S. M., Ernesto Méndez, V., Läderach, P. and Rice, R. A. 2014. Shade coffee: update on a disappearing refuge for biodiversity. *BioScience* 64(5), 416–28. doi:10.1093/biosci/biu038.

Johns, T., Powell, B., Maundu, P. and Eyzaguirre, P. B. 2013. Agricultural biodiversity as a link between traditional food systems and contemporary development, social integrity and ecological health. *Journal of the Science of Food and Agriculture* 93(14), 3433–42. doi:10.1002/jsfa.6351.

Johnson, J. A., Runge, C. F., Senauer, B., Foley, J. and Polasky, S. 2014. Global agriculture and carbon trade-offs. *Proceedings of the National Academy of Sciences of the United States of America* 111(34), 12342–7. doi:10.1073/pnas.1412835111.

Jose, S. 2009. Agroforestry for ecosystem services and environmental benefits: an overview. *Agroforestry Systems* 76(1), 1-10. doi:10.1007/s10457-009-9229-7.

Kennedy, C. M., Miteva, D. A., Baumgarten, L., Hawthorne, P. L., Sochi, K., Polasky, S., Oakleaf, J. R., Uhlhorn, M. E. and Kiesecker, J. 2016. Bigger is better: improved nature conservation and economic returns from landscape-level mitigation. *Science Advances* 2(7), e1501021. doi:10.1126/sciadv.1501021.

Kessler, M., Hertel, D., Jungkunst, H. F., Kluge, J., Abrahamczyk, S., Bos, M., Buchori, D., Gerold, G., Gradstein, S. R., Kohler, S., Leuschner, C., Moser, G., Pitopang, R., Saleh, S., Schulze, C. H., Sporn, S. G., Steffan-Dewenter, I., Tjitrosoedirdjo, S. S. and Tscharntke, T. 2012. Can joint carbon and biodiversity management in tropical agroforestry landscapes be optimized? *PLOS ONE* 7(10), e47192. doi:10.1371/journal.pone.0047192.

Kim, D.-G., Kirschbaum, M. U. F. and Beedy, T. L. 2016. Carbon sequestration and net emissions of CH_4 and N_2O under agroforestry: synthesizing available data and suggestions for future studies. *Agriculture, Ecosystems and Environment* 226, 65-78. doi:10.1016/j.agee.2016.04.011.

Koschke, L., Fürst, C., Frank, S. and Makeschin, F. 2012. A multi-criteria approach for an integrated land-cover-based assessment of ecosystem services provision to support landscape planning. *Ecological Indicators* 21, 54-66. doi:10.1016/j.ecolind.2011.12.010.

Kragt, M. E. and Robertson, M. J. 2014. Quantifying ecosystem services trade-offs from agricultural practices. *Ecological Economics* 102, 147-57. doi:10.1016/j.ecolecon.2014.04.001.

Kumar, B. M. 2011. Species richness and aboveground carbon stocks in the homegardens of central Kerala, India. *Agriculture, Ecosystems and Environment* 140(3-4), 430-40. doi:10.1016/j.agee.2011.01.006.

Kumar, B. M. and Nair, P. K. R. 2004. The enigma of tropical homegardens. *Agroforestry Systems* 61, 135-52.

Kuyah, S., Öborn, I., Jonsson, M., Dahlin, A. S., Barrios, E., Muthuri, C., Malmer, A., Nyaga, J., Magaju, C., Namirembe, S., Nyberg, Y. and Sinclair, F. L. 2016. Trees in agricultural landscapes enhance provision of ecosystem services in Sub-Saharan Africa. *International Journal of Biodiversity Science, Ecosystem Services and Management* 12, 255-73. doi:10.1080/21513732.2016.1214178.

Landuyt, D., Broekx, S., D'Hondt, R., Engelen, G., Aertsens, J. and Goethals, P. L. M. 2013. A review of Bayesian belief networks in ecosystem service modelling. *Environmental Modelling and Software* 46, 1-11. doi:10.1016/j.envsoft.2013.03.011.

Laurance, W. F., Sayer, J. and Cassman, K. G. 2014. Agricultural expansion and its impacts on tropical nature. *Trends in Ecology and Evolution* 29(2), 107-16. doi:10.1016/j.tree.2013.12.001.

Lin, B. B. 2010. The role of agroforestry in reducing water loss through soil evaporation and crop transpiration in coffee agroecosystems. *Agricultural and Forest Meteorology* 150(4), 510-8. doi:10.1016/j.agrformet.2009.11.010.

López-Bravo, D. F., Virginio-Filho, EdM. and Avelino, J. 2012. Shade is conducive to coffee rust as compared to full sun exposure under standardized fruit load conditions. *Crop Protection* 38, 21-9. doi:10.1016/j.cropro.2012.03.011.

Lopez-Sampson, A. and Somarriba, E. 2005. Árboles frutales en fincas de cacao orgánico del Alto Beni, Bolivia. *Agroforesteria en las Americas* 43-44, 38-45.

Lorenz, K. and Lal, R. 2014. Soil organic carbon sequestration in agroforestry systems. A review. *Agronomy for Sustainable Development* 34(2), 443–54. doi:10.1007/s13593-014-0212-y.

Lu, N., Fu, B., Jin, T. and Chang, R. 2014. Trade-off analyses of multiple ecosystem services by plantations along a precipitation gradient across Loess Plateau landscapes. *Landscape Ecology* 29(10), 1697–708. doi:10.1007/s10980-014-0101-4.

Mahapatra, A. K. and Panda, P. C. 2012. Wild edible fruit diversity and its significance in the livelihood of indigenous tribals: evidence from eastern India. *Food Security* 4(2), 219–34. doi:10.1007/s12571-012-0186-z.

Manners, R. and Varela-Ortega, C. 2018. The role of decision-making in ecosystem service trade-offs in Lowland Bolivia's Amazonian agricultural systems. *Ecological Economics* 153, 31–42. doi:10.1016/j.ecolecon.2018.06.021.

Marre, J. B., Pascoe, S., Thebaud, O., Jennings, S., Boncoeur, J. and Coglan, L. 2016. Information preferences for the evaluation of coastal development impacts on ecosystem services: a multi-criteria assessment in the Australian context. *Journal of Environmental Management* 173, 141–50. doi:10.1016/j.jenvman.2016.01.025.

Martin, A. R., Rapidel, B., Roupsard, O., Van den Meersche, K., de Melo Virginio Filho, E., Barrios, M. and Isaac, M. E., de Melo Virginio Filho, E., Barrios, M., Isaac, M. E. and Barton, K. 2017 Intraspecific trait variation across multiple scales: the leaf economics spectrum in coffee. *Functional Ecology* 31(3), 604–12. doi:10.1111/1365-2435.12790.

Mattsson, E., Ostwalt, M., Nissanka, S. P. and Marambe, B. 2013. Homegardens as a multi-functional land-use strategy in Sri Lanka with focus on carbon sequestration. *AMBIO* 42(7), 892–902. doi:10.1007/s13280-013-0390-x.

Mattsson, E., Ostwald, M., Nissanka, S. P. and Pushpakumara, D. K. N. G. 2015. Quantification of carbon stock and tree diversity of homegardens in a dry zone area of Moneragala District, Sri Lanka. *Agroforestry Systems* 89(3), 435–45. doi:10.1007/s10457-014-9780-8.

McKechnie, J., Colombo, S., Chen, J., Mabee, W. and MacLean, H. L. 2011. Forest bioenergy or forest carbon? Assessing trade-offs in greenhouse gas mitigation with wood-based fuels. *Environmental Science and Technology* 45(2), 789–95. doi:10.1021/es1024004.

Mendez, V. E., Lok, R. and Somarriba, E. 2001. Interdisciplinary analysis of homegardens in Nicaragua: micro-zonation, plant use and socioeconomic importance. *Agroforestry Systems* 51(2), 85–96. doi:10.1023/A:1010622430223.

Mohri, H., Lahoti, S., Saito, O., Mahalingam, A., Gunatilleke, N., Irham, Hoang, V. T., Hitinayake, G., Takeuchi, K. and Herath, S. 2013. Assessment of ecosystem services in homegarden systems in Indonesia, Sri Lanka, and Vietnam. *Ecosystem Services* 5, 124–36. doi:10.1016/j.ecoser.2013.07.006.

Monroe, P. H. M., Gama-Rodrigues, E. F., Gama-Rodrigues, A. C. and Marques, J. R. B. 2016. Soil carbon stocks and origin under different cacao agroforestry systems in Southern Bahia, Brazil. *Agriculture, Ecosystems and Environment* 221, 99–108. doi:10.1016/j.agee.2016.01.022.

Montagnini, F. and Nair, P. K. R. 2004. Carbon sequestration: an underexploited environmental benefit of agroforestry systems. *Agroforest. Syst.* 61, 281–95.

Montagnini, F., Ibrahim, M. and Murgueitio, E. R. 2013. Silvopastoral systems and climate change in Latin America. *Bois et Forêts des Tropiques* 316, 2–14.

Mora, F., Balvanera, P., García-Frapolli, E., Castillo, A., Trilleras, J. M., Cohen-Salgado, D. and Salmerón, O. 2016. Trade-offs between ecosystem services and alternative pathways toward sustainability in a tropical dry forest region. *Ecology and Society* 21(4). doi:10.5751/ES-08691-210445.

Mortimer, R., Saj, S. and David, C. 2017. Supporting and regulating ecosystem services in cacao agroforestry systems. *Agroforestry Systems* 92(6), 1639–57. doi:10.1007/s10457-017-0113-6.

Mueller, N. D., Gerber, J. S., Johnston, M., Ray, D. K., Ramankutty, N. and Foley, J. A. 2012. Closing yield gaps through nutrient and water management. *Nature* 490(7419), 254–7. doi:10.1038/nature11420.

Nair, P. K. R. 2001. Do tropical homegardens elude science, or is it the other way around? *Agroforestry Systems* 53(2), 239–45. doi:10.1023/A:1013388806993.

Nair, P. K. R. 2012. Carbon sequestration studies in agroforestry systems: a reality-check. *Agroforestry Systems* 86(2), 243–53. doi:10.1007/s10457-011-9434-z.

Nair, P. K. R., Mohan Kumar, B. and Nair, V. D. 2009. Agroforestry as a strategy for carbon sequestration. *Journal of Plant Nutrition and Soil Science* 172(1), 10–23. doi:10.1002/jpln.200800030.

Narango, D. L., Tallamy, D. W., Snyder, K. J. and Rice, R. A. 2019. Canopy tree preference by insectivorous birds in shade-coffee farms: implications for migratory bird conservation. *Biotropica*, 1–12. doi:10.1111/btp.12642.

Nelson, E., Mendoza, G., Regetz, J., Polasky, S., Tallis, H., Cameron, D., Chan, K. M. A., Daily, G. C., Goldstein, J., Kareiva, P. M., Lonsdorf, E., Naidoo, R., Ricketts, T. H. and Shaw, M. 2009. Modeling multiple ecosystem services, biodiversity conservation, commodity production, and tradeoffs at landscape scales. *Frontiers in Ecology and the Environment* 7(1), 4–11. doi:10.1890/080023.

Ngo Bieng, M. A., Gidoin, C., Avelino, J., Cilas, C., Deheuvels, O. and Wery, J. 2013. Diversity and spatial clustering of shade trees affect cacao yield and pathogen pressure in Costa Rican agroforests. *Basic and Applied Ecology* 14(4), 329–36. doi:10.1016/j.baae.2013.03.003.

Noponen, M. R. A., Healey, J. R., Soto, G. and Haggar, J. P. 2013. Sink or source–the potential of coffee agroforestry systems to sequester atmospheric CO_2 into soil organic carbon. *Agriculture, Ecosystems and Environment* 175, 60–8. doi:10.1016/j.agee.2013.04.012.

Opoku, I. Y., Akrofi, A. Y. and Appiah, A. A. 2002. Shade trees are alternative hosts of the cocoa pathogen Phytophthora megakarya. *Crop Protection* 21(8), 629–34. doi:10.1016/S0261-2194(02)00013-3.

Orozco-Aguilar, L. and Somarriba, E. 2005. Árboles maderables en fincas de cacao orgánico del Alto Beni, Bolivia. *Agroforesteria en las Americas* 43–44, 46–53.

Padovan, M. P., Brook, R. M., Barrios, M., Cruz-Castillo, J. B., Vilchez-Mendoza, S. J., Costa, A. N. and Rapidel, B. 2018. Water loss by transpiration and soil evaporation in coffee shaded by *Tabebuia rosea Bertol.* and *Simarouba glauca* dc. compared to unshaded coffee in sub-optimal environmental conditions. *Agricultural and Forest Meteorology* 248, 1–14. doi:10.1016/j.agrformet.2017.08.036.

Pérez-Flores, J., Pérez, A. A., Suárez, Y. P., Bolaina, V. C. and Quiroga, A. L. 2017. Leaf litter and its nutrient contribution in the cacao agroforestry system. *Agroforestry Systems*. doi:10.1007/s10457-017-0096-3.

Perfecto, I. and Vandermeer, J. 2008. Spatial pattern and ecological process in the coffee agroforestry system. *Ecology* 89(4), 915–20. doi:10.1890/06-2121.1.

Perfecto, I. and Vandermeer, J. 2010. The agroecological matrix as alternative to the land-sparing/agriculture intensification model. *Proceedings of the National Academy of Sciences of the United States of America* 107(13), 5786-91. doi:10.1073/pnas.0905455107.

Pfund, J. L., Watts, J. D., Boissiere, M., Boucard, A., Bullock, R. M., Ekadinata, A., Dewi, S., Feintrenie, L., Levang, P., Rantala, S., Sheil, D., Sunderland, T. and Urech, Z. L. 2011. Understanding and integrating local perceptions of trees and forests into incentives for sustainable landscape management. *Environmental Management* 48(2), 334-49. doi:10.1007/s00267-011-9689-1.

Pinho, R. C., Miller, R. P. and Alfaia, S. S. 2012. Agroforestry and the improvement of soil fertility: a view from Amazonia. *Applied and Environmental Soil Science* 2012, 1-11. doi:10.1155/2012/616383.

Pinoargote, M., Cerda, R., Mercado, L., Aguilar, A., Barrios, M. and Somarriba, E. 2017. Carbon stocks, net cash flow and family benefits from four small coffee plantation types in Nicaragua. *Forests, Trees and Livelihoods* 26(3), 183-98. doi:10.1080/14728028.2016.1268544.

Poppenborg, P. and Hölscher, D. 2009. The influence of emergent trees on rainfall distribution in a cacao agroforest (Sulawesi, Indonesia). *Flora - Morphology, Distribution, Functional Ecology of Plants* 204(10), 730-6. doi:10.1016/j.flora.2008.10.003.

Power, A. G. 2010. Ecosystem services and agriculture: tradeoffs and synergies. *Philosophical Transactions of the Royal Society of London. Series B, Biological Sciences* 365(1554), 2959-71. doi:10.1098/rstb.2010.0143.

Pumariño, L., Sileshi, G. W., Gripenberg, S., Kaartinen, R., Barrios, E., Muchane, M. N., Midega, C. and Jonsson, M. 2015. Effects of agroforestry on pest, disease and weed control: a meta-analysis. *Basic and Applied Ecology* 16(7), 573-82. doi:10.1016/j.baae.2015.08.006.

Rahman, S. A., Sunderland, T., Kshatriya, M., Roshetko, J. M., Pagella, T. and Healey, J. R. 2016. Towards productive landscapes: trade-offs in tree-cover and income across a matrix of smallholder agricultural land-use systems. *Land Use Policy* 58, 152-64. doi:10.1016/j.landusepol.2016.07.003.

Rapidel, B., Ripoche, A., Allinne, C., Metay, A., Deheuvels, O., Lamanda, N., Blazy, J.-M., Valdés-Gómez, H. and Gary, C. 2015. Analysis of ecosystem services trade-offs to design agroecosystems with perennial crops. *Agronomy for Sustainable Development* 35(4), 1373-90. doi:10.1007/s13593-015-0317-y.

Rice, R. A. 2011. Fruits from shade trees in coffee: how important are they? *Agroforestry Systems* 83(1), 41-9. doi:10.1007/s10457-011-9385-4.

Rios, N., Andrade, H. J. and Ibrahim, M. 2008. Evaluacion de la recarga hidrica en sistemas silvopastoriles en paisajes ganaderos. *Zootecnia Tropical* 26, 183-6.

Rockstrom, J., Williams, J., Daily, G., Noble, A., Matthews, N., Gordon, L., Wetterstrand, H., DeClerck, F., Shah, M., Steduto, P., de Fraiture, C., Hatibu, N., Unver, O., Bird, J., Sibanda, L. and Smith, J. 2017. Sustainable intensification of agriculture for human prosperity and global sustainability. *AMBIO* 46(1), 4-17. doi:10.1007/s13280-016-0793-6.

Rodriguez, J. P., Beard, T. D., Bennet, E. M., Cumming, G. S., Cork, S. J., Agard, J., Dobson, A. P. and Peterson, G. D. 2006. Trade-offs across space-time and ecosystem services. *Ecology and Society* 11(1), 1-28. doi:10.5751/ES-01667-110128.

Rousseau, G. X., Deheuvels, O., Rodriguez Arias, I. and Somarriba, E. 2012. Indicating soil quality in cacao-based agroforestry systems and old-growth forests: the potential of soil macrofauna assemblage. *Ecological Indicators* 23, 535–43. doi:10.1016/j.ecolind.2012.05.008.

Rudel, T. K., Bates, D. and Machinguashi, R. 2002. A tropical forest transition? Agricultural change, out-migration, and secondary forests in the Ecuadorian Amazon. *Annals of the Association of American Geographers* 92(1), 87–102. doi:10.1111/1467-8306.00281.

Saarikoski, H., Barton, D., Mustajoki, J., Keune, H., Gomez-Baggethun, E. and Langmayer, J. 2015. Multi-criteria decision analysis (MCDA) in ecosystem service valuation, p. 6.

Saaty, T. L. 1980. *The Analytic Hierarchy Process*. McGraw-Hill, New York.

Saha, S. K., Nair, P. K. R., Nair, V. D. and Kumar, B. M. 2009. Soil carbon stock in relation to plant diversity of homegardens in Kerala, India. *Agroforestry Systems* 76(1), 53–65. doi:10.1007/s10457-009-9228-8.

Salgado-Mora, M. G., Ibarra-Nunez, G., Macias-Samano, J. E. and Lopez-Baez, O. 2007. Diversity of shade trees in cacao plantations of Soconusco, Chiapas, Mexico. *Interciencia* 32, 763–8.

Salinas-Peba, L. and Parra-Tabla, V. 2007. Phenology and pollination of Manilkara zapota in forest and homegardens. *Forest Ecology and Management* 248(3), 136–42. doi:10.1016/j.foreco.2007.04.046.

Sauvadet, M., den Meersche, K. V., Allinne, C., Gay, F., de Melo Virginio Filho, E., Chauvat, M., Becquer, T., Tixier, P. and Harmand, J.-M. 2019. Shade trees have higher impact on soil nutrient availability and food web in organic than conventional coffee agroforestry. *Science of the Total Environment* 649, 1065–74.

Schroth, G. and da Mota, D. S. S. 2014. Agroforestry: complex multistrata agriculture. In: van Alfen, N. (Ed.), *Encyclopedia of Agriculture and Food Systems* (2nd edn., vol. 1), pp. 195–207. Elsevier. doi:10.1016/B978-0-444-52512-3.00030-9.

Schroth, G., Krauss, U., Gasparotto, L., Duarte-Aguilar, J. A. and Vohland, K. 2000. Pests and diseases in agroforestry systems of the humid tropics. *Agroforestry Systems* 50(3), 199–241. doi:10.1023/A:1006468103914.

Schroth, G., Lehmann, J., Rodrigues, M. R. L., Barros, E. and Macedo, J. L. V. 2001. Plant-soil interactions in multistrata agroforestry in the humid tropics. *Agroforestry Systems* 53(2), 85–102. doi:10.1023/A:1013360000633.

Schroth, G., Jeusset, A., Gomes, AdS., Florence, C. T., Coelho, N. A. P., Faria, D. and Läderach, P. 2014. Climate friendliness of cocoa agroforests is compatible with productivity increase. *Mitigation and Adaptation Strategies for Global Change* 21(1), 67–80. doi:10.1007/s11027-014-9570-7.

Schroth, G., Bede, L. C., Paiva, A. O., Cassano, C. R., Amorim, A. M., Faria, D., Mariano-Neto, E., Martini, A. M. Z., Sambuichi, R. H. R. and Lôbo, R. N. 2015. Contribution of agroforests to landscape carbon storage. *Mitigation and Adaptation Strategies for Global Change* 20(7), 1175–90. doi:10.1007/s11027-013-9530-7.

Seneviratne, G., Kuruppuarachchi, K. A. J., Somaratne, S. and Seneviratne, K. A. C. 2010. Nutrient cycling and safety-net mechanism in the tropical homegardens. *International Journal of Agricultural Resarch* 5, 529–42.

Sijtsma, F., Heide, C. and Van Hinsberg, A. 2012. Biodiversity and decision-support: integrating CBA and MCA. In: Hull, A., Alexander, E., Khakee, A. and Woltjer, J. (Eds), *Evaluation for Participation and Sustainability in Planning*. Routledge, London, pp. 197–218.

Silva Moço, M. K., Gama-Rodrigues, E. F., Gama-Rodrigues, A. C., Machado, R. C. R. and Baligar, V. C. 2008. Soil and litter fauna of cacao agroforestry systems in Bahia, Brazil. *Agroforestry Systems* 76, 127–38.

Somarriba, E. 2012. The population dynamics and productivity of Acacia pennatula in the pasturelands of the Nature Reserve Mesas de Moropotente, Estelí, Nicaragua. *Agroforestry Systems* 84(1), 1–9. doi:10.1007/s10457-011-9447-7.

Somarriba, E., Beer, J., Alegre-Orihuela, J., Andrade, H. J., Cerda, R., DeClerck, F., Detlefsen, G., Escalante, M., Giraldo, L. A., Ibrahim, M., Krishnamurthy, L., Mena-Mosquera, V. E., Mora-Degado, J. R., Orozco, L., Scheelje, M. and Campos, J. J. 2012. Mainstreaming agroforestry in Latin America. In: Nair, P. K. R. and Garrity, D. (Eds), *Agroforestry - the Future of Global Land Use*. Springer Netherlands, Dordrecht, pp. 429–53.

Somarriba, E., Cerda, R., Orozco, L., Cifuentes, M., Dávila, H., Espin, T., Mavisoy, H., Ávila, G., Alvarado, E., Poveda, V., Astorga, C., Say, E. and Deheuvels, O. 2013. Carbon stocks and cocoa yields in agroforestry systems of Central America. *Agriculture, Ecosystems and Environment* 173, 46–57. doi:10.1016/j.agee.2013.04.013.

Somarriba, E., Suárez-Islas, A., Calero-Borge, W., Villota, A., Castillo, C., Vílchez, S., Deheuvels, O. and Cerda, R. 2014. Cocoa-timber agroforestry systems: *Theobroma cacao-Cordia alliodora* in Central America. *Agroforestry Systems* 88(6), 1001–19. doi:10.1007/s10457-014-9692-7.

Somarriba, E., Carreño-Rocabado, G., Amores, F., Caicedo, W., Oblitas Gillés de Pélichy, S., Cerda, R. and Ordóñez, J. C. 2017. Trees on farms for livelihoods, conservation of biodiversity and carbon storage: evidence from Nicaragua on this 'invisible' resource. *Advances in Agroforestry* 12, 369–93. doi:10.1007/978-3-319-69371-2_15.

Somarriba, E., Orozco-Aguilar, L., Cerda, R. and López-Sampson, A. 2018. *Analysis and Design of the Shade Canopy of Cocoa-based Agroforestry Systems*. Burleigh Dodds Series in Agricultural Science, Cambridge, UK, pp. 469–99. doi:10.19103/AS.2017.0021.29.

Sonwa, D. J., Coulibaly, O., Adesina, A. A., Weise, S. F. and Tchatat, M. 2002. Integrated pest management in cocoa agroforests in southern Cameroon: constraints and overview. *Integrated Pest Management Reviews* 7(3), 191–9. doi:10.1023/B:IPMR.0000027499.52755.0b.

Soto-Pinto, L., Anzueto, M., Mendoza, J., Ferrer, G. J. and de Jong, B. 2010. Carbon sequestration through agroforestry in indigenous communities of Chiapas, Mexico. *Agroforestry Systems* 78(1), 39–51. doi:10.1007/s10457-009-9247-5.

Soto-Pinto, L., Perfecto, I. and Caballero-Nieto, J. 2012. Shade over coffee: its effects on berry borer, leaf rust and spontaneous herbs in Chiapas, Mexico. *Agroforestry Systems* 55, 37–45.

Spake, R., Lasseur, R., Crouzat, E., Bullock, J. M., Lavorel, S., Parks, K. E., Schaafsma, M., Bennett, E. M., Maes, J., Mulligan, M., Mouchet, M., Peterson, G. D., Schulp, C. J. E., Thuiller, W., Turner, M. G., Verburg, P. H. and Eigenbrod, F. 2017. Unpacking ecosystem service bundles: towards predictive mapping of synergies and trade-offs between ecosystem services. *Global Environmental Change* 47, 37–50. doi:10.1016/j.gloenvcha.2017.08.004.

Sperber, C. F., Nakayama, K., Valverde, M. J. and Neves, FdS. 2004. Tree species richness and density affect parasitoid diversity in cacao agroforestry. *Basic and Applied Ecology* 5(3), 241–51. doi:10.1016/j.baae.2004.04.001.

Staver, C., Guharay, F., Monterroso, D. and Muschler, R. G. 2001. Designing pest-suppressive multistrata perennial crop systems: shade-grown coffee in Central America. *Agroforestry Systems* 53(2), 151–70. doi:10.1023/A:1013372403359.

Tamburini, G., De Simone, S., Sigura, M., Boscutti, F. and Marini, L. 2016. Soil management shapes ecosystem service provision and trade-offs in agricultural landscapes. *Proceedings. Biological Sciences* 283(1837). doi:10.1098/rspb.2016.1369.

Tscharntke, T., Clough, Y., Bhagwat, S. A., Buchori, D., Faust, H., Hertel, D., Hölscher, D., Juhrbandt, J., Kessler, M., Perfecto, I., Scherber, C., Schroth, G., Veldkamp, E. and Wanger, T. C. 2011. Multifunctional shade-tree management in tropical agroforestry landscapes - a review. *Journal of Applied Ecology* 48(3), 619–29. doi:10.1111/j.1365-2664.2010.01939.x.

Tscharntke, T., Tylianakis, J. M., Rand, T. A., Didham, R. K., Fahrig, L., Batary, P., Bengtsson, J., Clough, Y., Crist, T. O., Dormann, C. F., Ewers, R. M., Frund, J., Holt, R. D., Holzschuh, A., Klein, A. M., Kleijn, D., Kremen, C., Landis, D. A., Laurance, W., Lindenmayer, D., Scherber, C., Sodhi, N., Steffan-Dewenter, I., Thies, C., van der Putten, W. H. and Westphal, C. 2012. Landscape moderation of biodiversity patterns and processes - eight hypotheses. *Biological Reviews of the Cambridge Philosophical Society* 87(3), 661–85. doi:10.1111/j.1469-185X.2011.00216.x.

Tully, K. L., Lawrence, D. and Scanlon, T. M. 2012. More trees less loss: nitrogen leaching losses decrease with increasing biomass in coffee agroforests. *Agriculture, Ecosystems and Environment* 161, 137–44. doi:10.1016/j.agee.2012.08.002.

Vaast, P. and Somarriba, E. 2014. Trade-offs between crop intensification and ecosystem services: the role of agroforestry in cocoa cultivation. *Agroforestry Systems* 88(6), 947–56. doi:10.1007/s10457-014-9762-x.

van Noordwijk, M. and Purnomosidhi, P. 1995. Root architecture in relation to tree-soil-crop interactions and shoot pruning in agroforestry. *Agroforestry Systems* 30(1–2), 161–73. doi:10.1007/BF00708919.

van Noordwijk, M., Rahayu, S., Hairiah, K., Wulan, Y. C., Farida, A. and Verbist, B. 2002. Carbon stock assessment for a forest-to-coffee conversion landscape in Sumber-Jaya (Lampung, Indonesia): from allometric equations to land use change analysis. *Science in China* 45, 75–85.

van Vliet, N., Mertz, O., Heinimann, A., Langanke, T., Pascual, U., Schmook, B., Adams, C., Schmidt-Vogt, D., Messerli, P., Leisz, S., Castella, J.-C., Jørgensen, L., Birch-Thomsen, T., Hett, C., Bech-Bruun, T., Ickowitz, A., Vu, K. C., Yasuyuki, K., Fox, J., Padoch, C., Dressler, W. and Ziegler, A. D. 2012. Trends, drivers and impacts of changes in swidden cultivation in tropical forest-agriculture frontiers: a global assessment. *Global Environmental Change* 22(2), 418–29. doi:10.1016/j.gloenvcha.2011.10.009.

Vebrova, H., Lojka, B., Husband, T. P., Zans, M. E. C., Van Damme, P., Rollo, A. and Kalousova, M. 2014. Tree diversity in cacao agroforests in San Alejandro, Peruvian Amazon. *Agroforestry Systems* 88(6), 1101–15. doi:10.1007/s10457-013-9654-5.

Villamagna, A. M., Angermeier, P. L. and Bennett, E. M. 2013. Capacity, pressure, demand, and flow: a conceptual framework for analyzing ecosystem service provision and delivery. *Ecological Complexity* 15, 114–21. doi:10.1016/j.ecocom.2013.07.004.

Vlek, P., Khamzina, A., Azadi, H., Bhaduri, A., Bharati, L., Braimoh, A., Martius, C., Sunderland, T. and Taheri, F. 2017. Trade-offs in multi-purpose land use under land degradation. *Sustainability* 9(12). doi:10.3390/su9122196.

Wade, A. S. I., Asase, A., Hadley, P., Mason, J., Ofori-Frimpong, K., Preece, D., Spring, N. and Norris, K. 2010. Management strategies for maximizing carbon storage and tree species diversity in cocoa-growing landscapes. *Agriculture, Ecosystems and Environment* 138(3-4), 324–34. doi:10.1016/j.agee.2010.06.007.

Wang, N., Jassogne, L., van Asten, P. J. A., Mukasa, D., Wanyama, I., Kagezi, G. and Giller, K. E. 2015. Evaluating coffee yield gaps and important biotic, abiotic, and management factors limiting coffee production in Uganda. *European Journal of Agronomy* 63, 1–11. doi:10.1016/j.eja.2014.11.003.

Wartenberga, A. C., Blasera, W. J., Gattingerc, A., Roshetkob, J. M., Van Noordwijkb, M. and Sixa, J. 2017. Does shade tree diversity increase soil fertility in cocoa plantations? *Agriculture, Ecosystems and Environment* 248, 190–9. doi:10.1016/j.agee.2017.07.033.

Whitney, C. W., Tabuti, J. R. S., Hensel, O., Yeh, C.-H., Gebauer, J. and Luedeling, E. 2017. Homegardens and the future of food and nutrition security in southwest Uganda. *Agricultural Systems* 154, 133–44. doi:10.1016/j.agsy.2017.03.009.

Wong, C. P., Jiang, B., Kinzig, A. P., Lee, K. N. and Ouyang, Z. 2015. Linking ecosystem characteristics to final ecosystem services for public policy. *Ecology Letters* 18(1), 108–18. doi:10.1111/ele.12389.

Wu, J., Liu, W., Chen, C. and Bennett, J. 2016. Can intercropping with the world's three major beverage plants help improve the water use of rubber trees? *Journal of Applied Ecology* 53(6), 1787–99. doi:10.1111/1365-2664.12730.

Xue, J., Gui, D., Lei, J., Zeng, F., Huang, R. and Mao, D. 2016. Development of a participatory Bayesian network model for integrating ecosystem services into catchment-scale water resources management. *Hydrology and Earth System Sciences Discussions*, 1–37. doi:10.5194/hess-2016-618.

Xue, J., Gui, D., Lei, J., Sun, H., Zeng, F. and Feng, X. 2017. A hybrid Bayesian network approach for trade-offs between environmental flows and agricultural water using dynamic discretization. *Advances in Water Resources* 110, 445–58. doi:10.1016/j.advwatres.2016.10.022.

Zheng, H., Wang, L. and Wu, T. 2019. Coordinating ecosystem service trade-offs to achieve win-win outcomes: a review of the approaches. *Journal of Environmental Sciences* 82, 103–12. doi:10.1016/j.jes.2019.02.030.

Zomer, R. J., Trabucco, A., Coe, R., Place, F., van Noordwijk, M. and Xu, J. C. 2014. Trees on farms: an update and reanalysis of agroforestry's global extent and socio-ecological characteristics. Working Paper 179. World Agroforestry Centre (ICRAF) Southeast Asia Regional Program, Bogor, Indonesia. doi:10.5716/WP14064.PDF.

Zomer, R. J., Neufeldt, H., Xu, J., Ahrends, A., Bossio, D., Trabucco, A., van Noordwijk, M. and Wang, M. 2016. Global tree cover and biomass carbon on agricultural land: the contribution of agroforestry to global and national carbon budgets. *Scientific Reports* 6, 29987. doi:10.1038/srep29987.

Zuidema, P. A., Leffelaar, P. A., Gerritsma, W., Mommer, L. and Anten, N. P. R. 2005. A physiological production model for cocoa (*Theobroma cacao*): model presentation, validation and application. *Agricultural Systems* 84(2), 195–225. doi:10.1016/j.agsy.2004.06.015.

Index

www.ingramcontent.com/pod-product-compliance
Lightning Source LLC
LaVergne TN
LVHW020628300926
849161LV00004B/643
* 9 7 8 1 7 8 6 7 6 2 2 0 7 *